Advances in Experimental Medicine and Biology

Volume 1174

More information about this series at http://www.springer.com/series/5584

Sarah Perrett • Alexander K. Buell
Tuomas P. J. Knowles
Editors

Biological and Bio-inspired Nanomaterials

Properties and Assembly Mechanisms

Springer

Editors
Sarah Perrett
National Laboratory of Biomacromolecules
Institute of Biophysics
Chinese Academy of Sciences
Beijing, China

Alexander K. Buell
Department of Biotechnology
and Biomedicine
Technical University of Denmark
DTU, Lyngby, Denmark

Tuomas P. J. Knowles
Department of Chemistry
University of Cambridge
Cambridge, UK

ISSN 0065-2598 ISSN 2214-8019 (electronic)
Advances in Experimental Medicine and Biology
ISBN 978-981-13-9790-5 ISBN 978-981-13-9791-2 (eBook)
https://doi.org/10.1007/978-981-13-9791-2

This Springer imprint is published by the registered company Springer Nature Singapore Pte Ltd.
The registered company address is: 152 Beach Road, #21-01/04 Gateway East, Singapore 189721, Singapore

Contents

Chapter 1
Dynamics and Control of Peptide Self-Assembly and Aggregation

Georg Meisl, Thomas C. T. Michaels, Paolo Arosio, Michele Vendruscolo, Christopher M. Dobson, and Tuomas P. J. Knowles

Abstract The aggregation of proteins into fibrillar structures is a central process implicated in the onset and development of several devastating neuro-degenerative diseases, but can, in contrast to these pathological roles, also fulfil important biological functions. In both scenarios, an understanding of the mechanisms by which soluble proteins convert to their fibrillar forms represents a fundamental objective for molecular sciences. This chapter details the different classes of microscopic processes responsible for this conversion and discusses how they can be described by a mathematical formulation of the aggregation kinetics. We present easily accessible experimental quantities that allow the determination of the dominant pathways of aggregation, as well as a general strategy to obtain detailed solutions to the kinetic rate laws that yield the microscopic rate constants of the individual processes of nucleation and growth. This chapter discusses a framework for a structured approach to address key questions regarding the dynamics of protein aggregation and shows how the use of chemical kinetics to tackle complex biophysical systems can lead to a deeper understanding of the underlying physical and chemical principles.

Keywords Chemical kinetics · Aggregation mechanisms · Scaling exponent · Global analysis

---

G. Meisl (✉) · T. C. T. Michaels · M. Vendruscolo · C. M. Dobson
Department of Chemistry, University of Cambridge, Cambridge, UK
e-mail: gm373@cam.ac.uk

P. Arosio
Department of Chemistry and Applied Bioscience, ETH Zurich, Zurich, Switzerland

T. P. J. Knowles
Centre for Misfolding Diseases, Department of Chemistry, University of Cambridge, Cambridge, UK

Cavendish Laboratory, University of Cambridge, Cambridge, UK
e-mail: tpjk2@cam.ac.uk

S. Perrett et al. (eds.), *Biological and Bio-inspired Nanomaterials*,
Advances in Experimental Medicine and Biology 1174,
https://doi.org/10.1007/978-981-13-9791-2_1

1.1 Introduction

The self-assembly of proteins into ordered linear structures is an important process for many living systems, for example in the context of the formation of the cytoskeletal filaments. When it occurs in a controlled manner, this process can therefore be central to the functionality of an organism, but conversely, unwanted filamentous aggregation of proteins can have devastating effects on an organism's health. One such process of particular significance is the aggregation of proteins into elongated structures, amyloid fibrils, which may consist of thousands or more copies of the same protein [1, 2]. Surprisingly, a large variety of unrelated proteins have the ability to form amyloid structures, and once formed, these entities possess an inherent propensity towards promoting the conversion of further proteins into the amyloid form [3, 4]. The study of protein aggregation has become an important area of research largely because the proliferation of amyloid fibrils is closely associated with several devastating and increasingly prevalent diseases, including type II diabetes, Parkinson's and Alzheimer's diseases [5–7]. However, there is also a number of proteins that self-assemble into fibrillar structures that are not associated with disease but are functional and essential for living organisms [8–12]. Important examples of such functional protein assemblies include for instance biofilaments of actin and tubulin, that are key parts of the eukaryotic cytoskeleton [9–11], as well as functional amyloid structures [13] that possess roles as catalytic scaffolds [14], as depots for hormones [15], in the functioning of pathogens [16] or as components of bacterial biofilms [17, 18]. The existence of functional amyloids has also inspired the use of such structures as functional biomaterials in various nanotechnological applications [19, 20], a factor that has further contributed to the interest in understanding how filamentous self-assembly works.

From a biophysical point of view, the formation of filamentous structures from dispersed proteins represents an elementary form of supra-molecular assembly since it is generally homo-molecular in nature [9, 21]. Yet, despite this apparent simplicity, many different molecular-level events contribute to the overall fibril formation process and the competition and interplay between these microscopic steps often results in rich dynamical behaviour. Obtaining a molecular-level kinetic description of self-assembling systems is thus a particularly challenging task which involves considering a complex interconnected network of several distinct microscopic steps, such as nucleation, growth or fragmentation processes [22, 23]. In this chapter we describe in detail how the use of chemical kinetics in the context of protein aggregation allows one to overcome this challenge. We demonstrate that this approach provides a general strategy for quantifying the rates of the individual microscopic steps of filamentous growth. This advance illuminates which parts of the full reaction network determine the aggregation behaviour in a given system and which ones can instead be neglected, thus providing a very useful mechanistic framework for designing strategies for controlling protein aggregation in technological applications or suppressing it for therapeutic purposes.

The chapter is organized as follows. In the first section, we discuss the microscopic-level processes that contribute to the overall protein aggregation reaction and outline a general approach for mathematically modelling the resulting reaction network. We then consider the interaction and competition between these individual processes and describe in detail how such complex scenarios can be understood within the framework of kinetic theory. The following section briefly looks at the application of these kinetic models of protein aggregation in the context of data analysis for discovering the dominant microscopic processes in action. Finally, we conclude with a discussion on how the resulting mechanistic understanding of protein aggregation forms the basis of devising rational strategies to employ inhibitory compounds or modulations of the environmental conditions to control the pathways by which the aggregation reaction proceeds.

1.2 Kinetic Theory of Protein Aggregation

Chemical kinetics provide the mathematical framework for predicting the time course of a chemical reaction. In general, a kinetic description of a chemical reaction is derived by breaking the overall process down into a sequence of one or more relevant steps. The law of mass action then yields differential equations (so called rate laws) that describe the rates of each one of these individual steps in terms of the concentrations of the species involved. The rate constants are the constants of proportionality entering such relationships and the reaction orders describe the power a particular concentration is raised to. Although for simple elementary reactions the reaction orders correspond to the actual number of species involved in the reaction, such a simple physical interpretation of reaction orders may not apply to reactions with many steps, such as encountered in the complex models discussed here. Rate laws emerge directly from a consideration of the various steps that constitute the overall reaction and thus represent the best tool for establishing unknown mechanisms. As we will discuss later, if the reaction mechanism is unknown, we can carry out experiments to determine the reaction orders with respect to each reactant and then try out various trial reaction mechanisms to see which one fits best with the experimental data. An important point to recognize here is that the rate constants and reaction orders need to be constrained by the experiments. In the following, we apply these concepts from chemical kinetics to the study of protein aggregation phenomena.

1.2.1 Fundamental Processes in Protein Aggregation

In order to develop a kinetic model of aggregation, it is first necessary to establish the species and the microscopic-level processes that are likely to be involved in the overall assembly reaction. In this context, it is important to keep the kinetic

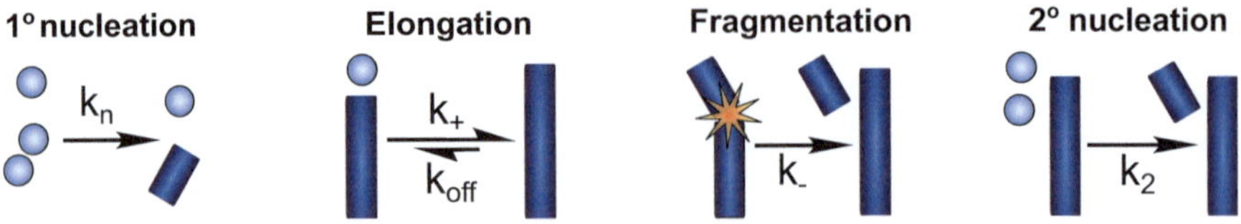

Fig. 1.1 Microscopic processes of aggregation. A schematic depiction of the fundamental microscopic processes of aggregation, also considered in the basic model discussed in Sects. 1.2.2 and 1.2.3. Spheres represent monomer, cylinders fibrillar species. Extensions to these processes to account for their multi-step nature, which may become evident under certain conditions, are discussed in Sect. 1.3

model as minimalistic as possible and only include those microscopic steps that are required to explain experimental observations or that are suggested by our physical intuition of the system. For this reason, we discuss here the simplest kinetic model of filamentous assembly [23], in which filamentous aggregates are described as linear chains of monomers that are formed, grow and multiply according to the following three basic categories of processes that form the core parts of the aggregation reaction (see Fig. 1.1 for a schematic visualization of the individual processes):

- **De novo formation of aggregates – Primary nucleation** – Primary nucleation is the spontaneous formation of a growth-competent aggregate (nucleus) from monomers alone, without the involvement of fibrillar species. As such, primary nucleation represents always the first event in an aggregation process starting with monomers only. The primary nucleation step reflects the fact that the formation of small filaments is unfavourable so that aggregates smaller than a certain critical size are unstable; the nucleus represents the smallest aggregate species that is more likely to grow than to dissociate back into monomers. Although it involves only monomers, it can, and in many cases does, happen heterogeneously, on surfaces such as for example the air-solution interface or lipid membranes [24, 25]. Removal of specific interfaces or experiments at different surface to volume ratios can give insights into the extent to which such interface effects alter the kinetics [26].
- **Growth of existing aggregates – Elongation and dissociation** – Existing aggregates are capable of further growth through elongation, but may also shrink through dissociation processes. During the elongation reaction, soluble protein adds onto the ends of an existing fibril thus leading to an increase of the overall aggregate mass. This process typically involves the attachment of monomeric protein to either end of an existing fibril followed by a conformational rearrangement of the added monomer into a structure with high β-sheet content. The addition step is typically rate-determining, but under certain conditions the multi-step nature of the elongation reaction can become apparent (see Sect. 1.3.2).

Dissociation is the reverse process of elongation and describes the removal of monomers from the ends of an existing aggregate.[1]

- **Multiplication of existing aggregates – Secondary processes** – With the term secondary processes we summarize fibril multiplication events that lead to the formation of new growth-competent aggregates in a manner that depends on the current fibril population [23, 28]. Fragmentation constitutes the simplest example of a secondary process, where the rate of fibril multiplication depends solely on the mass of existing aggregates and is independent of the monomers [29]. Fragmentation processes have been found experimentally to be active in the in-vitro aggregation of the Aβ peptide of Alzheimer's disease under mechanical agitation [30] or in the formation of prions [31–33]. Another relevant example of a secondary process is secondary (or surface-catalysed) nucleation. Secondary nucleation describes the formation of new growth-competent aggregates from the interaction between monomeric species and the surfaces of existing fibrils. A physical manifestation of this mechanism would be that the fibril surface catalyses the formation of a fibril nucleus, which, once detached from the aggregate surface, yields a new growth competent fibril. Thus, mathematically, secondary nucleation is distinguished from fragmentation by the fact that the fragmentation rate does not depend on the free monomer concentration whereas that of secondary nucleation does depend on the free monomer concentration. Secondary nucleation was first described by Ferrone, Eaton and co-workers in the context of their work on sickle haemoglobin aggregation [34–36]. It has since been found to be the major mechanism of aggregate multiplication in other protein systems including the aggregation of the Aβ peptide of Alzheimer's disease in-vitro under non-shaking conditions [28, 37, 38].

1.2.2 *The Master Equation: Quantifying the Kinetics of Aggregation*

By considering the rates at which each one of the individual processes listed above changes the concentration, $f(t, j)$, of aggregates of size j at time t, we obtain the following set of rate equations, known as the master equation [39]:

$$\begin{aligned}\frac{\partial f(t,j)}{\partial t} &= 2k_+ m(t) f(t, j-1) - 2k_+ m(t) f(t,j) \quad \text{(Elongation)}\\ &+2k_{\text{off}} f(t, j+1) - 2k_{\text{off}} f(t,j) \quad \text{(Dissociation)}\end{aligned}$$

[1]Dissociation ensures that fibril growth is reversible, in accordance with the principle of detailed balance, however, in most aggregation reactions which are performed at significant supersaturation this process does not have significant influence on the time course of the aggregate mass, which is the observable of main interest to our discussion (see Sect. 1.2.3, "Common Approximations"). It becomes relevant only at the very late stages of the aggregation process, when the aggregate mass has equilibrated and the aggregate length distribution tends towards an exponential distribution [27]. Equally the reverse of fragmentation is negligible and thus ignored.

$$
\begin{aligned}
&+2k_- \sum_{i=j+1}^{\infty} f(t,i) - k_-(j-1) f(t,j) \quad \text{(Fragmentation)} \\
&+k_n m(t)^{n_c} \delta_{j,n_c} \quad \text{(Primary nucleation)} \\
&+k_2 m(t)^{n_2} \sum_{i=n_c}^{\infty} i f(t,i)\, \delta_{j,n_2} \quad \text{(Secondary nucleation)} \qquad (1.1) \\
&\text{(valid for } j \geq n_c)
\end{aligned}
$$

where $\delta_{i,j}$ is the Kronecker delta and $m(t)$ is the concentration of monomers.

The terms on the right hand side of the master equation have straightforward interpretation and represent, in order, contributions from elongation, dissociation, fragmentation, primary nucleation and secondary nucleation with the respective rate constants as summarized in Table 1.1. For example, the rate of secondary nucleation depends on the monomer concentration raised to the reaction order n_2 and the available surface area of aggregated protein which is assumed to be proportional to the aggregate mass concentration, $\sum_{j=n_c}^{\infty} j f(t, j)$. The factor of 2 on the first and second lines of Eq. (1.1) appears because the rate constants for elongation and dissociation are defined per filament end and fibrils grow from both ends. Moreover, implicit in our formulation is the assumption that the rate constants for the various processes are size independent. In particular, the fragmentation rate is assumed to be uniform along the length of the aggregate, meaning that aggregates are equally likely to break at each bond between monomers. Physically this behaviour could be envisioned if fragmentation is governed by random fluctuations, leading to breakage of the bonds holding the fibril together. However, under mechanical stress one could imagine a fragmentation rate that depends on the total length of the aggregate [41, 42]. In such a case our approach constitutes a mean-field approximation, as long as the average fibril length is approximately constant in time.

A system in-vivo might be modelled using Eq. (1.1) assuming a constant concentration of monomeric protein, as the organism constantly replenishes protein consumed through the aggregation process. In the majority of cases, however, the aggregation kinetics is measured in an in-vitro context, where the total mass of protein and not the monomer concentration stays constant. Under these circumstances, the conservation of total protein mass $m_{\text{tot}} = m(t) + \sum_{j=n_c}^{\infty} j f(t, j)$ has to be enforced, yielding an additional rate equation for $m(t)$:

$$
\frac{\mathrm{d}m(t)}{\mathrm{d}t} = -\frac{\mathrm{d}}{\mathrm{d}t} \sum_{j=n_c}^{\infty} j f(t, j). \qquad (1.2)
$$

Finally, it is important to notice here that a reaction step (e.g. elongation step) that is effectively written as a single term in the master equation (1.1) may in fact consist of several steps, whose effect can become kinetically apparent under certain environmental conditions. This observation will become extremely important as we

Table 1.1 Parameters

Parameter/Units	description
m, (m_0)/c	**(Initial) monomer concentration.** $m(t)$ is the concentration of free, non-aggregated monomer, called m_0 at the beginning of the aggregation reaction.
M, (M_0)/c	**(Initial) fibril mass concentration.** $M(t)$ is the mass concentration of aggregates, i.e. the equivalent monomer concentration if the aggregates were re-dissolved. Its value at the beginning of the reaction is M_0, which is 0 in the case of an unseeded aggregation reaction.
P, (P_0)/c	**(Initial) fibril number concentration.** $P(t)$ is the number concentration of aggregates, proportional to the number concentration of growth competent ends, which are the points at which the aggregate can elongate. Its value at the beginning of the reaction is P_0, which is 0 in the case of an unseeded aggregation reaction. P is linked to M by the average fibril length, L, via $M/P = L$. P is difficult to measure directly but can be estimated from M by using the average fibril length.
k_n/t^{-1}c$^{-n_c+1}$	**Primary nucleation rate constant.** This appears as $k_n m^{n_c}$ in the rate of formation of primary nuclei. It has units of time^{-1} concentration$^{-n_c+1}$.
n_c/unit-less	**Reaction order of primary nucleation.** This appears as $k_n m^{n_c}$ in the rate of formation of primary nuclei. Its simple interpretation, for example in the context of classical nucleation theory, is that of a nucleus size, however this interpretation is only valid if the reaction is a simple single step process. It is unit-less and typically has a value between 0 and 5.
k_+/t^{-1} c^{-1}	**Elongation rate constant.** This appears as $2k_+mP$ in the rate of formation of new aggregate mass. It has units of time^{-1} concentration^{-1}.
k_{off}/t^{-1}	**Depolymerisation rate constant.** This appears as $-2k_{\text{off}}P$ in the rate of aggregate mass formation and is the rate at which monomers are lost from fibril ends. It has units of time^{-1}. This may be a global fitting parameter. However, in most cases it is negligibly small.
k_-/t^{-1}	**Fragmentation rate constant.** This appears as k_-M in the rate of formation of new growth competent ends from fragmentation. It has units of time^{-1}. This form of the fragmentation rate assumes that an aggregate is equally likely to break anywhere along its length, with the time-scale of breaking given by $1/k_-$.
k_2/t^{-1} c$^{-n_2}$	**Secondary nucleation rate constant.** This appears as $k_2 m^{n_2} M$ in the rate of formation of secondary nuclei. It has units of time^{-1} concentration$^{-n_2}$.
n_2/unit-less	**Reaction order of secondary nucleation.** This appears as $k_2 m^{n_2} M$ in the rate of formation of secondary nuclei. Its simple interpretation, for example in the context of classical nucleation theory, is that of a nucleus size, however this interpretation is only valid if the reaction is a simple single step process. It is unit-less and typically has a value between 0 and 4 [28, 40], although larger values are possible.

learn more about saturation effects in Sect. 1.3.2. In order to correctly interpret the results of our kinetic analysis and develop appropriate extensions of existing models, it is thus important to keep in mind what range of detailed mechanisms will map onto this coarse-grained description and what mechanisms require instead additional complexity. For this reason we have summarized the formal definitions of the kinetic model in Table 1.2.

Table 1.2 Definitions of the kinetic model of aggregation

Aggregates are linear chains of monomers, no branching occurs (i.e. new aggregates that form at the surface detach).
Fibrils can fracture at any point with rate k_-. This rate is unchanged throughout the fibril and does not depend on its length, i.e. every monomer-monomer bond has an equal probability of breaking.
Monomers can attach and detach at the ends of fibrils with rate constants k_+ and $(k_{\text{off}} + k_-)$ respectively (i.e. k_{off} determines how much more likely a monomer at the end is to break off compared to the fibril breaking at any other position). These rate constants are independent of the size of the aggregate.
Aggregates are unstable below a certain size, n_c, and the transient population of aggregates smaller than n_c is hence neglected.
Stable aggregates of size n_c can form from monomers with the nucleation rate constant k_n. This process is referred to as primary nucleation.
Stable aggregates of size n_2 can form from monomers, heterogeneously on the surface of existing aggregates, with the nucleation rate constant k_2. This process is referred to as secondary nucleation. It is assumed not to influence the elongation behaviour of the aggregates serving as a nucleation surface.
All rate constants are reaction limited, i.e. the effects of diffusion are not considered explicitly.
Interactions between aggregates are neglected.

1.2.3 *Principal Moments and Moment Equations*

Being a set of infinitely many coupled and non-linear differential equations, the master equation (1.1) seems very hard to solve. However, in a data analysis context this difficultly does not necessarily preclude its use. In most cases, the full aggregate size distribution $f(t, j)$ is in fact not easily accessible through experiments. Instead, measurements of average quantities, such as the total fibril mass, are made, which can already provide significant insights into the mechanisms of aggregation.

1.2.3.1 Principal Moments

A particularly useful strategy to reduce the complexity of the master equation is given by the introduction of the principal moments of the aggregate length distribution, defined by the equation:

$$Q_N(t) = \sum_{j=n_c}^{\infty} j^N f(t, j). \tag{1.3}$$

Of particular experimental interest are the zero-th moment

$$P(t) = Q_0(t) = \sum_{j=n_c}^{\infty} f(t, j) \tag{1.4}$$

and the first moment

$$M(t) = Q_1(t) = \sum_{j=n_c}^{\infty} jf(t,j). \tag{1.5}$$

These quantities, which correspond to the fibril number and mass concentrations respectively, are the most common experimentally measured observables (see Sect. 1.4).

1.2.3.2 Moment Equations

Differentiating the above equations for the lowest moments, Eqs. (1.4) and (1.5), in combination with the master equation (1.1) we obtain two differential equations for $P(t)$ and $M(t)$, commonly referred to as the moment equations [39]:

$$\frac{\mathrm{d}P(t)}{\mathrm{d}t} = k_n m(t)^{n_c} + [k_- + k_2 m(t)^{n_2}]\, M(t) - k_-(2n_c - 1)P(t) \tag{1.6}$$

$$\frac{\mathrm{d}M(t)}{\mathrm{d}t} = [2m(t)k_+ - 2k_{\mathrm{off}} - k_- n_c(n_c - 1)]\, P(t) + n_c k_n m(t)^{n_c} + n_2 k_2 m(t)^{n_2} M(t). \tag{1.7}$$

Importantly, these moment equations form a closed system of two differential equations so that $P(t)$ and $M(t)$ can be determined independently of the full aggregate size distribution $f(t,j)$. Thus, compared to the initial infinitely dimensional master equation, the introduction of principal moments not only provides a direct link to common experimental observables but also results in considerable simplification of the mathematics.[2]

1.2.3.3 Common Approximations

At this stage, it is advisable to introduce some approximations to further simplify the moment equations before attempting to solve them (see Table 1.3 for the specific

[2]When analysing the kinetics of aggregate formation, we simply fit rate laws and obtain rate constants and reaction orders, but we do not directly monitor the process. The microscopic mechanism is inferred by our interpretation of the fitted parameters. This leads to an interesting phenomenon: From a mathematical point of view, considering only the moment equations (1.8) and (1.9), fragmentation and secondary nucleation with $n_2 = 0$ (i.e. the rate determining step of secondary nucleation is monomer independent, see Sect. 1.3.2) are equivalent and hence indistinguishable in this kind of kinetic analysis. In order to distinguish between these two possibilities, experiments that yield information on the fibril distribution are necessary. This could constitute measurements of the full length distribution of fibrils, use trapping of fibrils in filters to test for seeding of monomer nucleation or employ the addition of specific labels to fibrils and monomer [43–45].

Table 1.3 Common approximations

Approximation	Description
Mass from nucleation events	If long fibrils are produced, i.e. on average significantly bigger than the nucleus sizes n_2 and n_c, the mass produced directly from nucleation events will be negligible. This means last two terms in Eq. (1.7) will be negligible. The aggregates observed in the context of most aggregating systems usually consist of several thousand monomers, making this a good approximation.
Loss through fragmentation	If a piece smaller than the nucleus size breaks off the ends it will re-dissolve and therefore lead to a decrease in both mass and number concentration. Because of the large average size and the usually small value of the fragmentation constant this effect is also negligible. The term last term in equation (1.6) becomes negligible and so does the "k_-" term in equation (1.7).
Loss through depolymerisation	Depolymerisation is the reverse of the elongation reaction. If m_∞ denotes the free monomer concentration in equilibrium, i.e. the solubility of the monomer, then the depolymerisation rate is given by $k_{\text{off}} = k_+ m_\infty$. Therefore the depolymerisation rate will only become significant, i.e. comparable to the elongation rate, at low monomer concentrations approaching m_∞. If the experiment is performed at concentrations well above the solubility, where most of the initial monomer concentration will be incorporated into aggregates, then the depolymerisation rate can be neglected for the description of the kinetics.

details of the various approximations employed here). These approximations are based on considerations of the relative magnitudes of the rates of certain microscopic processes. For example, under typical environmental conditions nucleation processes are much slower than elongation; this condition ensures that long fibrils will form from monomers. For this reason, elongation is the only process that depletes monomers significantly and the contribution of nucleation processes to the overall increase in aggregate mass can be considered to be negligible compared to that of elongation. These approximations result in the following simplified set of moment equations:

$$\frac{\mathrm{d}P(t)}{\mathrm{d}t} = k_n m(t)^{n_c} + \left[k_- + k_2 m(t)^{n_2}\right] M(t) \tag{1.8}$$

$$\frac{\mathrm{d}M(t)}{\mathrm{d}t} = 2k_+ m(t) P(t). \tag{1.9}$$

These equations are the fundamental kinetic equations for filamentous assembly with secondary pathways and will thus serve as the basis for all descriptions in the remainder of this chapter unless otherwise stated.

1.2.4 Solving the Moment Equations: The Fixed-Point Method

The moment equations (1.8) and (1.9) fully describe the time evolution of commonly measured quantities, the aggregate mass and number concentrations. From these equations, one could, at least in principle, calculate the aggregate number and mass at any point in time, given the values of the rate constants and the boundary conditions. In practice, the rate constants can be determined for example through kinetic experiments, whereas the boundary conditions are imposed by the type of system at hand, most commonly an in-vitro context where the total mass of protein stays constant. Unfortunately, exact solutions to the moment equations (1.8) and (1.9) exist only in very few cases, including Oosawa's solution to filamentous growth dominated by primary nucleation [9], and in general one has to rely on approximation methods in order to find accurate expressions for the aggregate mass concentration, $M(t)$, as a function of time.

One could ask why there is any need to derive analytic solutions to the moment equations in the first place: The closed system of differential equations, (1.8) and (1.9), can simply be integrated numerically to arbitrary accuracy, so what advantage is gained by deriving analytic solutions? A minor factor is that analytic solutions will significantly increase computation speed during data fitting, as numerical integration of differential equations can be both computationally expensive and less robust than evaluation of an analytical function. Much more importantly, the availability of analytic solutions also allows the derivation of other system representative quantities and provides the basis for a more in depth understanding of the origins of the system's behaviour. In particular, analytic solutions give a handle on the relative importance of the different microscopic processes and can be used to derive qualitative constraints on the mechanism, based on easily measurable macroscopic parameters [37].

A powerful approach to find an approximate solution to the moment equations is the fixed-point iteration method [46]. The underlying idea of this approach is to reformulate the moment equations as a fixed-point equation and subsequently apply the associated fixed-point operator repeatedly to an initial guess. The repeated application of the fixed-point operator will cause the so-constructed series of approximative solutions to converge to the exact solution of the moment equations. Making an appropriate choice for the initial guess is particularly important for the success of the fixed-point method, largely because the repeated application of the fixed point operator can easily lead to very complex and thus impractical expressions for $M(t)$. We now illustrate these principles on the moment equations (1.8) and (1.9).

To obtain the explicit form of the fixed-point relevant to our case, we integrate Eq. (1.9) formally and recast it in terms of an integral operator[3]:

---

[3]Note that the fixed-point operator takes the form of Eq. (1.10) as long as the equation for $\mathrm{d}M/\mathrm{d}t$, Eq. (1.9) remains unchanged.

$$M(t) = 2k_+m_0 e^{-2k_+\int_0^t P(\tau)d\tau} \int_0^t e^{-2k_+\int_0^\tau P(\bar{\tau})d\bar{\tau}} P(\tau)d\tau, \tag{1.10}$$

where, for simplicity, we have assumed that only monomers are present initially ($M_0 = P_0 = 0$).

To construct an appropriate initial guess for the fixed-point iteration, we make use of our physical intuition by realizing that the assumption of constant monomer concentration will be accurate during the early times of the reaction when the number of aggregates is small and the monomers are thus not significantly depleted. This solution, which was originally obtained by Eaton, Ferrone and co-workers, is found by linearising Eqs. (1.8) and (1.9) by enforcing a constant monomer concentration, $m(t) = m_0$, resulting in a set of linear differential equations that can be solved straightforwardly to yield:

$$P_{\text{init}}(t) = \frac{k_n m_0^{n_c}}{\kappa} \sinh(\kappa t) \tag{1.11}$$

$$M_{\text{init}}(t) = \frac{k_n m_0^{n_c}}{k_- + k_2 m_0^{n_2}} \left[\cosh(\kappa t) - 1\right], \tag{1.12}$$

where

$$\kappa = \sqrt{2k_+ m_0 \left(k_- + k_2 m_0^{n_2}\right)}. \tag{1.13}$$

Finally, performing one fixed-point iteration by inserting Eq. (1.11) into Eq. (1.10) yields the following closed-form expression for $M(t)$:

$$\frac{M(t)}{m_0} = 1 - \exp\left(-\frac{k_n m_0^{n_c}}{k_- + k_2 m_0^{n_2}} \left[\cosh(\kappa t) - 1\right]\right). \tag{1.14}$$

The comparison between Eq. (1.14) and the numerical solution to the moment equations in Fig. 1.2 demonstrates the power of the fixed-point iteration method for tackling the mathematical complexity inherent to the description of filamentous growth phenomena. With this strategy similar accurate approximate solutions have been obtained for various other mechanisms of aggregation; the explicit expressions for the various different aggregation mechanisms presented in the following can be found in Meisl et al. [47].

1.2.5 Implications from Integrated Rate Laws

One of the biggest advantages of obtaining analytic solutions to the moment equations is the ability to derive expressions for representative observables that provide the basis for an in-depth understanding of the origins of the system's

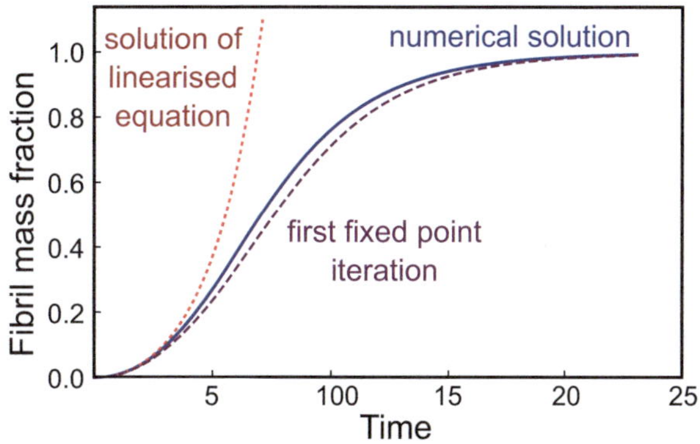

Fig. 1.2 Approximate solutions. The exact, numerically integrated solution is shown as the blue solid line. The linearised solution, dotted red line, is only valid at early times, as it does not take into account monomer depletion it increases exponentially, never reaching a plateau. The approximate solution, purple dashed line, obtained after one application of the fixed-point operator, approximates the accurate solution well over the full time-course of the reaction

behaviour. In particular, analytic solutions give a handle on the relative importance of the different microscopic processes and can thus be used to derive qualitative constraints on the mechanism, based on easily measurable macroscopic parameters [37]. In the following, we illustrate these points by exploring in detail how Eq. (1.14) contributes to our understanding of filamentous growth phenomena.

1.2.5.1 Early-Time Behaviour is Exponential

The first consequence of Eq. (1.14) is that for an aggregation reaction dominated by secondary processes, the fibrillar mass initially increases exponentially $M(t) \sim e^{\kappa t}$, where κ is an effective rate of multiplication that combines contributions from growth and the rate of all active secondary pathways:

$$\kappa = \sqrt{(\text{elongation rate}) \cdot (\text{sum of rates of secondary processes})}. \tag{1.15}$$

Hence, together with the elongation reaction, secondary processes form a positive feedback loop, where the growth of newly formed aggregates leads to an increased rate of the secondary processes, which in turn promotes the formation of further aggregates. The initial exponential increase of the aggregate mass when secondary processes are active is to be compared to the situation when secondary processes are absent and aggregates are formed only through primary nucleation. Under these circumstances, a more gradual polynomial increase $M(t) \sim t^2$ is predicted by Oosawa's exact solution [9]. This difference in behaviour can be useful as a rough guide for distinguishing systems dominated by primary or secondary nucleation through a simple inspection of the kinetic curve shape.

1.2.5.2 Half-Times and Scaling Exponents

There are many paths by which an aggregation reaction can proceed and a major challenge in the analysis of experimental data is to determine which aggregation mechanism is dominant. A fitting of the rate laws for the full network is often not feasible, therefore strategies to narrow down the number of possible mechanisms are required. One qualitative constraint to the underlying mechanism can be obtained by the analysis of the dependence of the half times on the monomer concentration. In the following, we will discuss how half times provide insights into the topology of the reaction network and thereby constrain the possible mechanisms in action.

The half time of aggregation, $t_{1/2}$, is defined as the time at which half the fibrillar mass present at the end of the reaction has been formed. The availability of Eq. (1.14) allows us to obtain an approximate expression for this half time by solving $M(t_{1/2}) = m_0/2$, yielding [48, 49]:

$$t_{1/2} = \frac{1}{\kappa} \log \left[\frac{\log(2)(k_- + k_2 m_0^{n_2})}{k_n m_0^{n_c}} \right]. \tag{1.16}$$

As expected from the dominance of secondary pathways, $t_{1/2}$ is found to be inversely proportional to the aggregate multiplication rate κ, even though a weak dependence of the half time on the primary nucleation emerges in form of a logarithmic correction. This result can be rationalized by the fact that the formation of new fibril ends is initially always controlled by primary nucleation even when secondary processes are dominant at later stages of the reaction. Note that the idea that in bulk experiments the observed lag time (i.e. the time until an increase in aggregate mass can be detected) is the time to formation of the first nucleus is a common misconception. In almost all bulk systems the first nucleus is formed effectively immediately, the lag time is instead the time until a sufficient amount of aggregated material has been formed to be detected. First, this is evident from the reproducibility of the lag time and lack of stochastic effects, second, a self-consistency check using the fitted values of the nucleation rates shows that a nucleation rate far away from that determined from the fitting would be required in order to put the system into a regime determined by stochastic nucleation effects [50]. However, stochastic behaviour, where the lag time is indeed determined by the time until formation of the first nucleus, can be observed if the reactions are performed in very small volumes, for example nl-sized droplets in a micro-fluidic device [51].

When either fragmentation or secondary nucleation is dominant, the dependence of $t_{1/2}$ on the monomer concentration, m_0, takes approximately the form of a power law $t_{1/2} \approx m_0^{\gamma}$, where γ is referred to as the scaling exponent. A double logarithmic plot of the half time versus monomer concentrations therefore gives a straight line with slope γ:

$$\gamma = \frac{d(\log(t_{1/2}))}{d(\log(m_0))} = m_0 \frac{d(\log(t_{1/2}))}{dm_0}. \tag{1.17}$$

The exact value of the scaling exponent depends on the monomer-dependence of the dominant pathway of aggregation and in general it takes the form:

$$\gamma \approx -\frac{(\text{reaction order of elongation}) + (\text{reaction order of dominant } 2^o \text{ process})}{2}, \tag{1.18}$$

where the factor of $1/2$ originates from the square root in κ. For example, fragmentation is monomer independent and so in a system which is dominated by fragmentation processes (i.e. the contribution from primary and secondary nucleation is negligible) we expect a scaling exponent of $\gamma = -1/2$. By contrast, the secondary nucleation step is characterized by a non-zero monomer dependence, hence a system dominated by secondary nucleation would result in a scaling exponent of $\gamma = -(n_2 + 1)/2$. Because the reaction order of secondary nucleation commonly takes low integer values around 2, the observation of a scaling exponents of a large magnitude, e.g. -1.5, is typically indicative of a secondary-nucleation-dominated aggregation mechanism. In the absence of secondary pathways, the scaling exponent is determined by the reaction order of the primary nucleation step alone, giving $\gamma = -n_c/2$. Hence, half-time plots as a function of monomer concentration and the associated scaling exponents are of central importance in establishing the main mechanism of aggregation of a given protein system.

1.3 The Full Aggregation Network: Interplay and Competition

In the previous section, we have demonstrated how the scaling behaviour of the half time with monomer concentration provides important insights into the nature of the dominant mechanism of filamentous assembly through the value of the scaling exponent γ. In the following, we demonstrate that, in addition to its actual value, the monomer concentration dependence of the scaling exponent can provide further insights into the full aggregation network, such as the existence of saturation effects or competing mechanisms [40].

1.3.1 Monomer Dependence of the Scaling Exponent as a Guide to Complex Mechanisms

The scaling exponent is determined by the reaction orders of the dominant processes on the aggregation pathway. Therefore, a scaling exponent which is independent of the monomer concentration is indicative of the fact that the dominant mechanism of aggregation is likely to remain unchanged over the range of monomer concentrations considered. By contrast, a scaling exponent that depends on the monomer concentration (i.e. the scaling exponent, not just the half times, depend on the monomer concentration) is indicative of a change of the reaction order of the

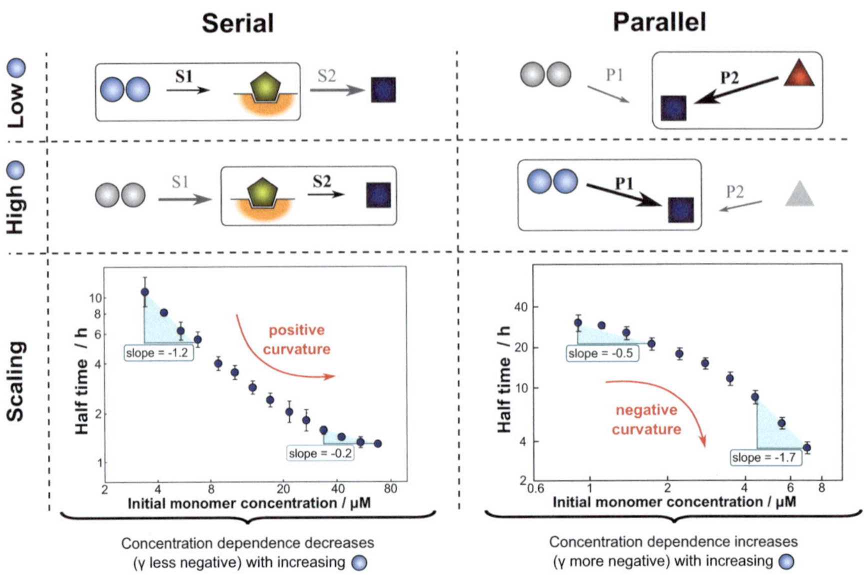

Fig. 1.3 Scaling in parallel and serial reaction networks. A schematic depiction of a simple parallel and a simple serial reaction network. Each one of the networks has a monomer-dependent step (S1 or P1) and monomer-independent step (S2 or P2). In the serial network S1 determines the rate at low concentrations, whereas S2 is rate-determining at high concentrations. By contrast, in the parallel system P2 dominates the reaction at low monomer concentrations, whereas P1 dominates at high concentrations. Experimental examples illustrate the two cases. For a serial system we observe a decrease in the monomer dependence of the reaction rate with increasing monomer concentration, resulting in a positive curvature of the half time plots (Aβ40 peptide [52]). A parallel system displays an increase in monomer dependence of the reaction rate with increasing monomer concentration, resulting in negative curvature in the half time plots (Aβ42 at low ionic strengths [53]). (Adapted from Ref. [40] with permission from the Royal Society of Chemistry)

dominant process and the direction (curvature) of this change (decrease or increase in monomer dependence) contains information about the topology of the reaction network [40].

To illustrate this principle, we consider the two simple reaction networks illustrated in Fig. 1.3. Both schemes describe the transformation of monomers (blue circles) into the same reaction product (blue squares). The first network, however, consists of two steps in parallel, whereas the second one is built up of two steps in series, with the conversion of intermediate to product occurring through the action of a catalyst.[4] Our focus is on the net rate of formation of product and on understanding its dependence on the monomer concentration.

[4]Note that for a simple serial reaction, without catalyst, the rate of formation of product depends on the rates of the individual reactions in a multiplicative fashion, so no saturation effects emerge and the overall monomer dependence will remain constant.

Since the reactions S1 and P1 are monomer-dependent but the reactions S2 and P2 are monomer-independent, we expect that lowering the monomer concentration sufficiently will lead to a regime where the processes S1 and P1 are significantly slower than S2 and P2. In the case of the serial network, this situation means that the process S1 becomes rate determining. By contrast, in the case of the parallel network, P1 is very slow and P2 is thus the main generator of product. Therefore, the overall rate of product formation in the serial network depends on the monomer concentration (as the monomer-dependent reaction S1 is rate-determining), whilst in the parallel network it is independent of the monomer concentration (effectively P2 alone is responsible for the formation of product).

In the opposite limit of high monomer concentration, the processes S1 and P1 are faster than S2 and P2. Under these circumstances, the net rate of product formation in the serial network is limited by the conversion of intermediate to product (S2) while the production of intermediate runs very fast. By contrast, in the parallel network, most of the product is formed through the monomer-dependent step P1. Hence, in the high monomer concentration regime, the rate of product formation in the serial network is determined by the monomer-independent reaction, whereas in the parallel network it is controlled by the monomer-dependent process.

By this simple example, we have illustrated how a parallel process possesses a weak monomer dependence at low monomer concentrations and a strong monomer dependence at high monomer concentrations and that the opposite is true for a saturating process. Therefore, an increase in the magnitude of the scaling exponent with increasing monomer concentration is characteristic of a parallel pathway (negative curvature in the half time plots), whereas a decrease of the scaling exponent is characteristic of a serial (i.e. saturating) pathway (positive curvature in the half time plots).

Applied to protein aggregation, these results suggest that a decrease in the magnitude of the scaling exponent with increasing monomer concentration requires a saturation of either elongation or secondary nucleation. Seeded experiments can directly sample the elongation reaction and hence distinguish between the two cases. An increase in the magnitude of the scaling exponent with increasing monomer concentration requires a competition between two processes in parallel. This can be either primary nucleation and one of the secondary processes or fragmentation and secondary nucleation competing with each other. In the following, we will discuss these two separate cases in more detail.

1.3.2 Saturation: Processes in Series

In its simplest interpretation, the steps in a kinetic scheme correspond to elementary steps of the reaction. Elementary steps cannot be broken down into sub-steps, as there are no further stable intermediates on the path from reactants to products. However, in the case of large complex molecules, such as proteins, reactions often involve conformational rearrangements and the formation of many interactions within a molecule and between species. It is usually not feasible to explicitly take into account all these processes and the kinetic description used to model the reaction therefore becomes coarse-grained, where each 'single step' may in fact consist of many elementary steps. In many situations this difficulty does not limit the applicability of single-step kinetic models, as these sub-steps are fast and not kinetically visible.[5] However, under certain environmental conditions, some of these sub-steps might become rate-determining so that the multi-step nature of the process becomes apparent. In particular this situation may arise if the individual sub-steps display different reaction orders and therefore respond differently to a change in concentration. In this case the kinetic descriptions have to be extended to take this effect into account.

As discussed above, such serial reactions, if the intermediate is catalytic, will display a decrease in monomer dependence as the monomer concentration increases, reflected in a positive curvature of the half time plots [40].

1.3.2.1 Multi-step Elongation

In the basic model discussed in Sect. 1.2.3 elongation was described as a single-step reaction, where monomeric species from solution directly add onto fibrils. In real amyloid forming systems the monomer is often present in a disordered state or folded differently to its conformation within the fibril [21, 54, 55]. Therefore, the elongation of amyloid fibrils will most likely involve monomers adsorbing onto the ends of fibrils and then adopting the β-sheet rich conformation required to maximise interactions within the amyloid fibril, in a process that will be determined by the intrinsic time-scale required for structural rearrangement. Indeed there is evidence for such a lock-dock mechanism in several systems [56–58]. As the first step of this process depends on the monomer concentration, whereas the second step is monomer-independent, this process resembles the serial network discussed above. Hence, we expect that its multi-step nature will become kinetically visible as a saturation effect at sufficiently high monomer concentrations.

The lock-dock mechanism can be modelled as a two-step reaction, with an initial, monomer-concentration-dependent attachment step followed a monomer-concentration-independent rearrangement step. This mechanism is formally equivalent to a Michaelis-Menten kinetics scheme from biochemistry [59], where the fibril end is the catalyst, the monomer the substrate and the product is a slightly longer fibril with a new growth competent free end. The corresponding scheme in terms of fibril number and mass is given in scheme 4 (Fig. 1.4).

[5]However, keep in mind that the rates and reaction orders of such coarse-grained processes are not as straightforward to interpret on a molecular level as in elementary reactions.

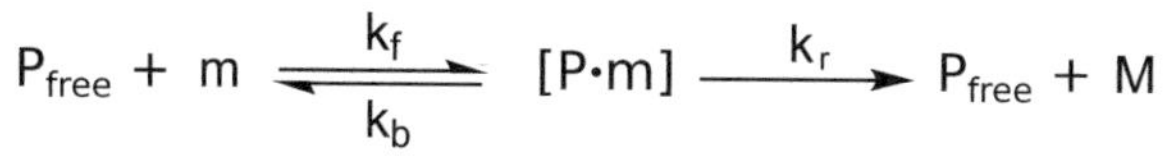

Fig. 1.4 **2-step elongation.** In a first step a monomer, m, binds to a free fibril end, P_{free}, with rate constant k_f, forming a complex, $[P \cdot m]$. The complex can dissociate again with rate constant k_b. Within this complex the monomer then rearranges to adopt the correct fibrillar conformation, forming a free end again with rate constant k_r and increasing the overall aggregate mass, M

Applying the standard Michaelis-Menten steady-state approach [60] results in an extended version of the equation for the aggregate mass as (compare to Eq. (1.9)):

$$\frac{\mathrm{d}M}{\mathrm{d}t} = 2k_+ \frac{m(t)}{1 + m(t)/K_E} P(t) \tag{1.19}$$

where $K_E = (k_b + k_r)/k_f$ and $k_+ = k_f k_r/(k_b + k_r)$. The physical interpretation of the Michaelis constant K_E gives the monomer concentration at which elongation is half saturated, i.e. the elongation step proceeds at half its maximum speed.

If the monomer concentration is low, $m(t) \ll K_E$, the initial monomer dependent step is slow and therefore determines the overall rate. This low monomer limit is identical to the simpler single step model assumed previously in Eq. (1.9). Therefore many systems can be successfully modelled with single step elongation kinetics, simply because they are far away from the saturation concentration, K_E, at which the rearrangement step becomes visible kinetically.

By contrast, if the monomer concentration is high, $m(t) \gg K_E$, the monomer dependent step is fast and the rearrangement step becomes rate-limiting. Under these circumstances Eq. (1.19) reduces to $\mathrm{d}M/\mathrm{d}t = 2k_+ K_E P(t)$. In this limit the overall elongation reaction does no longer depend on the monomer concentration; a further increase in monomer will not change the rate of elongation and the process is said to be saturated. Whilst in the low monomer limit, the simplified form of $\mathrm{d}M/\mathrm{d}t$ is valid for the entire time course, this is not true in the high monomer concentration case: Free monomer is depleted during the reaction and necessarily falls below K_E at some point, so the elongation rate slows and the aggregation reaches completion.

In general, upon saturation of elongation the scaling exponent will increase by $1/2$, but its absolute value also depends on the dominant nucleation mechanism:

$$\gamma \approx \gamma_{\text{nuc}} - \frac{1}{2(1 + m_0/K_E)} \tag{1.20}$$

where γ_{nuc} is the contribution of secondary nucleation processes to the scaling and γ_{elon} is the contribution from elongation, such that $\gamma = \gamma_{\text{nuc}} + \gamma_{\text{elon}}$. This expression interpolates between the limits of unsaturated elongation, $\gamma_{\text{elon}} = -1/2$, and fully saturated elongation, $\gamma_{\text{elon}} = 0$.

1.3.2.2 Multi-step Primary Nucleation

In the majority of cases it is sufficient to treat primary nucleation as a single step mechanism, with one rate constant and one reaction order, as described by the term $k_n m(t)^{n_c}$ in Eq. (1.9). In general, however, there may be several steps during the formation of a primary nucleus. If high quality data of the part of the aggregation reaction where primary nucleation is important exist, such as data on the concentrations of different oligomeric species [61], then these models can be used in order to gain insights into the details of the nucleation cascades. In either case, it is important to employ a minimal model approach to avoid over-fitting and indeed bulk experimental data can often be fit by a single-step primary nucleation model. This means that to within experimental accuracy the nucleation process may be described as a single-step reaction, but care needs to be taken when interpreting the microscopic parameters: for a true, elementary-step reaction, the reaction order of a species can indeed be interpreted as the number of molecules of that species taking part in the reaction, i.e. in this case the reaction order n_c would correspond to the nucleus size. By contrast, in this coarse grained model, n_c should not be naively interpreted as a nucleus size, but rather as the reaction order of the rate determining step. For example, if the nucleation reaction happened on an interface, the scaling can become significantly smaller than the nucleus. The multi-step nature of primary nucleation can also become apparent in a system where primary nucleation is the main process of aggregate multiplication (i.e. in the absence of secondary processes). In that case a multi-step primary nucleation can in fact change the time dependence of $M(t)$ to be $\approx t^{n+1}$ where n is the number of steps in the nucleation cascade. A detailed discussion of nucleation cascades can be found in Garcia et al. [62].

1.3.2.3 Multi-step Secondary Nucleation

We have already mentioned the possible complexity of primary nucleation, and the nucleation cascade models that exist to describe this process, and we have included a possible saturation effect into the elongation process. Now we will consider the details of the secondary processes and what extensions of our models may be necessary to account for their potential multi-step nature.[6]

It is hard to envision that secondary nucleation, a process where several monomers and fibrils have to meet, happens in a single step, as a three (or more) body collision. More likely fibril attachment and monomer encounter occur in separate steps. Kinetically, the multi-step nature of this process is relevant because the individual steps will have a different dependence on monomer concentration.

[6]Fragmentation is a first order reaction, dependent only on fibril mass and could reasonably be expected to follow single-step kinetics. No kinetic evidence for its multi-step nature exists to date, so it is not discussed here.

Similar to the effect discussed for elongation in Sect. 1.3.2.1, this may lead to saturation effects and a change in the reaction order as the monomer concentration is varied. Such a behaviour has been observed for several variants of the Aβ peptide, for a range of conditions [52, 63], see also Fig. 1.7.

One can envision different scenarios, depending on whether monomer encounter occurs on or off the fibril. In the former case, we model this effect by assuming that monomers bind to the surface of existing fibrils, where they may form a nucleus [64]. The rate of nucleus formation will then simply depend on the surface concentration (i.e. the coverage) to the power of the reaction order, n_2. Alternatively, monomers may meet directly, rather than first binding to the fibril surface, and the fibril surface serves as the catalytic site for conversion of the oligomeric species. In this case, following a similar line of argument as for elongation, we obtain a modified version of the moment equation (1.9) that takes into account the multi-step nature of secondary nucleation:

$$\frac{\mathrm{d}P}{\mathrm{d}t} = k_2 \frac{m(t)^{n_2}}{1 + m(t)^{n_2}/K_M} M(t) + k_n m(t)^{n_c} \tag{1.21}$$

where K_M is the Michaelis constant of secondary nucleation, and K_M^{1/n_2} corresponds to the monomer concentration at which secondary nucleation is half saturated, meaning that the nucleation step proceeds at half its maximum speed. In the limit of low monomer ($K_M^{1/n_2} \gg m$) we recover the expression for a single step nucleation, as in Eq. (1.8). At high concentrations ($m \gg K_M^{1/n_2}$) the reaction order of the secondary pathway in monomer is close to zero.

The scaling exponent is given by:

$$\gamma \approx -\frac{1}{2}\left(\frac{n_2}{1 + m_0/K_M} + 1\right). \tag{1.22}$$

This expression interpolates between $\gamma = -(n_2+1)/2$ and $\gamma = -1/2$ for the limits of low and high monomer, respectively.

1.3.3 Competition: Processes in Parallel

There are several processes that produce new growth competent aggregates: primary nucleation and the two secondary processes, fragmentation and secondary nucleation. These processes may all occur in parallel and can hence compete for being the dominant mechanism of aggregate formation. This competition can become kinetically visible through a shift in the dominant mechanism due to the differing monomer dependence of the three processes. Fragmentation is monomer independent, whereas primary and secondary nucleation depend on the monomer concentration to the powers of n_c and n_2, respectively.

As discussed previously, when two processes compete in parallel, the process with the higher reaction order will dominate at high concentrations. Therefore, we expect an increase in the monomer-dependence of the reaction rate with increasing monomer concentrations, reflected by negative curvature in the half time plots.

1.3.3.1 Competition Between Primary and Secondary Processes

A shift between a primary and a secondary dominated mechanism with a change in monomer concentration could be envisioned if the reaction orders of the two processes differ significantly. In all systems that display a secondary process studied thus far, this secondary process is so fast compared to the primary ones that it dominates the aggregation reaction at all accessible monomer concentrations. However, in computer simulations of aggregation [64] a competition between primary and secondary processes can be observed: Naturally nuclei that form from primary nucleation, homogeneously in solution, are larger than those formed on the surface of fibrils through secondary nucleation, therefore the reaction order for secondary nucleation is lower than that of primary, and primary nucleation dominates at sufficiently high monomer concentrations. The scaling exponent varies between the scaling for a secondary nucleation dominated mechanism, $\gamma = -(n_2 + 1)/2$, at low monomer, and that for a primary dominated mechanism, $\gamma = -n_c/2$, at high monomer. Moreover, a change in the curve shape, from the exponential shapes produced when secondary processes dominate to the polynomial shapes when primary nucleation dominates, is expected to occur.

1.3.3.2 Two Competing Secondary Processes

The contributions to the rate of growth competent end formation, $\mathrm{d}P/\mathrm{d}t$, from the secondary processes is $k_- M(t)$ for fragmentation and $k_2 m(t)^{n_2} M(t)$ for secondary nucleation. While both depend on the concentration of aggregates, $M(t)$, only secondary nucleation also depends on the monomer concentration. Therefore, if a monomer-dependent secondary pathway is in principle accessible in a given system, it should become dominant over fragmentation at sufficiently high monomer concentrations. At low concentrations, by contrast, fragmentation is expected to become dominant.

The relevant differential equation for the rate of new aggregate formation is given by:

$$\frac{\mathrm{d}P}{\mathrm{d}t} = k_n m(t)^{n_c} + (k_- + k_2 m(t)^{n_2}) M(t). \tag{1.23}$$

The approximate scaling exponent is determined as

$$\gamma \approx -\frac{1}{2}\left(\frac{n_2}{1 + K/m(0)^{n_2}} + 1\right), \tag{1.24}$$

Constant scaling (non-saturated)

Dominant pathway	Approximate scaling	Linear approximation to mass concentration
1° nucelation only	$-\frac{n_c}{2}$	$\approx 2k_+k_n m_0^{n_c} t^2$
fragmentation	$-\frac{1}{2}$	$\approx \mathrm{Exp}[\sqrt{2k_+k_- m_0}\, t]$
2° nucelation	$-\frac{1+n_2}{2}$	$\approx \mathrm{Exp}[\sqrt{2k_+k_2 m_0^{(n_2+1)}}\, t]$

Change in scaling (with increasing monomer conc.)

Dominant pathway	Change in scaling		Topology	Half time plots
	growth	nuc./mult.		
Saturating 2° nucleation	–	$+\frac{n_2}{2}$	serial	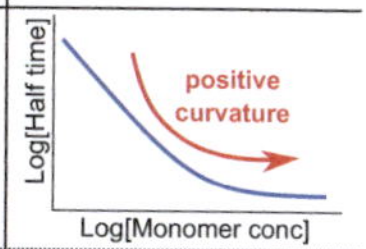
Saturating elongation	$+\frac{1}{2}$	–	serial	
Competing 2° processes	–	$-\frac{n_2}{2}$	parallel	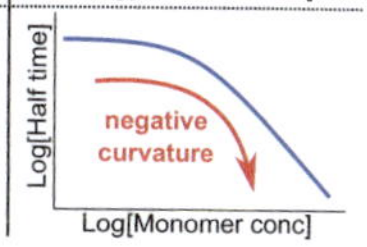
Competing 1° and 2° process	–	$\frac{n_2+1-n_c}{2}$	parallel	

Fig. 1.5 Scalings in different limits. This table summarizes the scaling and changes in scaling for the range of mechanisms discussed here. The top table gives the scaling when no competition or saturation is observed, as well as the approximate early time increase in aggregate mass concentration, $M(t)$, with time, t, as a function of the initial monomer concentration m_0. The bottom table gives the change in scaling exponent upon an increase in the monomer concentration, going from one extreme, such as the non-saturated regime, to the other extreme, such as the fully saturated regime. A value of $-1/2$ would thus correspond to a decrease in the scaling by $1/2$. (Adapted from Ref. [40] with permission from the Royal Society of Chemistry)

where $K = k_-/k_2$. As expected, this expression interpolates between the scaling for fragmentation, $\gamma = -1/2$, at low monomer concentrations and the scaling for secondary nucleation, $\gamma = -(n_2+1)/2$, at high monomer concentrations. The concentration at which the two processes contribute equally is determined by $K^{(1/n2)}$, where $K = k_-/k_2$. The scaling exponent decreases with increasing monomer and there is negative curvature in the half time plots (Fig. 1.5).

1.3.4 Representing the Reaction Network

So far, we have identified two quantities describing an aggregating system, $P(t)$ and $M(t)$, and have derived equations describing their time evolution in terms of the initial conditions and the physical properties of the system (the rate constants and the reaction orders). We have considered the individual processes and the effect of them acting in parallel and in series, now we combine them into a visual representation of the entire reaction network. Although the moment equations are very similar to mass action equations encountered in simple chemical reactions, the quantities $M(t)$ and $P(t)$ do not represent chemical species in the usual sense. As such, turning these equations into chemical schemes may at first seem unusual. Take for example the addition of a monomeric species to a fibril end with rate constant k_+: this leads to loss of a monomer, m, and production of a unit of mass, M. The number of ends P remains unchanged, so the corresponding chemical scheme would be

$$P + m \rightarrow P + M$$

although physically this simply corresponds to creating a fibril from a slightly shorter fibril and a monomer.

Applying this strategy, a reaction network in terms of the moments P and M can be formulated to visualise the aggregation reaction, see Fig. 1.6: Starting at the top left, primary processes use up monomer, m, and produce new fibrils ends P,

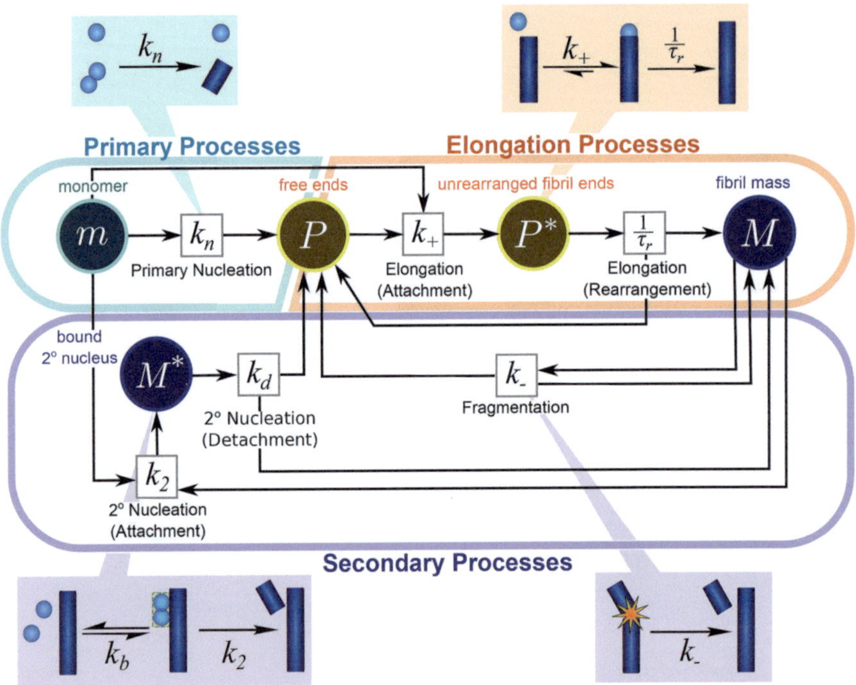

Fig. 1.6 Full reaction network. This reaction network explicitly shows all processes involved in the aggregation reaction discussed in this chapter. By following the arrows from monomer, m, one can trace the different possible paths to fibrils which the system can take. Note in particular the competition between the nucleation and fragmentation in parallel, (both form new aggregates P from aggregate mass M, but secondary nucleation also involves monomer m) as well as the positive feedback loop formed by elongation (from P to M) and the secondary processes (back from M to P). (Adapted from Ref. [40] with permission from the Royal Society of Chemistry)

and are independent of the concentration of fibrils present.[7] They are necessarily present in an aggregation reaction that starts from monomeric protein, as they are responsible for the formation of the initial aggregates from monomers alone. Elongation processes then also use up monomer m and produce fibril mass M. Although a growth competent end, P, is required for elongation, a new growth competent end is formed by the added monomer, so P is conserved in this reaction (note the arrow back towards P during the conversion of the intermediate P^*). Elongation is always present in an aggregating system of the kind studied here, as it is the process that produces the aggregate mass. Without elongation no fibrils would be formed, only small nuclei. The aggregate mass, M, can then catalyse the formation of further new aggregates, P, via one of two mechanisms: either fibrils fragment with rate k_-, or monomers nucleate on the fibril surface, via a bound intermediate M^*. Both secondary nucleation and fragmentation conserve aggregate mass M, as evident by the arrows back towards M.

This reaction network picture illustrates the basic principles of aggregation, not only how the different mechanisms compete in parallel but also why secondary nucleation is often of such central importance: Primary nucleation and elongation alone would be a simple linear reaction network (turquoise and orange, at the top of the network). However, if secondary processes are present they introduce a positive feedback loop: M is linked back to P to close the loop with elongation, either through fragmentation, directly with rate constant k_-, or via the intermediate, M^*, for secondary nucleation. As discussed previously this positive feedback results in a massive acceleration of the aggregation reaction, evident by the exponential increase in aggregate mass observed in experiment.

1.4 Application to Experiment: Global Fitting of Kinetic Data

A kinetic experiment of fibril formation usually involves measuring the concentration of aggregated material as a function of time, e.g. via a reporter dye such as Thioflavin-T (ThT), or recording the concentration of free monomer over time, e.g. via NMR techniques. In either case, the purity of the sample and the careful control of the experimental conditions are key. In fact, protein aggregation reactions tend to be very sensitive to impurities or surface effects and thus a meticulously optimized experimental procedure is a fundamental prerequisite for obtaining reproducible kinetic data [63].

As the mechanisms of aggregation are complex and thus involve many different parameters, the kinetic measurements have to be performed at a number of monomer concentrations in order to yield sufficient constraints for all the fitting parameters.

[7] As described in Sect. 1.2.3, "Common Approximations", the mass produced by nucleation can be neglected relative to that generated through elongation.

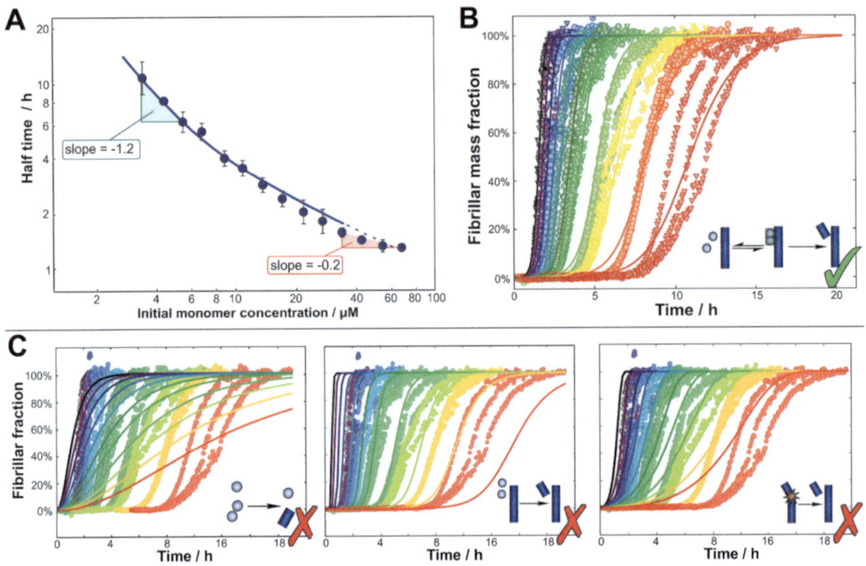

Fig. 1.7 Global fitting of aggregation data. (**a**) The half time plots are used to narrow down the mechanism. In this case strong monomer dependence at low monomer concentrations and the positive curvature indicate a saturation of secondary nucleation as the correct mechanism. (**b**) The mass concentration is fitted over the full time course to the approximate solution of the moment equations. This is a global fit, simultaneously at all monomer concentrations. (**c**) The fitting of a model that has only primary processes, a model that has fragmentation as its secondary process and a model that has a single-step secondary nucleation all fail. (Adapted from Ref. [52])

In a first step the half time plots can be used as an approximate guide to possible mechanisms (Fig. 1.7a). The solution for $M(t)$ for the mechanism in question is then fitted globally, i.e. to all data simultaneously, (Fig. 1.7b) to yield the rate constants and reaction orders of the individual processes [47]. It is important to also test other mechanism and ensure the data cannot be fitted by a simpler model (Fig. 1.7c).

Additional experiments in the presence of seeds can be used to provide further evidence for a given mechanism [63]. The variation of the aggregation behaviour upon addition of other compounds, such as possible inhibitors, can be used to determine at which point in the aggregation network this compound acts [65, 66].

1.5 Controlling Aggregation: Inhibitors and Solution Conditions

The understanding of the fundamental molecular processes that underlie the aggregation reaction opens the attractive possibility to rationally predict which processes have to be modified in order to affect the aggregation behaviour in a desired manner. This crucial application is relevant in many areas of biological and biotechnological sciences when there is the need to increase the yield of a desired product or to prevent aberrant aggregation. One important example is observed in the context of the aggregation of amyloidogenic peptides and proteins associated with the onset and development of neuro-degenerative disorders, where the inhibition of the aggregation process represents a promising strategy in the development of drugs for the prevention of these aggregation-related diseases. Another important application where aggregation must be strictly avoided is found in the industrial production of therapeutic proteins such as insulin and monoclonal antibodies.

The rate of every microscopic aggregation process can be expressed as the product of a reaction rate constant and the concentration of the reactive species. Typically, the reaction rate constant is strongly dependent on the intermolecular interactions that two colliding objects face while approaching each other. As a consequence, the kinetics of the aggregation process can be modulated by tuning the intermolecular forces via changes in the solution conditions. For instance, the screening of electrostatic interactions through an increase in the ionic strength of the solution can have significant effects on the aggregation behaviour: as the aggregating species are usually like-charged, higher ionic strength often results in reduced repulsion and increases all rates of assembly [53, 67, 68]. Moreover, different processes may be affected to a different degree, resulting in a shift of the dominant mechanism, for example through a saturation of the secondary nucleation process. In addition to shielding electrostatic interactions, another means to modify electrostatic interactions involves changing the charge of the protein: for example, α-synuclein displays a secondary nucleation-dominated process only under acidic conditions ($\mathrm{pH} < 5.8$). This effect may be due to a decrease in negative charge upon protonation of a side chain residue [38]. Equally, mutant proteins with altered charges can display significantly different aggregation behaviours and indeed, in the context of Alzheimer's and Parkinson's disease, many of the disease related mutants show an altered charge compared to the wild type [69, 70].

The methods described above affect the aggregation behaviour by modulating the interactions of the aggregating species in a complex manner, and often it is challenging to perturb specifically one single microscopic event by this modality. More targeted tools are offered by the use of species designed to bind a specific target in the aggregation network and thereby inhibit a specific process. This approach includes the addition into the reaction mixture of a large variety of modulators including small molecules, peptides, proteins and antibodies. In this case, the reaction rate is modified by changes in the concentration of the reactive species rather than by modifications of the reaction rate constants. In particular, a specific modification of a single microscopic event can be achieved by targeting a reactive species that participates only in the desired microscopic reaction.

For instance, one could envision the possibility of inhibiting elongation by adding compounds that bind to fibril ends preventing their further growth, i.e., reducing the concentration of fibril ends that are capable of adding monomers and elongate. Similarly, primary nucleation could be selectively suppressed by introducing molecules that interact with some pre-nucleus clusters, while secondary nucleation could be inhibited by adding species capable of binding to the surfaces of the fibrils. In Fig. 1.8 we outline the effect of inhibiting specifically several different processes on the time evolution of the fibrillar mass. In order to illustrate this concept, we simulate the inhibition of a reaction rate by decreasing the corresponding apparent reaction rate constant.[8] Depending on the process being inhibited, the half time or the slope of the aggregation curves can be modified in a different manner [65]. The characteristic changes in the shape of the reaction profiles in the absence and presence of the modulators can be analysed to identify inhibitors capable of modulating different processes [65].

In addition to modifying the aggregation kinetics, the inhibition of the aggregation process can have significant effects on the fibril length distribution at the completion of the reaction, as indicated by the simulated change in the fibril number concentrations shown in Fig. 1.8. At a fixed total protein concentration, a reduction in the number of fibrils at the completion of the reaction corresponds clearly to longer fibrils.

This observation is particularly important in the context of the search for drugs against diseases that are associated with protein aggregation, since increasing evidence indicates that different fibril lengths may display significant differences in toxicity. In particular, clusters of a few monomeric units are currently thought to be likely to represent the most toxic species. A key consequence highlighted by the kinetic analysis is the fact that for a successful therapeutic intervention it is crucial to achieve a targeted inhibition of specific steps rather than aiming at a generic arrest of the aggregation process [65]. Indeed, a decrease in the overall aggregation rate achieved for instance by suppressing the elongation rate may result in an increase in toxicity due to a shift of the fibril population towards shorter, more toxic species (Fig. 1.8b) [71]. By contrast, a specific suppression of the microscopic reactions which are most responsible for the generation of oligomers is desired. For this purpose, the kinetic platform described in this chapter provides an extremely powerful tool for the screening and the identification of compounds capable of achieving this goal. Indeed, this analysis has been applied to identify the variety of microscopic mechanisms through which biologically relevant components are capable of supressing aggregation under normal physiological conditions [72]. In particular, a molecular chaperone has been recognized to suppress specifically and dramatically the rate of secondary nucleation events that are responsible for the generation of toxic species during the aggregation of the Aβ42 peptide, the peptide associated with the onset and development of Alzheimer's disease [71]. From the lessons learnt from nature, we envision the possibility to design molecules that mimic the same effects and could represent promising drug candidates against neurodegenerative diseases [73].

[8]This is a treatment of inhibition as a perturbation to the models for aggregation of pure protein and does not explicitly include reactions of the inhibitor with the various species in the aggregation reaction network. Therefore, it does not reproduce intricate effects, for example due to the kinetics of inhibitor binding, but it does to a very good approximation yield the same overall behaviour as the more complex approach and is thus sufficient for the purposes of illustrating the effect or establishing which species is targeted by the inhibitor.

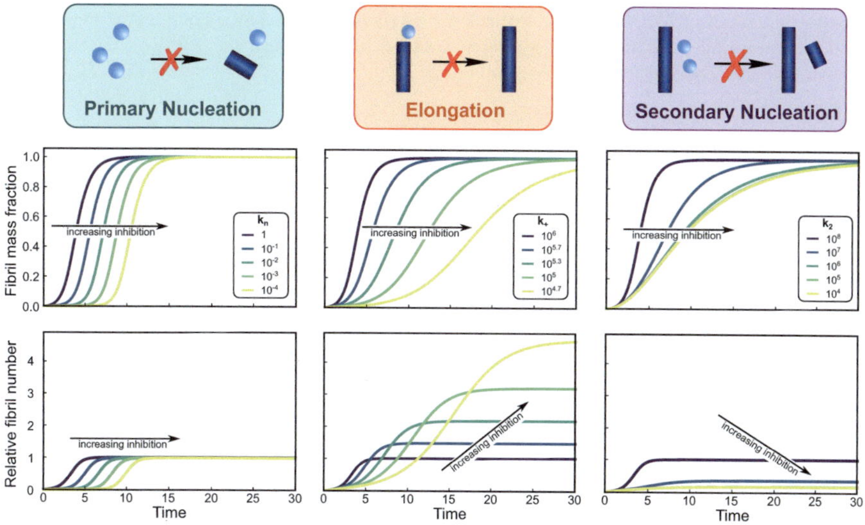

Fig. 1.8 Effect of inhibition of different processes. The effect of inhibition of different processes on the mass and number concentrations of fibrils is modelled. The system modelled is based approximately on the rate constants for Aβ42 from Alzheimer's disease. Blue curves have the highest rate constants (least inhibition), yellow curves the lowest rate constants (highest inhibition). An inhibition of primary nucleation results in a shift of the curves along the time axis; the fibril number at completion is not affected. An inhibition of elongation is more effective in delaying the formation of aggregate mass than an inhibition of primary nucleation, but it also significantly increases the fibril number at completion because nucleation processes are favoured over elongation processes, resulting in shorter fibrils. Last, an inhibition of secondary nucleation delays aggregate formation and significantly decreases the number of fibrils at equilibrium, because now elongation processes are favoured over nucleation processes, giving longer fibrils. Note that the effect of inhibiting secondary nucleation saturates because once secondary nucleation is sufficiently inhibited, primary nucleation will become the dominant process of aggregate multiplication and a further inhibition of secondary nucleation has no more effect on the kinetics. The y-axes are normalised with respect to the plateau value for the uninhibited case

1.6 Conclusions

An in depth understanding of the mechanisms of protein self-assembly is a fundamental prerequisite for controlling aggregation, both in the development of therapies to aggregation-related diseases and in the design of self-assembling materials. In this context, chemical kinetics is the central tool for determining the mechanisms of aggregation as it provides the means of gaining a molecular level understanding through measurements of easily accessible macroscopic quantities. Once established, a kinetic model can be then used to determine key steps in the reaction network and predict the effects of altering the mechanism.

Acknowledgements We would like to thank the Swiss National Science Foundation, Peterhouse College Cambridge, the European Research Council, the BBSRC, the EPSRC, the Newman Foundation and Sidney Sussex College Cambridge.

References

1. Sawaya MR, Sambashivan S, Nelson R, Ivanova MI, Sievers SA, Apostol MI, Thompson MJ, Balbirnie M, Wiltzius JJW, McFarlane HT, Madsen A, Riekel C, Eisenberg D (2007) Atomic structures of amyloid cross-beta spines reveal varied steric zippers. Nature 447(7143):453–457
2. Sunde M, Blake C (1997) The structure of amyloid fibrils by electron microscopy and x-ray diffraction. Adv Protein Chem 50:123–159
3. Knowles TPJ, Vendruscolo M, Dobson CM (2014) The amyloid state and its association with protein misfolding diseases. Nat Rev Mol Cell Biol 15:384–396
4. Ciryam P, Tartaglia GG, Morimoto RI, Dobson CM, Vendruscolo M (2013) Widespread aggregation and neurodegenerative diseases are associated with supersaturated proteins. Cell Rep 5(3):781–790
5. Boller F, Mizutani T, Roessmann U, Gambetti P (1980) Parkinson disease, dementia, and alzheimer disease: clinicopathological correlations. Ann Neurol 7(4):329–335
6. Lansbury PT, Lashuel HA (2006) A century-old debate on protein aggregation and neurodegeneration enters the clinic. Nature 443(7113):774–779
7. Bemporad F, Chiti F (2012) Protein misfolded oligomers: experimental approaches, mechanism of formation, and structure-toxicity relationships. Chem Biol 19(3):315–327
8. Oosawa F, Kasai M (1962) A theory of linear and helical aggregations of macromolecules. J Mol Biol 4:10–21
9. Oosawa F, Asakura S (1975) Thermodynamics of the polymerization of protein. Academic, New York
10. Alberts B, Johnson A, Lewis J, Raff M, Roberts K, Walter P (2002) Molecular biology of the cell, 4th edn. Garland Publishing, New York
11. Hill TL (1987) Linear aggregation theory in cell biology. Springer, New York
12. Fowler DM, Koulov AV, Balch WE, Kelly JW (2007) Functional amyloid–from bacteria to humans. Trends Biochem Sci 32(5):217–224
13. Kelly JW, Balch WE (2003) Amyloid as a natural product. J Cell Biol 161:461–462
14. Fowler DM, Koulov AV, Alory-Jost C, Marks MS, Balch WE, Kelly JW (2006) Functional amyloid formation within mammalian tissue. PLoS Biol 4(1):e6
15. Maji SK, Perrin MH, Sawaya MR, Jessberger S, Vadodaria K, Rissman RA, Singru PS, Nilsson KPR, Simon R, Schubert D, Eisenberg D, Rivier J, Sawchenko P, Vale W, Riek R (2009) Functional amyloids as natural storage of peptide hormones in pituitary secretory granules. Science 325(5938):328–332
16. Talbot NJ, Kershaw MJ, Wakley GE, De Vries OMH, Wessels JGH, Hamer JE (1996) Mpg1 encodes a fungal hydrophobin involved in surface interactions during infection-related development of magnaporthe grisea. Plant Cell 8(6):985–999
17. Romero D, Aguilar C, Losick R, Kolter R (2010) Amyloid fibers provide structural integrity to bacillus subtilis biofilms. Proc Natl Acad Sci U S A 107(5):2230–2234
18. Chapman MR, Robinson LS, Pinkner JS, Roth R, Heuser J, Hammar M, Normark S, Hultgren SJ (2002) Role of escherichia coli curli operons in directing amyloid fiber formation. Science 295(5556):851–855
19. Mankar S, Anoop A, Sen S, Maji SK (2011) Nanomaterials: amyloids reflect their brighter side. Nano Rev 2:6032. https://doi.org/10.3402/nano.v2i0.6032
20. Bolisetty S, Mezzenga R (2015) Amyloid-carbon hybrid membranes for universal water purification. Nat Nano 11:365–371

21. Dobson CM (1999) Protein misfolding, evolution and disease. Trends Biochem Sci 24(9):329–332
22. Ferrone F (1999) Analysis of protein aggregation kinetics. Methods Enzymol 309:256–274
23. Knowles TPJ, Waudby CA, Devlin GL, Cohen SIA, Aguzzi A, Vendruscolo M, Terentjev EM, Welland ME, Dobson CM (2009) An analytical solution to the kinetics of breakable filament assembly. Science 326(5959):1533–1537
24. Galvagnion C, Buell AK, Meisl G, Michaels TCT, Vendruscolo M, Knowles TPJ, Dobson CM (2015) Lipid vesicles trigger a-synuclein aggregation by stimulating primary nucleation. Nat Chem Biol 11:229–234
25. Grigolato F, Colombo C, Ferrari R, Rezabkova L, Arosio P (2017) Mechanistic origin of the combined effect of surfaces and mechanical agitation on amyloid formation. ACS Nano 11(11):11358–11367. PMID: 29045787
26. Pham CLL, Rey A, Lo V, Soulès M, Ren Q, Meisl G, Knowles TPJ, Kwan AH, Sunde M (2016) Self-assembly of MPG1, a hydrophobin protein from the rice blast fungus that forms functional amyloid coatings, occurs by a surface-driven mechanism. Sci Rep 6:25288
27. Michaels TCT, Garcia GA, Knowles TPJ (2014) Asymptotic solutions of the oosawa model for the length distribution of biofilaments. J Chem Phys 140(19):194906
28. Ruschak AM, Miranker AD (2007) Fiber-dependent amyloid formation as catalysis of an existing reaction pathway. Proc Natl Acad Sci U S A 104(30):12341–12346
29. Xue W-F, Hellewell AL, Gosal WS, Homans SW, Hewitt EW, Radford SE (2009) Fibril fragmentation enhances amyloid cytotoxicity. J Biol Chem 284(49):34272–34282
30. Cohen SIA, Linse S, Luheshi LM, Hellstrand E, White DA, Rajah L, Otzen DE, Vendruscolo M, Dobson CM, Knowles TPJ (2013) Proliferation of amyloid-beta42 aggregates occurs through a secondary nucleation mechanism. Proc Natl Acad Sci 110:9758–9763
31. Collins SR, Douglass A, Vale RD, Weissman JS (2004) Mechanism of prion propagation: amyloid growth occurs by monomer addition. PLoS Biol 2(10):e321
32. Kunes KC, Cox DL, Singh RRP (2005) One-dimensional model of yeast prion aggregation. Phys Rev E Stat Nonlin Soft Matter Phys 72(5 Pt 1):051915
33. Stöhr J, Weinmann N, Wille H, Kaimann T, Nagel-Steger L, Birkmann E, Panza G, Prusiner SB, Eigen M, Riesner D (2008) Mechanisms of prion protein assembly into amyloid. Proc Natl Acad Sci U S A 105(7):2409–2414
34. Ferrone FA, Hofrichter J, Eaton WA (1985) Kinetics of sickle hemoglobin polymerization. I. Studies using temperature-jump and laser photolysis techniques. J Mol Biol 183(4):591–610
35. Ferrone FA, Hofrichter J, Eaton WA (1985) Kinetics of sickle hemoglobin polymerization. II. A double nucleation mechanism. J Mol Biol 183(4):611–631
36. Bishop MF, Ferrone FA (1984) Kinetics of nucleation-controlled polymerization. A perturbation treatment for use with a secondary pathway. Biophys J 46(5):631–644
37. Cohen SIA, Vendruscolo M, Dobson CM, Knowles TPJ (2012) From macroscopic measurements to microscopic mechanisms of protein aggregation. J Mol Biol 421(2–3):160–171
38. Buell AK, Galvagnion C, Gaspar R, Sparr E, Vendruscolo M, Knowles TPJ, Linse S, Dobson CM (2014) Solution conditions determine the relative importance of nucleation and growth processes in α-synuclein aggregation. Proc Natl Acad Sci 111(21):7671–7676
39. Knowles TPJ, Buehler MJ (2011) Nanomechanics of functional and pathological amyloid materials. Nat Nanotechnol 6(8):469–479
40. Meisl G, Rajah L, Cohen SAI, Pfammatter M, Saric A, Hellstrand E, Buell AK, Aguzzi A, Linse S, Vendruscolo M, Dobson CM, Knowles TPJ (2017) Scaling behaviour and rate-determining steps in filamentous self-assembly. Chem Sci 8:7087–7097
41. Krapivsky PL, Redner S, Ben-Naim E (2010) A kinetic view of statistical physics. Cambridge University Press, Leiden
42. Xue W-F, Homans SW, Radford SE (2008) Systematic analysis of nucleation-dependent polymerization reveals new insights into the mechanism of amyloid self-assembly. Proc Natl Acad Sci U S A 105(26):8926–8931
43. Arosio P, Cedervall T, Knowles TPJ, Linse S (2016) Analysis of the length distribution of amyloid fibrils by centrifugal sedimentation. Anal Biochem 504:7–13

44. Nasir I, Linse S, Cabaleiro-Lago C (2015) Fluorescent filter-trap assay for amyloid fibril formation kinetics in complex solutions. ACS Chem Neurosci (8):1436–1444. PMID: 25946560
45. Gaspar R, Meisl G, Buell AK, Young L, Kaminski CF, Knowles TPJ, Sparr E, Linse S (2017) Secondary nucleation of monomers on fibril surface dominates α-synuclein aggregation and provides autocatalytic amyloid amplification. Q Rev Biophys 50:E6
46. Bender CM, Orszag SA (1999) Advanced mathematical methods for scientists and engineers. Springer, New York
47. Meisl G, Kirkegaard JB, Arosio P, Michaels TTC, Vendruscolo M, Dobson CM, Linse S, Knowles TPJ (2016) Molecular mechanisms of protein aggregation from global fitting of kinetic models. Nat Protoc 11(2):252–272
48. Cohen SIA, Vendruscolo M, Welland ME, Dobson CM, Terentjev EM, Knowles TPJ (2011) Nucleated polymerization with secondary pathways. I. Time evolution of the principal moments. J Chem Phys 135(6):065105
49. Cohen SIA, Vendruscolo M, Dobson CM, Knowles TPJ (2011) Nucleated polymerization with secondary pathways. II. Determination of self-consistent solutions to growth processes described by non-linear master equations. J Chem Phys 135(6):065106
50. Arosio P, Knowles TPJ, Linse S (2015) On the lag phase in amyloid fibril formation. Phys Chem Chem Phys 17:7606–7618
51. Knowles TPJ, White DA, Abate AR, Agresti JJ, Cohen SIA, Sperling RA, De Genst EJ, Dobson CM, Weitz DA (2011) Observation of spatial propagation of amyloid assembly from single nuclei. Proc Natl Acad Sci U S A 108(36):14746–14751
52. Meisl G, Yang X, Hellstrand E, Frohm B, Kirkegaard JB, Cohen SIA, Dobson CM, Linse S, Knowles TPJ (2014) Differences in nucleation behavior underlie the contrasting aggregation kinetics of the aβ40 and aβ42 peptides. Proc Natl Acad Sci 111:9384–9389
53. Meisl G, Yang X, Dobson CM, Linse S, Knowles TPJ (2017) Modulation of electrostatic interactions to reveal a reaction network unifying the aggregation behaviour of the aβ42 peptide and its variants. Chem Sci 8:4352–4362
54. Knowles TPJ, Vendruscolo M, Dobson CM (2015) The physical basis of protein misfolding disorders. Phys Today 68(3):36
55. Lee J, Culyba EK, Powers ET, Kelly JW (2011) Amyloid-beta forms fibrils by nucleated conformational conversion of oligomers. Nat Chem Biol 7(9):602–609
56. Scheibel T, Bloom J, Lindquist SL (2004) The elongation of yeast prion fibers involves separable steps of association and conversion. Proc Natl Acad Sci U S A 101(8):2287–2292
57. Esler WP, Stimson ER, Jennings JM, Vinters HV, Ghilardi JR, Lee JP, Mantyh PW, Maggio JE (2000) Alzheimer's disease amyloid propagation by a template-dependent dock-lock mechanism. Biochemistry 39(21):6288–6295
58. Buell AK, Blundell JR, Dobson CM, Welland ME, Terentjev EM, Knowles TPJ (2010) Frequency factors in a landscape model of filamentous protein aggregation. Phys Rev Lett 104(22):228101
59. Michaelis L, Menten M (1913) Die kinetik der invertinwirkung. Biochem Z 49:333
60. Connors KA (1990) Chemical kinetics: study of reaction rates in solution. Wiley, New York
61. Orte A, Clarke R, Balasubramanian S, Klenerman D (2006) Determination of the fraction and stoichiometry of femtomolar levels of biomolecular complexes in an excess of monomer using single-molecule, two-color coincidence detection. Anal Chem 78(22):7707–7715
62. Garcia GA, Cohen SIA, Dobson CM, Knowles TPJ (2014) Nucleation-conversion-polymerization reactions of biological macromolecules with prenucleation clusters. Phys Rev E 89:032712
63. Meisl G, Yang X, Frohm B, Knowles TPJ, Linse S (2016) Quantitative analysis of intrinsic and extrinsic factors in the aggregation mechanism of alzheimer-associated aβ-peptide. Sci Rep 6:18728
64. Saric A, Buell A, Meisl G, Michaels TCT, Dobson CM, Linse S, Knowles TPJ, Frenkel D (2016) Physical determinants of the self-replication of protein fibrils. Nat Phys 12:874–880
65. Arosio P, Vendruscolo M, Dobson CM, Knowles TPJ (2014) Chemical kinetics for drug discovery to combat protein aggregation diseases. Trends Pharmacol Sci 35(3):127–135

66. Abelein A, Graslund A, Danielsson J (2015) Zinc as chaperone-mimicking agent for retardation of amyloid β peptide fibril formation. Proc Natl Acad Sci U S A 112(17):5407–5412
67. Klement K, Wieligmann K, Meinhardt J, Hortschansky P, Richter W, Fändrich M (2007) Effect of different salt ions on the propensity of aggregation and on the structure of Alzheimer's abeta(1–40) amyloid fibrils. J Mol Biol 373(5):1321–1333
68. Buell AK, Hung P, Salvatella X, Welland ME, Dobson CM, Knowles TPJ (2013) Electrostatic effects in filamentous protein aggregation. Biophys J 104:1116–1126
69. Flagmeier P, Meisl G, Vendruscolo M, Knowles TPJ, Dobson CM, Buell AK, Galvagnion C (2016) Mutations associated with familial parkinson's disease alter the initiation and amplification steps of α-synuclein aggregation. Proc Natl Acad Sci U S A 113(37):10328–10333
70. Yang X, Meisl G, Frohm B, Thulin E, Knowles TPJ, Linse S (2018) On the role of sidechain size and charge in the aggregation of aβ42 with familial mutations. Proc Natl Acad Sci U S A 115(26):E5849–E5858
71. Cohen SIA, Arosio P, Presto J, Kurudenkandy FR, Biverstal H, Dolfe L, Dunning C, Yang X, Frohm B, Vendruscolo M, Johansson J, Dobson CM, Fisahn A, Knowles TPJ, Linse S (2015) The molecular chaperone brichos breaks the catalytic cycle that generates toxic ab oligomers. Nat Struct Mol Biol 22:207–213
72. Arosio P, Michaels TCT, Linse S, Månsson C, Emanuelsson C, Presto J, Johansson J, Vendruscolo M, Dobson C, Knowles TPJ (2016) Kinetic analysis reveals the diversity of microscopic mechanisms through which molecular chaperones suppress amyloid formation. Nat Commun 7:10948
73. Habchi J, Arosio P, Perni M, Costa AR, Yagi-Utsumi M, Joshi P, Chia S, Cohen SIA, Müller MBD, Linse S, Nollen EAA, Dobson CM, Knowles TPJ, Vendruscolo M (2016) An anticancer drug suppresses the primary nucleation reaction that initiates the production of the toxic aβ42 aggregates linked with alzheimer's disease. Sci Adv 2(2):e1501244

Chapter 2
Peptide Self-Assembly and Its Modulation: Imaging on the Nanoscale

Lanlan Yu, Yanlian Yang, and Chen Wang

Abstract This chapter intends to review the progress in obtaining site-specific structural information for peptide assemblies using scanning tunneling microscopy. The effects on assembly propensity due to mutations and modifications in peptide sequences, small organic molecules and conformational transitions of peptides are identified. The obtained structural insights into the sequence-dependent assembly propensity could inspire rational design of peptide architectures at the molecular level.

Keywords Peptide assembly · Modulation · Scanning tunneling microscopy

Abbreviations

4Bpy	4,4′-bipyridyl
AD	Alzheimer's disease
AFM	Atomic force microscopy
ALS	Amyotrophic lateral sclerosis
Aβ	Beta-amyloid peptides
CD	Circular dichroism
DPE	1,2-di(4-pyridyl)-ethylene
FTIR	Fourier transform infrared spectroscopy
FTLD	Frontotemporal lobar degeneration
hIAPP	Human islet amyloid polypeptide
HOPG	Highly oriented pyrolytic graphite
MD	Molecular dynamics
NMR	Nuclear magnetic resonance
$PcCu(SO_3Na)_4$	Copper phthalocyanine tetrasulfonate sodium

L. Yu · Y. Yang · C. Wang (✉)
National Center for Nanoscience and Technology, Chinese Academy of Sciences, Beijing, China
e-mail: wangch@nanoctr.cn

S. Perrett et al. (eds.), *Biological and Bio-inspired Nanomaterials*,
Advances in Experimental Medicine and Biology 1174,
https://doi.org/10.1007/978-981-13-9791-2_2

STM	Scanning tunneling microscopy
TDP-43	TAR DNA-binding protein 43
TEM	Transmission electron microscopy
ThT	Thioflavin T
TSE	Transmissible spongiform encephalopathy
UHV	Ultrahigh vacuum
VNTRs	Variable number of tandem repeats
XRD	X-ray diffraction

2.1 Introduction

Peptide assemblies have been drawing extensive interest due to their vital roles in biological recognition associated with interactions between proteins, interactions between peptide drugs and targets, disease-associated amyloid aggregation, and the design of novel functional biomaterials. Common to the key elements of molecular assembly, inter-peptide interactions include non-covalent interactions such as hydrogen bonds, electrostatic interactions (or salt bridges), the hydrophobic effect, van der Waals interactions and water-mediated hydrogen bonds [1, 2]. Representative examples of assembly processes are seen among amyloidal peptides which have been extensively investigated due to their relevance to over 20 diseases including Alzheimer's disease (AD) and Parkinson's disease (PD). The assembly of amyloidal peptides generally evolves from oligomers to protofibrils, and protofibrils subsequently twist together into mature fibrils and plaques [3]. Efforts to unveil the hierarchical assembly pathways and to decode the detailed structural information would be beneficial to modulate and manipulate them for the rational design of biological materials.

The advances in unraveling molecular mechanisms have contributed to a broad range of potential applications in areas including controlled release and drug delivery, cell culture, tissue engineering, antimicrobial and biomineralization materials [2, 4–15]. Furthermore, peptide assemblies on surfaces have also been widely explored as a reflection of biological and physiological processes *in vivo*, which are relevant to membranes and potential applications for biosensors and heterogeneous biocatalysis, drug delivery, and affinity chromatography [16–18]. It is worth noting that peptide assemblies on surfaces are typically stabilized through peptide-peptide and peptide-substrate interactions, similar to other widely studied surface-mediated molecular assemblies.

To analyze the assembly pathways and structural information, various experimental approaches have been attempted. The aggregation propensity and stability can be studied by kinetic measurements such as the thioflavin T (ThT) binding assay [19, 20]. Various spectroscopies, such as circular dichroism (CD) spectroscopy, Fourier transform infrared spectroscopy (FTIR) and ultraviolet Raman spectroscopy can provide information regarding secondary structure [21–24]. Imaging methods, including atomic force microscopy (AFM) and transmission electron microscopy (TEM) can be employed to obtain the morphology and size evolution of the peptide aggregates [25–27]. In terms of subtle details of peptide assembly structures, X-ray diffraction (XRD) [28–32], nuclear magnetic resonance (NMR) [25, 33–39] and cryo-electron microscopy [31] have been generally applied to obtain high-resolution structural information. However, these techniques are restricted by either the complicated procedures of sample preparation or the nature of peptides such as solubility and crystallinity. More significantly, the results are confined to an ensemble averaging regime and only provide signals from the convolutions of all residues.

Notably, scanning tunneling microscopy (STM) has been widely exploited to probe the assembly behavior and single-molecular structure of small molecules [40–48] on surfaces due to its submolecular structural resolution and adaptability to various conditions including ambient conditions and ultrahigh vacuum (UHV) conditions. STM offers the advantage of easy sample preparation procedure, and more importantly, its structural resolution is theoretically down to single molecule, and even amino acid residue-specific resolution. Thus far, STM has successfully demonstrated its capacity for investigating peptide assemblies ranging from simple model peptides [49–51] to complex disease-related peptides [27, 52, 53].

In this chapter, we will firstly introduce the STM investigations on surface-bound peptide assembly structures with residue-specific resolution. Subsequently, the influence of mutations and modifications (phosphorylation/glycosylation) in peptide assemblies and the aggregation propensities are presented, as well as interactions between the peptides with small molecules including terminus modulators and side group modulators. Finally, the correlation of peptide assemblies on surface and in solution and the peptide conformational transitions when adsorbed on surfaces are also discussed.

2.2 Peptide Self-Assembly Structures on Surfaces

Structural analysis is central to gaining insight into inter-peptide interactions and the relationship between peptide sequence and assembly structures. As a technique with submolecular-level resolution, the potential of STM in structural analysis of peptide molecules and peptide assemblies under various environments including UHV as well as ambient conditions has been established [49–53].

For UHV-STM investigations, peptide molecules are usually adsorbed onto metal surfaces defined by surface indexes (such as Cu(110) and Au(111)) and at relatively low temperatures to minimize thermally induced fluctuation effects on high-resolution imaging. The smallest peptide molecule comprised of only two amino acid residues (the chiral dipeptide di-*L*-alanine) was first observed to self-assemble on a metal surface using UHV-STM. Sub-monolayer growth of di-*L*-alanine on Cu(110) was observed and the peptide was found to nucleate into small islands with single or double molecule rows at low coverage (Fig. 2.1a) [51]. The dialanines appear as pairs of a smaller and a larger protrusion in STM images, aligned along the [332] direction. Within a single row, the alignment of molecules with the same orientation indicates that the –COOH moiety of one molecule binds to

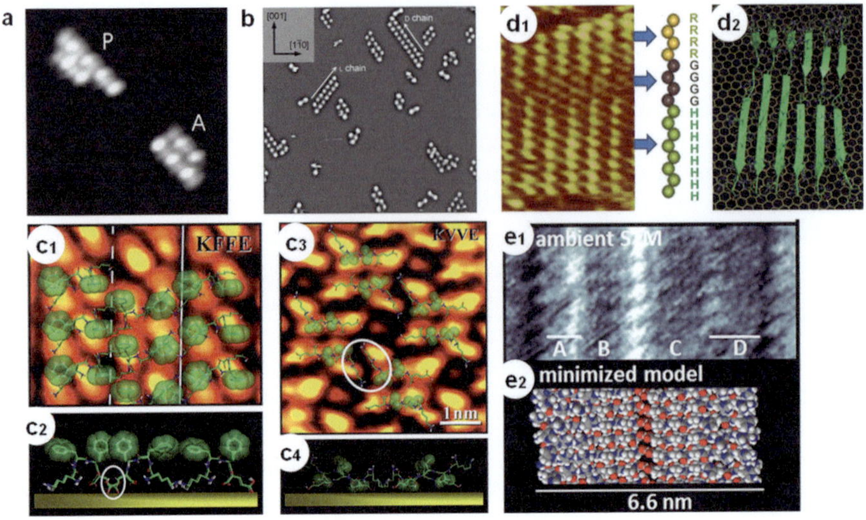

Fig. 2.1 (**a**) STM image of di-*L*-alanine on Cu(110) at low coverage. The molecules were evaporated at 248 K and scanning took place at 208 K. Two islands show parallel (P) or anti-parallel (A) peptide molecules in adjacent rows. (**b**) STM image of co-adsorbed *L*-Phe-*L*-Phe and *D*-Phe-*D*-Phe on Cu(110) surface at room temperature. The arrows indicate the growth direction of the homochiral chains. Image size: 36 nm × 34 nm. (**c1**, **c3**) STM image of KFFE and KVVE on the Au(111) surface with parallel and antiparallel structures, respectively. The representative model with Phe side chains highlighted in green. The oval indicates antiparallel organization of the Lys side chain in (**c3**). (**c2**) The side view of the models with Au surface indicated by the horizontal bar. The oval indicates putative surface binding in (**c2**). Val side chain carbons highlighted by transparent spheres for (**c3**) and (**c4**). (**d1**) STM image of $R_4G_4H_8$ self-assembly. (**d2**) Schematic of the molecular structure resulting from MD simulations. (**e1**) Ambient STM image of HA showing symmetry with UHV image and contrast difference in the feature near the center. (**b2**) Energy-minimized model of HA peptides showing alignment of peptide termini and block boundaries with images above. (Reprinted and adapted with permission from the *Surface Science* http://www.sciencedirect.com/science/article/pii/S0039602803011300 [51] (**a**); *Angewandte Chemie International Edition* http://onlinelibrary.wiley.com/doi/10.1002/ange.200700194/full [54] (**b**); *ACS Nano* http://pubs.acs.org/doi/abs/10.1021/nn301708d [55] (**c**); *Journal of the American Chemical Society* http://pubs.acs.org/doi/abs/10.1021/ja307198u [50] (**d**); *Journal of the American Chemical Society* http://pubs.acs.org/doi/abs/10.1021/ja408550a [57] (**e**))

the –NH2 moiety of the neighboring molecule, very likely by a hydrogen bond. At low evaporation temperature, the adjacent rows in a particular island could be either parallel or anti-parallel depending on the thermal history of the surface. At higher temperature, all molecules within the same island have the same orientation. This indicates the presence of an energy barrier between the parallel and the antiparallel adsorption directions, of which the parallel configuration has the lower energy.

Moreover, the chiral dipeptide, diphenylalanine with *L*-Phe-*L*-Phe and *D*-Phe-*D*-Phe forms can be distinguished on Cu(110) in the UHV-STM images, as shown in Fig. 2.1b [54]. The topography of the peptide molecule is represented by two high contrast features and a central low contrast part corresponding to the two electron-rich phenyl rings and the peptide backbone, respectively. The discernible molecular chirality of the adsorbed Phe-Phe molecules indicates that their stereogenic centers are involved in the molecule-surface interactions. Meanwhile, the STM video recording of two isolated *D* chirality molecules was performed to attain a dynamic description of the chiral-recognition phenomenon, which reveals that the molecules form an initial pair by experiencing first a change in the adsorption geometry and then go through several further rearrangements to reach the final stable state. Since this chiral recognition occurs only when the two molecules are positioned at a close distance, it implies that these conformational changes could be induced by mutual intermolecular interaction. The results reveal the dynamic process of enantioselective molecular interactions resulting from conformational adjustments.

The above illustrated structural resolutions of dipeptides can be also demonstrated in longer peptides on surfaces. The tetrapeptides KFFE and KVVE have been reported to be among the minimal amyloid fibril-forming peptide units *in vitro*. Highly ordered adsorption structures (Fig. 2.1c) were observed on an inert Au(111) surface by STM under UHV conditions [55]. The observations reveal that KFFE and KVVE adopt parallel and antiparallel structures, respectively, which are similar to their arrangements in fibrillar structures. This investigation offers interesting perspectives for gaining single-molecule insights into amyloid fibril formation by using a surface science approach with STM.

For ambient STM investigations, peptide assemblies can be clearly resolved under atmospheric conditions, commonly at room temperature. Highly oriented pyrolytic graphite (HOPG) is usually chosen as the substrate due to its chemical inertness. More importantly, peptide assemblies at the liquid-solid interface are more representative of natural states in biological systems in comparison to UHV conditions. Studies have been carried out on the assembly behavior of cyclic peptides [56], model peptides [49] and disease-related peptides [27, 53] to unveil the mechanisms of peptide assembly and aggregation. The design and study of residue-specific recognition between peptide sequences has been pursued because of its significance in understanding of the peptide assembly mechanism and peptide-peptide interactions. Despite the high resolution capabilities of the STM technique, it remains challenging to distinguish individual residues in the peptide sequence because the STM image brightness contrast is a convolution of both the electronic and topographic structures and affected by the adsorption stability and conformations of the adsorbed molecules. However, assuming that backbones of peptides contribute nearly the same in the STM observations, it is feasible to explore the dependence of the brightness contrast on the side groups.

Such efforts have been reflected in the ambient STM study of homogeneous assemblies of two model peptides, $R_4G_4H_8$ and $F_4G_4H_8$ (Fig. 2.1d) [50]. The separation between the two adjacent peptide strands was measured to be 4.6 ± 0.2 Å, indicating the formation of β-sheet structure. Combined with the length distributions of the two peptides, it suggests that both model peptides form parallel β-sheet assembly structures on the HOPG surface. Meanwhile, these amino acid residues could be distinguished qualitatively depending on the observed brightness in the order Phe > His > Arg > Gly. Furthermore, all-atom molecular dynamic (MD) simulations were performed to explore the conformational dynamics of the peptide assemblies adsorbed onto a graphene sheet, which shows that the interaction energy of four different residues with the underlying graphite surface follows the same order as the STM brightness contrast. This correlation between the brightness contrast and the related interaction energy of the different residues may shed light upon the sequence effects and conformation effects on the peptide assemblies. In another related study, differentiating between histidine and alanine residues (Fig. 2.1e), and side chain orientations in individual histidine residues were also successfully achieved in the block peptide with the sequence HHHHHAAAAA, by correlating features in STM images with those in energy-optimized models in molecular modeling [57]. These pioneering works pave the way for studies of site-specific interactions between molecular species and peptides, and between heterogeneous peptides and protein structures at the single-molecule level.

2.3 Mutation/Modification Effects on Peptide Assemblies

Mutations in peptides, and modifications such as phosphorylation and glycosylation exist widely and play essential roles in biological systems. For surface-bound peptide assemblies, the chemical nature of the amino acids is crucial for the peptide-peptide and peptide-substrate interactions. It is reasonable that the alteration of an amino acid side chain or the introduction of phosphoryl or saccharide groups (*i.e.* phosphorylation or glycosylation) would significantly influence the binding affinities and assembly characteristics of peptides.

Mutation effects on disease-associated peptides have drawn considerable interest, especially for the amyloidal peptides related to degenerative diseases and some cancers. For instance, the beta-amyloid peptide (Aβ), the human islet amyloid polypeptide (hIAPP), the TAR DNA-binding protein 43 (TDP-43) and the prion protein have been extensively explored with respect to various aspects using miscellaneous methods due to their close relevance to the following currently intractable diseases: Alzheimer's disease (AD), type II diabetes mellitus, amyotrophic lateral sclerosis (ALS) and frontotemporal lobar degeneration (FTLD), and transmissible spongiform encephalopathy (TSE) [58–63]. Mutations found in nature related to AD include Dutch (E22Q) [64], Flemish (A21G) [65], Arctic (E22G) [66], Iowa (D23N) [67], English (H6R) [68], Tottori (D7N) [69], and A2V [70] variants. These mutations are associated with the early development of AD and the peptides have high amyloid aggregation propensities. Over 30 mutations have also been detected in the TDP-43 protein among ALS patients [71, 72]. Furthermore, many efforts have been made to understand the assembly behavior of mutant analogs of hIAPP as well. However, more studies are still needed to fully understand the complex mechanisms of amyloid assembly and the related neurotoxicity.

Mutational analysis directly compares the differences in behaviour caused by the replacement of residues at the same site. This approach can provide information of the effect of certain residues on the assembly kinetics, aggregation propensity, stability and fibril structure. For instance, a particular mutant which has a lower aggregation propensity could suggest that the original residue contributes to this property to a large extent. Furthermore, single point mutations provide a bridge between the nature of a specific amino acid and its role in the physico-chemical and functional properties of the peptide.

Herein, the assembly structures of hIAPP-related mutants have been investigated by using STM, including $hIAPP_{8–37}$, rat $IAPP_{8–37}$ ($rIAPP_{8–37}$), $rIAPP_{8–37}R18H$ on a HOPG surface [73]. In order to identify the folding sites in the sequence, a chaperone molecule, 4,4′-bipyridyl (4Bpy), was introduced to co-assemble with the peptide to allow identification of the C-terminus of the peptide, due to the strong O–H···N hydrogen bond between the carboxyl terminus of the peptide molecule and the nitrogen atom of the 4Bpy molecule. With one terminus fixed and labeled, the exact C-terminal region participating in the surface-bound peptide assembly can be determined by the number of residues, which is obtained from dividing the length of C-terminal strand by 0.325 nm, the typical distance of two neighboring residues in parallel β-sheet structures. Likewise, the N-terminal region can be deduced by repeating the experiment with the reversed sequence, whereby the sequence is identical but with reversed N and C termini. This strategy has been readily applied to investigate the key sites of beta-structural assembly consisting of multiple beta motifs. The high resolution STM images of IAPP analogs present typical lamellar features, as shown in Fig. 2.2. The results of a statistical analysis show that the $IAPP_{8–37}$ analogs have the common motifs of $IAPP_{8–17}$ and $IAPP_{26–37}$ with the most probable folding sites at Ser19/Ser20 and Gly24. The peptides manifest similar tendencies to form amyloid fibrils in the N and C termini owing to the similarity of the peptide sequences, while the differences caused by changes in specific key sites demonstrate the effect of sequence variations. The length distributions of $rIAPP_{8–37}$ are wider compared with $hIAPP_{8–37}$. This indicates the multiplicity of the $rIAPP_{8–37}$ beta-structure motif because of the six different amino acid residues in the sequence.

To unravel the role of "hot spots" in the beta-like-structure stability, the folding structure of $rIAPP_{8–37}$ R18H was investigated by creating a single point mutation. The length distributions of both $rIAPP_{8–37}$ R18H and $rIAPP_{37–8}$ R18H show bimodal characteristics, which shows the direct difference caused by the mutation of His18 to Arg. The smaller components of the length distributions in $rIAPP_{8–37}$ R18H and $rIAPP_{37–8}$ R18H suggest one more possible key site at Asn21 and Gly24, respectively, which indicates the possible extension of the beta-structure motif at both N and C termini. There is an overlapping possible beta-structure motif $rIAPP_{20–23}$ R18H, which could be attributed to the three beta segments in $rIAPP_{8–37}$ R18H linked by turns at His18-Ser19 and Gly24-Pro25. The extension of the beta-structure motif and the possible third beta-structure motif may result from the enhanced tendency of beta-structure formation and fibrillogenesis induced by the mutation R18H in the peptide rIAPP. This single-point mutation provides evidence of possible collateral effects on the folding structure, in which the residue His18 may have a long-range effect in modulating peptide stacking interactions. These results could open a new window for investigating the assembly polymorphism of peptides with multiple β-structure motifs.

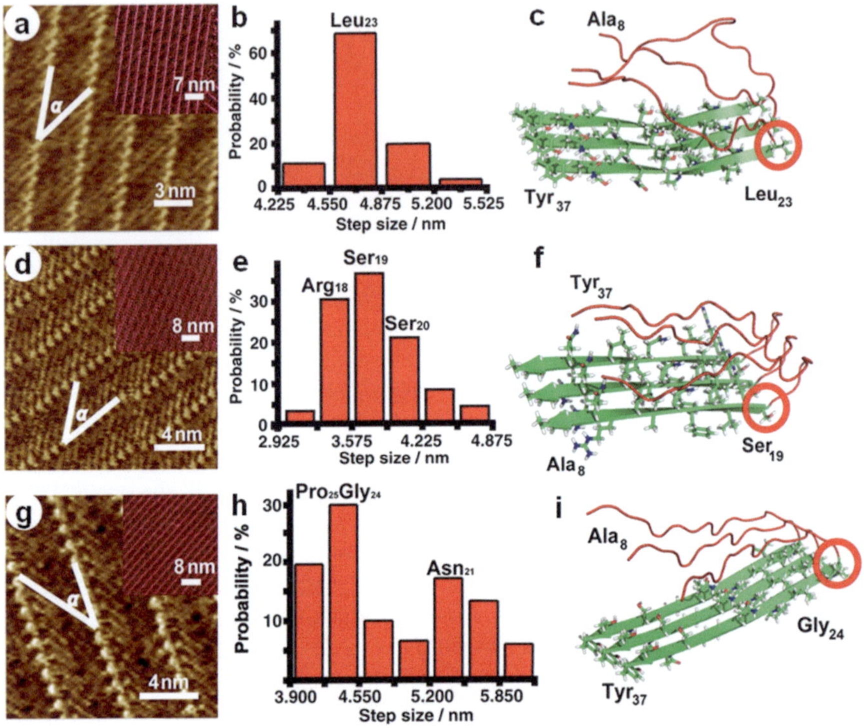

Fig. 2.2 (**a**, **d**, **g**) High resolution STM images of rIAPP$_{8-37}$ (**a**), rIAPP$_{37-8}$ (**d**), rIAPP$_{8-37}$ R18H (**g**) co-assembled with 4Bpy molecules. Insets show the corresponding large scale STM images. (**b**, **e**, **h**) Length distribution histograms of rIAPP$_{8-37}$ (**b**), rIAPP$_{37-8}$ (**e**) and rIAPP$_{8-37}$ R18H (**h**) C-strand motifs measured from the related images. The step size for the histograms are 0.325 nm. (**c**, **f**, **i**) Proposed models for the folding sites in rIAPP$_{8-37}$, rIAPP$_{37-8}$ and rIAPP$_{8-37}$ R18H beta-structure motifs, respectively. (Reprinted and adapted with permission from the *Proceedings of the National Academy of Sciences of the United States of America* http://www.pnas.org/content/108/49/19605.full [73])

The effects of mutations in the β-motif of amyloidogenic TDP-43 mutants has also been explored by combining STM with ThT fluorescence assays [74]. The β-domains of the wild-type TDP-43 (Wt) and three mutant TDP-43 peptides, including an ALS-related mutant peptide: phosphorylated A315T mutant TDP-43 (A315T(p)) and two model mutant peptides: A315T, A315E, were carefully compared by using STM. They demonstrate an order of the length of the core region according to the statistical analysis of the results: Wt = A315T<A315T(p)<A315E, which agrees with the order of aggregation propensity as determined by the ThT assay (Fig. 2.3a). It was hypothesised that the stronger aggregation of A315T(p) and A315E could be ascribed to the electrostatic interactions between the negatively charged phosphoryl moiety/glutamic acid and the positively charged Arg293 residue in the N-terminus, which was subsequently confirmed through studying the R293G mutation in the

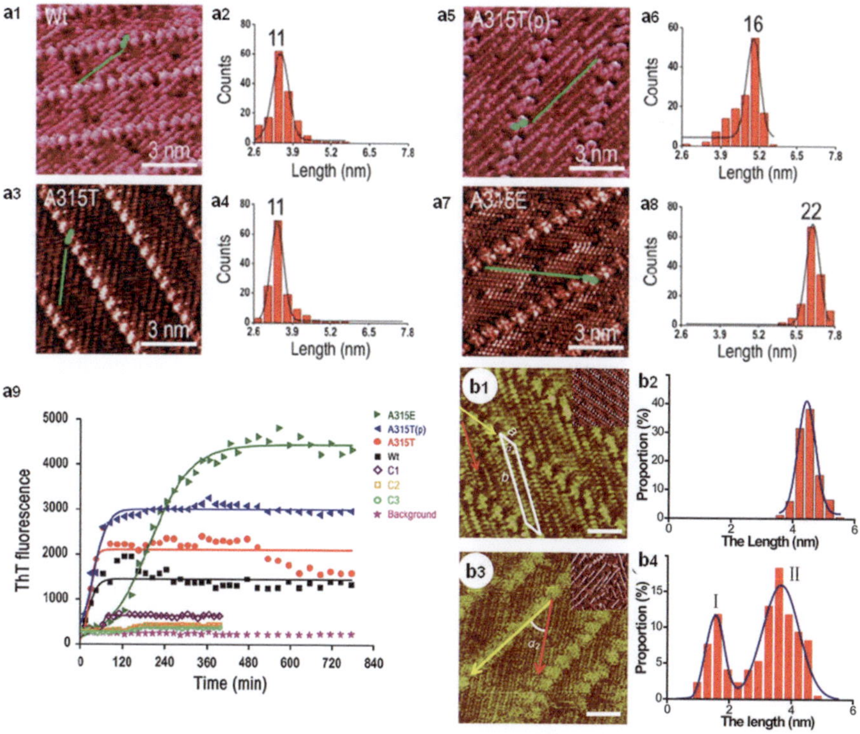

Fig. 2.3 (**a**) Different lengths of β-sheet C-strands of the Wt, A315T, A315T(p) and A315E TDP-43 peptide forms observed by STM experiments. High resolution STM images of 2D assemblies of 4Bpy-labeled Wt, A315T, A315T(p) and A315E peptides are shown in panels (**a1**, **a3**, **a5** and **a7**), respectively. The histograms of the length distributions of the core β-domain with Gaussian fitting for Wt (**a2**), A315T (**a4**), A315T(p) (**a6**) and A315E (**a8**) peptides. (**a9**) The time courses of ThT fluorescence experiments of TDP-43 peptides A315E, A315T(p), A315T, Wt TDP-43 peptide and C1, C2, C3. (**b**) The effect of glycosylation on peptide assemblies identified with STM. (**b1**, **b3**) High-resolution STM images of p-VNTR/4Bpy and g-VNTR/4Bpy co-assembly, respectively. Scale bar: 2 nm. The inset images are the corresponding large scale STM images. (**b2**, **b4**) The histograms of the length distribution of C-terminal strands of p-VNTR and g-VNTR, respectively. (Reprinted and adapted with permission from the *Journal of structural Biology* http://www.sciencedirect.com/science/article/pii/S1047847712002936 [74] (**a**) and the *Journal of Physical Chemistry C* http://pubs.acs.org/doi/abs/10.1021/acs.jpcc.5b12357 [75] (**b**))

above mentioned peptides in the same way. Importantly, this positive correlation between β-domain length on the surface and the aggregation propensity in solution reveals the feasibility of understanding the aggregation process and subsequent cytotoxicity of peptides by β-motif structural analysis on surfaces at the molecular level.

All coassembly structures of peptide/4Bpy on the HOPG surface above show typical lamellar features, as shown in Fig. 2.3a, while the key sites differ between the different peptide sequences. Notably, the key site in the C-terminal β-motif of

A315T(p), an ALS phosphorylated mutant, is located at Phe316, which is the correct neighboring residue (closer to the C-terminus) to the phosphorylated Thr315. In another study, STM investigation of a glycosylated peptide was performed to reveal the glycosylation effect on the peptide assemblies. Glycoprotein MUC1 with a variable number of tandem repeats (VNTRs), serves as a promising target for cancer immunotherapy. The VNTR is a 20-amino acid extracellular domain with the sequence HGVTSAPDTRPAPGSTAPPA. The assembly structures of pristine VNTR (p-VNTR) and the glycosylated VNTR (g-VNTR) with the glycosylation of disaccharide (T antigen) on the Thr9 residue and monosaccharide (Tn antigen) on the Ser15 residue were directly compared by using STM (Fig. 2.3b) [75]. The p-VNTR peptides form highly homogeneous lamellar-like assemblies on the HOPG surface, while the g-VNTR peptide assembly shows apparent heterogeneity. This difference in the assembly characteristics provides direct evidence for a glycosylation induced destabilization effect on the glycopeptide assemblies. Furthermore, the histogram of the length distribution of C-terminal strands of g-VNTR shows two peaks (I and II), assigned to key sites Thr16 and Arg10. Notably, key residues Thr16 and Arg10 are located exactly adjacent to the glycosylated residues Ser15 and Thr9. This demonstrates that glycosylation can induce a pronounced site-specific destabilization effect on peptide assemblies. Moreover, disaccharides exert greater influence on the assembly stability than monosaccharides. These results provide insights into the effects of phosphorylation/glycosylation on peptide assemblies and sequence effects on the local structural stability of proteins.

2.4 Coassembly of Peptides with Small Molecules

A number of small molecules have been exploited as probes to track amyloid aggregation or for modulating the peptide assembly structures and aggregation processes. They demonstrate potential applications for the diagnosis and treatment of conformational diseases caused by protein misfolding. For example, surfactants, copper/zinc ion chelators, well-known bioactive molecules (such as apomorphine, rifamycine, curcumin, porphyrins), and dyes (such as Congo red (CR), thioflavin T (ThT)) and their derivatives have been reported to inhibit amyloid aggregation [76–78]. Unveiling the mechanism of interactions between these molecules and related peptides could provide a theoretical basis for the structure-based drug design and facilitate the rational design of peptide-organic architectures.

2.4.1 Small Molecules Interacting with the Termini of Peptides

It is well recognized that peptide-peptide interactions are the driving force for peptide assembly. Based on similar molecular interactions, introducing peptide-organic molecule interactions can also provide complementary strategies for studying peptide assemblies and their interactions and thereby aid in constructing peptide-organic architectures. Such efforts have been reflected in many investigations on the effects of small molecular modulators of peptide assembly and aggregation processes. In particular, it is feasible to tune peptide aggregation by introducing hydrogen bonding interactions with molecular modulators of the peptide termini, such as chaperone-like molecules 4Bpy and 1,2-di(4-pyridyl)-ethylene (DPE) [26, 52, 73, 79–81].

Firstly, it is of great importance to clarify the effect of chaperone-like molecules on the peptide assembly structures by real-time observation. Introducing tip manipulation has been demonstrated to be an effective way to rearrange or remove molecules adsorbed onto surfaces and liquid-solid interfaces [82–84]. An amyloidal peptide Q11 (QQKFQFQFEQQ) was studied as a model peptide. Chaperone-mediated peptide assemblies (Q11/4Bpy) adopt ordered lamellar structures with a homogeneous distribution of the surface-bound peptide lengths, which could be attributed to the core segment of the Q11 peptide [85], as shown in Fig. 2.4a. With a low-bias-voltage scan, the 4Bpy molecules were removed and the assembled peptides were agitated simultaneously and subsequently formed a new pristine peptide assembly at equilibrium *in situ*. This pristine peptide assembly presents relatively disordered structures with a broad and bimodal distribution of core segment lengths. This direct observation reveals the effect of chaperone molecules on the peptide assembly structures, which could aid construction of peptide architectures with various functions.

In another related study, the self-assembly behavior of a critical amyloidal peptide segment KLVFF (Aβ (16–20)) and its co-assembly behavior with pyridine modulators (4Bpy and 4′-chloro-2,2′:6′,2′′-terpyridine (Cl-Ter)) was observed and secondary structure transformation was detected (Fig. 2.4b) [86]. The uniform KLVFF/4Bpy co-assembly was observed to present a parallel β-sheet-like conformation with sandwich-like striped structures. However, a new close-packed ladder-like structure was formed in the KLVFF/Cl-Ter co-assembly. Based on the conformation, configuration and interaction sites of the modulators, the secondary structure of KLVFF is transformed from a parallel to an antiparallel β-sheet conformation, which was further verified by FTIR spectroscopy. Moreover, molecular mechanistic investigations on this peptide-terpyridine coassembly further reveal the associated contributions from multiple interactions in polymorphic peptide assemblies with distinct geometrical arrangements. Most importantly, the peptide-molecule co-assemblies demonstrate a synergistic inhibitory effect on Aβ42-induced cytotoxicity [87].

Modulation effects of 4Bpy on AD-related peptide assemblies were also identified with STM. The variation of amino acid residues in the peptide sequence shows significant effects on the modulation of peptide assemblies as mentioned above. In particular, the critical effect of the C-terminus of the peptide Aβ42 on its amyloid aggregation has been addressed [88]. Herein, the mechanistic changes of the assembly behaviour of variants of Aβ42 caused by the 4Bpy molecular tether at its C-terminus was explored (Fig. 2.5) [89]. In the self-assembly of the Aβ42 peptide, each β-strand is visible as a bright double-dotted ribbon and its periodicity

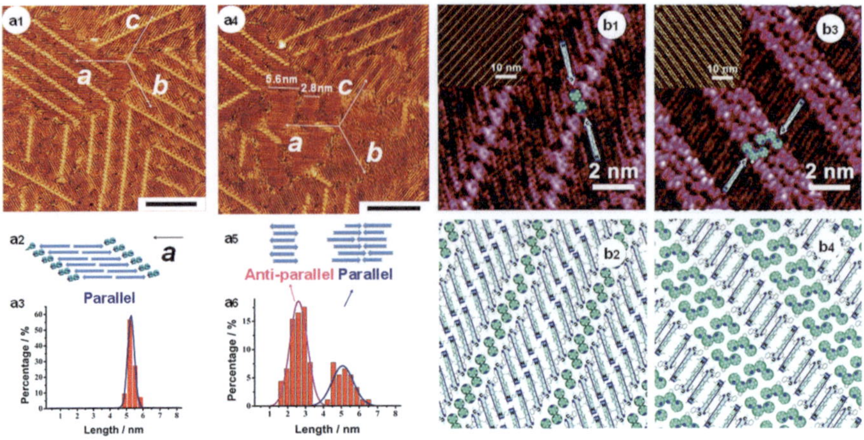

Fig. 2.4 (**a1**, **a4**) STM images of the Q11/4Bpy co-assembly on the HOPG surface and pristine peptide Q11 assembly after removing the 4Bpy molecules in the central region. Arrows in the image present three main directions of peptide molecules. Tunneling conditions for imaging in (**a1**, **a4**): $I = 300$ pA, $V = 600$ mV. Scale bar = 10 nm. Tunneling conditions for manipulation, the tip scans in *a* direction, $I = 600$ pA, $V = 10$ mV, scan rate = 15 Hz, scan size = 25 × 25 nm. (**a2**, **a5**) Schematic models of peptide co-assemblies in *a* direction with an ordered parallel β-sheet structure (**a2**) and pristine peptide assemblies in antiparallel and parallel β-sheet structures (**a5**). (**a3**, **a6**) The histograms of the peptide length distribution fitted with a Gaussian distribution. (**b1**, **b3**) High-resolution STM images (with schematic models) of KLVFF/4Bpy and KLVFF/Cl-Ter co-assemblies demonstrate the modulation of KLVFF secondary structures by pyridine molecules. (**b2**, **b4**) Tentative models of the KLVFF/4Bpy and KLVFF/Cl-Ter coassemblies on the HOPG surface, respectively. (Reprinted and adapted with permission from the *ChemPhysChem* http://onlinelibrary.wiley.com/wol1/doi/10.1002/cphc.201500340/full [85] (**a**), and the *Chemical Communications* http://pubs.rsc.org/en/content/articlehtml/2014/cc/c4cc02748e [86] (**b**))

in a single β-strand was revealed by Fourier transform analysis, which demonstrates that each peak, *i.e.* one pair of bright spots, represents one amino acid residue. Based on the lengths of surface-bound motifs, the number of residues in the core regions of the Aβ42 hairpins is assigned to be around 9–15 residues, and the angle between Aβ42 molecules and the stripe axis in pure peptide assemblies was approximately 90°. By contrast, in the presence of the tethered molecule 4Bpy, two characteristic types (type I and II) of assembled structures are visible in the STM images. The residue number in the C-terminal β-strands of the Aβ42 hairpins was calculated to be mainly 10/11 residues (accounting for 44.1%) in type I co-assembly, and 14 residues (accounting for 54.1%) in type II co-assembly. Furthermore, differences between the two structures also exist in the angles between the molecular axes of the Aβ42 and the axes of the stripe, which are 25° and 33°, respectively. Therefore, the peptide strands with lengths other than 10/11 and 14 may be energetically unstable, and not involved in the co-assemblies after molecular tethering. Membrane permeability and cell viability assays showed that the tuned Aβ42 assemblies with the tethered 4Bpy molecules possess weak membrane disruption ability, resulting in reduced cytotoxicity. This study indicates that tethering small molecules at the termini of

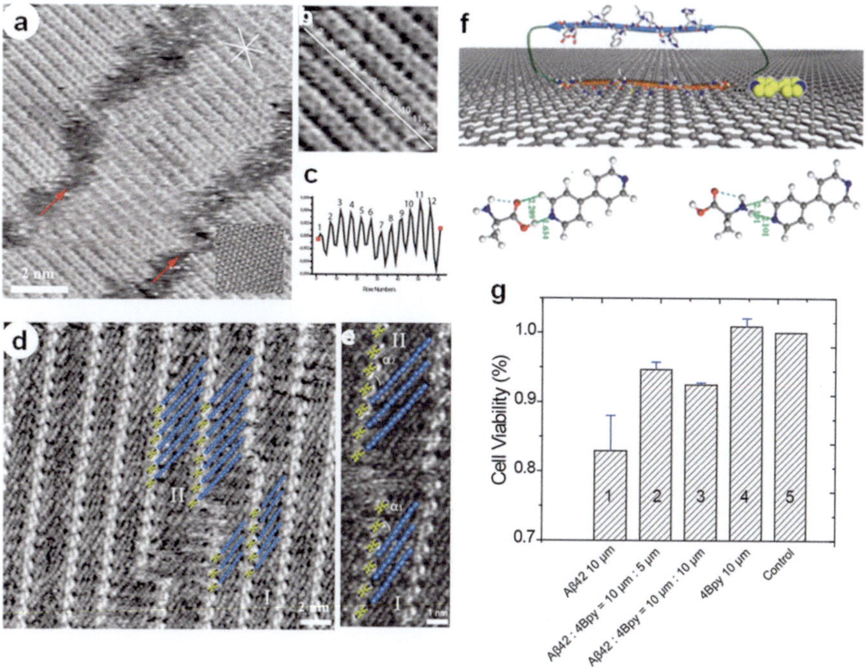

Fig. 2.5 (**a**) A large-scale STM image of Aβ42 assemblies on HOPG surface. (**b**) High-resolution STM image of Aβ42. (**c**) The Fourier transform analysis of corresponding peptide strands indicated in (**b**). (**d** and **e**) STM images of Aβ42/4Bpy co-assemblies with basic building models. (**f**) The basic building model for Aβ42/4Bpy co-assemblies on the HOPG surface. The binding energy of two peptide termini with the nitrogen atom of 4Bpy obtained from theoretical calculations: (Lower left) C-terminus of peptide with nitrogen atom of 4Bpy, $E = 0.78$ eV; (Lower right) N-terminus of peptide with nitrogen atom of 4Bpy molecule, $E = 0.42$ eV. In (**d** and **e**), the blue and yellow stripes denote the surface-bound β-strands and the pyridyl rings, respectively. The 4Bpy binds to the carboxyl group of Aβ42. (**g**) The cytotoxicity of SH-SY5Y cells induced by Aβ42 and Aβ42/4Bpy solutions measured by WST-8 toxicity assay. (Reprinted and adapted with permission from *ACS Nano* http://pubs.acs.org/doi/abs/10.1021/nn503737r [89])

amyloidal proteins may be a promising strategy for therapeutic intervention of AD and other degenerative diseases.

2.4.2 Small Molecules Interacting with the Side Groups of Peptides

A number of small molecules have been exploited as molecular probes to study amyloid aggregation processes [90, 91], suggested to influence fibrillization [92–94], and thus are considered as potential candidates for diagnostic and therapeutic applications. Unveiling the mechanism of t interactions between the protein and the binding molecules (drugs, dyes and ligands) is of critical importance for understanding the biological processes and applications. STM has the capability to image both single small molecules and single peptides, and so can be adopted for real-space imaging of adsorption sites and configurations of small molecules on the target peptides. For instance, the binding behavior of ThT on the prion peptide segment, GNNQQNY, was identified with four different binding modes [95].

Furthermore, investigation of the relative binding affinities of labeling molecules for specific amino acid residues was also performed (Fig. 2.6) [96]. Four representative polyamino acids were selected as model peptides including octaphenylalanine (polyF_8) with a hydrophobic aromatic group, octahistidine (polyH_8) with a positively charged group, octatyrosine (polyY_8) with a hydrophilic aromatic group, and heptaglutamine (polyQ_7) with a hydrophilic group. The adsorption and distribution of a labeling dye molecule and a photodynamic therapy drug, copper phthalocyanine tetrasulfonate sodium ($PcCu(SO_3Na)_4$) on the four model peptides was directly visualized using STM. By frequency counting, the relative

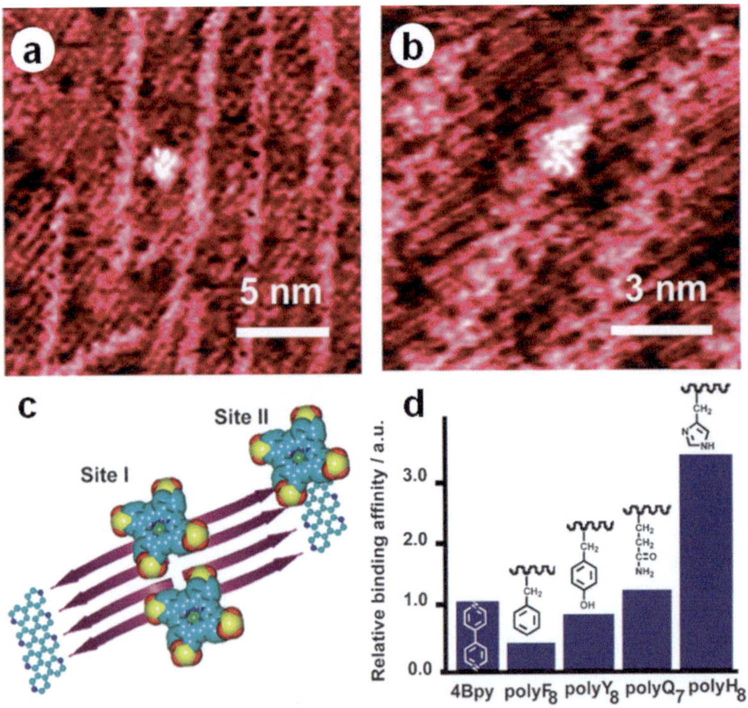

Fig. 2.6 (**a**, **b**) High resolution STM images of $PcCu(SO_3Na)_4$ adsorption atop polyQ_7 and atop 4Bpy. (**c**) Proposed model for the $PcCu(SO_3Na)_4$ adsorption sites. Sites I and II correspond to the adsorption of $PcCu(SO_3Na)_4$ atop polypeptides and 4Bpy molecules, respectively. (**d**) The relative binding affinities of $PcCu(SO_3Na)_4$ for polyF_8, polyQ_7, polyH_8, and polyY_8. (Reprinted and adapted with permission from *Chemical Communications* http://pubs.rsc.org/en/content/articlehtml/2011/cc/c1cc12380g [96])

binding affinity of the polypeptide and 4Bpy molecules can be obtained. Since the affinity of $PcCu(SO_3Na)_4$ towards 4Bpy can be considered constant in these peptide-organic co-assemblies, the relative binding affinities of $PcCu(SO_3Na)_4$ for the four polypeptides were found to have the ratio ($polyH_8$:$polyQ_7$:$polyY_8$:$polyF_8$ = 3.32:1.29:0.79:0.34). The high affinity of $PcCu(SO_3Na)_4$ on $polyH_8$ could be ascribed to the electrostatic interaction between the positively charged imidazole moiety with the negatively charged sulfonic acid group, while the low affinity on $polyF_8$ could be due to the effect of the solvent-accessible surface area of the side chains in water. The binding of dye or drug molecules on model peptides also provides important insight into the mechanism of chemical and structural dependence on specific amino acid residues.

Here, we present a summary of the experimentally observed peptide lengths in surface-bound assemblies, with and without co-assembled small molecules interacting with the peptide termini. The data reveal the relationship between the number of residues in the peptide sequence and the experimentally observed number of residues in the peptide assemblies in STM observations (Table 2.1). Short peptides comprising 2–7 residues can be fully extended in the assembly structures, as shown in Fig. 2.7. It appears that for both homogeneous and heterogeneous sequences with less than seven residues, the peptides are fully extended on the surface. This observation indicates the effect of persistence length in peptide structures. In terms of relatively long peptides with more than seven residues, the number of residues involved in assemblies may vary, which suggests pronounced sequence dependence. These observations demonstrate that peptide structures on surfaces could provide insights into the fundamental mechanisms underlying the sequence dependence of the persistent lengths of peptide chains.

Table 2.1 The number of residues in surface-bound peptide assemblies identified by STM

No.	Peptide name	Full[a]	Observed[b]	References
1	AA	2	2	Stensgaard [51]
2	FF	2	2	Lingenfelder [54]
3	KFFE	4	4	Kalashnyk [55]
4	KVVE	4	4	Kalashnyk [55]
5	KLVFF (Aβ16-20)	5	5	Niu et al. [86]
6	A5	5	5	Niu et al. [97]
7	GNNQQNY	7	7	Mao et al. [95]
8	Q7	7	7	Guo et al. [49]
9	Q8	8	7/8	Guo et al. [49]
10	Q7	7	7	Wang et al. [96]
11	F8	8	8	Wang et al. [96]
12	Y8	8	8	Wang et al. [96]
13	H8	8	8	Wang et al. [96]

(continued)

Table 2.1 (continued)

No.	Peptide name	Full[a]	Observed[b]	References
14	H5A5	10	10	Claridge et al. [57]
15	Aβ33-42	10	10	Liu et al. [52]
16	H5F5	10	6	Guo et al. [98]
17	F5H5	10	10	Guo et al. [98]
18	N5F5	10	8	Guo et al. [98]
19	F5 N5	10	10	Guo et al. [98]
20	A5F5	10	7	Guo et al. [98]
21	F5A5	10	10	Guo et al. [98]
22	A5N5	10	7	Guo et al. [98]
23	N5A5	10	9	Guo et al. [98]
24	H5D5	10	7	Guo et al. [98]
25	D5H5	10	6	Guo et al. [98]
26	H5A5	10	8	Guo et al. [98]
27	A5H5	10	9	Guo et al. [98]
28	Q11	11	8	Yu et al. [85]
29	Aβ10-20	11	11	Liu et al. [26]
30	KKKFAFAFAFAKKK	14	14	Liu et al. [99]
31	DELERRIRELEARIK	15	15	Mao et al. [100]
32	R4G4H8	16	16	Mao et al. [50]
33	F4G4H8	16	16	Mao et al. [50]
34	VNTR	20	14	Yu et al. [75]
35	hIAPP8-37	30	14/15	Mao et al. [73]
36	hIAPP37-8	30	13	Mao et al. [73]
37	rIAPP8-37	30	15	Mao et al. [73]
38	rIAPP37-8	30	12	Mao et al. [73]
39	rIAPP8-37 R18H	30	13/14/17	Mao et al. [73]
40	rIAPP37-8 R18H	30	11/12/17	Mao et al. [73]
41	hIAPP	37	11	Mao et al. [27]
42	rIAPP	37	12	Mao et al. [27]
43	Aβ42	42	12	Ma et al. [53]
44	Aβ42	42	10/11/14	Liu et al. [89]
45	Wt (TDP-43(286-331))	46	11	Xu et al. [101]
46	A315T	46	11	Xu et al. [101]
47	A315T(p)	46	16	Xu et al. [101]
48	A315E	46	22	Xu et al. [101]
49	Re-Wt	46	14	Xu et al. [101]
50	Re-A315T	46	14	Xu et al. [101]
51	Re-A315T(p)	46	14	Xu et al. [101]
52	Re-A315E	46	14	Xu et al. [101]

[a]Full: the number of residues in the peptide sequence
[b]Observed: the number of residues in the surface-bound peptide assemblies observed by STM

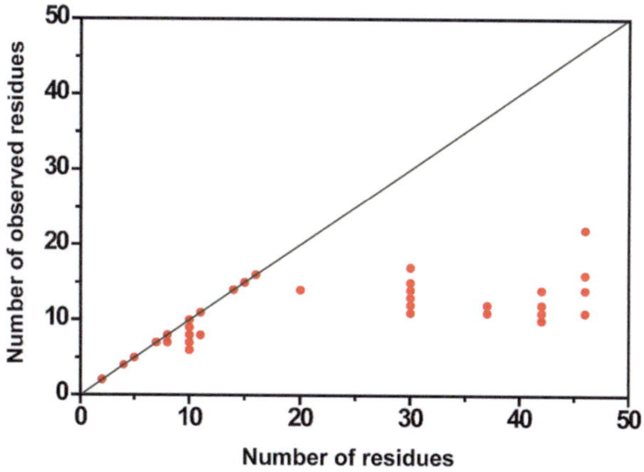

Fig. 2.7 The relationship between the number of residues in the full peptide sequence and the number of residues observed by STM involved in the peptide assemblies on surfaces. The data points on the grey line (y = x) mean that the number of observed residues is equal to its number of residues in the full peptide sequence, i.e., all the residues in the sequence are participated in the peptide assembly

2.5 Correlation of Peptide Assemblies on Surfaces and in Solution

Studies of peptide assemblies on surfaces imaged using STM provide structural information and unveil the formation mechanisms, providing constructive guidance for modulating peptide aggregation and therapeutic applications associated with hydrophobic membranes, including in biological aqueous systems. However, the two-dimensional liquid-solid system, which introduces a hydrophobic interface, is distinct from the three-dimensional solution system. Peptide assembly behavior on surfaces results from the delicate balance between peptide-peptide, peptide-solvent and peptide-substrate interactions [1, 3, 102]. Nevertheless, the main driving forces for the assembly behavior could still be peptide-peptide interactions determined by the physico-chemical properties of amino acid residues in the specific sequence. As mentioned above, the β-motif length on surfaces and the aggregation propensity in bulk systems have been revealed to be positively correlated.

It is also well recognized from studies on adsorption processes that proteins can undergo conformational rearrangements when adsorbed at liquid-solid interfaces from solutions [103–106]. More importantly, the pathogenesis of some degenerative diseases, such as AD, prion diseases, and type 2 diabetes, is associated with an α-to-β conformational transition of part of the protein structure [107–110], which eventually results in the formation of amyloid deposits. The unnatural conformations in the deposits could involve the formation of elastic layers at interfaces and also insoluble amyloid aggregates. In the following, the detailed conformational transformation and relationship between secondary structure and assembly behavior in these two systems are reviewed.

The structural changes of a *de novo* designed polypeptide during peptide adsorption at liquid-solid interfaces were investigated to explore the relationship between the conformation in solution and on surfaces [100]. The polypeptide with sequence DELERRIRELEARIK is initially stable in solution and in the crystalline state, adopting a mostly α-helical conformation. When adsorbed on a graphite surface, the polypeptide molecules readily form homogeneous β-sheet-like assemblies which can be observed by STM (Fig. 2.8a). In addition, supporting evidence was provided by CD analysis that the polypeptide conformation transforms from α-helix to β-sheet upon the addition of graphite particles to the polypeptide solution. Further all-atom molecular dynamics (MD) simulations were carried out to explore the dynamics of its early stage conformation transition at the water-graphite interface and the driving force underlying this transition [111]. The results reveal that unfolding of the α-helix and assembly into an amorphous dimer of the peptide occurs on the graphene surface and the adsorption and the unfolding of the peptide is initiated from the C-terminal region resulting from strong interactions between residues Arg_{13}-Ile_{14}-Lys_{15} and the graphene surface. The studies of such conformational changes could benefit our understanding of the deposition process of disease-associated peptides and also protein-surface interactions.

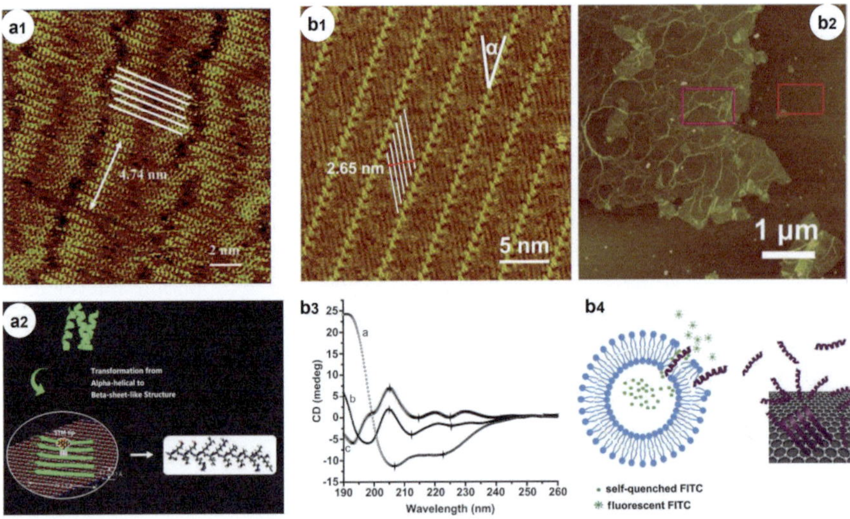

Fig. 2.8 (**a1**) High-resolution STM image of assembly of the polypeptide DELERRIRELEARIK. The white arrow covers ten peptide molecules with a length of 4.74 nm. (**a2**) Schematic illustration of the structural transformation of the polypeptide. The polypeptide structural change induced by the surface is proposed to be α-helical in solution to a β-sheet-like structure on the HOPG surface. (**b1**) STM image of Vpr13-33 peptide assembly on HOPG surface. (**b2**) Tapping mode AFM image of 0.1 mg/mL Vpr13-33 on mica with previously deposited GO. (**b3**) CD spectra of Vpr13-33 in the presence of GO. (a) Vpr13-33 in 1: 1 (v/v) mixture of trifluoroethanol and water, adopting α-helical secondary structure; (b) Vpr13-33 with graphene oxide, showing β-turn and β-sheet secondary structure induced by graphene oxide; (c) the centrifugal mother liquid. (**b4**) Schematic model for the mechanism of how GO reduces Vpr13-33-induced cytotoxicity. (Reprinted and adapted with permission from *Langmuir* http://pubs.acs.org/doi/abs/10.1021/la901342r [100] (a) and *Biomaterials* http://www.sciencedirect.com/science/article/pii/S0142961212012185 [112] (b))

A similar α-to-β conformational transition was also observed in a fragment of the viral protein R (Vpr), Vpr13-33, which plays a vital role in regulating nuclear transport of human immunodeficiency virus (HIV) [112]. Preferential adsorption of Vpr13-33 peptide on hydrophobic graphene oxide (GO) accompanied by a conformational transition was observed by AFM and CD. The high-resolution assembly structures of Vpr13-33 on a graphite surface observed by STM further confirms that β-sheet structures form on the GO surface (Fig. 2.8b). It is therefore deduced that the presence of GO, with a similar hydrophobic surface to the interior of lipid bilayers, causes the conformational change and accelerated aggregation of Vpr13-33 peptide owing to hydrophobic forces. Moreover, it further leads to reduced cytotoxicity towards neuroblastoma cells and T cells, which could be related to inhibition of the "pore forming" process of Vpr13-33 indicated by a fluorescence leakage assay. This finding suggests that the carbon material GO is a potential candidate for therapeutic applications in biological systems. This work also sheds light on peptide-surface interactions and membrane-associated proteins.

In addition, aiming at revealing aggregation kinetics of peptides, electrochemical analyses were combined with AFM imaging of Aβ42 fibril formation on the same basal plane HOPG [113]. The results illustrate that conformational changes during the consecutive aggregation stages from monomers, oligomers, protofibrils to mature amyloid fibrils, can be directly monitored by following the electrochemical signal of Tyr oxidation that consistently decreases during the course of Aβ42 aggregation (Fig. 2.9). This method simultaneously provides kinetic and morphological characterization of the peptide aggregation process and thus may be used as a

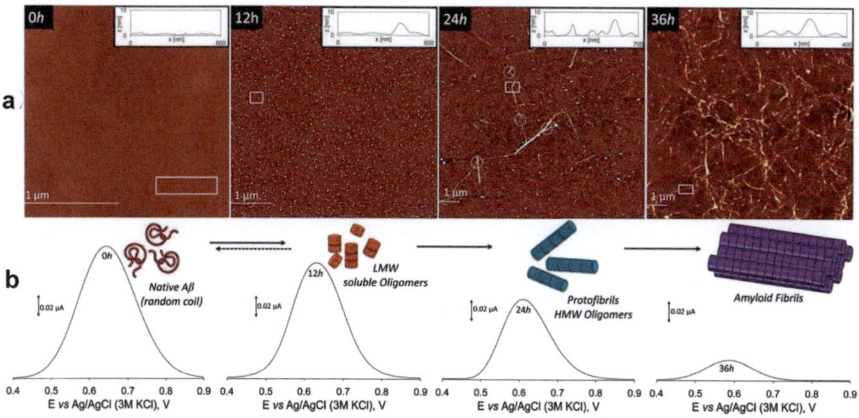

Fig. 2.9 (**a**) Representative AFM images and (**b**) baseline-corrected DPVs recorded with Aβ42 adsorbed on freshly cleaved HOPG in different aggregation states. Aβ42 incubation times are indicated at the upper left corner. Insets in (**a**): height profiles corresponding to the white rectangle sections in the AFM images. Insets in (**b**): schematic representation of the Aβ42 aggregation process (LMW and HMW: low and high molecular weight, respectively). (Reprinted and adapted with permission from *Nanoscale Publishing* http://pubs.rsc.org/en/content/articlehtml/2014/nr/c4nr02413c [113])

potential screening platform for analysis of aggregation kinetics and drug response, which would be beneficial for efficient monitoring of the progression and treatments of AD.

2.6 Conclusions and Perspectives

Unveiling the mechanisms of peptide assembly and aggregation is essential to deepen our insights into peptide assembly pathways and rational design of peptide-based biomaterials. This chapter has summarized the efforts in studying site-specific assemblies of simple model peptides and disease-associated peptides on surfaces by STM. Subtle structural information such as chirality and the conformations of side groups has been successfully obtained in model peptide assemblies. In terms of disease-related peptides, labeling molecules are introduced to benefit the identification of the key folding sites in amyloid aggregates, which deepens the understanding of amyloidal assemblies and inspires the modulation of the aggregation processes. In addition, recent work on peptide-organic molecule interactions provide complementary avenues for studying peptide assemblies and constructing peptide-organic architectures and also points to potential diagnostic and therapeutic applications. Related investigations further reveal the positive correlation between the core lengths of peptides on surfaces and their aggregation propensity in solution as well as their conformational transitions when absorbed on surfaces. Further endeavors should be made towards recognition of amino acid residues in the surface-bound peptide sequences, decoding the effect of mutations/modifications and environmental conditions, and understanding peptide-organic and peptide-surface interactions.

Acknowledgement This work was supported by the National Natural Science Foundation of China (91127043, 21332006, 21273051) and the National Basic Research Program of China (2013CB934200) and the Chinese Academy of Sciences (XDA09030306, YZ201317). Financial support from the CAS Key Laboratory for Biological Effects of Nanomaterials and Nanosafety and the Key Laboratory of Standardization and Measurement for Nanotechnology are also gratefully acknowledged.

References

1. Whitesides GM, Mathias JP, Seto CT (1991) Molecular self-assembly and nanochemistry: a chemical strategy for the synthesis of nanostructures. Science 254:1312–1319
2. Zhao X, Zhang S (2006) Molecular designer self-assembling peptides. Chem Soc Rev 35:1105–1110
3. Yang YL, Wang C (2009) Hierarchical construction of self-assembled low-dimensional molecular architectures observed by using scanning tunneling microscopy. Chem Soc Rev 38:2576–2589
4. Gazit E (2010) Bioinspired chemistry diversity for self-assembly. Nat Chem 2:1010–1011

5. Stupp SI (2010) Self-assembly and biomaterials. Nano Lett 10:4783–4786
6. Santoso S, Hwang W, Hartman H, Zhang SG (2002) Self-assembly of surfactant-like peptides with variable glycine tails to form nanotubes and nanovesicles. Nano Lett 2:687–691
7. Von Maltzahn G, Vauthey S, Santoso S, Zhang SU (2003) Positively charged surfactant-like peptides self-assemble into nanostructures. Langmuir 19:4332–4337
8. Zhang SG (2002) Emerging biological materials through molecular self-assembly. Biotechnol Adv 20:321–339
9. Ellis-Behnke RG, Liang YX, You SW, Tay DKC, Zhang SG, So KF, Schneider GE (2006) Nano neuro knitting: peptide nanofiber scaffold for brain repair and axon regeneration with functional return of vision. Proc Natl Acad Sci U S A 103:5054–5059
10. Narmoneva DA, Oni O, Sieminski AL, Zhang SG, Gertler JP, Kamm RD, Lee RT (2005) Self-assembling short oligopeptides and the promotion of angiogenesis. Biomaterials 26:4837–4846
11. Yang YL, Khoe U, Wang XM, Horii A, Yokoi H, Zhang SG (2009) Designer self-assembling peptide nanomaterials. Nano Today 4:193–210
12. Goldberg M, Langer R, Jia XQ (2007) Nanostructured materials for applications in drug delivery and tissue engineering. J Biomater Sci Polym Ed 18:241–268
13. Fairman R, Akerfeldt KS (2005) Peptides as novel smart materials. Curr Opin Struct Biol 15:453–463
14. Scanlon S, Aggeli A (2008) Self-assembling peptide nanotubes. Nano Today 3:22–30
15. Fleming S, Ulijn RV (2014) Design of nanostructures based on aromatic peptide amphiphiles. Chem Soc Rev 43:8150–8177
16. Sethuraman A, Belfort G (2005) Protein structural perturbation and aggregation on homogeneous surfaces. Biophys J 88:1322–1333
17. Bhakta SA, Evans E, Benavidez TE, Garcia CD (2015) Protein adsorption onto nanomaterials for the development of biosensors and analytical devices: a review. Anal Chim Acta 872:7–25
18. Peng Q, Mu HL (2016) The potential of protein-nanomaterial interaction for advanced drug delivery. J Control Release 225:121–132
19. Morimoto A, Irie K, Murakami K, Masuda Y, Ohigashi H, Nagao M, Fukuda H, Shimizu T, Shirasawa T (2004) Analysis of the secondary structure of beta-amyloid (A beta 42) fibrils by systematic proline replacement. J Biol Chem 279:52781–52788
20. Senguen FT, Lee NR, Gu X, Ryan DM, Doran TM, Anderson EA, Nilsson BL (2011) Probing aromatic, hydrophobic, and steric effects on the self-assembly of an amyloid-beta fragment peptide. Mol BioSyst 7:486–496
21. Zandomeneghi G, Krebs MRH, Mccammon MG, Fandrich M (2004) FTIR reveals structural differences between native beta-sheet proteins and amyloid fibrils. Protein Sci 13:3314–3321
22. Makin OS, Serpell LC (2005) Structures for amyloid fibrils. FEBS J 272:5950–5961
23. Sachse C, Xu C, Wieligmann K, Diekmann S, Grigorieff N, Faendrich M (2006) Quaternary structure of a mature amyloid fibril from Alzheimer's A beta(1-40) peptide. J Mol Biol 362:347–354
24. Kim YS, Liu L, Axelsen PH, Hochstrasser RM (2008) Two-dimensional infrared spectra of isotopically diluted amyloid fibrils from A beta 40. Proc Natl Acad Sci U S A 105:7720–7725
25. Petkova AT, Ishii Y, Balbach JJ, Antzutkin ON, Leapman RD, Delaglio F, Tycko R (2002) A structural model for Alzheimer's beta-amyloid fibrils based on experimental constraints from solid state NMR. Proc Natl Acad Sci U S A 99:16742–16747
26. Liu L, Zhang L, Niu L, Xu M, Mao X, Yang Y, Wang C (2011) Observation of reduced cytotoxicity of aggregated amyloidogenic peptides with chaperone-like molecules. ACS Nano 5:6001–6007
27. Mao X, Ma X, Liu L, Niu L, Yang Y, Wang C (2009) Structural characteristics of the beta-sheet-like human and rat islet amyloid polypeptides as determined by scanning tunneling microscopy. J Struct Biol 167:209–215
28. Makin OS, Atkins E, Sikorski P, Johansson J, Serpell LC (2005) Molecular basis for amyloid fibril formation and stability. Proc Natl Acad Sci U S A 102:315–320

29. Nelson R, Sawaya MR, Balbirnie M, Madsen AO, Riekel C, Grothe R, Eisenberg D (2005) Structure of the cross-beta spine of amyloid-like fibrils. Nature 435:773–778
30. Sunde M, Serpell LC, Bartlam M, Fraser PE, Pepys MB, Blake CCF (1997) Common core structure of amyloid fibrils by synchrotron X-ray diffraction. J Mol Biol 273:729–739
31. Makin OS, Serpell LC (2004) Structural characterisation of islet amyloid polypeptide fibrils. J Mol Biol 335:1279–1288
32. Serpell LC, Blake CCF, Fraser PE (2000) Molecular structure of a fibrillar Alzheimer's A beta fragment. Biochemistry 39:13269–13275
33. Jaroniec CP, Macphee CE, Bajaj VS, Mcmahon MT, Dobson CM, Griffin RG (2004) High-resolution molecular structure of a peptide in an amyloid fibril determined by magic angle spinning NMR spectroscopy. Proc Natl Acad Sci U S A 101:711–716
34. Petkova AT, Leapman RD, Guo ZH, Yau WM, Mattson MP, Tycko R (2005) Self-propagating, molecular-level polymorphism in Alzheimer's beta-amyloid fibrils. Science 307:262–265
35. Iwata K, Fujiwara T, Matsuki Y, Akutsu H, Takahashi S, Naiki H, Goto Y (2006) 3D structure of amyloid protofilaments of beta(2)-microglobulin fragment probed by solid-state NMR. Proc Natl Acad Sci U S A 103:18119–18124
36. Tycko R (2006) Solid-state NMR as a probe of amyloid structure. Protein Peptide Lett 13:229–234
37. Wasmer C, Lange A, Van Melckebeke H, Siemer AB, Riek R, Meier BH (2008) Amyloid fibrils of the Het-s(218-289) prion form a beta solenoid with a triangular hydrophobic core. Science 319:1523–1526
38. Nielsen JT, Bjerring M, Jeppesen MD, Pedersen RO, Pedersen JM, Hein KL, Vosegaard T, Skrydstrup T, Otzen DE, Nielsen NC (2009) Unique identification of supramolecular structures in amyloid fibrils by solid-state NMR spectroscopy. Angew Chem Int Ed 48:2118–2121
39. Luca S, Yau W-M, Leapman R, Tycko R (2007) Peptide conformation and supramolecular organization in amylin fibrils: constraints from solid-state NMR. Biochemistry 46:13505–13522
40. Qiu XH, Wang C, Zeng QD, Xu B, Yin SX, Wang HN, Xu SD, Bai CL (2000) Alkane-assisted adsorption and assembly of phthalocyanines and porphyrins. J Am Chem Soc 122:5550–5556
41. Chang S, Liu R, Wang L, Li M, Deng K, Zheng Q, Zeng Q (2016) Formation of ordered coronene clusters in template utilizing the structural transformation of hexaphenylbenzene derivative networks on graphite surface. ACS Nano 10:342–348
42. Cyr DM, Venkataraman B, Flynn GW (1996) STM investigations of organic molecules physisorbed at the liquid-solid interface. Chem Mater 8:1600–1615
43. De Feyter S, De Schryver FC (2003) Two-dimensional supramolecular self-assembly probed by scanning tunneling microscopy. Chem Soc Rev 32:139–150
44. Dhirani AA, Zehner RW, Hsung RP, Guyotsionnest P, Sita LR (1996) Self-assembly of conjugated molecular rods: a high-resolution STM study. J Am Chem Soc 118:3319–3320
45. Yang GH, Liu GY (2003) New insights for self-assembled monolayers of organothiols on Au(111) revealed by scanning tunneling microscopy. J Phys Chem B 107:8746–8759
46. Xu J, Xiao X, Deng K, Zeng Q (2016) Transformation of self-assembly of a TTF derivative at the 1-phenyloctane/HOPG interface studied by STM-from a nanoporous network to a linear structure. Nanoscale 8:1652–1657
47. Barth JV, Costantini G, Kern K (2005) Engineering atomic and molecular nanostructures at surfaces. Nature 437:671–679
48. Xu B, Yin SX, Wang C, Zeng QD, Qiu XH, Bai CL (2001) Identification of hydrogen bond characterizations of isomeric 4Bpy and 2Bpy by STM. Surf Interface Anal 32:245–247
49. Guo Y, Wang C, Hou J, Yang A, Zhang X, Wang Y, Zhang M, Yang Y, Wang C (2012) Odd-even sequence effect of surface-mediated peptide assemblies observed by scanning tunneling microscopy. Chin J Chem 30:1987–1991
50. Mao XB, Guo YY, Luo Y, Niu L, Liu L, Ma XJ, Wang HB, Yang YL, Wei GH, Wang C (2013) Sequence effects on peptide assembly characteristics observed by using scanning tunneling microscopy. J Am Chem Soc 135:2181–2187

51. Stensgaard I (2003) Adsorption of di-*L*-alanine on Cu(110) investigated with scanning tunneling microscopy. Surf Sci 545:L747–L752
52. Liu L, Zhang L, Mao XB, Niu L, Yang YL, Wang C (2009) Chaperon-mediated single molecular approach toward modulating A beta peptide aggregation. Nano Lett 9:4066–4072
53. Ma X, Liu L, Mao X, Niu L, Deng K, Wu W, Li Y, Yang Y, Wang C (2009) Amyloid beta (1-42) folding multiplicity and single-molecule binding behavior studied with STM. J Mol Biol 388:894–901
54. Lingenfelder M, Tomba G, Costantini G, Ciacchi LC, De Vita A, Kern K (2007) Tracking the chiral recognition of adsorbed dipeptides at the single-molecule level. Angew Chem Int Ed 46:4492–4495
55. Kalashnyk N, Nielsen JT, Nielsen EH, Skrydstrup T, Otzen DE, Laegsgaard E, Wang C, Besenbacher F, Nielsen NC, Linderoth TR (2012) Scanning tunneling microscopy reveals single-molecule insights into the self-assembly of amyloid fibrils. ACS Nano 6:6882–6889
56. Wang YB, Niu L, Li YB, Mao XB, Yang YL, Wang C (2010) Single molecule studies of cyclic peptides using molecular matrix at liquid/solid interface by scanning tunneling microscopy. Langmuir 26:16305–16311
57. Claridge SA, Thomas JC, Silverman MA, Schwartz JJ, Yang YL, Wang C, Weiss PS (2013) Differentiating amino acid residues and side chain orientations in peptides using scanning tunneling microscopy. J Am Chem Soc 135:18528–18535
58. Forloni G, Angeretti N, Chiesa R, Monzani E, Salmona M, Bugiani O, Tagliavini F (1993) Neurotoxicity of a prion protein-fragment. Nature 362:543–546
59. Lorenzo A, Razzaboni B, Weir GC, Yankner BA (1994) Pancreatic-islet cell toxicity of amylin associated with type-2 diabetes-mellitus. Nature 368:756–760
60. Goedert M, Spillantini MG (2006) A century of Alzheimer's disease. Science 314:777–781
61. Roberson ED, Mucke L (2006) 100 years and counting: prospects for defeating Alzheimer's disease. Science 314:781–784
62. Brody DL, Magnoni S, Schwetye KE, Spinner ML, Esparza TJ, Stocchetti N, Zipfel GJ, Holtzman DM (2008) Amyloid-beta dynamics correlate with neurological status in the injured human brain. Science 321:1221–1224
63. Chen-Plotkin AS, Lee VMY, Trojanowski JQ (2010) TAR DNA-binding protein 43 in neurodegenerative disease. Nat Rev Neurol 6:211–220
64. Levy E, Carman MD, Fernandezmadrid IJ, Power MD, Lieberburg I, Vanduinen SG, Bots G, Luyendijk W, Frangione B (1990) Mutation of the Alzheimer's-disease amyloid gene in hereditary cerebral-hemorrhage, Dutch type. Science 248:1124–1126
65. Hendriks L, Vanduijn CM, Cras P, Cruts M, Vanhul W, Vanharskamp F, Warren A, Mcinnis MG, Antonarakis SE, Martin JJ, Hofman A, Van Broeckhoven C (1992) Presenile-dementia and cerebral-hemorrhage linked to a mutation at Codon-692 of the beta-amyloid precursor protein gene. Nat Genet 1:218–221
66. Nilsberth C, Westlind-Danielsson A, Eckman CB, Condron MM, Axelman K, Forsell C, Stenh C, Luthman J, Teplow DB, Younkin SG, Naslund J, Lannfelt L (2001) The 'arctic' APP mutation (E693G) causes Alzheimer's disease by enhanced A beta protofibril formation. Nat Neurosci 4:887–893
67. Grabowski TJ, Cho HS, Vonsattel JPG, Rebeck GW, Greenberg SM (2001) Novel amyloid precursor protein mutation in an Iowa family with dementia and severe cerebral amyloid angiopathy. Ann Neurol 49:697–705
68. Janssen JC, Beck JA, Campbell TA, Dickinson A, Fox NC, Harvey RJ, Houlden H, Rossor MN, Collinge J (2003) Early onset familial Alzheimer's disease – mutation frequency in 31 families. Neurology 60:235–239
69. Wakutani Y, Watanabe K, Adachi Y, Wada-Isoe K, Urakami K, Ninomiya H, Saido TC, Hashimoto T, Iwatsubo T, Nakashima K (2004) Novel amyloid precursor protein gene missense mutation (D678N) in probable familial Alzheimer's disease. J Neurol Neurosurg Psychiatry 75:1039–1042
70. Di Fede G, Catania M, Morbin M, Rossi G, Suardi S, Mazzoleni G, Merlin M, Giovagnoli AR, Prioni S, Erbetta A, Falcone C, Gobbi M, Colombo L, Bastone A, Beeg M, Manzoni C,

Francescucci B, Spagnoli A, Cantu L, Del Favero E, Levy E, Salmona M, Tagliavini F (2009) A recessive mutation in the APP gene with dominant-negative effect on amyloidogenesis. Science 323:1473–1477

71. Neumann M, Sampathu DM, Kwong LK, Truax AC, Micsenyi MC, Chou TT, Bruce J, Schuck T, Grossman M, Clark CM, Mccluskey LF, Miller BL, Masliah E, Mackenzie IR, Feldman H, Feiden W, Kretzschmar HA, Trojanowski JQ, Lee VMY (2006) Ubiquitinated TDP-43 in frontotemporal lobar degeneration and amyotrophic lateral sclerosis. Science 314:130–133
72. Lagier-Tourenne C, Polymenidou M, Cleveland DW (2010) TDP-43 and fus/tls: emerging roles in RNA processing and neurodegeneration. Hum Mol Genet 19:R46–R64
73. Mao XB, Wang CX, Wu XK, Ma XJ, Liu L, Zhang L, Niu L, Guo YY, Li DH, Yang YL, Wang C (2011) Beta structure motifs of islet amyloid polypeptides identified through surface-mediated assemblies. Proc Natl Acad Sci U S A 108:19605–19610
74. Xu M, Zhu L, Liu J, Yang Y, Wu JY, Wang C (2013) Characterization of beta-domains in C-terminal fragments of TDP-43 by scanning tunneling microscopy. J Struct Biol 181:11–16
75. Yu LL, Sun ZY, Yu Y, Qu FY, Yang YL, Li YM, Wang C (2016) Molecular evidence of glycosylation effect on the peptide assemblies identified with scanning tunneling microscopy. J Phys Chem C 120:6577–6582
76. Nie Q, Du X-G, Geng M-Y (2011) Small molecule inhibitors of amyloid beta peptide aggregation as a potential therapeutic strategy for Alzheimer's disease. Acta Pharmacol Sin 32:545–551
77. Porat Y, Abramowitz A, Gazit E (2006) Inhibition of amyloid fibril formation by polyphenols: structural similarity and aromatic interactions as a common inhibition mechanism. Chem Biol Drug Des 67:27–37
78. Stains CI, Mondal K, Ghosh I (2007) Molecules that target beta-amyloid. ChemMedChem 2:1674–1692
79. Wang CX, Yang AH, Li X, Li DH, Zhang M, Du HW, Li C, Guo YY, Mao XB, Dong MD, Besenbacher F, Yang YL, Wang C (2012) Observation of molecular inhibition and binding structures of amyloid peptides. Nanoscale 4:1895–1909
80. Yu Y, Yang YL, Wang C (2015) Site-specific analysis of amyloid assemblies by using scanning tunneling microscopy. Chin J Chem 33:24–34
81. Yang YL, Wang C (2013) Single-molecule studies on individual peptides and peptide assemblies on surfaces. Philos Trans A Math Phys Eng Sci 371:20
82. Dujardin G, Mayne A, Robert O, Rose F, Joachim C, Tang H (1998) Vertical manipulation of individual atoms by a direct STM tip-surface contact on Ge(111). Phys Rev Lett 80:3085–3088
83. Hwang IS, Chang SH, Fang CK, Chen LJ, Tsong TT (2004) Probing dynamics of a phase transition of two-dimensional nano-domains with STM imaging and manipulation. Surf Sci 572:L331–L337
84. Uchida H, Huang D, Grey F, Aono M (1993) Site-specific measurement of adatom binding-energy differences by atom extraction with the STM. Phys Rev Lett 70:2040–2043
85. Yu Y, Yang Y, Wang C (2015) Identification of core segment of amyloidal peptide mediated by chaperone molecules by using scanning tunneling microscopy. ChemPhysChem 16:2995–2999
86. Niu L, Liu L, Xu M, Cramer J, Gothelf KV, Dong MD, Besenbacher F, Zeng QD, Yang YL, Wang C (2014) Transformation of beta-sheet structures of the amyloid peptide induced by molecular modulators. Chem Commun 50:8923–8926
87. Niu L, Liu L, Xi WH, Han QS, Li Q, Yu Y, Huang QX, Qu FY, Xu M, Li YB, Du HW, Yang R, Cramer J, Gothelf KV, Dong MD, Besenbacher F, Zeng QD, Wang C, Wei GH, Yang YL (2016) Synergistic inhibitory effect of peptide-organic coassemblies on amyloid aggregation. ACS Nano 10:4143–4153
88. Lansbury PT, Costa PR, Griffiths JM, Simon EJ, Auger M, Halverson KJ, Kocisko DA, Hendsch ZS, Ashburn TT, Spencer RGS, Tidor B, Griffin RG (1995) Structural model for the beta-amyloid fibril based on interstrand alignment of an antiparallel-sheet comprising a C-terminal peptide. Nat Struct Biol 2:990–998

89. Liu L, Niu L, Xu M, Han QS, Duan HY, Dong MD, Besenbacher F, Wang C, Yang YL (2014) Molecular tethering effect of C-terminus of amyloid peptide A beta 42. ACS Nano 8:9503–9510
90. Ivanova MI, Sawaya MR, Gingery M, Attinger A, Eisenberg D (2004) An amyloid-forming segment of beta 2-microglobulin suggests a molecular model for the fibril. Proc Natl Acad Sci U S A 101:10584–10589
91. Fowler DM, Koulov AV, Balch WE, Kelly JW (2007) Functional amyloid – from bacteria to humans. Trends Biochem Sci 32:217–224
92. Sato T, Kienlen-Campard P, Ahmed M, Liu W, Li HL, Elliott JI, Aimoto S, Constantinescu SN, Octave JN, Smith SO (2006) Inhibitors of amyloid toxicity based on beta-sheet packing of A beta 40 and A beta 42. Biochemistry 45:5503–5516
93. Blanchard BJ, Chen A, Rozeboom LM, Stafford KA, Weigele P, Ingram VM (2004) Efficient reversal of Alzheimer's disease fibril formation and elimination of neurotoxicity by a small molecule. Proc Natl Acad Sci U S A 101:14326–14332
94. Heiser V, Scherzinger E, Boeddrich A, Nordhoff E, Lurz R, Schugardt N, Lehrach H, Wanker EE (2000) Inhibition of huntingtin fibrillogenesis by specific antibodies and small molecules: implications for Huntington's disease therapy. Proc Natl Acad Sci U S A 97:6739–6744
95. Mao XB, Guo YY, Wang CX, Zhang M, Ma XJ, Liu L, Niu L, Zeng QD, Yang YL, Wang C (2011) Binding modes of thioflavin t molecules to prion peptide assemblies identified by using scanning tunneling microscopy. ACS Chem Neurosci 2:281–287
96. Wang CX, Mao XB, Yang AH, Niu L, Wang SN, Li DH, Guo YY, Wang YB, Yang YL, Wang C (2011) Determination of relative binding affinities of labeling molecules with amino acids by using scanning tunneling microscopy. Chem Commun 47:10638–10640
97. Niu L, Ma XJ, Liu L, Mao XB, Wu DX, Yang YL, Zeng QD, Wang C (2010) Molecularly tuned peptide assemblies at the liquid-solid interface studied by scanning tunneling microscopy. Phys Chem Chem Phys 12:11683–11687
98. Guo YY, Hou JF, Zhang XM, Yang YL, Wang C (2017) Stabilization effect of amino acid side chains in peptide assemblies on graphite studied by scanning tunneling microscopy. Chemphyschem 18:1–10
99. Liu L, Li YB, Xia D, Bortolini C, Zhang S, Yang YL, Pedersen JS, Wang C, Besenbacher F, Dong MD (2015) A self-assembled nanopatch with peptide-organic multilayers and mechanical properties. Nanoscale 7:2250–2254
100. Mao XB, Wang YB, Liu L, Niu L, Yang YL, Wang C (2009) Molecular-level evidence of the surface-induced transformation of peptide structures revealed by scanning tunneling microscopy. Langmuir 25:8849–8853
101. Xu M, Zhu L, Liu J, Yang Y, Wu JY, Wang C (2013) Characterization of beta-domains in C-terminal fragments of TDP-43 by scanning tunneling microscopy. J Struct Biol 181:11–16
102. Yang YL, Wang C (2009) Solvent effects on two-dimensional molecular self-assemblies investigated by using scanning tunneling microscopy. Curr Opin Colloid Interface Sci 14:135–147
103. Latour RA, Rini CJ (2002) Theoretical analysis of adsorption thermodynamics for hydrophobic peptide residues on SAM surfaces of varying functionality. J Biomed Mater Res 60:564–577
104. Ostuni E, Grzybowski BA, Mrksich M, Roberts CS, Whitesides GM (2003) Adsorption of proteins to hydrophobic sites on mixed self-assembled monolayers. Langmuir 19:1861–1872
105. Chakarova SD, Carlsson AE (2004) Model study of protein unfolding by interfaces. Phys Rev E 69:021907
106. Sethuraman A, Vedantham G, Imoto T, Przybycien T, Belfort G (2004) Protein unfolding at interfaces: slow dynamics of alpha-helix to beta-sheet transition. Proteins 56:669–678
107. Jayawickrama D, Zink S, Vandervelde D, Effiong RI, Larive CK (1995) Conformational-analysis of the beta-amyloid peptide fragment, beta(12-28). J Biomol Struct Dyn 13:229–244
108. Mihara H, Takahashi Y, Ueno A (1998) Design of peptides undergoing self-catalytic alpha-to-beta transition and amyloidogenesis. Biopolymers 47:83–92

109. Pan KM, Baldwin M, Nguyen J, Gasset M, Serban A, Groth D, Mehlhorn I, Huang ZW, Fletterick RJ, Cohen FE, Prusiner SB (1993) Conversion of alpha-helices into beta-sheets features in the formation of the scrapie prion proteins. Proc Natl Acad Sci U S A 90:10962–10966
110. Peretz D, Williamson RA, Matsunaga Y, Serban H, Pinilla C, Bastidas RB, Rozenshteyn R, James TL, Houghten RA, Cohen FE, Prusiner SB, Burton DR (1997) A conformational transition at the N terminus of the prion protein features in formation of the scrapie isoform. J Mol Biol 273:614–622
111. Ou L, Luo Y, Wei G (2011) Atomic-level study of adsorption, conformational change, and dimerization of an alpha-helical peptide at graphene surface. J Phys Chem B 115:9813–9822
112. Zhang M, Mao XB, Wang CX, Zeng WF, Zhang CL, Li ZJ, Fang Y, Yang YL, Liang W, Wang C (2013) The effect of graphene oxide on conformation change, aggregation and cytotoxicity of HIV-1 regulatory protein (Vpr). Biomaterials 34:1383–1390
113. Lopes P, Xu M, Zhang M, Zhou T, Yang YL, Wang C, Ferapontova EE (2014) Direct electrochemical and AFM detection of amyloid-beta peptide aggregation on basal plane HOPG. Nanoscale 6:7853–7857

Chapter 3
The Kinetics, Thermodynamics and Mechanisms of Short Aromatic Peptide Self-Assembly

Thomas O. Mason and Alexander K. Buell

Abstract The self-assembly of short aromatic peptides and peptide derivatives into a variety of different nano- and microstructures (fibrillar gels, crystals, spheres, plates) is a promising route toward the creation of bio-compatible materials with often unexpected and useful properties. Furthermore, such simple self-assembling systems have been proposed as model systems for the self-assembly of longer peptides, a process that can be linked to biological function and malfunction. Much effort has been made in the last 15 years to explore the space of peptide sequences, chemical modifications and solvent conditions in order to maximise the diversity of assembly morphologies and properties. However, quantitative studies of the corresponding mechanisms of, and driving forces for, peptide self-assembly have remained relatively scarce until recently. In this chapter we review the current state of understanding of the thermodynamic driving forces and self-assembly mechanisms of short aromatic peptides into supramolecular structures. We will focus on experimental studies of the assembly process and our perspective will be centered around diphenylalanine (FF), a key motif of the amyloid β sequence and a paradigmatic self-assembly building block. Our main focus is the basic physical chemistry and key structural aspects of such systems, and we will also compare the mechanism of dipeptide aggregation with that of longer peptide sequences into amyloid fibrils, with discussion on how these mechanisms may be revealed through detailed analysis of growth kinetics, thermodynamics and other fundamental properties of the aggregation process.

Keywords FF · Aromaticity · Amyloid · Self-assembly · Crystal · Fibril · Gel · Nucleation · Microfluidics · Biomaterials

---

T. O. Mason
Department of Materials and Interfaces, Weizmann Institute of Science, Rehovot, Israel

A. K. Buell (✉)
Department of Biotechnology and Biomedicine, Technical University of Denmark, DTU, Lyngby, Denmark
e-mail: alebu@dtu.dk

S. Perrett et al. (eds.), *Biological and Bio-inspired Nanomaterials*,
Advances in Experimental Medicine and Biology 1174,
https://doi.org/10.1007/978-981-13-9791-2_3

3.1 Introduction

It has been known for decades that (poly)peptides can assemble into ordered supramolecular structures, such as crystals and fibres [1]. Such assemblies were mostly studied in the context of structural biology (e.g. protein crystals [2]) or cellular biology (e.g. filaments of the cytoskeleton). Only in the second half of the twentieth century it became clear that filamentous protein assemblies can also be linked to disease processes [3]. A particular type of filamentous assembly, so-called amyloid fibrils, were found in a range of different disorders, ranging from Alzheimer's disease to type 2 diabetes [4]. Amyloid fibrils can be formed by a large variety of unrelated polypeptides and share a common morphology as linear, mostly unbranched structures of several nanometers in diameter and several micrometers in length. These assemblies are characterised by a high β-sheet content and characteristic tinctorial properties and X-ray diffraction pattern [5]. Until recently, mostly low resolution structural information was available on assemblies from sequences longer than about 10 residues, e.g. from limited proteolysis experiments [6] whereas the possibility to crystallise short (<10 aa) peptides allowed access to atomically resolved structures [7]. The recent progress in the production of homogeneous samples of amyloid fibrils (in vitro [8, 9] and ex vivo [10]) as well as technical progress in solid state NMR spectroscopy and cryo-electron microscopy now allows to define the structures of amyloid fibrils from longer peptide sequences at the individual amino acid residue level of detail.

The finding that such supramolecular protein structures play a role in diseases triggered a large scale effort to elucidate the mechanisms and driving forces that lead to the formation of amyloid fibrils, as well as a search for potential inhibitors of their formation. In the course of such studies, Gazit and co-workers aimed at identifying minimal amino acid sequence motifs that would be responsible for the self-assembly of the proteins implicated in disease [11]. These investigations ultimately led to the isolated study of the diphenylalanine (FF) peptide [12], a central motif (position 19 and 20) of the sequence of the amyloid β peptide, the aggregation of which is a hallmark of Alzheimer's disease [13]. It was found that FF, rather than assemble into amyloid fibrils, formed crystalline hollow tubes [12], which were later shown to have the same molecular arrangement as the crystals of the FF peptide the structure of which had been solved a few years earlier [14]. The study of these structures really gained momentum when it was shown that the resulting nano- to microscale structures could be functionalised in a variety of ways and displayed interesting mechanical, optical and electrical properties [15–18]. Ultimately, this line of research converged with the field of short peptide gelators. It had been known for some time that, in addition to many other small molecules, short peptides or even amino acids (aromatic or aliphatic), often terminally capped or functionalised with aromatic protection groups, were capable of self-assembly and gelation [19–21]. To-date, hundreds of studies have been published that report on the assembly and structure formation of short (aromatic) peptides in order to create assemblies with potentially useful properties. Most of the experimental work in this area in the last

15 years has focused on the creation of new types of materials, whereas fundamental mechanistic studies have received less attention. The lack of basic physico-chemical characterisation has in some cases led to misinterpretation of observed phenomena and, more generally, inhibited quantitative studies of the dipeptide systems. This situation has changed in the last five years or so, where significant efforts and progress have been made in the elucidation of the fundamental driving forces responsible for the assembly of aromatic dipeptides and their mechanisms of assembly. We attempt to give a summary of these studies and the advances in the understanding of this class of assembly processes. In addition, we will address the question as to what we can learn from the self-assembly of short aromatic peptides about the assembly characteristics of longer peptide sequences into (disease-related or functional) amyloid fibrils. We will start the chapter by reviewing the most important types of intermolecular interactions responsible for peptide self-assembly. The reader familiar with these basic physico-chemical concepts is invited to skip the first section.

3.2 The Nature of the Interactions Responsible for Peptide Assembly

A key concept in biologically-relevant self-assembly processes is the reversibility of the interactions [22, 23] – in contrast to reactions where bonds are formed or broken by external stimulus, self-assembly relies on the sum total of a number of lower energy interactions being favourable in order to achieve a stable form. The interactions in a self-assembly process depend on multi-polar interactions, hydrogen bonding, steric effects, dispersion forces, in addition to 'forces' arising from the environment in which assembly takes place, most notably for the purposes of this chapter, the hydrophobic effect.

3.2.1 Hydrogen Bonding

Energetically, the strongest interactions beyond the covalent bonds in biological systems are electrostatic in nature – variations in electron density giving rise to a force between two building blocks that may be attractive or repulsive, in contrast to the always-attractive dispersion forces. In polypetides, major structural elements are defined by the pattern of hydrogen bonding between amide linkages they display [24, 25]. The hydrogen bond is a peculiar case of an electrostatic attraction for both its ubiquity and its strength relative to other intermolecular forces. It occurs between a dipole of the form $^{\delta^-}A - H^{\delta^+}$ and an acceptor B, a region of high electron density typically (but not always) associated with an electronegative atom in a dipole or an anion [26]. B donates electron density to the partially positively charged hydrogen atom, resulting in an attractive potential. Steiner in a 2002 review highlighted the concept of a hydrogen bond as being describable as a "frozen" proton transfer [27]. The wide range of possible species A and B, together with geometric restraints imposed by the local covalent bonds and resultant steric effects leads to a range of energies for the hydrogen bond, but in biomolecules these rarely exceed 10 $k_B T$[28] as isolated interactions. Importantly, in aqueous solution, the competition from hydrogen-bonding solvent water lowers the overall free energy of formation of intramolecular hydrogen bonds [29] to a range in which they may be reversible in vivo, as expected and required in self-assembly processes [23, 30].

An important implication of hydrogen bonding systems in structure formation is geometry [24, 31]. The hydrogen bond, as an electrostatic multipole, has an orientation dependent energy, with a minimum for the three nuclei involved being collinear. In systems where external constraints on the geometry of the hydrogen bond exist (as is the case in biomacromolecules), the hydrogen bond can be observed at less-favourable angles. The second geometric implication of the hydrogen bond is its range – the interaction energy between the dipole (A-H) and monopolar acceptors scales as r^{-2}, and the dipole-dipole contributions scale as r^{-3}. Monopole-monopole interactions are most important in the case of interactions between formal charges, such as in salt bridges and these scale as r^{-1} [32]. In short peptides terminated by the free amine and carboxylic acid, these interactions commonly occur at the termini, and also occur between charged side chains, the high energy of the attraction allowing self-assembling peptides to be designed based simply on the predictability of the proton transfer occurring [33–35]. Indeed, in native proteins, in the vast majority of cases, species capable of hydrogen bonding are observed to be involved in a hydrogen bond [36], and these bonding interactions have favoured bond angles and bond lengths. The hydrogen bond tends towards a co-linear geometry for the three atoms – the $A - H \cdots B$ angle tending to 180° with increasing bond strength [37], and the angle $H \cdots CO$ at carbonyl acceptors has an energy minimum at 120°, around which there is clustering in the distribution of observed angles in the solid state [38]. Alkyl protons, both simple and activated, can also play a role as secondary hydrogen bond donors in protein self-assembly and molecular recognition. These weak interactions are explained relatively conventionally for what has been said thus far – the feebly polarised C-H bonds can, in the presence of a relatively strong acceptor, form a hydrogen bond. No capacity for acceptance of hydrogen bonds is observed for alkanes, and hence the side chains of the amino acids alanine, valine, leucine, isoleucine and proline do not operate as hydrogen bond acceptors. The hierarchy of hydrogen bond strength has greater structural implication in short peptides, with each interaction contributing a proportionally greater of the total than in large biomolecules, where the weak bonds tend to have a greater significance in structure and recognition [39]. Of particular present interest is the capacity of the aromatic phenylalanine and tyrosine to act as both aromatic C-H donors and as acceptors of hydrogen bonds perpendicular to the ring centre [39].

Backbone-backbone interactions represent a significant proportion of hydrogen bonding interactions in natively-folded proteins [36], with other interactions of the conventional form such as (N, NH, NH_2^+, O)H$\cdots$(OH, O=, N). However, hydrogen bonding is not limited to hydrogen covalently bonded to one of two closely-approaching nitrogen or oxygen atoms [40]. The interaction is fairly general provided there exists an A-H dipole with a partial positive charge on the hydrogen, and a lone pair on B [41, 42]. Phenylalanine for example can, and frequently does act as a hydrogen bond acceptor [43]. The reason for the observed dual functionality is the aromaticity phenomenon (see below), which leads to non-isotropic electron density in much the same manner as more familiar dipolar species, but with different symmetry and hence interaction geometry.

3.2.2 Hydrophobicity

The previously described effects are inherent to the molecules themselves – hydrogen bonding and electrostatic interactions will occur between suitable species in the gas phase, and similarly steric effects limit the allowable conformations of isolated molecules and those contacting others. In dealing with biologically-relevant systems, however, consideration of the solvent environment is essential to any analysis of the mechanism and thermodynamics of the assembly process. The hydrophobic effect is the name given to the familiar phenomenon whereby non-polar species display far lower solubility in water than they do in organic solvents, tending to phase separate despite only feeble attractive forces between the non-polar species that form the new phase, and operates on both apolar and amphiphilic solutes, the latter giving rise to micellar and bilayer self-assembly [44, 45]. Water structure, characterised by extensive and highly fluxional chains of (directional) hydrogen bonds, is frustrated by the presence of the solute, which acts as a 'cavity' to which no hydrogen bonds may be made [46, 47]. Small solutes, those with radii of curvature on the molecular scale such as linear hydrocarbons, are less disruptive to water structure than are flatter/larger species which do not afford the water molecules the ability to form hydrogen bonding chains around the solute, instead forming an interface [48–50]. The contribution of this process to the free energy of solvation naturally has a linear dependence on surface area, and for solute radii larger than a few nanometres, the surface energy is the dominant term and tends towards the value observed in bulk [51, 52]. The small solutes, through their effects as cavities in the hydrogen bond network, affect the solvent in the hydration layer most strongly, the number of water molecules in this layer scaling with solute volume [52].

The entropic and enthalpic signature of small-molecule hydrophobic solvation at low temperature is a free energy composed of a favourable enthalpic (from intermolecular interactions) component and an unfavourable entropic (reduction in the number of energetically favourable conformations available to the water molecules in the presence of the hydrophobe) component. The ordering of the water around the solute is a compensated process. The restrictions on the conformations available to the water molecules results in there being a significant change in the heat capacity of the solvation, where the effects due to this localised ordering diminish rapidly with temperature [53]. The compensated ordering process, yielding a small negative free energy, can be seen as a mechanism that enhances the solubility of hydrophobic species at low temperatures [54]. The magnitude of fluctuations in water density, a relevant concern for the creation of a cavity, at the microscale are fairly independent of temperature [55], certainly when compared to less 'structured' organic media, and a cavity-insertion study demonstrates that this invariance in compressibility of solvent determines both the dominance of excluded-volume effects in the entropy of solvation for small species, and the observed high partial molar heat capacity associated with hydrophobic solvation [52, 56]. Desolvation and the formation of hydrophobic aggregates in a new phase is driven by the differential scaling of the entropic excluded-volume type contributions and the enthalpic interface contributions. Above a critical radius, which for short hydrocarbons in water is around 1 nm, clusters display surface-area energy scaling (sublinear in aggregation number), lower in energy than the separate excluded-volume cavities [51, 57], and so the new phase tends to grow. The early emergence of rudimentary structure in protein folding has been ascribed to the desolvation and interior segregation of hydrophobic residues in the "hydrophobic collapse" model of protein folding, in which desolvation, with or without secondary structure formation is on the main folding pathway [58, 59]. Buried residues typically pack extremely closely in their native state- the packing efficiency in some cases being found to be as high as in organic crystalline solids [60].

As a structural determinant in short-peptide self-assembled aggregates, hydrophobicity is similarly a phenomenon associated with the absence of strong hydrogen bonding or other polar interaction. Segregation of the polar and apolar components on adoption of an ordered structure is usually achieved in the solid state, and those dipeptides which reliably crystallise have remarkable commonalities in the arrangement of their hydrogen bonded and charged moieties, permitting groupings into a comparatively small number of structural classes [61]. The higher-energy electrostatic intermolecular interactions are present in all the surveyed crystalline structures, but their connectivity and spatial arrangement are moderated by the packing requirements of the side chains.

3.2.3 Aromaticity in Proteins and Short Peptides

Aromaticity is a property of cyclic species with delocalised electron density above and below the plane of the ring. All bonding molecular orbitals of the ring system are filled with electron pairs, while nonbonding and antibonding orbitals are unfilled. It is a stable phenomenon, and reactions commonly favour its retention rather more than comparable systems (for example, linear conjugated bonds) without aromatic character. The electron density distribution has a significant quadrupolar component. The symmetry is that of a spherical harmonic of degree 2 (vs. 1 for the dipole, 0 for the isolated charge) and so, as in the case of the dipole, the quadrupole interaction is directional, its sign depending on the approaching species and the orientation of the quadrupole. In proteins, of course, even phenylalanine will display lower order multipoles due to the asymmetry induced by the bonds to the backbone, this being obviously true for the substituted tyrosine and the heterocyclic tryptophan side chains. The interactions of the quadrupole have a steeper fall-off with distance than comparable dipolar interactions, the dipole-quadrupole having r^{-4} dependence and inter-quadrupole interactions varying as r^{-5}. Despite the tendency of quadrupolar interactions to be significant only at short ranges, the ubiquity of inter-aromatic interactions between the side chains of the amino acids phenylalanine, tyrosine and tryptophan suggests that the peculiar directionality [62] of the quadrupolar interactions may play an important role in the generation or stability of protein secondary and tertiary structure. The aromatic residues have a significant tendency to co-locate in folded sequences, comparable to that of oppositely-charged side chains [63], and aromatic pairing was found in 89% of a representative sample of proteins (including the aromatic neutral histidine side chain) for which high-resolution data was available in the Protein Data Bank [64]. Aromatic-aromatic pairs adopted both stacked-offset or 'herringbone' geometry and T-shaped C-H$\cdots\pi$ 'hydrogen bond' pairs [64]. A distinction from general hydrophobic interactions in the folding process is suggested by analysis of the sequence separation of the interacting pairs. 74% of aromatic-aromatic interactions in a statistical sample were found to occur between distant pairs, suggesting that formation of the π interactions follows folding of secondary structure [65]. Both of the favoured geometries involve favourable quadrupole-quadrupole interactions, while the T-shaped arrangement has a significant dipole-quadrupole character [66, 67]. The effect of the quadrupole is to slightly favour the geometry where the C-H bond vector points to the acceptor ring centre, whereas non-aromatic donors are found to 'aim' at centres and edges with little energy difference [27]. Both appear in native protein structures, with stacked-offset arrangements more common in isolated pairs not part of a network [68]. A tendency has been observed for aromatics and certain other hydrophobic amino acids to be encoded in β-sheet forming regions of native structure, a phenomenon that has been linked to the steric effect of branching on carbon 3 of the side chain [69] or from the presence of the bulky (and conformationally rigid) aromatic side chains of Trp, Phe and Tyr [70]. The large accessible surface area of the side chains is also significant in their pairing. An exponential relationship was found between side chain solvent accessible surface area and the average number of atoms within a contact radius of the atoms of the side chain. The aromatic species, with a large and hydrophobic side chain surface area, were found to have significantly more neighbouring atoms in folded structures than any other amino acid [71].

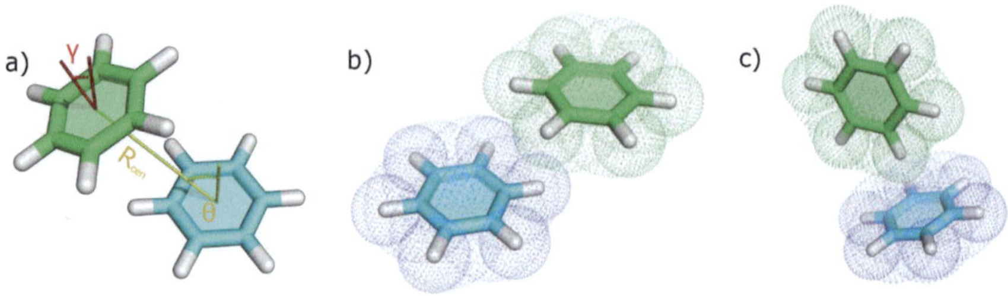

Fig. 3.1 The geometry of the π-π interactions can be described by a variety of parameters, in (**a**) two angles and a distance are defined, the angle θ being from the normal of one ring to the inter-centroid vector R_{cen}, and γ is the angle between the plane normals. (**b**) A representative favourable [72] stacked-offset interaction is shown – $20° < \theta < 50°$, $\gamma < 30°$. In (**c**) the T-shaped edge-to-face geometry is shown, these being classified by $60° < \theta, \gamma < 90°$. The intercentroid distance is below 6.5 Å, peaking around 5.3 Å for F-F pairs [68], naturally with stacked offset dominating at lower R_{cen} and T-shaped at greater range [64]. Detection of interaction by finding the closest intermolecular carbon-carbon contact (R_{CC}) gives a sharper range distribution, peaking around R_{CC}=3.8 Å

3.3 The Role of Phenylalanine Residues in Peptide Self-Assembly into Amyloid Fibrils

The hexapeptide human islet amyloid polypeptide fragment hIAPP (22–27), sequence NFGAIL, was found to assemble in vitro into amyloid fibrils in a similar fashion to its parent polypeptide [73]. An alanine-scanning study showed the importance of the phenylalanine residue to the formation of fibrils by the fragment and it was found that only the mutation F2A completely abolished the capacity of the species to form fibrils [11]. Other studies on fragments of the Aβ peptide showed that KLVFF (Aβ (16–20)) was capable of inhibiting Aβ fibril growth in vitro [74], as were cholylated sequences cho-QKLVFF, cho-KLVFF and cho-LVFFA [75]. This prompted further research into the possible role of phenylalanine specifically [76] in the processes of amyloid formation and deposition, with a focus on π-stacking interactions as a driving force for aggregation. Still shorter polypeptides, the tetrapeptides KFFE and KVVE, were also shown to be capable of self-organisation into amyloid-like β-sheet structures [77]. This remarkable aggregation behaviour, ascribed to sequence-dependent β-sheet propensity and terminal charge-charge interactions suggested that the central dipeptide FF, Aβ (19–20), could display aggregation properties and processes relevant to those of longer chains. The impact of two consecutive phenylalanine residues was later highlighted in a study that compared the assembly behaviour of extended versions of KLVFF/KLVF (capped and uncapped analogs) [78]. However, the special role of aromatic residues in peptide self-assembly has also been challenged in a study that demonstrated similar self-assembly of aliphatic and disease-amyloid derived short peptides [79]. The debate as to the role and importance of aromatic residues in the formation of amyloid fibrils is still ongoing. Two recent studies have applied very similar strategies to probe specifically the roles of phenylalanine in the amyloid fibril formation of the Aβ (1–40 peptide) [80, 81], with strikingly opposing results. While in one study the replacement of the phenylalanine residue at position 19 was found to lead to a completely abolished aggregation [80], the identical substitution in the other study leads to significantly accelerated aggregation with respect to the wild type [81]. This remarkable discrepancy highlights the difficulty that answering such seemingly simple questions entails. The control of a multitude of experimental factors, in particular peptide origin (recombinant vs. synthetic) and purity, solution conditions and experimental assay design, are all absolutely crucial in any attempt to answer this type of question. If any of these factors differs between studies, a direct comparison is no longer straightforward.

3.4 Experimental Methods to Study Short Peptide Assembly

The self-assembly of short aromatic peptides can be induced in a variety of ways and the self-assembly process monitored in situ or ex situ by a large array of experimental methods, that we summarise in this section, with a particular focus on studies of diphenylalanine (FF).

3.4.1 The Choice of the Assembly Conditions for Self-Assembly

The self-assembly of FF and its derivatives into ordered structures is most commonly, but not exclusively, induced and investigated in solution. In many studies, FF was dissolved at high concentration in an organic solvent (mostly Hexafluoroisopropanol, HFIP) and the self-assembly was then induced by injecting the concentrated stock solution into a low solubility solvent (mostly water) [12]. A large number of different combinations of high solubility and low solubility solvents have been investigated [82–88] and it has been shown that very different morphologies can be obtained, depending on the choice of solvents. Indeed, it was shown that even small impurities of solvents can have a significant effect on the self-assembly process [28]. Mason et al. have determined the solubilities of FF in a variety of solvents at room temperature [88], allowing a rational choice of solvent combination based on their relative FF-solubilising abilities. In the context of the creation and characterisation of novel materials, the choice of appropriate solvent for FF assembly is mostly determined by the desired type of structure. However, assembly in water has a privileged role, due to the relevance of self-assembly studies in water for biology and also biocompatibility of the so-created structures. Indeed, in some of the earliest studies of FF, the peptide was dissolved at elevated temperatures in pure water, followed by cooling down of the solution, which led to self-assembly [14, 89]. It has also been demonstrated that a change in pH is a suitable method to induce FF assembly, based on the strong pH-dependence of FF solubility [90]. A concentrated

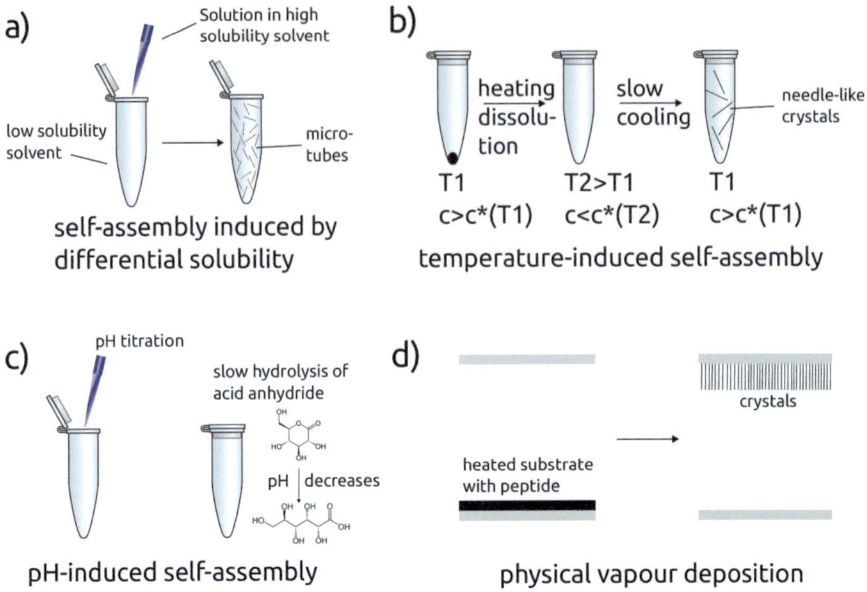

Fig. 3.2 Different methods to induce small aromatic peptide assembly (**a**) Introduction of a concentrated solution of dipeptide in a high solubility solvent, such as acetic acid or HFIP into a low solubility solvent, such as water or methanol [12, 88]. (**b**) Heating of a dipeptide solution at a concentration c to a temperature T2 where the solubility of the peptide, c*, is higher than the peptide concentration, followed by slow cooling [89, 96]. (**c**) pH titration [91] or in situ pH change through hydrolysis of an acid anhydride (here illustrated with glucono-δ-lactone, [92]) exploit the pH-dependence of peptide solubility [90] to induce self-assembly. (**d**) Self-assembly can also be induced by physical vapour deposition in vacuum [97–99], leading to the formation of arrays of crystals oriented perpendicularly to the surface

solution of FF in trifluoroacetic acid (TFA) can be titrated with ammonia solution, which leads to the formation of FF nanowires [91]. A particularly elegant method of pH-change-induced self-assembly has been developed in the context of short peptide hydrogel formation, whereby the slow hydrolysis of an acid anhydride (most notably glucono-δ-lactone, GdL) to the corresponding acid (gluconic acid) slowly and homogeneously acidifies the peptide solution, leading to significantly more homogeneous gel appearance compared to titration with a strong acid [92]. An alternative method for homogeneous induction of self-assembly is enzymatic activation that has been presented for FF derivatives [93, 94], whereby the precursor molecule is designed to be the substrate of an enzyme, such as a kinase. Self-assembly can also be induced through enzymatic action if the dipeptide is created in situ from an activated amino acid derivative (methanoate ester) and an amino acid amide, whereby self-assembly of the formed dipeptide immediately follows [95] (Fig. 3.2).

Assembly of FF from the gas phase has also been investigated, using physical vapour deposition (PVD) [97] and it has been demonstrated that at sublimation temperatures above 190°C, the FF peptide cyclizes [98]. The use of plasma enhanced CVD at 200°C was also reported [99], and most likely also yields assemblies of chemically modified species rather than intact FF. Vapour deposition at moderate temperatures initially only yields amorphous films. Hollow tubes of intact FF can be obtained by hydrating the amorphous layer or by depositing the FF at lower grade vacuum, illustrating the importance of water in the formation of the crystalline tubes [98].

3.4.2 Microscopic Methods

The observation of the peptide assemblies with different forms of microscopy is the most direct method of detection and is often used to define and distinguish different morphologies [88, 100]. In many cases, in particular in the case of gel formation, the self-assembled structures are too small to be observed by optical brightfield microscopy, and therefore atomic force microscopy (AFM) [16, 101, 102], scanning electron microscopy (SEM) [101, 103, 104] or transmission electron microscopy (TEM), either in the form of negative staining TEM [103, 105] or in the form of cryo-TEM [106] can be used. The use of atomic force microscopy has the additional benefit of allowing to probe the mechanical properties of the assemblies [16, 104]. For fibrillar structures of diameters in the nm range and lengths in the μm range, various fluorescence microscopy techniques can also be employed, such as confocal laser scanning [107–109] or super-resolution microscopy [108]. In particular, these direct nanoscopic imaging techniques have enabled highly detailed mechanistic studies of the assembly mechanisms [108] (see below).

3.4.3 Spectroscopic Methods

A wide range of different spectroscopic methods have been used to detect and follow the self-assembly process. In particular spectroscopic tools have been employed in order to monitor the change in peptide conformation between the soluble and assembled states. Fluorescence spectroscopy can be used, both in the form of measurements of the intrinsic fluorescence of the aromatic residues [110], or through the binding of a fluorescent molecule, such as the amyloid fibril-binding dye Thioflavin-T [103, 105] or fluorescently labeled DNA [111]. Other types of spectroscopy that have been shown to be useful for structural characterisations are circular dichroism (CD) [101, 105, 106] and Fourier transform infrared (FT-IR) [103, 105, 107] spectroscopy, that are able to inform about secondary structure motifs formed by the peptides. Furthermore, NMR spectroscopy has also been used, either to follow the decrease in soluble peptide [100, 112], to monitor structural changes of assemblies [106], to characterize the charge state of the peptide, e.g. as a function of concentration [78], or to study the interactions between assemblies and other species, such as metal ions [113].

Despite not being a spectroscopic method, we include here also mass spectrometry, in the form of ion mobility mass spectrometry, as it has been used in order to detect the presence of clusters of aromatic dipeptides in solution [114].

3.4.4 Scattering, Rheological, Calorimetric and Conductivity-Based Methods

The often dramatic change in size upon self-assembly makes it possible to follow the self-assembly process by a variety of scattering methods. Dynamic light scattering (DLS) [90, 100], small angle X-ray scattering (SAXS) [115], as well as small angle neutron scattering (SANS) [107, 116, 117] have been employed to follow the assembly process. In the case of small angle scattering (SAXS and SANS), the data can sometimes be fitted to a model corresponding to the emergence of rod-like micelles/fibrils upon the induction of self-assembly [107].

The self-assembly is often associated with a considerable change in mechanical properties, in particular in the case of gel-formation. This can be followed by the measurement of the rheological properties, in particular the ratio between elastic and dissipative behaviour [94, 105, 107, 117]. In the case of orthogonally assembling, self-sorting multi-component gels, the rheological properties provide a measure of the assembly state of the individual components and can be selectively tuned through external stimuli [118].

In some cases, the self-assembly reaction can also be directly carried out inside a calorimeter, with the aim to study the thermodynamic signatures of the assembly reaction. In the case of Boc-FF, the peptide initially forms amorphous spherical structures when a concentrated solution in ethanol is diluted into water (see below). These structures can be metastable against transformation into crystals for sufficient amounts of time such that the heat exchange linked to this conversion can be measured with a differential scanning calorimeter [100].

Inspired by classical methods for the determination of the critical micelle concentrations of surfactants, it is also possible to follow different levels of peptide self-assembly by measuring the concentration-dependence of the conductivity of the peptide solution [107]. This type of method is based on the fact that the electrophoretic mobility of assemblies can differ significantly from those of their building blocks.

3.4.5 *Microfluidics*

A relatively recent addition to the toolbox of methods for the study of short aromatic peptide assembly is microfluidics, which is particularly well-suited for self-assembled structures that can be directly observed by optical microscopy, see Fig. 3.3. Microfluidics allows the study of peptide assembly in real time by optical time lapse microscopy (Fig. 3.3c, e) while maintaining well-defined solution conditions, by constant flow of a solution with defined concentration over the growing assemblies, eliminating depletion effects [96]. The concentration can be adjusted by changing the degree of supersaturation of the stock solution that is drawn into the device (Fig. 3.3b and reference [96]) or else by changing the relative proportions of supersaturated peptide solution and water/buffer in an on-chip mixing device (Fig. 3.3d and reference [119]). This methodology has recently been applied in a series of studies that investigated the growth rate of FF microcrystals as a

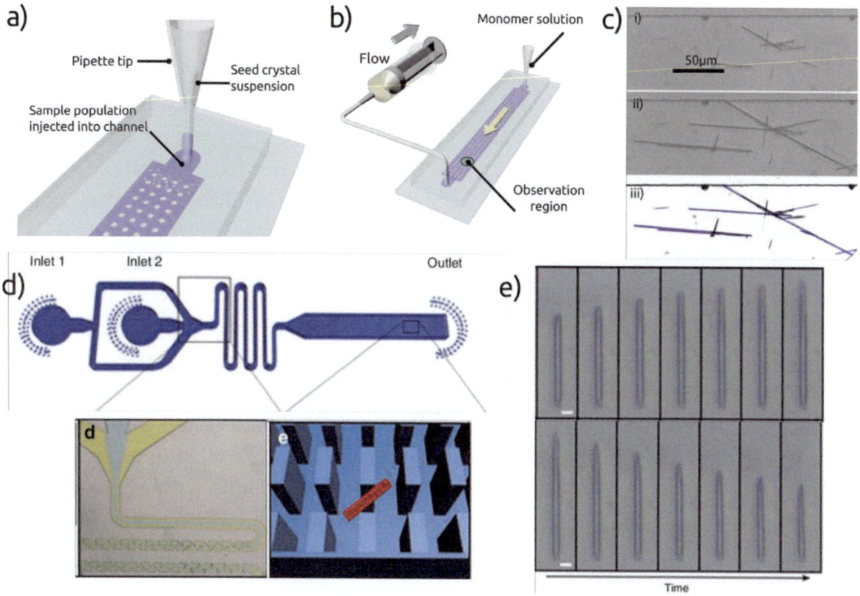

Fig. 3.3 Microfluidic methods to study diphenylalanine self-assembly (**a**) Pre-formed FF crystals can be injected into a wide microfluidic channel and trapped by support columns that avoid collapse of the channel. (**b**) The trapped seed crystals can be flushed with a constant stream of a peptide solution at well-defined concentration. (**c**) Optical time lapse microscopy can be used to follow the growth of the crystals. (**d**) A mixing device can also be used, whereby the relative proportions of supersaturated peptide solution and water/buffer can be varied, leading to an adjustable degree of saturation of the peptide solution in contact with the seed crystals. (**e**) If the seed crystals are flushed with a supersaturated solution, they are observed to grow (top), whereas they shrink when incubated with a sub-saturated solution (bottom). (**a**)–(**c**) reproduced with permission from reference [96] ©2016 American Chemical Society, and (**d**)–(**e**) reproduced from reference [119] under the Creative Commons Licence (CC-BY-4.0)

function of concentration [96, 119] and temperature [90]. This approach allows extensive and accurate measurements to be made and the resulting data has contributed significantly to a better understanding of assembly mechanism of FF (see below).

3.5 Thermodynamic Stability of Peptide Assemblies

The thermodynamics of short aromatic peptide assembly depends crucially on the exact sequence of the peptide, including capping and protection groups, as well as the solution conditions, i.e. temperature, pH and salt concentration. In the following, we will discuss the thermodynamic stability of crystalline and fibrillar structures and how they are influenced by various factors.

3.5.1 Thermal Stability of FF Crystals

It has been reported that FF microcrystals are very stable against heating in the dry state [101]. In the dry state, these structures can withstand temperatures above 100°C and thermogravimetric analysis reveals the loss of water from the crystal structure [101]. However, at temperatures above 150°C, it was shown that chemical decomposition sets in that can lead to the loss of phenylalanine fragments [102]. Furthermore, the FF peptide can cyclize under loss of water at such high temperatures, which leads to an irreversible collapse of the tubular crystals [120, 121]. Full thermal decomposition of the peptide film finally occurs at temperatures above 300°C [101, 122]. On the other hand, amorphous films of FF can be converted into fibrillar morphology through dry heating at 150°C [122], possibly through rearrangement or, more likely, sublimation of chemically degraded/cyclized FF, similar to the observed formation of deposits on AFM cantilever observed in an earlier study [102]. This is corroborated by the finding that pre-cyclized, phenylalanine-containing dipeptides can form fibrils and gels [123].

High thermal stability of FF assemblies above 100°C (in an autoclave) has also been reported in aqueous solution [101]. Only recently, the solubility of FF in water has been systematically studied as a function of temperature, by measuring the FF concentration in the supernatant in equilibrium with FF crystals [90]. It was found that the solubility of FF dramatically increases with temperature, from about 0.5 g/L (1.6 mM) at 10°C to about 1.9 g/L (6 mM) at 70°C. Interestingly, phenylalanine shows a similar temperature dependence of its solubility, the absolute solubility being much higher [124], whereas benzene shows a very weak temperature dependence of its solubility in water [125]. It is interesting to note that benzene, being a prototypical hydrophobic molecule, has a higher solubility in water than the diphenylalanine molecule. The temperature-dependent solubilities of F, FF and benzene are compared in Fig. 3.4. In the light of the reported temperature

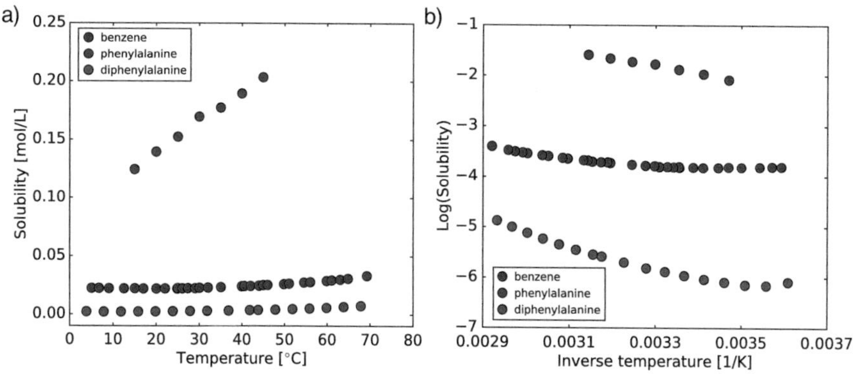

Fig. 3.4 Temperature-dependent solubilities of benzene, phenylalanine and diphenylalanine. (**a**) The equilibrium solubility of benzene [125], phenylalanine [124] and diphenylalanine [90] are plotted against temperature. (**b**) The logarithms of the solubility data are plotted against the inverse absolute temperature in a van't Hoff-like plot. The sign of the slope at any one point corresponds to the sign of the enthalpy of self-assembly (F, FF) or phase separation (benzene)

dependence of FF solubility, it is likely that FF crystals at 2 g/L dissolve completely upon autoclaving and reform upon cooling of the sample [101]. At the same time it is plausible that the individual FF molecules will be more stable against chemical degradation in aqueous solution compared to the dry state, given that the major degradation process is the cyclization reaction which, being associated with the loss of water, is unfavourable in aqueous solution.

Systematic analysis of the temperature-dependent solubility of FF is not only useful for rationalising experiments that probe the thermal stability of its assemblies, but also allows to decompose the thermodynamic driving force of the assembly into enthalpic and entropic contributions, through a van't Hoff analysis (see Fig. 3.4 for van't Hoff plots of F, FF and benzene). This type of analysis has been carried out for FF and it was found that while the free energy of assembly is virtually invariant throughout the investigated temperature range (4–68°C), the balance of enthalpy and entropy changes significantly [90]. While the assembly of FF into crystalline structures is purely entropy-driven below 10°C, it becomes enthalpically favourable above 10°C and entropically unfavourable above 30°C, a manifestation of entropy-enthalpy compensation. Overall, this temperature dependent solubility follows the behaviour expected for a peptide system dominated by the hydrophobic effect [126].

3.5.2 Chemical Stability of FF Crystals

The stability of FF crystals against non-polar solvents has also been highlighted [101]. Systematic analysis of the solubilities of FF in a wide variety of solvents reveals that the solubility varies by a factor of more than 200,000 between the lowest solubility solvent (hexane, 0.002 g/L) and the highest solubility solvent (acetic acid, 430 g/L) investigated [88]. It was also shown that the thermodynamic stability of FF crystals is strongly pH-dependent, with the solubility minimum coinciding with the pH range where FF occurs as a globally neutral zwitterion [90], and the solubility at pH 1 being approximately ten times higher than at the solubility minimum [127]. Interestingly, addition of NaCl only marginally changed the solubility of FF in pure water; only at NaCl concentrations of 1 M and above, the solubility decreases appreciably [90], suggesting that electrostatic screening effects do not play important roles in the assembly thermodynamics. The effect of high NaCl concentrations is likely related to the well-known phenomenon of salting out, that can for example be observed for aqueous solutions of benzene [128].

3.5.3 Non-crystalline Short Aromatic Peptide Assemblies

3.5.3.1 Fibrils and Gels

Structurally, one of the most interesting features of short aromatic peptides is their ability to form fibrils and gels in addition to crystalline structures. In this context, fibrils are defined as highly anisotropic structures with two dimensions of the order of (a few times) the molecular size and one dimension that can be of the order of μm, and which do not show long range crystalline order. Fibril/crystal dimorphism is of course not restricted to (aromatic) dipeptides, but is also observed for longer sequences, in particular those derived from amyloid fibrils [129, 130]. While uncapped zwitterionic dipeptides easily assemble into crystalline structures, capped versions display a clear preference for assembly into fibrils, even though they can be crystallised under some conditions [93, 131]. In particular large, bulky aromatic capping groups, such as fluorenylmethyloxycarbonyl (Fmoc) [132], carboxybenzyl (Cbz) and naphtalene, direct the FF peptide into fibrillar structures [133–135], whereas FF with smaller capping groups (Boc, acetate, amide) is still able to form crystals [100, 133]. The fibril formation and subsequent gelation of N-terminally capped aromatic dipeptides has been studied particularly extensively [20, 116, 134–136]. However, whether a gel or a crystal is formed is not exclusively determined by chemical modifications of the FF peptide, but also strongly depends on the solvent conditions. In the case of uncapped FF, polar, hydrogen-bonding solvents are found to favour the formation of crystals [14, 88] and non-polar solvents favouring the formation of gels [82, 85, 86]. A theoretical thermodynamic framework to decide whether the gel or the crystal is the equilibrium structure has been developed and is based on the geometry of the self-assembling molecule and its solvophilic or solvophobic properties [137]. It is predicted by this model that the gel, usually considered a metastable phase compared to the crystal, can in fact correspond to the thermodynamic minimum, depending on the solution conditions. In cases, however, where a system under the same conditions can form both a gel and a crystal, it is usually found that the crystalline state appears after the gel state and corresponds to the final thermodynamic minimum [100, 106, 135], in agreement with Ostwald's rule of stages [100]. Structurally, gel and crystal are often distinct [100, 135], suggesting that they do not interconvert directly, but that the thermodynamically more stable crystal phase nucleates either independently or catalysed by the gel (heterogeneous nucleation).

The most prominent example of an N-terminally capped FF peptide that forms fibrillar gels in aqueous solution is the Fmoc-FF system, which has been characterised in detail [20]. It was proposed that the fibrils are stabilised through aromatic stacking interactions, a design principle that has been successfully mimicked with molecules containing both a phenyl and fluorophenyl group [138]. Notwithstanding the fact that Fmoc-FF displays very robust fibril formation, it was shown that peptoid substitutions (displacement of the phenyl ring from the side chain to the amide nitrogen) of one or both of the phenylalanine residues impede (if one F is changed) or completely abolish (if both F are changed) gel formation [139].

Gel formation in N-terminally capped dipeptides is usually induced by dissolving the peptide at elevated pH and then lowering the pH, either through titration with acid or in situ through the hydrolysis of an anhydride (Fig. 3.2 and references [92] and [131]). Gel formation occurs as soon as the pK_a value of the C-terminal carboxy-group is reached, suggesting that it is mainly the globally neutral species that is incorporated into the fibrils. However, through pioneering and detailed pH-titration studies, it could be shown that the initial fibrillar assemblies are not globally neutral, but consist of molecules with an average degree of ionisation of 0.66 [136]. Upon further acidification, subsequent transitions into higher order structures occur that accompany increasing neutralisation of the peptides. Through such titration studies [131, 136], as well as solution state NMR experiments [78] it became clear that the process of self-assembly actually modulates the pK_a of the peptide, with the pK_a of the assembled structures being significantly higher than that of the isolated peptide, whereby the difference can correspond to up to 5 pH units [136]. The protonation state of the peptides is also strongly dependent on the total peptide concentration [78]. If a pH titration is performed on such a peptide system, the self-assembly sets in as soon as appropriate pH values are reached, but the process of self-assembly is slow on the time scale of the titration experiments and therefore the system will not be at thermodynamic equilibrium during the titration. Indeed, the titration of self-assembling peptide systems is an ingenious way to access the thermodynamics of gel-forming systems that are otherwise difficult to study. This method was applied to characterize the thermodynamics of fibril and gel formation of the FFDD tetrapeptide and its sequence permutations [110]. Free energies of self-assembly were determined from titration curves under the assumption that assemblies only form from neutral species, an assumption, however, which might not be fully valid [113, 115, 136]. Nevertheless, from a comparison of the titration behaviour of a series of closely related peptides, a relative ordering of the driving forces for assembly can be deduced [110]. This type of data would otherwise be difficult to obtain, given the high stability of these fibrils and the resulting low equilibrium concentration of soluble peptide that are challenging to measure directly. In the same study [110], one of the very few quantitative physico-chemical investigations of aromatic peptide fibril formation published to-date, the kinetics of assembly was followed by NMR and fluorescence spectroscopy, as well as small angle neutron scattering and the important conclusion was drawn that the kinetics and the thermodynamics of the assembly of the six homologous peptides are not correlated [110].

The thermodynamic stability of different gel-like N-terminally naphthoxyaetyl (NAP)-protected dipeptides was also evaluated qualitatively in the context of a study that aimed at the creation of a self-assembling system that displays dynamic instability [95]. The motivation behind this fascinating type of study is to artificially create a system that displays properties similar to biological polymers, in particular tubulin [140]. The competition between enzymatic dipeptide formation, peptide self-assembly and enzymatic peptide hydrolysis was investigated and it was found that only in the case of the YF peptide, a gel persisted at equilibrium, due to the high thermodynamic stability of the assemblies of this peptide, whereas YY and YL dissolved completely [95].

In addition to pH-drop induced fibril- and gel-formation, it was also shown that N-terminally naphtalene (as well as Fmoc)-capped peptides can be gelled at high pH by adding salt ions that cross-link the rod-like structures, present at elevated pH ($>$10) [115]. It was proposed that at high pH, it is not fibrils that form, but rather worm-like micelles with the deprotonated carboxylic acid groups pointing towards the water. These micelles can then be cross-linked by the addition of metal ions (such as Ca^{2+}). The direct interactions of calcium ions with such peptide micelles is confirmed with solution state NMR spectroscopy [113]. The concentration-dependent formation of such micelles at pH 10.5 and different peptide concentrations has been characterized in detail, using a variety of experimental techniques [107]. In this context it should be noted that the distinction between fibrils and elongated micelles is not simply a matter of semantics; indeed it has recently been shown that while peptide fibrils display a formation mechanism characterised by nucleated growth, micelles (of a different self-assembling system based on lipid-like molecules) form without an appreciable nucleation step and do also not seed further growth [108]. In has been demonstrated that it is possible to trigger the conversion of a micellar aromatic peptide system into a fibrillar system through an external stimulus, such as the dephosphorylation of a tyrosine residue in the peptide [141]. The removal of the phosphate group renders the peptide derivative less amphiphilic and hence the micelles less stable. Detailed spectroscopic characterisation of this system suggests that the micelles are not directly converted into fibrils, but that they rather dissolve first, followed by a rapid de novo nucleation of fibrils [141].

While most of the hydrogel-forming peptides and peptide derivatives so far reported have been found serendipitously, it has also been shown that coarse grained simulations allow to explore the sequence space, at least of tripeptides [142]. In this way, several previously unknown gel-forming (in the absence of organic solvent) tripeptides were predicted and experimentally verified (KYF, KYY, KFF and KYW). The important role of aromatic residues is obvious from this list.

3.5.3.2 Amorphous Materials

In addition to the formation of ordered crystalline and fibrillar structures, short aromatic peptides and peptide derivatives can also form disordered structures. When screening a range of aromatic dipeptides, such as FW, WY, WF and WW, Reches and Gazit noticed that only FW was capable of ordered crystalline assembly, in agreement with a later large scale survey of published crystal structures [61], even though the crystals were mixed with amorphous assemblies [12]. Interestingly, it has been found that FF itself at highly acidic pH values, where its solubility is very high (see above), also does not form crystalline, but rather dendritic assemblies that are likely to be disordered at the molecular level [127]. Both the high solubility and the inability to form crystalline assemblies is likely to be linked to the net charge carried by the peptide at this pH, which would need to be neutralised by the incorporation of a counter ion into the crystal. Intriguingly, the individual phenylalanine molecule can be crystallised from acidic solution (dilute formic acid) by incorporating two chemically distinct F molecules into the asymmetric unit cell: a neutral F zwitterion and a positively charged F ion associated with a formate molecule [143].

The formation of other types of regularly shaped assemblies that are molecularly disordered can be observed for a range of N-terminally capped FF derivatives (Boc-FF and Azobenzene-FF (PPA)). The introduction of an ethanol solution of Boc-FF into water leads to the immediate formation of rather uniform spherical aggregates (Fig. 3.5 and references [104] and [100]), probably as a result of a de-mixing process. The dependence of the formation of these spherical aggregates on the ethanol and peptide content of the solution was studied systematically (Fig. 3.5c). It was shown that the spheres are amorphous in structure (Fig. 3.5d) and represent metastable assemblies that evolve further into fibrils and finally into crystals (Fig. 3.5e, f and reference [100]). The transformation of the spheres can be monitored directly through optical microscopy, dynamic light scattering, NMR spectroscopy, and differential scanning calorimetry (DSC) [100] and the ensemble of observations suggests that the spheres are dynamic structures in rapid exchange with monomeric peptide that dissolve once the thermodynamically more stable fibrillar phase has formed, which then grows at the expense of the spheres (Fig. 3.5e). Thermodynamic analysis shows that the system evolves down a gradient in free energy (Fig. 3.5f), as expected for a spontaneous process and that the enthalpy change associated with the dissolution of spheres and simultaneous formation of the gel or crystals shows little temperature dependence [100]. How the net negative charge of the Boc-FF is accommodated in the final crystal structure is at present unknown. A related system is the N-terminally-azobenzene-capped FF (PPA), which forms amorphous spherical structures upon evaporation from an HFIP solution [144]. Interestingly, the layer of spherical structures converts into fibrils upon incubation with water, as was studied by AFM [144]. It is unclear whether the spheres assemble directly into fibrils or whether the latter nucleate de novo and grow at the expense of the dynamically dissolving spheres, as in the case of Boc-FF [100]. The observation that upon contact with water the spheres become significantly smaller [144] suggests the latter possibility as the likely mechanism.

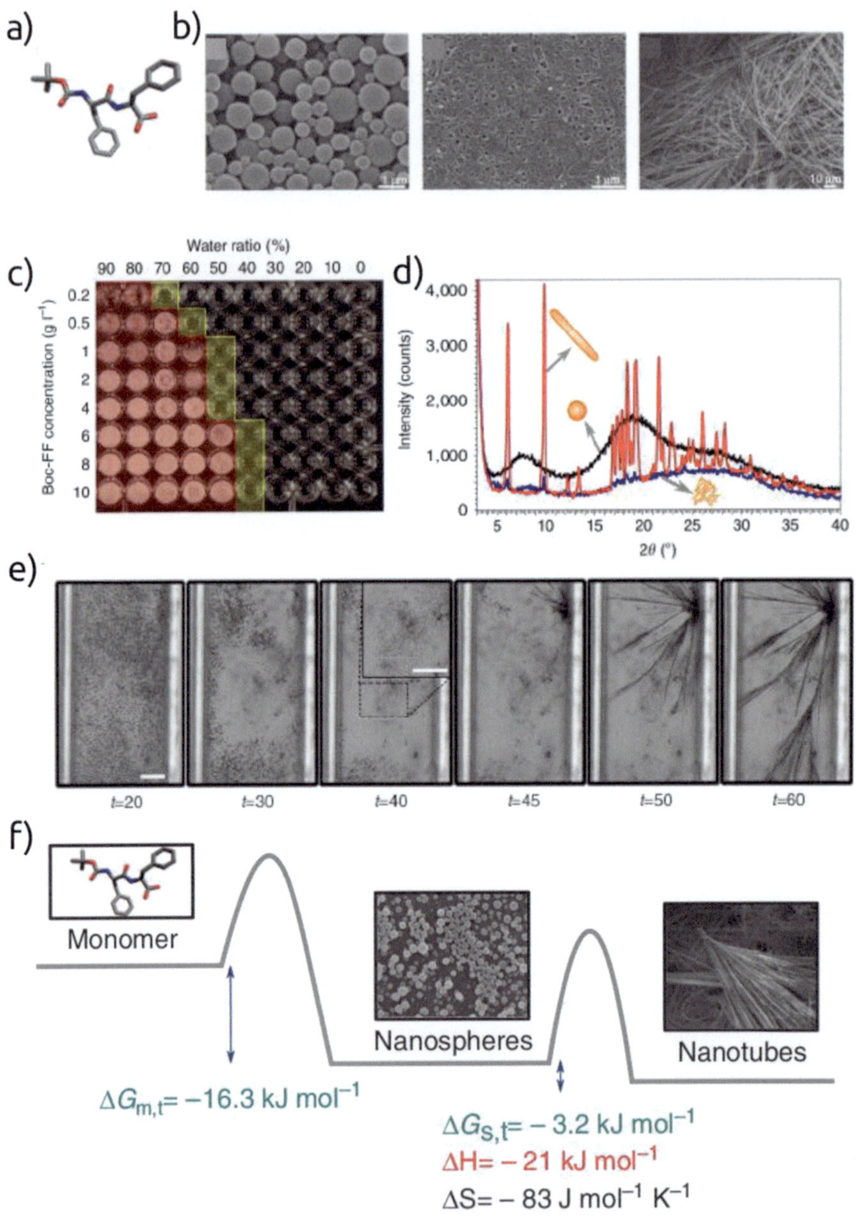

Fig. 3.5 The thermodynamics of self-assembly of Boc-FF (**a**) The structure of the Boc-FF molecule. (**b**) SEM images of the different self-assembled phases that Boc-FF can form. Left: spheres, centre: gel, right: crystals (**c**) Illustration of the solubility of Boc-FF as a function of peptide concentration and ethanol content. The red zone corresponds to the crystalline phase, the yellow zone to spherical structures and the black zone to complete dissolution of the Boc-FF. (**d**) Powder diffraction spectra demonstrate that the spheres are amorphous and the gel is mostly amorphous as well. (**e**) Self-assembly experiments inside glass microcapillaries illustrate the step-

Related observations have been made for layers of FF that are spin-coated onto various substrates directly from HFIP solution, leading to the formation of dendritic fibrillar structures, which upon prolonged incubation at 100% humidity convert into more ordered, crystalline-appearing structures [145]. Recently, it was shown that also the Fmoc-DOPA molecule undergoes a series of transitions from spherical amorphous structures to a fibrillar gel and finally to a crystal upon rapid dilution of an ethanol stock solution into water [106]. In this study, the various phases of the system were structurally characterised in detail through a range of spectroscopic (CD, NMR, IR) and microscopic (cryo-TEM) methods, but no thermodynamic characterisation of the relative stability of the different phases was given.

A very interesting phenomenon in the framework of the thermodynamic stability of FF-derived assemblies is the observation that fibrils formed from C-terminally capped, and therefore cationic FF peptide transform into spherical structures upon dilution of the solution below 7 g/L [111, 146]. This structural transition is partly reversible upon re-concentration of the sample [146]. A theoretical description has been presented that attempts to explain this behaviour as a result of a competition between unfavourable interfacial and favourable volume terms that are concentration dependent [146]. It is tempting to contrast this fluid-like model of C-terminally capped FF spherical structures (in solution) with the observation that (dried) N-terminally capped FF displays metal-like stiffness [104]. In our view, these observations of seemingly contradictory mechanical properties of very similar types of structures illustrate the necessity for direct measurements of the mechanical properties in solution.

Remarkably, it has also been reported that uncapped FF, albeit the D-enantiomeric variant, transforms into spherical or vesicular structures upon dilution [89]. This behaviour is very different from what is observed for the L-enantiomeric form of FF, tubes of which are simply found to dissolve upon dilution below the critical concentration [90, 119]. This difference in observed behaviour could be due to the difference in chirality, but that seems rather unlikely. Therefore, this conflicting behaviour remains at present unresolved, but future detailed microfluidic studies [96] have the potential to resolve this question.

Fig. 3.5 (continued) wise assembly mechanism from spheres over gel to crystals and suggest that the formation of a new phase leads to the dissolution of the less stable previous phase. (**f**) Proposed energy landscape of the Boc-FF peptide, with determined free energy and enthalpy differences indicated. The system follows a down-hill free energy trajectory, as described by Ostwald's rule of stages. (Adapted with permission from reference [100], ©2014 Springer Nature)

3.6 Mechanistic and Kinetic Description of Aromatic Peptide Assembly

Ordered self-assembly of aromatic dipeptides leads to the formation of elongated structures, either fibrils or needle-like crystals, the latter with diameters in the range of hundreds of nanometer to several μm and lengths that can attain many μm. In the following, we will discuss the current state of knowledge of how such structures form, with, as in the remainder of this chapter, a particular focus on crystalline assemblies of FF.

3.6.1 Growth Processes

An early model of how microcrystalline FF assemblies might form is based on the idea that sheet structures are first formed which then either roll up along one dimension to form tubes, or along two dimensions to form spherical structures [147]. However, recent experimental results from the direct, real-time study of FF microcrystals in contact with monomer solutions of varying concentrations inside microfluidic channels, using time lapse optical microscopy, clearly demonstrate the growth of the crystals through the addition of soluble peptide (Fig. 3.5 and references [96], [119] and [90]). These observations render the formation, at least of fully formed microcrystals, through the rolling up of sheets extremely unlikely. The dependence of the axial growth rate of the assemblies on the concentration of soluble FF was measured and it was found to be approximately linear at low degrees of supersaturation (see footnote[1]), while the assemblies were observed to shrink when washed with a sub-saturated solution [119]. A study which extended into a regime of higher supersaturation found that the supersaturation dependence of both the axial and radial growth rates, in a regime limited by surface processes, is exponential (Fig. 3.6a and reference [90]). The fact that the concentration dependence of the growth rate in both dimensions was precisely known, combined with the ability to maintain the soluble FF concentration constant at a specifically adjusted value inside the microfluidic flow reactor, allowed tuning of the aspect ratio of the FF structures [96]. This was achieved by allowing the structures to evolve at a concentration where they grow almost exclusively longer and not wider (compare absolute rates of axial and radial growth in Fig. 3.6a, b), and represents one of the first examples where detailed knowledge of the assembly kinetics has been used for morphological control of assembled peptide structures. The exponential dependence of the axial growth rate of the crystals on the concentration of soluble peptide reflects the fact that the growth reaction is a nucleated process (Fig. 3.6c),

[1]The dimensionless supersaturation σ of the peptide solution is defined as $\sigma = \frac{c-c^*}{c^*}$, where c^* is the critical concentration, see Fig. 3.11 a for an illustration of how the critical concentration can be determined.

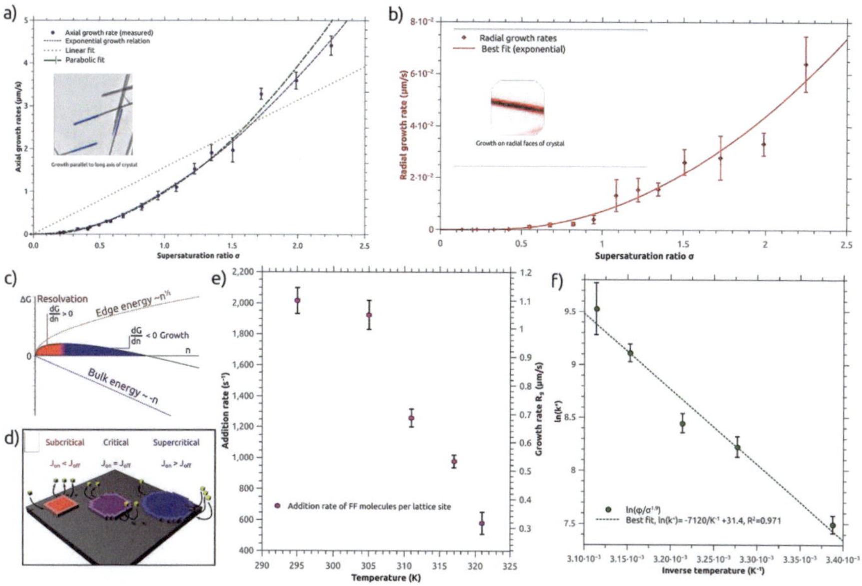

Fig. 3.6 The growth kinetics of FF crystals from microfluidics experiments (**a**) The axial growth kinetics as a function of solution supersaturation. Fits to different models (exponential, polynomic, linear) are shown. (**b**) The radial growth kinetics as a function of solution supersaturation. An exponential fit is shown. (**c**) Illustration of the origin of the energy barrier for two-dimensional nucleation, i.e. the competition between favourable bulk energy and the unfavourable edge energy of the two-dimensional nucleus. (**d**) Illustration of sub-critical, critical and supercritical sizes of nuclei. (**e**) The axial growth rates of FF crystals measured at different temperatures at a constant concentration of soluble FF (1.2 g/L). (**f**) The growth rate is corrected for the different degrees of supersaturation and the growth rate constant has been determined and plotted in the form of an Arrhenius plot, which illustrates that the crystal growth is an activated process. (**a**)–(**d**) adapted with permission from reference [96] ©2016 American Chemical Society, and (**e**)–(**f**) adapted with permission from reference [90] ©2017 American Chemical Society

whereby new crystal layers nucleate on the advancing phase (2D-nucleation) and spread across the face (Fig. 3.6d and reference [96]). The absolute growth rates reported in these two microfluidic studies differ significantly; at an equivalent degree of supersaturation of ca. 0.7, the absolute growth rates are reported to be 5 μm/min [119] or 24 μm/min [96]. The reason for the discrepancy is likely to lie in the differing limiting regimes that operated in each study. Arnon et al. operated in a flow regime where arrival at the growth face was transport limited. Both studies used flow reactor chambers of comparable cross section, but the flow rates varied by a factor of 200. At low flow rates (below 100 μl/h), the rate of crystal face growth was observed to be approximately linearly dependent on flow rate [96], in accordance with diffusion-limited growth [148].

In addition to the growth of assemblies formed from unmodified FF, the growth of cyclized FF from DMSO is also reported [119]. Interestingly, it was noted that while cyclo-FF assemblies grow bidirectionally, crystalline assemblies formed from unmodified FF display only unidirectional growth, a difference that is explained through the different crystal symmetry.

In order to gain insight into the energy barriers of the assembly process, the growth of FF assemblies was measured at different temperatures [90]. When preformed FF assemblies are incubated with a constant soluble peptide concentration at different temperatures, a decreasing rate of growth with increasing temperature is observed (Fig. 3.6e). This unexpected result can be understood when the strong increase in solubility of FF with increasing temperature is considered (Fig. 3.4 and reference [90]). The driving force for net growth of the assemblies is the supersaturation σ, i.e. the difference between actual concentration of the soluble peptide and the critical concentration [90, 119]. Therefore, at constant total concentration, the net driving force for growth will strongly decrease with increasing temperature. Given the exponential dependence of the growth rate on the supersaturation [96], it can be understood why the observed growth rate decreases. However, if this effect is taken into account, it is found that the crystal growth process is actually a thermally activated process (Fig. 3.4f), as has been found for the growth of amyloid fibrils [149].

It was recently shown that FF and Boc-FF can co-assemble and that the presence of Boc-FF slows down the assembly compared to pure FF [150]. Furthermore, the structures formed in the presence of Boc-FF are shorter, which was explained through their higher frangibility under the assembly conditions of vigorous stirring. Co-assembly, rather than separate assembly was indicated by spatially resolved time-of-flight secondary ion mass spectrometry (TOF-SIMS) measurements. Furthermore, in another study by the same authors, the mechanical properties (stiffness and Young's modulus) of FF-Boc-FF co-assembled structures were determined by AFM and it was found that the co-assemblies display a significantly lower Young's modulus than the pure FF assemblies [151]. In this study, the assemblies from different ratios of the two peptide variants were also spectroscopically characterised and a change in π-stacking and hydrogen bonding was proposed. These results show that despite the crystalline nature of the assemblies, the incorporation of foreign molecules is possible and likely leads to the formation of defects in the crystal structure.

The growth kinetics of fibrils formed by the six sequence permutations of the DDFF peptide have recently been investigated by fluorescence spectroscopy and SANS [110] and the second order growth rate constants have been determined. It was shown that depending on the sequence, the fibril growth rates differed by up to a factor of 10, assuming the same number of growing fibrils for all sequences. Given the uncertainty on the absolute nucleation rate (see below), the latter assumption could be the source of some error. However, this study [110] represents one of the first attempts to quantitatively describe the growth rates of short aromatic peptide assemblies in a size range where bright field time lapse microscopy [90, 96, 119] is not possible, and where therefore ensemble techniques are more easily employed. It has, however, recently been shown that even the growth of thin aromatic peptide fibrils can be studied by confocal laser scanning microscopy (CLSM) when the fibrils can be appropriately fluorescently stained [108]. In this study it was demonstrated that an N-terminally protected triphenylalanine (FFF) peptide displayed seeded fibril growth, whereas no seeding was observed for fibrils made from a lipid-like micelle-forming hydrogelator [108].

3.6.2 Nucleation Processes

Nucleation is in general more difficult to study than growth, because nucleation is a rare process and because the nucleated structures rapidly evolve through growth [152], leading in the case of short aromatic dipeptides typically to high aspect ratio structures, such as fibrils and needle-like crystals. Nucleation can not easily be directly observed, due to the often nanoscopic nature of the nuclei and the stochastic occurrence of nucleation. A phenomenological approach to describing the relative rates of nucleation and growth of crystalline structures could consider the distributions of absolute lengths, diameters and wall thicknesses (in the case of tubular crystals) of the final structures. It has been shown that these properties can be tuned in the case of FF tubes by adjusting the relative proportions of different solvents [153], probably reflecting different ratios of nucleation and growth rates (kinetic factors), but probably also varying free energies of the different crystal-solution interfaces (thermodynamic factors). Along similar lines, the introduction of concentrated stock solutions of FF in high solubility solvents, such as HFIP [12] or acetic acid [88], into water leads to many small crystals, whereas the slow cooling to room temperature of an aqueous solution of FF that was saturated at 80°C leads to the formation of much fewer and larger crystals [90]. The nucleation rate is therefore, as expected, highly dependent on the (local) degree of supersaturation. This could explain why at high salt concentration, where the solubility of FF is reduced (see above and reference [90]), smaller crystals are formed, but this could also be partly due to an increased nucleation rate constant in the presence of salt or indeed inhibition of growth by association of counterions at growth faces. While, therefore, general and qualitative conclusions about nucleation processes can be drawn from such observations, nucleation itself remains difficult to capture. The elusive nature of the nucleation process might explain why a range of competing models have been proposed for how aromatic dipeptides and related molecules nucleate their self-assembly. As mentioned above, it has been proposed that FF tube-like crystals and spherical structures might form through the rolling up of sheets [147]. While this is unlikely to explain the formation of fully growth structures, it cannot be excluded that the formation of the initial nanoscopic nucleus resembles such a process.

Among the different types of nucleation scenarios, the formation of dimers has been proposed on various occasions to play a prominent role [110, 114, 154]. Especially mass spectrometry, in the form of time-of-flight or ion mobility mass spectrometry, has been employed to demonstrate the presence of dimeric structures in solution [114, 154]. Ion mobility mass spectrometry is probably the highest resolution molecular size measurement technique (short of directly determining a crystal-, NMR- or cryo-TEM-structure), and is able to distinguish between different conformations of small molecules, such as dipeptides, as well as between different degrees of water clustering around such molecules, as shown in the case of end-capped FF [114]. On the other hand, this method has the caveat of requiring transfer of the molecule or complex of interest into the gas phase. Depending on the dominant type of interaction in the complex, transfer to the gas phase can change the relative interaction strength compared to measurement in the condensed phase, which can lead to a biased distribution of species. Indeed, the question as to whether dipeptides or even single amino acids occur in solution in the form of dimers or larger clusters has been a long standing one and a summary and discussion of the debate surrounding this question can be found in [155]. In a recent study employing dynamic light scattering (DLS), it has been shown that at least in the case of FF, no larger pre-nucleation clusters exist, as the scattering intensity-weighted size distribution was found to be dominated by a species of approximately 1 nm hydrodynamic diameter [90]. The resolution of these measurements is not high enough to decide whether this dominant population corresponds to monomers, dimers or even slightly larger species. However, it is interesting to note that the size distributions from DLS are essentially identical for strongly sub-saturated (in methanol) and strongly super-saturated solutions (in water) [90]. If an equilibrium population of multimers with a finite binding affinity existed, it would be expected that its population depends strongly on the total peptide concentration. The strongly sub-saturated conditions (2 g/L in methanol) correspond to conditions where it has previously been proposed that FF occurs in the form of dimers [154], which is compatible with the size distribution measured in reference [90].

On a more conceptual level, studies that show a significant fraction of the peptide to occur in the form of dimers or multimers [114, 154] are unlikely to describe nuclei, given that nuclei, by their very definition, are transient, rare species because they correspond to the species with highest free energy along the reaction coordinate. In a study that investigated the kinetics and thermodynamics of six tetrapeptides (all permutations of the FFDD peptide), it was proposed that dimer formation is slow and therefore corresponds to the rate-determining step [110]. This conclusion is based on the observation that the fluorescence intensity of the phenylalanine residues initially increases steeply after a lag time and then slowly decreases, whereby the decrease corresponds to the formation of fibrils. However, the fluorescence signal in this case cannot unambiguously be ascribed to dimeric species and it is difficult to estimate how large the population of dimers/oligomers is, given that their fluorescence properties cannot be studied in isolation. Nevertheless, the data do strongly suggest that this system is initially dominated by the formation of unstable/metastable smaller species, followed by the further evolution of these species through growth, which corresponds to an energetic down-hill process.

Information about the thermodynamic stability of small species can also be obtained by analysing the dependence of soluble peptide on the total peptide concentration [90]. It is found that in the case of FF, the soluble concentration corresponds to the total concentration up to the critical concentration and remains constant at higher total concentrations. Such a behaviour, analysed within the framework of linear polymerisation, suggests a large critical nucleus size and hence a highly cooperative nucleation process [90].

A very different approach to studying the nucleation of FF microtubes has recently been taken, whereby amorphous films of FF, formed from HFIP, have been hydrated and studied by highly time-resolved Raman spectroscopy [156]. It was thereby found that the transition from amorphous to fully crystalline material involves the formation of an intermediate species. It is proposed that the entire peptide population is converted into this intermediate initially, followed by full conversion into the final state. This model is compatible with the observed growth of FF microtubes by addition of soluble peptide [90, 96, 119], if one assumes a dissolution of the intermediate, followed by growth of the final structures, rather than a direct interconversion.

Overall, the currently available data suggests that gel-forming systems might display rather small nucleus sizes of the order of a dimer [110], leading to rapid nucleation (time scales of seconds to minutes). Another aromatic peptide system that supports this conclusion is given by the N-terminally protected triphenylalanine peptides developed by Hamachi [108]. Here, a solution that is left to cool to room temperature displays rapid de novo nucleation of fibrils within minutes [108]. On the other hand, crystallizing systems are more likely to feature significantly larger nuclei, as indicated by the ability to create supersaturated solutions of FF that can be stable for hours [96]. Such systems are more likely to be appropriately described within the framework of classical nucleation theory [152]. A note of caution is appropriate here, however, as these statements are likely to only hold for very short sequences. For significantly longer sequences, such as the amyloid fibril-forming intrinsically disordered protein α-synclein, monomer solutions can be kinetically stable for days [157]. While this behaviour could also be indicative of a large nucleus size, it is more likely that it stems from a high energetic/entropic cost of forming even a small nucleus.

3.7 Structure of Dipeptide Crystals with Particular Emphasis on FF

The structural analysis of assemblies of dipeptides, in particular also aromatic ones, was significantly advanced by Carl-Henrik Gørbitz [14, 61, 158]. In particular, he solved the crystal structure of FF crystallized from water and highlighted the importance of water inside nanochannels for the structural integrity of the crystal. In the case of FF, the occupancy of the water sites is found to be between 0.1 and 0.5, whereas in other dipeptide crystals, water molecules can play an even more well-defined structural role [14]. The importance of water in such crystal structures is rather general, but the water is not always found inside channels – it can also play a bridging role, such as in YW crystals [158]. This layer structure is also interesting in that it displays hydrogen bonding (albeit weak) between the ammonium terminus and the aromatic tyrosine ring. The establishment of continuous hydrogen-bonded networks in short peptide structures is a hallmark of their self-assembly, and notably all backbone hydrogen bonding interactions of uncharged dipeptides can easily adopt planar conformation, leading to layered structures and the potential for polar/nonpolar segregation [159, 160]. In cases where the drive to segregate nonpolar species is strong, and, not unconnected, where the species are notably bulky, the hydrogen-bond 'planes' may adopt inward curvature, maintaining the critical connectivity while allowing the large R-groups to segregate and pack efficiently [14, 161]. This can lead to the formation of nanochannels, unit-cell-level voids occupied by both solvation species and indeed non-associated solvent or other species. The energetic cost of porous structures due to loss of van der Waals interactions can be steep, so it is perhaps notable that the hydrophobic dipeptide family of molecules display such arrangements with a degree of frequency.

3.7.1 Hydrophobic Structures in Aromatic Dipeptides

Two major structural classes of porous hydrophobic dipeptide crystals are known, each named for the dipeptide forming the archetypical structure. The Val-Ala class [160, 162] displays hexagonal hydrophobic pores ringed by six left-handed dipeptide double helices, interacting with each other by an unusual bridging interaction involving hydrogen bond donation by the amide and C_α to a carboxylate group pointed end-on at the two donors. The structure is known for seven of the nine dipeptides containing Val, Ala and Ile, with di-alanine and di-isoleucine being the exceptions. The second major structural family is the Phe-Phe family, and it is naturally this family on which this section focuses. The key property of this family is the total segregation of the hydrogen bonding chains into mutually non-interacting tubular regions within the structure, the tubular regions being described by helices of hydrogen bonded, hydrophilic dipeptide backbones surrounding a solvated void. The tubes are completely surrounded by interacting 'sleeves' formed by the bulky side chains, and the terminal ammonium projects a proton into the channel to a solvent hydrogen bond acceptor. It is known for the dipeptides IL, FL, LL, LF, FW, WG and of course FF, the most dramatic demonstration [14, 160] (Fig. 3.7).

3.7.2 Hydrogen Bond Connectivity in Aromatic Dipeptides

A statistical study [61] of hydrogen bond connectivity in dipeptide structures showed that out of the 160 investigated crystal structures of dipeptides, 152 are uncharged zwitterions, and only 8 are charged and crystallize with associated counter ions. This finding shows that the presence of charges on zwitterionic peptides can be well-accommodated in crystals in a head-to-tail arrangement. In the survey, the hydrogen bonding is classed in terms of the unbounded chains

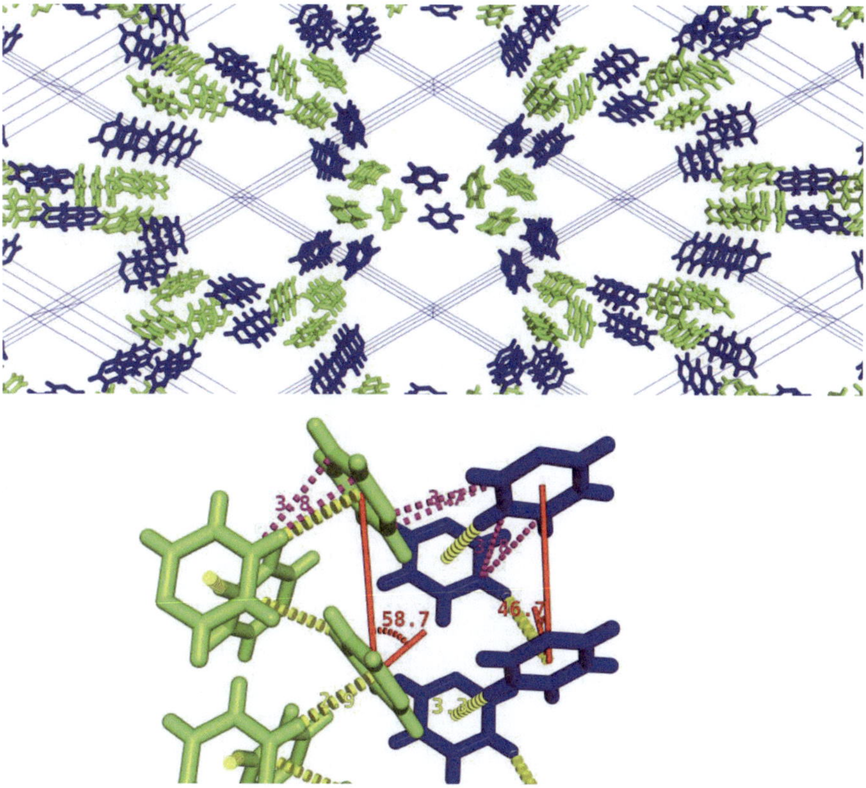

Fig. 3.7 Top: The continuous hydrophobic domain within FF arises partly from the bulk of the side chains, shown here within the hexagonal unit cell. The polar backbones line the channel, and form essentially one-dimensional structures. F1, the N-terminal amino acid, is in blue, and F2 is in green. The asymmetric unit cell contains only one FF, and so there is one F1 environment and one F2 environment, each F1 or F2 site being interchangeable with other F1 or F2 sites by symmetry operations of the $P6_1$ space group. Bottom: Geometry of interactions within the hydrophobic domain shows a remarkable helix structure for F2 sites akin to McGaughey's energy-minimised pinwheel for the benzene trimer [68], linked by T-shape type aromatic-aromatic interactions (R_{cen} = 5.1 Å, close to the empirical Phe-Phe pair minimum [68] and the calculated minimum [66]), with closest approach through the carbon 4 C-H$\cdots$ π bond (yellow) at a distance from H to the ring centroid of 2.9 Å. The comparable interaction for F1 does not have the proton pointed directly at the ring, but retains an offset T-shape geometry. The measurement for closest carbon-carbon approach (magenta), often used in automated systems for the discovery of interaromatic contact, is 3.8 Å on both the F1 and F2 sites. The F1-F2 interactions are shown at 3.7 Å between C2 of F1 and C1 and C2 of F2. The vertical spacing of each site is naturally the lattice parameter c, 5.46 Å, at angle θ (in red) 58.7° on F2 and 46.7° on F1. This geometry suggests a lack of $\pi-\pi$ interaction along the long axis of the crystal [163–165]

formed by hydrogen bonding in the crystal – these chains consist of a repeating unit that can be mapped on to itself by a simple translation along unit cell vectors [166, 167], and are expressed in terms of graph set terminology [168], in this case specifically the chain patterns denoted as $C(n)$, with n the number of atoms in the repeating unit. The eight-membered repeat unit is by a long way the most common motif, and in this case the eight atoms involved are those along the dipeptide backbone, corresponding to the "head-to-tail" charge-charge interaction pattern, with surveyed dipeptides showing up to three independent eight-membered chains. Of the 160 structures surveyed, only 15 did not display a C(8) chain. FF displays two, remarkable for their helical shape – one helix is right handed, and one unit cell in length, the other left handed and five unit cells in length. A critical characteristic of FF in terms of its relevance to amyloid fibrils is that its crystal grown in aqueous solution *does not show* a C(4) chain corresponding to inter-amide hydrogen bonding. The central amide is instead involved in the more common C(5) pattern, with the amide proton closely approaching the carboxylate terminus rather than a neighbouring amide. A more conventional 'amyloid-like' hydrogen bond structure is achieved by the FF-methanol solvate [88], which displays C(4) inter-amide and C(8) head-to-tail patterns (the C(8) displaying a twofold rotational axis normal to the backbone plane). Notably, the short $C\alpha \cdots O{=}C$ distance, demonstrating a hydrogen bond [169, 170], is a familiar feature in parallel β-sheet structures. This connectivity is, interestingly, exactly that observed for the dipeptide di(phenylglycine) in its inclusion compounds with certain chiral sulphoxides [171]. Its aqueous self-assembly process yields a spherical aggregate, suggesting a lack of long-range crystalline structure [147]. Conversely, successive prolonging of the aliphatic linker between the phenyl ring and the peptide backbone yields non-hollow elongated structures, as well as plate-like structures, which reflects presumably the different crystal packing, as well as a different balance of crystal nucleation and growth rates [172]. The chaotic response of dipeptide structures to even small changes in molecular identity has been observed frequently [160, 173], and a number of ring-substituted FF analogues are known: substitution at the 4 positions of the ring with fluoro, iodo and nitro groups yield extremely narrow tubular structures, similar in appearance but remarkably different in their structure. FTIR indicates antiparallel β-sheet structure for the halo-substituted dipeptides, while di-(4-nitrophenyl)alanine has a distinct signal consistent with random coil connectivity, possibly suggesting involvement of the nitro group (a strong hydrogen bond acceptor) in the polar/hydrogen bonded network of the aggregate. Di-(1-naphthyl)alanine and di-(2-naphthyl)alanine yield flexible tubular structures showing β-sheet FTIR signals, and di-(1-biphenyl)alanine crystallises as square plates [174]. The formation of plate-like structures has also been observed for the FFF tripeptide [103]. Based on molecular simulations, those structures are reported to be thermodynamically more stable than FF tubes, but this result is not experimentally verified. It is, however, plausible, given the trend in solubility observed between F and FF (Fig. 3.4).

Importantly, by comparing experimental and simulated powder diffraction data, it was also shown by Gørbitz that the molecular arrangement of large FF crystals are identical to the much smaller ones obtained by introducing concentrated solutions in organic solvents, such as HFIP [12], into water [175]. This conclusion gained further support when FF assemblies of different sizes (sub μm thickness and wider than μm) were structurally analysed by AFM and polarized Raman microspectroscopy and no differences were found [176].

3.7.3 Macroscale Aggregate Structure

The overall tubular appearance of FF crystals grown from water is one of the incredible traits of the diphenylalanine aggregate, almost seeming as though its habit recapitulates its unit cell down to the incorporation of a solvated void! Naturally, this phenomenon was the cause of much early discussion [12, 14] and was exploited by, for example, filling the core with elemental silver [12]. It is currently still unresolved why FF forms hollow crystals, but a potential explanation can be found in reference [177], where it was proposed that hollow crystals from small molecules can be formed when the growth rate of the crystal is faster than the arrival of peptide building block through diffusion. In this case the hollow core simply reflects the insufficient delivery of peptide building blocks in the centre of the growing crystal. In the light of this hypothesis, it can also be understood that under certain solution conditions, solid rods rather than hollow tubes can be formed from FF for example by titrating a solution of FF in TFA with ammonia solution [91, 178], or by sonication [178, 179]. The sonication process is likely to yield a more homogeneous peptide solution and prevents the built-up of concentration gradients that can lead to the formation of hollow tubes [177]. Interestingly, it was reported that the molecular arrangement of tubes and rods is not identical [178]. It is suggested that the different types of structures can be inter-converted [178], but a more detailed look at the protocols involved ("annealing" at 80°C, 2 g/L) suggests, based on the solubility of FF (see above) that the rods are dissolving and tubes are subsequently nucleating de novo. That differences in solution conditions should lead to the nucleation of different structures and morphologies (high peptide concentration and high ionic strength in the case of the rods and lower concentration and slow increase in supersaturation through cooling in the case of the tubes) seems overall more likely than a large scale, concerted rearrangement of an entire crystal. That said, it has been proposed elsewhere that transient structural changes can be induced by ultrasonication [180, 181], and these changes are not attributed to heating of the sample. It has to be noted, however, that sonication can lead to significant local heating and that the overall temperature increase of the sample is not necessarily representative of the maximum temperatures locally reached inside the sample [182].

3.8 Comparison with the Assembly of Longer Sequences into Amyloid Fibrils

3.8.1 Structural Comparison

In light of the remarkable ability of the tetrapeptides KFFE and KVVE to form unambiguous amyloid fibrils [77], a property shared with a number of short, hydrophobic peptides containing phenylalanine [183–185], and the presence of neighbouring phenylalanine residues in an aggregation-critical sequence of the amyloidogenic Aβ peptides, the possibility that FF represents an instructive model system in which interactions key to the early stages of amyloid aggregation could be simulated is a tantalising one, and indeed was a major driving force in much of the research into FF and similar short aromatic peptides.

There are certain superficial structural similarities between the water-grown crystal of FF and particular amyloid systems, outside of their compositional similarities – side chain interaction and close packing/interdigitation is a common property of cross-β amyloid structures [186], and is clearly a strong influence in FF self-assembly. We have already discussed the tendency of phenylalanine to occur in β-sheet-forming regions of native chains and the β-sheet is, of course, the fundamental structural element of the amyloid fibril, and any tendency to promote β-sheet formation can, with a few caveats, be seen as a tendency to stabilise amyloid structures. Inspection of the Ramachandran plots for the aromatic amino acids shows the extent of the steric restrictions imposed by the involvement of the γ-carbons in a rigid ring system. Application of the Ramachandran plot to dipeptide conformation is problematic, however – the angles are dihedral angles between amide CO-NH bonds and the two backbone bonds of the C_α; ψ is defined as the C_o^{i-1}-N-C_α-C_o dihedral angle, while ϕ is the N-C_α-C_o-N^{i+1} dihedral angle (Fig. 3.9a, b). Both are measured clockwise around the central bond in the N-terminal to C-terminal direction. Obviously, FF only has one amide bond, and the C-terminal is a carboxylate and the N-terminal an ammonium, strictly meaning that the N-terminal Phe only has a defined ψ (156.9°), and the C-terminal only a defined ϕ (54.7°). At the cost of some assumptions, replacing a carboxylate oxygen with an imaginary amide yields two potential ψ values for F2, at 44.5° and −138.8°. A ϕ angle of 156.9° (F1) is associated nearly exclusively with β-sheet secondary structure, while the situation for F2 is more unusual – the positive ϕ angle, a consequence of the cis arrangement of side chains, is more rarely seen in native protein structures and is typically associated with a left-handed α-helix [187] (for $\psi = 44.5°$) or an unclassified structure [188] for $\psi = -138.8°$. Inspection of Fig. 3.8 shows that the conformation where the first F is in a sheet-like conformation and the second in a left-hand helix or the unclassified conformation is relatively rare – of the more than 16,000 phenylalanine residues in FF or FFF sequences

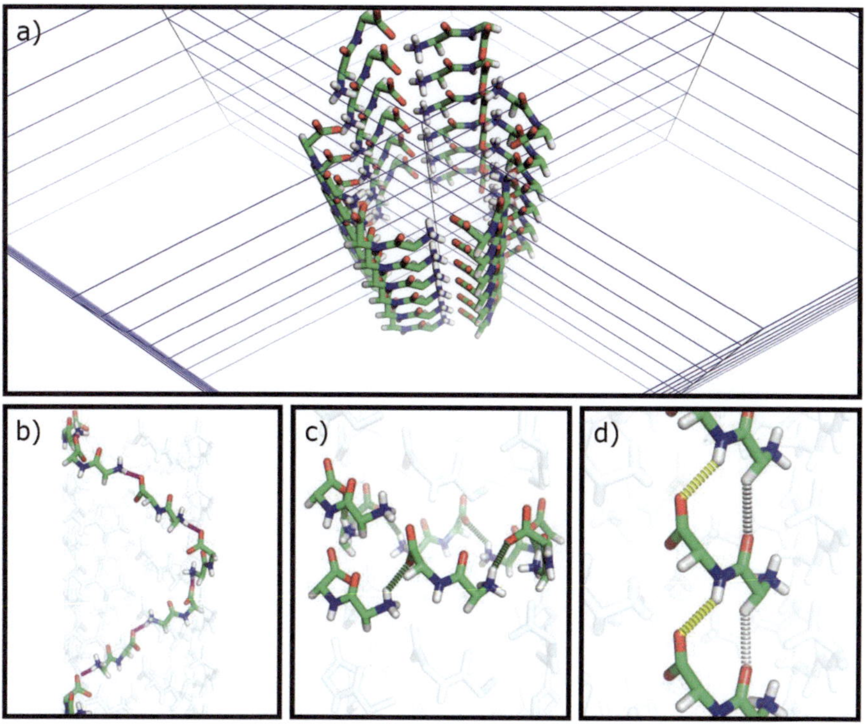

Fig. 3.8 (**a**) Hydrogen bonded network in FF, superimposed on the unit cell. The hydrogen bonds form closed structures due to the curvature imposed by the bulky side chains. (**b**) Six-membered left-handed helix, C(8). (**c**) Six-membered right-handed helix, C(8). (**d**) Axial chains, the strongly-bonded C(5) chain in yellow and the weaker Cα-carbonyl, a C(4)

sampled (see footnote[2]), only 40 show geometries approaching the ones observed in diphenylalanine crystals (Fig. 3.9c). A cyclic peptide RACAFFC containing the FF motif was the subject of a study into its activity as an antagonist at a chemokine receptor [191], and measurement of the ϕ, ψ angles of the two phenylalanines, constrained as part of a five-membered ring closed by a disulphide bond shows they are similarly not in a conformation associated with secondary structure (F6 does show angles typical of a β-sheet, but F5 is well outside the conventional limits). The amyloids formed by KLVFFA, however, show that both phenylalanines are part of a continuous β-sheet structure in all polymorphs and have dihedral angles well within the expected bounds [192], and the same applies for the oligomerising β-hairpin structures formed by longer core fragments of Aβ [193, 194].

[2]A search was performed on the PDB [189] for the "FF" sequence motif, where it occurs as part of an unmodified sequence, at resolution < 2.5 Å and without ligands. The sample was analysed using Pymol 2.1.0 [190] via the get_phipsi command and the output plotted using Gnuplot.

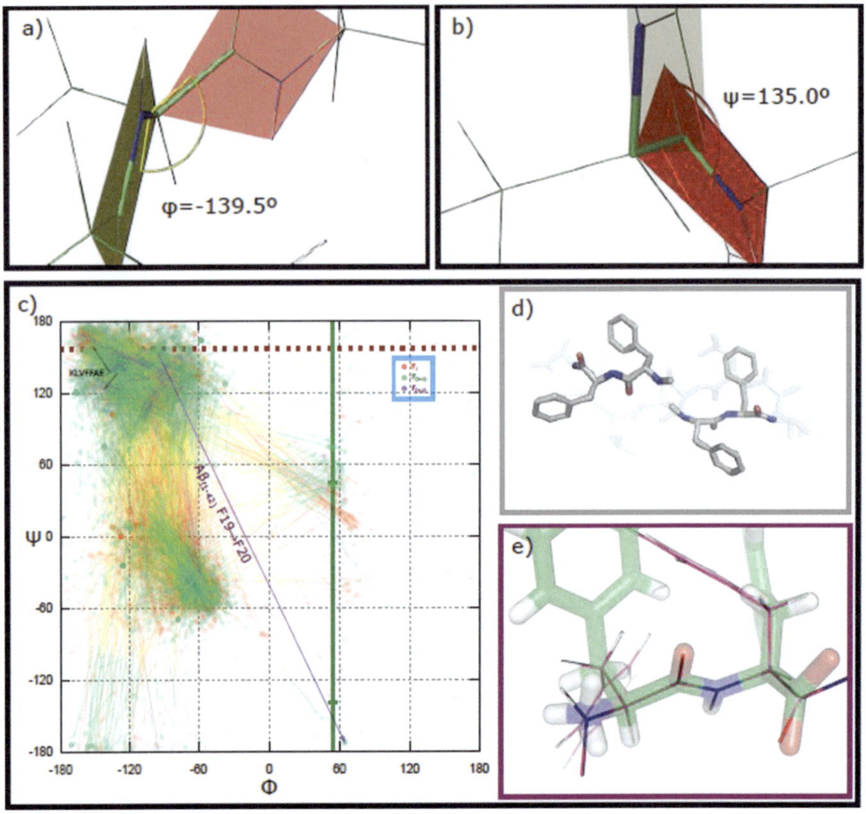

Fig. 3.9 The Ramachandran angles of the FF sequence in a statistical sample (**a**) Ramachandran angle ϕ, measured in a right-handed fashion in the N-to-C terminal direction. The angle shown is typical of an antiparallel β-sheet. (**b**) The angle ψ, again with a value typically seen in the antiparallel β-sheet. ϕ and ψ values are restricted by steric effects to ranges of 'allowed' values. (**c**) A statistical sample from the Protein Data Bank of 8044 occurrences of the unmodified FF sequence in 3087 PDB entries, collected by the authors for this work. Direction of the chain is shown by the colour gradation. In nature, the sequence has a relatively typical Ramachandran distribution, with a tendency for a first residue in a helical-type structure to result in a second residue with sheet-like conformation. Two black arrows represent the pair of FF sequences in PDB 3ow9 [192], a polymorph of KLVFFA with typical anti-parallel β-sheet conformation (Panel **d**) The dipeptide crystal has ϕ_1=156.9°, with ψ_1 undefined. The dotted red line shows the ϕ angle on F1. The carboxylate terminus of F2 allows the guessing of two possible "ψ angles", for $\phi_2 = 64.7°$. Recognition of a similar structure with a sheet-like first residue, and a second in a left-handed conformation, reveals the curious fact that FF (Aβ_{19-20}) adopts a similar backbone structure to PDB entry 5NAO, Aβ_{1-42}, as solved by Wälti [195]. (**e**) Superposition of the FF crystalline conformation (green) over the conformation of the sequence in Aβ_{1-42}

The helical connectivity of the hydrogen bonding chains has some similarities to a proposed model of polyglutamine aggregation advanced by Perutz et al. [196] contemporaneously with early, but highly influential, reports on the properties of FF [12], though the β-helix model for polyglutamine species has not stood the test of time [197, 198]. The parallel arrangement of the two-membered chains might be seen as something of a departure from expected behaviour for *short* polypeptides forming amyloid fibrils, due to the increased importance of terminal charge complementarity, which forces antiparallel organisation in KFFE and other asymmetric short peptides with charged residues [77, 199]. A critical factor in the topology of the aggregate structure is the twist observed for amyloid fibrils – this phenomenon is due to the steric effects of the chiral backbone [200]. Only if $\phi + \psi = 0$ on average along the extended chain in the sheet are the amide groups parallel. In general, the average has a small finite value, resulting in a tendency towards a progressive deflection of the polar direction of the amide groups along the chain, and hence to a helical twist around the axial direction of the polymeric fibril or β-sheet [201]. This helical twist effectively limits the radial dimension of the fibril, as fibril addition sites further from the central position experience progressively greater elastic strain. The equilibrium radius of the fibrils arises from the balance between this elastic deformation cost and the free energy of radial addition to the fibril [202]. Polymorphism in radial dimension, helical twist pitch, handedness and indeed in coiling of the fibril into nanotube-like ribbons have been observed [203, 204]. The limited width, due to mechanical effects, is a key property of amyloid fibrils; indeed, it is a major reason why they have a fibrillar morphology and critically it distinguishes the fibril from the crystal. This was shown in a theoretical study, where the free energy as a function of the twist angle was treated as a perturbation in accordance with the Landau theory of phase transitions, a mean-field theory whereby the perturbation can be expanded as a Taylor series [205]. In the case of the twisting of amyloid fibrils, the chirality of the individual chains requires the retention of the odd-powered terms in the expansion which would otherwise cause asymmetry about the transition point, though the linear term is shown to have a zero coefficient. The analysis in the paper proceeds to identify the stationary points of the series expansion, with a minimum at zero twist (a crystalline state), and a minimum at a specific twist angle, this twist angle being dependent on the equilibrium twist angle (the twist angle at zero fibril thickness), and the ratio of the torsional spring constant of the fibril (a function of the square of fibril thickness) to the spring constant for rotations between individual chains in the fibril (going as the cube of the chain length for long chains). A critical fibril thickness was established, above which the crystalline form would be stable [205]. This critical width scales linearly with chain length, and is of course also dependent on other material parameters, many of which are common to all amyloid structures. It is to be noted that the origin of the asymmetry, the equilibrium chiral twist of the amide backbone, cannot exert influence in the dipeptides – only one amide bond exists, and so it is impossible for there to be an equilibrium series of amide rotations about the backbone. The ammonium and carboxylate termini are freely rotated, and interact by a monopole-monopole force that, inherently, only has a distance and no directional dependence.

A crystal is defined by its repeating, three-dimensional motif and in principle, any translation along an integer number of unit cell vectors will lead to an identical environment. This property has been confirmed for two distinct FF solvomorphs [88, 175] among many other short peptides [61] by virtue of the analysis of X-ray diffractograms. Diphenylalanine crystallises from water into the space group $P6_1$, which on the face of it has a helical appearance, borne out by the angled C(8) chains which spiral the inside of the nanochannels. What is not helical, though, is the vertical (axial) connectivity, the C(5) chain- each step along a multiple of the short axial unit cell vector brings you to an identical diphenylalanine. This argument cannot be made for the amyloid axial translation unless the pitch of the helix was a precise multiple of the hydrogen bond separation. Even though the repeat distance is highly consistent along a fibril [206], it cannot be said to represent a crystalline repeat due to the far lower energy of minute fluctuations over the repeat distance. Still harder to equate to the crystalline regime are the radial 'vectors' in amyloid fibrils. For these sites, as has been discussed, energy and concomitantly equilibrium shape change with distance from the fibril centre, and there exist defined upper limits for the radial dimension. Nevertheless, the attachment of monomeric peptide onto the surfaces of amyloid fibrils of the Aβ peptide has been demonstrated to be the first step of the autocatalytic secondary nucleation process and the affinity has been determined from quantitative surface plasmon resonance (SPR) biosensing studies to be up to two orders of magnitude weaker compared to the affinity for the fibril end [207, 208].

3.8.2 Comparison of Assembly Kinetics and Thermodynamics of Short Aromatic Peptides and Longer Amyloid Forming Sequences

It has been shown that in diphenylalanine, the radial growth rate is a continuous, monotonic function of excess concentration of the dipeptide in water (Fig. 3.6b and reference [96]), while in the amyloid fibril continuous radial growth is not normally observed [209, 210]. The aggregation process of FF, in common with other crystalline species, is highly cooperative. Concerns of edge and surface energies determine the stability of 'islands' of a new layer at a surface (Fig. 3.6d) and give rise to complex, higher-order dependence of the incorporation rate on the excess arrivals of monomer at the surface over the off-rate of resolvating species (Figs. 3.6a and 3.10a). In contrast, the kinetics of elongation of amyloid fibrils have been shown in a range of studies to depend linearly on the concentration of soluble peptide at low concentrations and saturate at higher concentrations (Fig. 3.10b and references [211, 212]), and this distinct concentration dependence has been interpreted as strong evidence for a mechanism whereby the fibrils elongate through the addition of monomeric, rather than oligomeric building blocks [211]. The effect of monomer concentration on axial growth rate is a readily-identified difference between crystalline and fibrillar systems- crystalline systems display cooperativity at moderate supersaturation, while fibrillar systems transition from non-cooperative (linear in supersaturation) to anti-cooperative (sublinear) regimes at

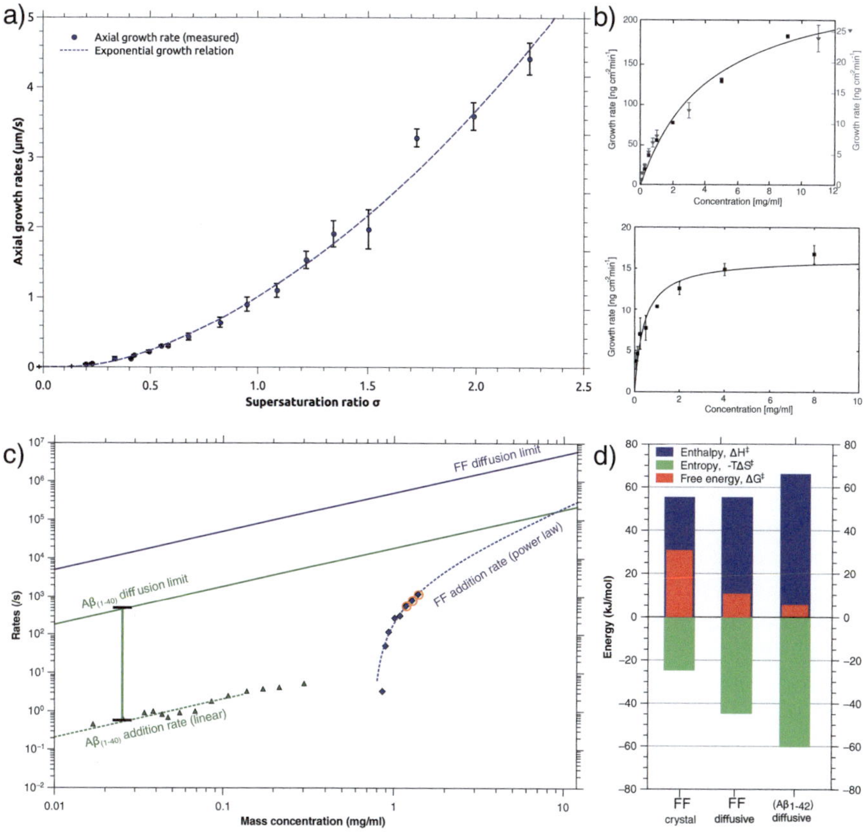

Fig. 3.10 Comparison of the concentration-dependence and free energy barriers of peptide self-assembly into amyloid fibrils and crystals (**a**) The axial growth rates of FF crystals as a function of solution supersaturation σ displays an exponential dependence. (**b**) The dependence of amyloid fibril elongation rate on the solution concentration of insulin under two different solution conditions (top)- the growth rates in 100 mM NaCl/20% acetic acid (squares) and in 10 mM HCl with no added salt [212] can be scaled to the same master curve, despite an order of magnitude difference in absolute rate and the concentration dependence of fibril elongation rate in the longer chain α-lactalbumin amyloid system (bottom), showing onset of saturation at lower concentration [212]. (**c**) Log–log graph comparing the aggregate growth rates of amyloid fibrils of the Aβ(1–40) peptide with those of FF crystals. The solid lines represent the diffusional fluxes of monomers into a reaction volume on the end of the aggregate and correspond to the theoretical maximum rates. (**d**) Comparison of the free energies, entropies and enthalpies of activation of FF crystal growth with those of amyloid fibril growth of the Aβ(1–42) peptide. The kinetic data of FF has been analysed in two different ways, using a diffusive model for direct comparability with amyloid fibril growth and also by taking the crystalline nature of the assemblies explicitly into account. Details of this analysis can be found in reference [90]. (**a**), (**c**) and (**d**) adapted with permission from reference [90] ©2017 American Chemical Society and (**b**) adapted with permission from reference [212] ©2010 American Physical Society

a supersaturation determined by internal dynamics of the attached monomer during incorporation [212].

The difference in growth mechanisms between amyloid fibrils and crystals renders a direct comparison of the absolute growth rates and in particular of the underlying free energy barriers not straightforward. Nevertheless, a first idea can be gained by comparing the absolute rates of growth of these structures with the theoretical maximum of the assembly rate, the diffusional arrival of the peptide into a reaction volume of size comparable to the soluble building block. In Fig. 3.10c such a comparison is shown for the FF peptide and the Aβ(1–40) peptide. The difference between the actual rate and the theoretical maximum is approximately a factor of 10^3 in the case of Aβ(1–40) across a range of concentrations, and due to the saturation of fibril elongation this factor increases at high concentrations. On the other hand, the exponential dependence of the axial growth rate of FF leads to a strongly decreasing difference between the rates of diffusional arrival and growth. If the difference between maximal and actual rates is interpreted as stemming from a free energy barrier, the latter is therefore concentration dependent [90].

All faces of the crystal have been shown to grow continuously at finite rates in suitable conditions, but another comparison with the amyloid system may be made in that the crystals of FF have extreme aspect ratios, and through careful control of solution conditions, can be made to grow axially with extremely slow radial growth [96], a phenomenon arising from the exponential dependence of growth rate on supersaturation on each face. The observation is explained in terms of the intermolecular interaction chains coplanar with the initial deposition of a new face. In FF, the radial faces may either intersect the helical hydrogen bonded channels, which will readily stabilise incoming FF, or they may present a simple hydrophobic surface, compare Figs. 3.7 and 3.8. The close-packed hydrophobic surface is a relatively low energy face, and so the borders of new layers on the axial face come at a rather lower cost than their counterparts on the radial face, which must always expose unsatisfied pendant hydrogen bonds. As a result of the greater nucleation rate of new layers, the axial face grows faster. Certain theoretical models have seen success in explaining amyloid nucleation and growth rates in terms of pseudo-one-dimensional crystallisation [213, 214] – crystals where the energy difference between the faces is such as to render radial growth impossible for wide ranges of conditions, though it should be mentioned that the Gibbs-Wulff theorem, to which analogy is made for the equilibrium aggregate shape, is only demonstrable in aggregates capable of rapid equilibration, i.e. very small ones. Kinetic processes are dominant determinants of crystal shape at the macroscale, and the early stages of polypeptide aggregation frequently involve non-fibrillar oligomeric species, for which the crystal theories of amyloid formation do not account.

For shorter species, the line between a crystal and an amyloid fibril can be a narrow one, and the classic examples are the Sup35 fragments GNNQQNY and NNQQNY, a heptapeptide and hexapeptide which, like their parent chain, readily form amyloid fibrils [129, 215]. Unusually, though, a crystalline polymorph is also observed in each case at slightly higher monomer concentration, this crystal having very similar structure to the fibril as determined by X-ray diffractometry (the 4.7 Å meridional reflection being particularly notable), and an interesting point is made that the crystals were also highly acicular and despite the efforts of the researchers, crystals could not be grown continuously in at least the radial dimension. A microcrystal of NNQQNY is shown, with a width slightly over 1 μm, corresponding to approximately 500 unit cells, and appears to have a somewhat irregular surface, suggesting a high density of grain boundaries parallel to the long axis [7]. Other short peptide polymorphic systems have been prepared, and they have a notably wide range of side chains [216] – NNQQ, a tetrapeptide with purely polar residues, fibrillises, as does penta-glutamine, results in keeping with observations of the pathology of Huntington's chorea, a neuropathological amyloidosis triggered by extended polyglutamine repeats [217]. The same work established a fibrillar state for the hexapeptide SSTSAA, a fairly hydrophilic fragment with short side chains, as well as larger, more hydrophobic species. A key finding of this extensive survey into amyloidogenic fragments is the concept of "steric zippers". These fragments of longer chains are computationally predicted and experimentally confirmed to be amyloidogenic in their own right, and as such are considered to be potential candidates for "trigger" regions of the parent chains, whose intermolecular interaction can lead to the nucleation of the fibrils [218, 219]. The fragments proved capable in many cases of enhancing fibril formation of the parent polypeptide, and vice versa. In the diphenylalanine case, the identity of the dipeptide as $A\beta_{19-20}$ is frequently remarked upon in studies of the species, and recent NMR-derived structures of $A\beta_{1-42}$ show the motif buried in a hydrophobic "pocket", interestingly in a similar 'cis' configuration of the side chains [195, 220]. However, a recent cryo-EM study showed F20 oriented out of the hydrophobic 'core' containing F19 [9]. In this study, the fibrils were formed in 30% acetonitrile solution. This alteration of the solvent environment may render the projection of F20 at the surface less unfavourable than in the aqueous/salt systems of references [195] and [220]. The Ramachandran angles are $\phi_{F1} = -94.7°$, $\psi_{F1} = 152.7°$, $\phi_{F2} = 64.7°$, $\psi_{F2} = -172.5°$ in the structure as given by Waelti et al. Interestingly, these correspond very closely to an equivalent FF conformation in the crystal, and a glance at the structural solution shows the effect of the same forces. The hydrophobic residues are buried, and indeed interact in a clear herringbone-type chain typical of the π-π interaction, even at significant cost to the continuity of the hydrogen bonding between successive chains [195]. The effects of hydrophobicity are readily discerned from the convoluted structure of the chain in the fibril. Two separate hydrophobic pockets are formed by each of the chains forming the dimeric 'unit cell' in the axial direction, one involving the residues 16–22 (KLVFFA), and the second involving interdigitating aliphatic side chains from residue 31 onwards.

Comparison of the thermodynamics of dipeptide and polypeptide aggregation is an important step towards rationalising or discounting the employment of short peptides as model systems for amyloid propagation. In particular, the desolvation of the aromatic benzyl groups is predicted to contribute a significant negative free energy to the process of amyloid fibril formation. In polypeptide systems with hydrophobic residues, the enthalpic component of the free energy of desolvation (upon folding and burial of the hydrophobic residues) is positive at low temperature, transitioning through zero to give a negative ΔH at high temperature [221]. The free energy of solvation remains relatively unchanged, however, indicating an entropic contribution to the free energy that is nearly equal in magnitude and opposite in sign. This compensated relationship is diagnostic of the influence of the hydrophobic effect [222]. In the case of FF, the signature of the free energy, as well as its compensated nature correspond very well to that of a system dominated by the hydrophobic effect (Fig. 3.11b).

The dependence of the free energy of fibril elongation, per peptide unit, on the length of the polypeptide has been investigated for a range of amyloid-forming systems, and an empirical power-law relationship governed by contact area between chains was proposed to describe the scaling of this property (Fig. 3.11c and reference [223]). The free energy was established by depolymerisation studies with a strong denaturant and analysis of the data with the linear polymerisation model [224], where the equilibrium constant for the addition to fibrils of all lengths reduces to that between fibril ends and monomers. Diphenylalanine displays markedly greater non-polar side chain surface area per peptide than typical biologically relevant polypeptides, and the π-stacking interactions will represent a significant portion of the side chain interaction energy, explaining the comparably higher stability of FF assemblies shown in Fig. 3.11c.

Finally, the temperature dependence of the axial growth rate of crystals formed by the FF peptide (Fig. 3.6f), combined with theoretical analysis in the framework of both diffusive and surface nucleation models has also been used in order to define the enthalpic and entropic character of the free energy barriers of this process [90]. As mentioned above, within the framework of crystal growth theory, the free energy barrier is concentration dependent. The free energy barrier shown in Fig. 3.10d has been calculated from the measured absolute growth rate at 295 K shown in Fig. 3.6e (1.2 g/L soluble FF, corresponding to $\sigma \sim 1$ at this temperature). Interestingly, despite the fact that the overall mechanism of assembly is clearly distinct between amyloid fibrils and crystals (see above), the magnitude and energetic signature of the free energy barriers are remarkably similar, with both types of barriers displaying a significant unfavourable enthalpic contribution, which is largely compensated by a favourable entropy of activation (Fig. 3.10d). For amyloid fibrils this structure of the free energy barriers has been explained through the interactions that need to be broken in order to reach the transition state (intramolecular secondary structure elements, as well as hydrogen bonds with the solvent) [149]. On the other hand, the desolvation of hydrophobic regions of the peptide upon the first contacts with the fibril end is associated with a favourable entropy of activation [149].

The comparison between amyloid fibrils and crystalline peptide systems therefore shows that while the assembly mechanisms can be quite distinct, the fundamental forces responsible for the assembly, and which determine both its thermodynamics and kinetics, are highly similar.

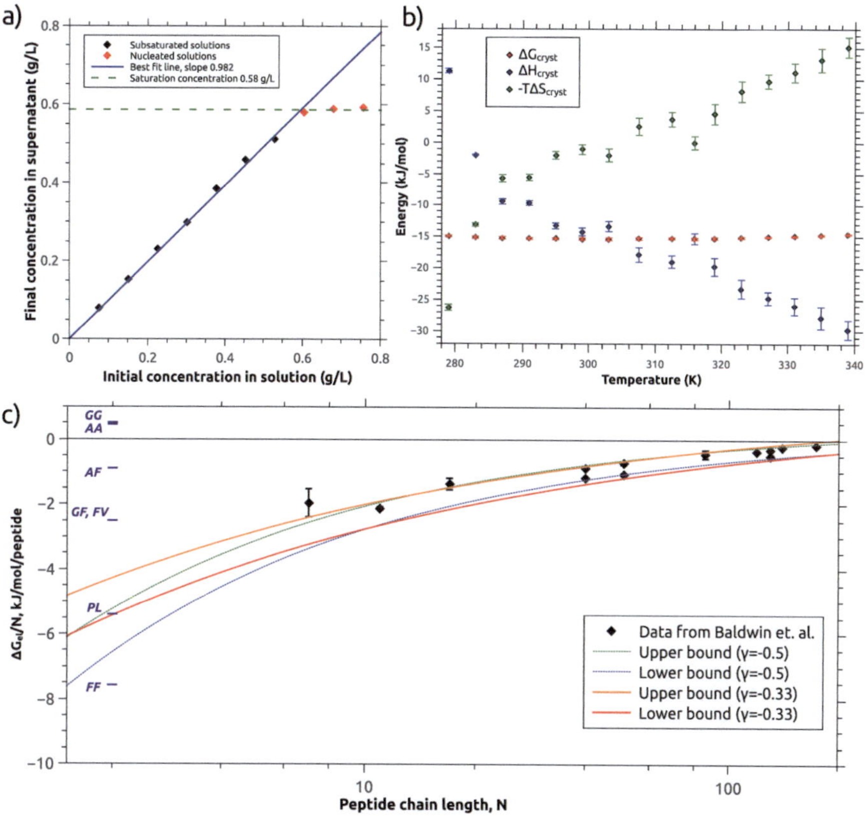

Fig. 3.11 Comparison of the thermodynamics of crystallisation with that of amyloid fibril formation (**a**) Supernatant concentrations, following centrifugation of samples equilibrated at 293 K, are plotted as a function of varying initial peptide concentration. The abrupt change in behavior, from linear with slope 1 to independent of initial concentration, at a value of 0.58 g/L demonstrates the absence of association below the critical concentration. (**b**) Overall thermodynamics of FF crystallization as a function of temperature, derived from the solubility data in Fig. 3.4. The entropy-enthalpy compensation of the crystallisation process is clearly apparent. (**c**) Peptide self-assembly thermodynamics and the limit of short sequences. Data from reference [223] with addition of the dipeptides FF, AF, FV, AA, GF, GG and PL. (Figure adapted with permission from reference [90] ©2017 American Chemical Society)

3.9 Conclusions and Future Perspectives

In this chapter, we have given a comprehensive overview over the state of knowledge of the physico-chemical and mechanistic aspects of short aromatic peptide assembly into fibrils, crystals and various other morphologies. It is rewarding to see that quantitative mechanistic studies have seen a steep rise in the last 5 years or so, and that our understanding of the driving forces responsible for these fascinating self-assembly processes is slowly, but surely, catching up with the enormous body of work that has so far mostly been based on empirical findings. A crucial remaining question is if, and how, the newly gained mechanistic understanding can be translated into rational control of the morphology and improvement of the material properties of (aromatic) peptide assemblies. The motivation to achieve this aim has the potential to be the major driving force for further fundamental studies in the coming years. While we think that the fully de novo design of assembly mechanisms and properties is not yet routinely possible, the experimental and theoretical toolbox now at our disposal allows a highly detailed characterization of any given self-assembling peptide system, which in turn enables systematic control of its assembly process. The chaotic nature of the relationship between molecular structure of the short peptide monomer and the properties of its aggregate as yet precludes direct prediction of the latter from the former, although advances in areas outside traditional biophysics, for example crystal structure prediction, continue to expand the range of methods by which short peptide structures can be rationally designed. The amyloid cross-β structure, where it is experimentally found to occur in the short peptides, provides a biomedically-relevant model system amenable to detailed computation.

The finding that such simple peptide systems can be shown to adopt such a wide range of structures and properties is both a driving force for new research and a long-standing challenge to researchers seeking to rationalise their findings and base predictions on them. Recent quantitative and theoretical studies, seeking to establish the basic parameters of the assembly processes, should provide a base for such predictions, and for the intelligent design of new self-assembling systems based on these monomers. The field, more than 20 years on from the first organised studies into aromatic dipeptides, continues to provide new and often surprising results. We have also aimed to provide an overview of the relationship between these fascinating systems and the self-assembly of full-length polypeptides, and the extent to which understanding of one system – the forces driving and the mechanisms describing its aggregation, the properties and applications of the self-assembled structure, and the experimental techniques for its study – can be used to understand the other.

Acknowledgements TOM thanks the Newman Foundation and the Weizmann Institute for funding. AKB thanks the Turnberg Foundation for a travel grant to Tel Aviv (2011), that enabled to start the mechanistic studies of short aromatic peptide self-assembly.

References

1. Waugh DF (1944) The linkage of corpuscular protein molecules. I. A fibrous modification of insulin. J Am Chem Soc 66:663–663
2. Kendrew JC, Bodo G, Dintzis HM, Parrish R, Wyckoff H, Phillips DC (1958) A three-dimensional model of the myoglobin molecule obtained by x-ray analysis. Nature 181:662–666
3. Kidd M (1963) Paired helical filaments in electron microscopy of Alzheimers disease. Nature 197:192–193

4. Dobson CM (2003) Protein folding and misfolding. Nature 426:884–890
5. Sunde M, Serpell LC, Bartlam M, Fraser PE, Pepys MB, Blake CC (1997) Common core structure of amyloid fibrils by synchrotron X-ray diffraction. J Mol Biol 273:729–739
6. Polverino de Laureto P, Taddei N, Frare E, Capanni C, Costantini S, Zurdo J, Chiti F, Dobson CM, Fontana A (2003) Protein aggregation and amyloid fibril formation by an SH3 domain probed by limited proteolysis. J Mol Biol 334:129–141
7. Nelson R, Sawaya MR, Balbirnie M, Madsen A, Riekel C, Grothe R, Eisenberg D (2005) Structure of the cross-beta spine of amyloid-like fibrils. Nature 435:773–778
8. Colvin MT, Silvers R, Frohm B, Su Y, Linse S, Griffin RG (2015) High resolution structural characterization of Aβ42 amyloid fibrils by magic angle spinning NMR. J Am Chem Soc 137:7509–7518
9. Gremer L, Schölzel D, Schenk C, Reinartz E, Labahn J, Ravelli RBG, Tusche M, Lopez-Iglesias C, Hoyer W, Heise H, Willbold D, Schröder GF (2017) Fibril structure of amyloid-β(1–42) by cryo-electron microscopy. Science 358:116–119
10. Fitzpatrick AWP, Falcon B, He S, Murzin AG, Murshudov G, Garringer HJ, Crowther RA, Ghetti B, Goedert M, Scheres SHW (2017) Cryo-EM structures of tau filaments from Alzheimer's disease. Nature 547:185–190
11. Azriel R, Gazit E (2001) Analysis of the minimal amyloid-forming fragment of the islet amyloid polypeptide. An experimental support for the key role of the phenylalanine residue in amyloid formation. J Biol Chem 276:34156–34161
12. Reches M, Gazit E (2003) Casting metal nanowires within discrete self-assembled peptide nanotubes. Science 300:625–627
13. Masters CL, Simms G, Weinman NA, Multhaup G, McDonald BL, Beyreuther K (1985) Amyloid plaque core protein in Alzheimer disease and down syndrome. Proc Natl Acad Sci U S A 82:4245–4249
14. Görbitz CH (2001) Nanotube formation by hydrophobic dipeptides. Chemistry 7:5153–5159
15. Yemini M, Reches M, Gazit E, Rishpon (2005) Peptide nanotube-modified electrodes for enzyme-biosensor applications. J Anal Chem 77:5155–5159
16. Kol N, Adler-Abramovich L, Barlam D, Shneck RZ, Gazit E, Rousso I (2005) Selfassembled peptide nanotubes are uniquely rigid bioinspired supramolecular structures. Nano Lett 5:1343–1346
17. Kholkin A, Amdursky N, Bdikin I, Gazit E, Rosenman G (2010) Strong piezoelectricity in bioinspired peptide nanotubes. ACS Nano 4:610–614
18. Wang M, Xiong S, Wu X, Chu PK (2011) Effects of water molecules on photoluminescence from hierarchical peptide nanotubes and water probing capability. Small 7:2801–2807
19. Mitra RN, Das D, Roy S, Das PK (2007) Structure and properties of low molecular weight amphiphilic peptide hydrogelators. J Phys Chem B 111:14107–14113
20. Smith AM, Williams RJ, Tang C, Coppo P, Collins RF, Turner ML, Saiani A, Ulijn RV (2008) Fmoc-Diphenylalanine self assembles to a hydrogel via a novel architecture based on π–π interlocked β-sheets. Adv Mater 20:37–41
21. Johnson EK, Adams DJ, Cameron PJ (2011) Peptide based low molecular weight gelators. J Mater Chem 21:2024–2027
22. Whitesides G, Mathias J, Seto C (1991) Molecular self-assembly and nanochemistry: a chemical strategy for the synthesis of nanostructures. Science 254:1312–1319
23. Whitesides GM, Grzybowski B (2002) Self-assembly at all scales. Science 295:2418–2421
24. Kabsch W, Sander C (1983) Dictionary of protein secondary structure: pattern recognition of hydrogen-bonded and geometrical features. Biopolymers 22:2577–2637
25. Frishman D, Argos P (1995) Knowledge-based protein secondary structure assignment. Proteins Struct Funct Bioinf 23:566–579
26. Perrin CL, Nielson JB (1997) Strong hydrogen bonds in chemistry and biology. Annu Rev Phys Chem 48:511–544
27. Steiner T (2002) The hydrogen bond in the solid state. Angew Chem Int Ed 41:48–76
28. Wang J, Liu K, Xing R, Yan X (2016) Peptide self-assembly: thermodynamics and kinetics. Chem Soc Rev 45:5589–5604

29. Bowie JU (2011) Membrane protein folding: how important are hydrogen bonds? Curr Opin Struct Biol 21:42–49
30. Sheu S-Y, Yang D-Y, Selzle H, Schlag E (2003) Energetics of hydrogen bonds in peptides. Proc Natl Acad Sci 100:12683–12687
31. Grzybowski BA, Ishchenko AV, DeWitte RS, Whitesides GM, Shakhnovich EI (2000) Development of a knowledge-based potential for crystals of small organic molecules: calculation of energy surfaces for C=O·HN hydrogen bonds. J Phys Chem B 104:7293–7298
32. Stone A (2013) The theory of intermolecular forces. Oxford University Press, Oxford
33. Gazit E (2007) Self-assembled peptide nanostructures: the design of molecular building blocks and their technological utilization. Chem Soc Rev 36:1263–1269
34. Holmes TC, de Lacalle S, Su X, Liu G, Rich A, Zhang S (2000) Extensive neurite outgrowth and active synapse formation on self-assembling peptide scaffolds. Proc Natl Acad Sci U S A 97:6728–6733
35. Caplan MR, Schwartzfarb EM, Zhang S, Kamm RD, Lauffenburger DA (2002) Control of self-assembling oligopeptide matrix formation through systematic variation of amino acid sequence. Biomaterials 23:219–227
36. McDonald IK, Thornton JM (1994) Satisfying hydrogen bonding potential in proteins. J Mol Biol 238:777–793
37. Kroon J, Kanters J (1974) Non-linearity of hydrogen bonds in molecular crystals. Nature 248:667
38. Allen FH, Bird CM, Rowland RS, Raithby PR (1997) Hydrogen-bond acceptor and donor properties of divalent sulfur (*Y*-S-*Z* and *R*-S-H). Acta Crystallogr B Struct Sci Cryst Eng Mater 53:696–701
39. Sarkhel S, Desiraju GR (2004) N-H···O, O-H···O, and C-H···O hydrogen bonds in protein–ligand complexes: strong and weak interactions in molecular recognition. Proteins Struct Funct Bioinf 54:247–259
40. Pogorelyi VK (1977) Weak hydrogen bonds. Russ Chem Rev 46:316–336
41. Desiraju GR (1991) Hydration in organic crystals: prediction from molecular structure. J Chem Soc Chem Commun 6:426–428
42. Derewenda ZS, Lee L, Derewenda U (1995) The occurrence of C–H··· O hydrogen bonds in proteins. J Mol Biol 252:248–262
43. Perutz M (1993) The role of aromatic rings as hydrogen-bond acceptors in molecular recognition. Phil Trans R Soc A 345:105–112
44. Kauzmann W (1959) Some factors in the interpretation of protein denaturation. In: Anfinsen C, Anson M, Bailey K, Edsall JT (Eds) Advances in protein chemistry, vol 14. Academic, New York, pp 1–63
45. Tanford C (1978) The hydrophobic effect and the organization of living matter. Science 200:1012–1018
46. Pratt LR, Chandler D (1977) Theory of the hydrophobic effect. J Chem Phys 67:3683–3704
47. Pratt LR, Chandler D (1980) Hydrophobic solvation of nonspherical solutes. J Chem Phys 73:3430–3433
48. Stillinger FH (1973) In: Kay RL (ed) Structure in aqueous solutions of nonpolar solutes from the standpoint of scaled-particle theory. The physical chemistry of aqueous system: a symposium in honor of Henry S. Frank on his seventieth birthday. Springer US, Boston, pp 43–60
49. Lum K, Chandler D, Weeks JD (1999) Hydrophobicity at small and large length scales. J Phys Chem B 103:4570–4577
50. Huang DM, Chandler D (2002) The hydrophobic effect and the influence of solute-solvent attractions. J Phys Chem B 106:2047–2053
51. Huang DM, Geissler PL, Chandler D (2001) Scaling of hydrophobic solvation free energies. J Phys Chem B 105:6704–6709
52. Chandler D (2005) Interfaces and the driving force of hydrophobic assembly. Nature 437:640–647

53. Patterson D, Barbe M (1976) Enthalpy-entropy compensation and order in alkane and aqueous systems. J Phys Chem 80:2435–2436
54. Shinoda K, Fujihira M (1968) The analysis of the solubility of hydrocarbons in water. Bull Chem Soc Jpn 41:2612–2615
55. Fine RA, Millero FJJ (1973) Compressibility of water as a function of temperature and pressure. Chem Phys 59:5529–5536
56. Garde S, Hummer G, García AE, Paulaitis ME, Pratt LR (1996) Origin of entropy convergence in hydrophobic hydration and protein folding. Phys Rev Lett 77:4966
57. Raschke TM, Tsai J, Levitt M (2001) Quantification of the hydrophobic interaction by simulations of the aggregation of small hydrophobic solutes in water. Proc Natl Acad Sci U S A 98:5965–5969
58. Dill KA (1985) Theory for the folding and stability of globular proteins. Biochemistry 24:1501–1509
59. Agashe VR, Shastry M, Udgaonkar JB (1995) Initial hydrophobic collapse in the folding of barstar. Nature 377:754
60. Richards FM (1977) Areas, volumes, packing and protein structure. Annu Rev Biophys Bioeng 6:151–176
61. Görbitz CH (2010) Structures of dipeptides: the head-to-tail story. Acta Crystallogr B 66:84–93
62. Samanta U, Pal D, Chakrabarti P (1999) Packing of aromatic rings against tryptophan residues in proteins. Acta Crystallogr D Biol Crystallogr 55:1421–1427
63. Chakrabarti P, Bhattacharyya R (2007) Geometry of nonbonded interactions involving planar groups in proteins. Prog Biophys Mol Biol 95:83–137
64. Chourasia M, Sastry GM, Sastry, GN (2011) Aromatic-aromatic interactions database, A2ID: an analysis of aromatic networks in proteins. Int J Biol Macromol 48:540–552
65. Thomas A, Meurisse R, Brasseur R (2002) Aromatic side-chain interactions in proteins. II. Near- and far-sequence Phe-X pairs. Proteins 48:635–644
66. Hobza P, Selzle HL, Schlag EW (1996) Potential energy surface for the Benzene Dimer. Results of ab initio CCSD(T) calculations show two nearly isoenergetic structures: T-shaped and parallel-displaced. J Phys Chem 100:18790–18794
67. Ninkovic DB, Andric JM, Malkov SN, Zaric SD (2014) What are the preferred horizontal displacements of aromatic-aromatic interactions in proteins? Comparison with the calculated benzene-benzene potential energy surface. Phys Chem Chem Phys 16:11173–11177
68. McGaughey GB, Gagné M, Rappé AK (1998) pi-Stacking interactions. Alive and well in proteins. J Biol Chem 273:15458–15463
69. Street AG, Mayo SL (1999) Intrinsic β-sheet propensities result from van der Waals interactions between side chains and the local backbone. Proc Natl Acad Sci U S A 96:9074–9076
70. Malkov SN, Živković MV; Beljanski MV, Hall MB, Zarić SD (2008) A reexamination of the propensities of amino acids towards a particular secondary structure: classification of amino acids based on their chemical structure. J Mol Model 14:769–775
71. Samanta U, Bahadur RP, Chakrabarti P (2002) Quantifying the accessible surface area of protein residues in their local environment. Protein Eng Des Sel 15:659–667
72. Hunter CA, Singh J, Thornton JM (1991) Pi-pi interactions: the geometry and energetics of phenylalanine-phenylalanine interactions in proteins. J Mol Biol 218:837–846
73. Tenidis K, Waldner M, Bernhagen J, Fischle W, Bergmann M, Weber M, Merkle M-L, Voelter W, Brunner H, Kapurniotu A (2000) Identification of a penta- and hexapeptide of islet amyloid polypeptide (IAPP) with amyloidogenic and cytotoxic properties. J Mol Biol 295:1055–1071
74. Tjernberg LO, Näslund J, Lindqvist F, Johansson J, Karlström AR, Thyberg J, Terenius L, Nordstedt C (1996) Arrest of beta-amyloid fibril formation by a pentapeptide ligand. J Biolog Chem 271:8545–8548
75. Findeis MA, Musso GM, Arico-Muendel CC, Benjamin HW, Hundal AM, Lee, J-J, Chin J, Kelley M, Wakefield J, Hayward NJ, Molineaux SM (1999) Modified-peptide inhibitors of amyloid β-peptide polymerization. Biochemistry (Mosc) 38:6791–6800, PMID: 10346900

76. Gazit E (2002) A possible role for p-stacking in the self-assembly of amyloid fibrils. FASEB J 16:77–83
77. Tjernberg L, Hosia W, Bark N, Thyberg J, Johansson J (2002) Charge attraction and beta propensity are necessary for amyloid fibril formation from tetrapeptides. J Biol Chem 277:43243–43246
78. Castelletto V, Hamley IW, Cenker C, Olsson U, Adamcik J, Mezzenga R, Miravet JF, Escuder B, Rodriguez-Llansola F (2011) Influence of end-capping on the self-assembly of model amyloid peptide fragments. J Phys Chem B 115:2107–2116
79. Lakshmanan A, Cheong DW, Accardo A, Di Fabrizio E, Riekel C, Hauser CA (2013) Aliphatic peptides show similar self-assembly to amyloid core sequences, challenging the importance of aromatic interactions in amyloidosis. Proc Natl Acad Sci U S A 110:519–524
80. Genji M, Yano Y, Hoshino M, Matsuzaki K (2017) Aromaticity of phenylalanine residues is essential for amyloid formation by Alzheimer's amyloid β-peptide. Chem Pharm Bull 65:668–673
81. Korn A, Surendran D, Krueger M, Maiti S, Huster D (2018) Ring structure modifications of phenylalanine 19 increase fibrillation kinetics and reduce toxicity of amyloid β (1–40). Chem Commun (Camb) 54:5430–5433
82. Yan X, Cui Y, He Q, Wang K, Li J (2008) Organogels based on self-assembly of diphenylalanine peptide and their application to immobilize quantum dots. Chem Mater 20:1522–1526
83. Demirel G, Malvadkar N, Demirel MC (2009) Control of protein adsorption onto core- shell tubular and vesicular structures of diphenylalanine/parylene. Langmuir 26:1460–1463
84. Su Y, Yan X, Wang A, Fei J, Cui Y, He Q, Li J (2010) A peony-flower-like hierarchical mesocrystal formed by diphenylalanine. J Mater Chem 20:6734–6740
85. Zhu P, Yan X, Su Y, Yang Y, Li J (2010) Solvent-induced structural transition of self-assembled dipeptide: from organogels to microcrystals. Chem Eur J 16:3176–3183
86. Huang R, Qi W, Su R, Zhao J, He Z (2011) Solvent and surface controlled self-assembly of diphenylalanine peptide: from microtubes to nanofibers. Soft Matter 7:6418–6421
87. Huang R, Wang Y, Qi W, Su R, He Z (2014) Temperature-induced reversible self-assembly of diphenylalanine peptide and the structural transition from organogel to crystalline nanowires. Nanoscale Res Lett 9:653
88. Mason TO, Chirgadze DY, Levin A, Adler-Abramovich L, Gazit E, Knowles TPJ, Buell AK (2014) Expanding the solvent chemical space for self-assembly of dipeptide nanostructures. ACS Nano 8:1243–53
89. Song Y, Challa SR, Medforth CJ, Qiu Y, Watt RK, Pena D, Miller JE, van Swol F, Shelnutt JA (2004) Synthesis of peptide-nanotube platinum-nanoparticle composites. Chem Commun (Camb) 1044–1045
90. Mason TO, Michaels TCT, Levin A, Dobson CM, Gazit E, Knowles TPJ, Buell AK (2017) Thermodynamics of polypeptide supramolecular assembly in the short-chain limit. J Am Chem Soc 139:16134–16142
91. Park JS, Han TH, Oh JK, Kim SO (2010) Capillarity induced large area patterning of peptide nanowires. J Nanosci Nanotech 10:6954–6957
92. Adams DJ, Butler MF, Frith WJ, Kirkland M, Mullen L, Sanderson P (2009) A new method for maintaining homogeneity during liquid–hydrogel transitions using low molecular weight hydrogelators. Soft Matter 5:1856–1862
93. Zhang Y, Kuang Y, Gao Y, Xu B (2010) Versatile small-molecule motifs for self-assembly in water and the formation of biofunctional supramolecular hydrogels. Langmuir 27:529–537
94. Zhou J, Du X, Gao Y, Shi J, Xu B (2014) Aromatic-aromatic interactions enhance interfiber contacts for enzymatic formation of a spontaneously aligned supramolecular hydrogel. J Am Chem Soc 136:2970–2973
95. Debnath S, Roy S, Ulijn RVJ (2013) Peptide nanofibers with dynamic instability through nonequilibrium biocatalytic assembly. Am Chem Soc 135:16789–16792
96. Mason TO, Michaels TCT, Levin A, Gazit E, Dobson CM, Buell AK, Knowles TPJ (2016) Synthesis of Nonequilibrium Supramolecular Peptide Polymers on a Microfluidic Platform. J Am Chem Soc 138:9589–9596

97. Adler-Abramovich L, Aronov D, Beker P, Yevnin M, Stempler S, Buzhansky L, Rosenman G, Gazit E (2009) Self-assembled arrays of peptide nanotubes by vapour deposition. Nat Nanotechnol 4:849
98. Bank-Srour B, Becker P, Krasovitsky L, Gladkikh A, Rosenberg Y, Barkay Z, Rosenman G (2013) Physical vapor deposition of peptide nanostructures. Polymer J 45:494
99. Vasudev MC, Koerner H, Singh KM, Partlow BP, Kaplan DL, Gazit E, Bunning TJ, Naik RR (2014) Vertically aligned peptide nanostructures using plasmaenhanced chemical vapor deposition. Biomacromolecules 15:533–540
100. Levin A, Mason TO, Adler-Abramovich L, Buell AK, Meisl G, Galvagnion C, Bram Y, Stratford SA, Dobson CM, Knowles TPJ, Gazit E (2014) Ostwalds rule of stages governs structural transitions and morphology of dipeptide supramolecular polymers. Nat Commun 5:5219
101. Adler-Abramovich L, Reches M, Sedman VL, Allen S, Tendler SJB, Gazit E (2006) Thermal and chemical stability of diphenylalanine peptide nanotubes: implications for nanotechnological applications. Langmuir 22:1313–1320
102. Sedman VL, Adler-Abramovich L, Allen S, Gazit E, Tendler SJB (2006) Direct observation of the release of phenylalanine from diphenylalanine nanotubes. J Am Chem Soc 128:6903–6908
103. Tamamis P, Adler-Abramovich L, Reches M, Marshall K, Sikorski P, Serpell L, Gazit E, Archontis G (2009) Self-assembly of phenylalanine oligopeptides: insights from experiments and simulations. Biophys J 96:5020–5029
104. Adler-Abramovich L, Kol N, Yanai I, Barlam D, Shneck RZ, Gazit E, Rousso I (2010) Self-assembled organic nanostructures with metallic-like stiffness. Angew Chem Int Ed Engl 49:9939–9942
105. Chen L, Morris K, Laybourn A, Elias D, Hicks MR, Rodger A, Serpell L, Adams DJ (2010) Self-assembly mechanism for a naphthalene-dipeptide leading to hydrogelation. Langmuir 26:5232–5242
106. Fichman G, Guterman T, Damron J, Adler-Abramovich L, Schmidt J, Kesselman E, Shimon LJ, Ramamoorthy A, Talmon Y, Gazit E (2016) Spontaneous structural transition and crystal formation in minimal supramolecular polymer model. Sci Adv 2:e1500827
107. Cardoso AZ, Mears LLE, Cattoz BN, Griffiths PC, Schweins R, Adams DJ (2016) Linking micellar structures to hydrogelation for salt-triggered dipeptide gelators. Soft Matter 12:3612–3621
108. Onogi S, Shigemitsu H, Yoshii T, Tanida T, Ikeda M, Kubota R, Hamachi I (2016) In situ real-time imaging of self-sorted supramolecular nanofibres. Nat Chem 8:743–752
109. Kubota R, Liu S, Shigemitsu H, Nakamura K, Tanaka W, Ikeda M, Hamachi I (2018) Imaging-based study on control factors over self-sorting of supramolecular nanofibers formed from peptide-and lipid-type hydrogelators. Bioconjug Chem 29(6):2058–2067
110. Tena-Solsona M, Escuder B, Miravet JF, Casttelleto V, Hamley IW, Dehsorkhi A (2015) Thermodynamic and kinetic study of the fibrillization of a family of tetrapeptides and its application to self-sorting. What takes so long? Chem Mater 27:3358–3365
111. Yan X, He Q, Wang K, Duan L, Cui Y, Li J (2007) Transition of cationic dipeptide nanotubes into vesicles and oligonucleotide delivery. Angew Chem Int Ed Engl 119:2483–2486
112. Wallace M, Iggo JA, Adams DJ (2015) Using solution state NMR spectroscopy to probe NMR invisible gelators. Soft Matter 11:7739–7747
113. Wallace M, Iggo JA, Adams DJ (2017) Probing the surface chemistry of self-assembled peptide hydrogels using solution-state NMR spectroscopy. Soft Matter 13:1716–1727
114. Do TD, Bowers MT (2015) Diphenylalanine self assembly: novel ion mobility methods showing the essential role of water. Anal Chem 87:4245–4252
115. Chen L, Pont G, Morris K, Lotze G, Squires A, Serpell LC, Adams DJ (2011) Salt-induced hydrogelation of functionalised-dipeptides at high pH. Chem Commun 47:12071–12073
116. Martin AD, Wojciechowski JP, Robinson AB, Heu C, Garvey CJ, Ratcliffe J, Waddington LJ, Gardiner J, Thordarson P (2017) Controlling self-assembly of diphenylalanine peptides at high pH using heterocyclic capping groups. Sci Rep 7:43947

117. Colquhoun C, Draper ER, Schweins R, Marcello M, Vadukul D, Serpell LC, Adams DJ (2017) Controlling the network type in self-assembled dipeptide hydrogels. Soft Matter 13:1914–1919
118. Shigemitsu H, Fujisaku T, Tanaka W, Kubota R, Minami S, Urayama K, Hamachi I (2018) An adaptive supramolecular hydrogel comprising self-sorting double nanofibre networks. Nat Nanotechnol 1(13):165–172
119. Arnon ZA, Vitalis A, Levin A, Michaels TC, Caflisch A, Knowles TP, Adler-Abramovich L, Gazit E (2016) Dynamic microfluidic control of supramolecular peptide self-assembly. Nat Commun 7:13190
120. Amdursky N, Beker P, Koren I, Bank-Srour B, Mishina E, Semin S, Rasing T, Rosenberg Y, Barkay Z, Gazit E, Rosenman G (2011) Structural transition in peptide nanotubes. Biomacromolecules 12:1349–1354
121. Handelman A, Natan A, Rosenman G (2014) Structural and optical properties of short peptides: nanotubes-to-nanofibers phase transformation. J Pept Sci 20:487–493
122. Ryu J, Park CB (2008) Solid-phase growth of nanostructures from amorphous peptide thin film: effect of water activity and temperature. Chem Mater 20:4284–4290
123. Kleinsmann AJ, Nachtsheim BJ (2013) Phenylalanine-containing cyclic dipeptides–the lowest molecular weight hydrogelators based on unmodified proteinogenic amino acids. Chem Commun 49:7818–7820
124. Zhou X, Fan J, Li N, Du Z, Ying H, Wu J, Xiong J, Bai J (2012) Solubility of l-phenylalanine in water and different binary mixtures from 288.15 to 318.15 K. Fluid Phase Equilibria 316:26–33
125. Franks F, Gent M, Johnson H (1963) 505. The solubility of benzene in water. J Chem Soc (Resumed) 2716–2723
126. Baldwin RL (1986) Temperature dependence of the hydrophobic interaction in protein folding. Proc Natl Acad Sci U S A 83:8069–8072
127. Han TH, Oh JK, Lee G-J, Pyun S-I, Kim SO (2010) Hierarchical assembly of diphenylalanine into dendritic nanoarchitectures. Colloids Surf B Biointerfaces 79:440–445
128. McDevit W, Long F (1952) The activity coefficient of benzene in aqueous salt solutions. J Am Chem Soc 74:1773–1777
129. Marshall KE, Hicks MR, Williams TL, Hoffmann SRVN, Rodger A, Dafforn TR, Serpell LC (2010) Characterizing the assembly of the Sup35 yeast prion fragment, GNNQQNY: structural changes accompany a fiber-to-crystal switch. Biophys J 98:330–338
130. Reynolds NP, Adamcik J, Berryman JT, Handschin S, Zanjani AAH, Li W, Liu K, Zhang A, Mezzenga R (2017) Competition between crystal and fibril formation in molecular mutations of amyloidogenic peptides. Nat Commun 8:1338
131. Chen L, Revel S, Morris K, C Serpell L, Adams DJ (2010) Effect of molecular structure on the properties of naphthalene- dipeptide hydrogelators. Langmuir 26:13466–13471
132. Fleming S, Debnath S, Frederix PW, Tuttle T, Ulijn RV (2013) Aromatic peptide amphiphiles: significance of the Fmoc moiety. Chem Commun 49:10587–10589
133. Reches M, Gazit E (2005) Self-assembly of peptide nanotubes and amyloid-like structures by charged-termini-capped diphenylalanine peptide analogues. Isr J Chem 45:363–371
134. Tang C, Ulijn RV, Saiani A (2011) Effect of glycine substitution on Fmocdiphenylalanine self-assembly and gelation properties. Langmuir 27:14438–14449
135. Houton KA, Morris KL, Chen L, Schmidtmann M, Jones JTA, Serpell LC, Lloyd GO, Adams DJ (2012) On crystal versus fiber formation in dipeptide hydrogelator systems. Langmuir 28:9797–9806
136. Tang C, Smith AM, Collins RF, Ulijn RV, Saiani A (2009) Fmoc-diphenylalanine self-assembly mechanism induces apparent p K a shifts. Langmuir 25:9447–9453
137. Ramos Sasselli I, Halling PJ, Ulijn RV, Tuttle T (2016) Supramolecular fibers in gels can be at thermodynamic equilibrium: a simple packing model reveals preferential fibril formation versus crystallization. ACS Nano 10:2661–2668
138. Hsu S-M, Lin Y-C, Chang J-W, Liu Y-H, Lin H-C (2014) Intramolecular interactions of a phenyl/perfluorophenyl pair in the formation of supramolecular nanofibers and hydrogels. Angew Chem Intl Ed Engl 53:1921–1927

139. Rajbhandary A, Nilsson BL (2017) Investigating the effects of peptoid substitutions in self-assembly of Fmoc-diphenylalanine derivatives. Pept Sci 108:1–11
140. Brouhard GJ (2015) Dynamic instability 30 years later: complexities in microtubule growth and catastrophe. Mol Biol Cell 26:1207–1210
141. Sadownik JW, Leckie J, Ulijn RV (2011) Micelle to fibre biocatalytic supramolecular transformation of an aromatic peptide amphiphile. Chem Commun (Camb) 47:728–730
142. Frederix PW, Scott GG, Abul-Haija YM, Kalafatovic D, Pappas CG, Javid N, Hunt NT, Ulijn RV, Tuttle T (2015) Exploring the sequence space for (tri-) peptide self-assembly to design and discover new hydrogels. Nat Chem 7:30–7
143. Görbitz C, Etter M (1992) Structure of l-phenylalanine l-phenylalaninium formate. Acta Crystallogr C 48:1317–1320
144. Zeng G, Liu L, Xia D, Li Q, Xin Z, Wang J, Besenbacher F, Skrydstrup T, Dong M (2014) Transition of chemically modified diphenylalanine peptide assemblies revealed by atomic force microscopy. RSC Adv 4:7516–7520
145. Korolkov VV, Allen S, Roberts CJ, Tendler SJ (2013) Surface mediated L-phenylalanyl-L-phenylalanine assembly into large dendritic structures. Faraday Discuss 166:257–267
146. Yan X, Cui Y, He Q, Wang K, Li J, Mu W, Wang B, Ou-yang Z-C (2008) Reversible transitions between peptide nanotubes and vesicle-like structures including theoretical modeling studies. Chem Eur J 14:5974–5980
147. Reches M, Gazit E (2004) Formation of closed-cage nanostructures by self-assembly of aromatic dipeptides. Nano Lett 4:581–585
148. Nielsen AE (1984) Electrolyte crystal growth mechanisms. J Cryst Growth 67:289–310
149. Buell AK, Dhulesia A, White DA, Knowles TP, Dobson CM, Welland ME (2012) Detailed analysis of the energy barriers for amyloid fibril growth. Angew Chem Intl Ed Engl 51:5247–5251
150. Adler-Abramovich L, Marco P, Arnon ZA, Creasey RCG, Michaels TCT, Levin A, Scurr DJ, Roberts CJ, Knowles TPJ, Tendler SJB, Gazit E (2016) Controlling the physical dimensions of peptide nanotubes by supramolecular polymer coassembly. ACS Nano 10:7436–7442
151. Creasey RCG, Louzao I, Arnon ZA, Marco P, Adler-Abramovich L, Roberts CJ, Gazit E, Tendler SJB (2016) Disruption of diphenylalanine assembly by a Boc-modified variant. Soft Matter 12:9451–9457
152. Buell AK (2017) The Nucleation of protein aggregates-from crystals to amyloid fibrils. Int Rev Cell Mol Biol 329:187–226
153. Wang Y, Huang R, Qi W, Xie Y, Wang M, Su R, He Z (2015) Capillary force-driven, hierarchical co-assembly of dandelion-like peptide microstructures. Small 11:2893–2902
154. Amdursky N, Molotskii M, Gazit E, Rosenman G (2010) Elementary building blocks of self-assembled peptide nanotubes. J Am Chem Soc 132:15632–15636
155. Gebauer D, Kellermeier M, Gale JD, Bergström L, Cölfen H (2014) Pre-nucleation clusters as solute precursors in crystallisation. Chem Soc Rev 43:2348–2371
156. Ishikawa M, Busch C, Motzkus M, Martinho H, Buckup T (2017) Two-step kinetic model as self-assembling mechanism for diphenylalanine micro/nanotubes formation. Phys Chem Chem Phys 19:31647–31654
157. Buell AK, Galvagnion C, Gaspar R, Sparr E, Vendruscolo M, Knowles TPJ, Linse S, Dobson CM (2014) Solution conditions determine the relative importance of nucleation and growth processes in α-synuclein aggregation. Proc Natl Acad Sci U S A 111(21):7671–7676
158. Görbitz CH, Hartviksen LM (2008) The monohydrates of the four polar dipeptides L-seryl-L-asparagine, L-seryl-L-tyrosine, L-tryptophanyl-L-serine and L-tyrosyl-L-tryptophan. Acta Crystallogr C 64:171–6
159. Görbitz CH, Etter MC (1992) Hydrogen bond connectivity patterns and hydrophobic interactions in crystal structures of small, acyclic peptides. Int J Pept Protein Res 39:93–110
160. Görbitz CH (2002) Turns, water cage formation and hydrogen bonding in the structures of l-valyl-l-phenylalanine. Acta Crystallogr B Struct Sci Cryst Eng Mater 58:512–518
161. Emge TJ, Agrawal A, Dalessio JP, Dukovic G, Inghrim JA, Janjua K, Macaluso M, Robertson LL, Stiglic TJ, Volovik Y, Georgiadis MM (2000) Alaninyltryptophan hydrate,

glycyltryptophan dihydrate and tryptophylglycine hydrate. Acta Crystallogr C Cryst Struct Commun 56:e469–e471
162. Görbitz CH, Gundersen E (1996) L-Valyl-L-alanine. Acta Crystallogr C Cryst Struct Commun 52:1764–1767
163. Sinnokrot MO, Valeev EF, Sherrill CD (2002) Estimates of the ab initio limit for pi-pi interactions: the benzene dimer. J Am Chem Soc 124:10887–10893
164. Lee EC, Kim D, Jurecka P, Tarakeshwar P, Hobza P, Kim KS (2007) Understanding of assembly phenomena by aromatic-aromatic interactions: benzene dimer and the substituted systems. J Phys Chem A 111:3446–3457
165. Prampolini G, Livotto PR, Cacelli I (2015) Accuracy of quantum mechanically derived force-fields parameterized from dispersion-corrected DFT data: the benzene dimer as a prototype for aromatic interactions. J Chem Theory Comput 11:5182–5196
166. Suresh C, Vijayan M (1987) X-ray studies on crystalline complexes involving amino acids and peptides. Part XIV: closed conformation and head-to-tail arrangement in a new crystal form of L-histidine L-aspartate monohydrate. J Biosci 12:13–21
167. Bernstein J, Davis RE, Shimoni L, Chang N-L (1995) Patterns in hydrogen bonding: functionality and graph set analysis in crystals. Angew Chem Int Ed Engl 34:1555–1573
168. Etter MC, MacDonald JC, Bernstein J (1990) Graph-set analysis of hydrogen-bond patterns in organic crystals. Acta Crystallogr B Struct Sci Cryst Eng Mater 46:256–262
169. Vargas R, Garza J, Dixon DA, Hay BP (2000) How strong is the CH···OC hydrogen bond? J Am Chem Soc 122:4750–4755
170. Fabiola GF, Krishnaswamy S, Nagarajan V, Pattabhi V (1997) C-H···O hydrogen bonds in β-sheets. Acta Crystallogr D Biol Crystallogr 53:316–320
171. Akazome M, Ueno Y, Ooiso H, Ogura K (2000) Enantioselective inclusion of methyl phenyl sulfoxides and benzyl methyl sulfoxides by (R)-phenylglycyl-(R)-phenylglycine and the crystal structures of the inclusion cavities. J Org Chem 65:68–76
172. Pellach M, Mondal S, Shimon LJ, Adler-Abramovich L, Buzhansky L, Gazit E (2016) Molecular engineering of self-assembling diphenylalanine analogues results in the formation of distinctive microstructures. Chem Mater 28:4341–4348
173. Eggleston DS, Hodgson DJ (1985) Intramolecular water bridge and a distorted trans peptide bond in the crystal structure of α-L-glutamyl-L-aspartic acid hydrate. Chem Biol Drug Des 26:509–517
174. Reches M, Gazit E (2006) Controlled patterning of aligned self-assembled peptide nanotubes. Nat Nanotechnol 1:195–200
175. Görbitz CH (2006) The structure of nanotubes formed by diphenylalanine, the core recognition motif of Alzheimer's beta-amyloid polypeptide. Chem Commun (Camb) 2332–2334
176. Lekprasert B, Korolkov V, Falamas A, Chis V, Roberts CJ, Tendler SJB, Notingher I (2012) Investigations of the supramolecular structure of individual diphenylalanine nano- and microtubes by polarized Raman microspectroscopy. Biomacromolecules 13:2181–2187
177. Eddleston MD, Jones W (2009) Formation of tubular crystals of pharmaceutical compounds. Cryst Growth Des 10:365–370
178. Kim J, Han TH, Kim Y-I, Park JS, Choi J, Churchill DG, Kim SO, Ihee H (2010) Role of water in directing diphenylalanine assembly into nanotubes and nanowires. Adv Mater 22:583–587
179. Li Q, Jia Y, Dai L, Yang Y, Li J (2015) Controlled rod nanostructured assembly of diphenylalanine and their optical waveguide properties. ACS Nano 9:2689–2695
180. Pappas CG, Frederix PW, Mutasa T, Fleming S, Abul-Haija YM, Kelly SM, Gachagan A, Kalafatovic D, Trevino J, Ulijn R, Bai S (2015) Alignment of nanostructured tripeptide gels by directional ultrasonication. Chem Commun 51:8465–8468
181. Pappas CG, Mutasa T, Frederix PW, Fleming S, Bia S, Debnath S, Kelly SM, Gachagan A, Ulijn RV (2015) Transient supramolecular reconfiguration of peptide nanostructures using ultrasound. Mater Horiz 2:198–202
182. Moholkar VS, Sable SP, Pandit AB (2000) Mapping the cavitation intensity in an ultrasonic bath using the acoustic emission. AIChE J 46:684–694

183. Tjernberg LO, Callaway DJ, Tjernberg A, Hahne S, Lilliehöök C, Terenius L, Thyberg J, Nordstedt C (1999) A molecular model of Alzheimer amyloid β-peptide fibril formation. J Biol Chem 274:12619–12625
184. Mazor Y, Gilead S, Benhar I, Gazit E (2002) Identification and characterization of a novel molecular-recognition and self-assembly domain within the islet amyloid polypeptide. J Mol Biol 322:1013–1024
185. Gazit E (2005) Mechanisms of amyloid fibril self-assembly and inhibition. Model short peptides as a key research tool. FEBS J 272:5971–5978
186. Jahn TR, Makin OS, Morris KL, Marshall KE, Tian P, Sikorski P, Serpell LC (2010) The common architecture of cross-β amyloid. J Mol Biol 395:717–727
187. Carugo O, Djinović-Carugo K (2013) Half a century of Ramachandran plots. Acta Crystallogr D Biol Crystallogr 69:1333–1341
188. Hollingsworth SA, Karplus PA (2010) A fresh look at the Ramachandran plot and the occurrence of standard structures in proteins. Biomol Concepts 1:271–283
189. Berman HM, Westbrook J, Feng Z, Gilliland G, Bhat TN, Weissig H, Shindyalov IN, Bourne PE (2000) The protein data bank. Nucleic Acids Res 28:235–242
190. Schrödinger LLC (2015) The PyMOL Molecular Graphics System, Version 2.1.0
191. Di Maro S et al (2016) Exploring the N-terminal region of C-X-C motif chemokine 12 (CXCL12): identification of plasma-stable cyclic peptides as novel, potent C-X-C chemokine receptor type 4 (CXCR4) antagonists. J Med Chem 59:8369–8380, PMID: 27571038
192. Colletier J-P, Laganowsky A, Landau M, Zhao M, Soriaga AB, Goldschmidt L, Flot D, Cascio D, Sawaya MR, Eisenberg D (2011) Molecular basis for amyloid-β polymorphism. Proc Natl Acad Sci U S A 108:16938–16943
193. Spencer RK, Li H, Nowick JS (2014) X-ray crystallographic structures of trimers and higher-order oligomeric assemblies of a peptide derived from Aβ 17–36. J Am Chem Soc 136:5595–5598
194. Kreutzer AG, Hamza IL, Spencer RK, Nowick JS (2016) X-ray crystallographic structures of a trimer, dodecamer, and annular pore formed by an Aβ17–36 β-hairpin. J Am Chem Soc 138:4634–4642, PMID: 26967810
195. Wälti MA, Ravotti F, Arai H, Glabe CG, Wall JS, Böckmann A, Güntert P, Meier BH, Riek R (2016) Atomic-resolution structure of a disease-relevant Aβ (1–42) amyloid fibril. Proc Natl Acad Sci U S A 113:E4976–E4984
196. Perutz MF, Finch JT, Berriman J, Lesk A (2002) Amyloid fibers are water-filled nanotubes. Proc Natl Acad Sci U S A 99:5591–5595
197. Sikorski P, Atkins E (2005) New model for crystalline polyglutamine assemblies and their connection with amyloid fibrils. Biomacromolecules 6:425–432
198. Buchanan LE, Carr JK, Fluitt AM, Hoganson AJ, Moran SD, de Pablo JJ, Skinner JL, Zanni MT (2014) Structural motif of polyglutamine amyloid fibrils discerned with mixed-isotope infrared spectroscopy. Proc Natl Acad Sci 111:5796–5801
199. Makin OS, Serpell LC (2005) Structures for amyloid fibrils. FEBS J 272:5950–5961
200. Harper JD, Lieber CM, Lansbury PT (1997) Atomic force microscopic imaging of seeded fibril formation and fibril branching by the Alzheimer's disease amyloid-β protein. Chem Biol 4:951–959
201. Chothia C (1973) Conformation of twisted β-pleated sheets in proteins. J Mol Biol 75:295–302
202. Aggeli A, Nyrkova IA, Bell M, Harding R, Carrick L, McLeish TC, Semenov AN, Boden N (2001) Hierarchical self-assembly of chiral rod-like molecules as a model for peptide β-sheet tapes, ribbons, fibrils, and fibers. Proc Natl Acad Sci USA 98:11857–11862
203. Usov I, Adamcik J, Mezzenga R (2013) Polymorphism complexity and handedness inversion in serum albumin amyloid fibrils. ACS Nano 7:10465–10474
204. Wang S-T, Lin Y, Spencer RK, Thomas MR, Nguyen AI, Amdursky N, Pashuck ET, Skaalure SC, Song CY, Parmar PA, Morgan RM, Ercius P, Aloni S, Zuckermann RN, Stevens MM (2017) Sequence-dependent self-assembly and structural diversity of Islet amyloid polypeptide-derived β-sheet fibrils. ACS Nano 11:8579–8589, PMID: 28771324

205. Knowles TPJ, Simone AD, Fitzpatrick AW, Baldwin A, Meehan S, Rajah L, Vendruscolo M, Welland ME, Dobson CM, Terentjev EM (2012) Twisting transition between crystalline and fibrillar phases of aggregated peptides. Phys Rev Lett 109:158101
206. Knowles TP, Smith JF, Craig A, Dobson CM, Welland ME (2006) Spatial persistence of angular correlations in amyloid fibrils. Phys Rev Lett 96:238301
207. Šarić, A.; Buell AK, Meisl G, Michaels TC, Dobson CM, Linse S, Knowles TP, Frenkel D Physical determinants of the self-replication of protein fibrils. (2016) Nat Phys 12:874–880
208. Cohen SIA, Cukalevski R, Michaels TCT, Saric A, Törnquist M, Vendruscolo M, Dobson CM, Buell AK, Knowles TPJ, Linse S (2018) Distinct thermodynamic signatures of oligomer generation in the aggregation of the amyloid-β peptide. Nat Chem 10:523–531
209. Cabriolu R, Auer S (2011) Amyloid fibrillation kinetics: insight from atomistic nucleation theory. J Mol Biol 411:275–285
210. Fitzpatrick AWP et al (2013) Atomic structure and hierarchical assembly of a crossβ amyloid fibril. Proc Natl Acad Sci USA 110:5468–5473
211. Collins SR, Douglass A, Vale RD, Weissman JS (2004) Mechanism of prion propagation: amyloid growth occurs by monomer addition. PLoS Biol 2:e321
212. Buell AK, Blundell JR, Dobson CM, Welland ME, Terentjev EM, Knowles TPJ (2010) Frequency factors in a landscape model of filamentous protein aggregation. Phys Rev Lett 104:228101
213. Kashchiev D, Auer S (2010) Nucleation of amyloid fibrils. J Chem Phys 132:215101
214. Cabriolu R, Kashchiev D, Auer S (2010) Atomistic theory of amyloid fibril nucleation. J Chem Phys 133:12B602
215. Balbirnie M, Grothe R, Eisenberg DS (2001) An amyloid-forming peptide from the yeast prion Sup35 reveals a dehydrated β-sheet structure for amyloid. Proc Natl Acad Sci U S A 98:2375–2380
216. Sawaya MR, Sambashivan S, Nelson R, Ivanova MI, Sievers SA, Apostol MI, Thompson MJ, Balbirnie M, Wiltzius JJW, McFarlane HT, Madsen AØ, Riekel C, Eisenberg D (2007) Atomic structures of amyloid cross-b spines reveal varied steric zippers. Nature 447:453 EP
217. Scherzinger E, Lurz R, Turmaine M, Mangiarini L, Hollenbach B, Hasenbank R, Bates GP, Davies SW, Lehrach H, Wanker EE (1997) Huntingtin-encoded polyglutamine expansions form amyloid-like protein aggregates in vitro and in vivo. Cell 90:549–558
218. Han H, Weinreb PH, Lansbury PT Jr (1995) The core alzheimers peptide NAC forms amyloid fibrils which seed and are seeded by β-amyloid: is NAC a common trigger or target in neurodegenerative disease? Chem Biol 2:163–169
219. Rodriguez JA et al (2015) Structure of the toxic core of a-synuclein from invisible crystals. Nature 525:486 EP
220. Colvin MT, Silvers R, Ni QZ, Can TV, Sergeyev I, Rosay M, Donovan KJ, Michael B, Wall J, Linse S, Griffin RG (2016) Atomic resolution structure of monomorphic Aβ42 amyloid fibrils. J Am Chem Soc 138:9663–9674, PMID: 27355699
221. Makhatadze GI, Privalov PL (1993) Contribution of hydration to protein folding thermodynamics: I. The enthalpy of hydration. J Mol Biol 232:639–659
222. Muller N (1988) Is there a region of highly structured water around a nonpolar solute molecule? J Solution Chem 17:661–672
223. Baldwin AJ, Knowles TPJ, Tartaglia GG, Fitzpatrick AW, Devlin GL, Shammas SL, Waudby CA, Mossuto MF, Meehan S, Gras SL, Christodoulou J, Anthony-Cahill SJ, Barker PD, Vendruscolo M, Dobson CM (2011) Metastability of native proteins and the phenomenon of amyloid formation. J Am Chem Soc 133:14160–14163, PMID: 21650202
224. Oosawa F, Kasai M (1962) A theory of linear and helical aggregations of macromolecules. J Mol Biol 4:10–21

Chapter 4
Bacterial Amyloids: Biogenesis and Biomaterials

Line Friis Bakmann Christensen, Nicholas Schafer, Adriana Wolf-Perez, Daniel Jhaf Madsen, and Daniel E. Otzen

Abstract Functional amyloid (FuBA) is produced by a large fraction of all bacterial species and represents a constructive use of the stable amyloid fold, in contrast to the pathological amyloid seen in neurodegenerative diseases. When assembled into amyloid, FuBA is unusually robust and withstands most chemicals including denaturants and SDS. Uses include strengthening of bacterial biofilms, cell-to-cell communication, cell wall construction and even bacterial warfare. Biogenesis is under tight spatio-temporal control, thanks to a simple but efficient secretion system which in *E. coli*, *Pseudomonas* and other well-studied bacteria includes a major amyloid component that is kept unfolded in the periplasm thanks to chaperones, threaded through the outer membrane via a pore protein and anchored to the cell surface through a nucleator and possibly other helper proteins. In these systems, amyloid formation is promoted through imperfect repeats, but other evolutionarily unrelated proteins either have no or only partially conserved repeats or simply consist of small peptides with multiple structural roles. This makes bioinformatics analysis challenging, though the sophisticated amyloid prediction tools developed from research in pathological amyloid together with the steady increase in identification of further examples of amyloid will strengthen genomic data mining. Functional amyloid represents an intriguing source of robust yet biodegradable materials with new properties, when combining the optimized self-assembly properties of the amyloid component with *e.g.* peptides with different binding properties or surface-reactive protein binders. Sophisticated patterns can also be obtained by co-incubating bacteria producing different types of amyloid, while amyloid inclusion bodies may lead to slow-release nanopills.

Keywords Curli · Functional amyloid in *Pseudomonas* · Sequence analysis · Fusion proteins with binding properties · Screening systems

---

L. F. B. Christensen · N. Schafer · A. Wolf-Perez · D. J. Madsen · D. E. Otzen (✉)
iNANO and Department of Molecular Biology and Genetics, Aarhus University, Aarhus, Denmark
e-mail: dao@inano.au.dk

S. Perrett et al. (eds.), *Biological and Bio-inspired Nanomaterials*,
Advances in Experimental Medicine and Biology 1174,
https://doi.org/10.1007/978-981-13-9791-2_4

4.1 Introduction

The amyloid β-sheet consists of a repetitive pattern of β-strands stacking on top of each other to form long fibrillary structure, whose axis is orthogonal to the β-strand (hence "cross- β"). The β-sheet secondary structure arranges the peptide backbone in an extended conformation, rich in inter-strand hydrogen bonds. The faces of the β-sheets can then stack and form very tightly interdigitated and dry interfaces [1]. Amyloid has long been associated with neurodegenerative diseases. However, it has been known for some time that many bacteria are able to utilize the amyloid structure for functional purposes such as providing strength to biofilms or as adhesins or toxins. Particularly within the last few decades, the prevalence of these functional bacterial amyloids (FuBAs) in natural habitats and across bacterial phyla has attracted much attention. In this chapter, we will provide a survey of the different FuBA systems described in the literature, discuss ways of identifying them from genomic data and present examples of their potential as a source of materials with new and versatile properties.

4.2 The Curli System: Quality-Conscious and Made to Last

Fibrillar structures, staining positive with Thioflavin T (ThT) and conformationally specific antibodies against amyloid fibrils, have been identified in many different environments, ranging from drinking water reservoirs to activated sludge plants [2, 3]. One of the most well-characterized of the FuBAs are the curli fibrils produced by *E. coli*. Curli appear as tangled fibers, 3–6 nm wide and up to several μm long [4–6]. Curli fibrils are extremely stable – often requiring concentrated formic acid or 1,1,1,3,3,3-hexafluoro-2-propanol (HFIP) to dissolve into monomers – and rich in β-sheets structure [6, 7]. These are characteristics that apply for many different FuBAs [8–11]. The curli fibrils are composed of one major and one minor protein component called CsgA and CsgB, respectively, which are present in an approximate ratio of 20:1 in the mature fibrils [12]. Wild-type (WT) CsgB is directed to the cell surface while WT CsgA is secreted from the cells [5]. The mature curli fibrils are especially involved in biofilm formation, where they account for a large portion of the matrix material, but also in general bacteria physiology [6, 13–16].

The machinery for curli production is dependent on seven proteins encoded by two operons, *csgBAC* and *csgDEFG* [17] (Fig. 4.1). This curli system was initially identified in some genera of *Enterobacteriales* [16, 19, 20], but has now been identified across four different phyla [21, 22]. Functional amyloid fibrils are known to be produced under tight regulation in many different systems [23] to avoid the toxic effects of the oligomers that form as an intermediate in amyloid fibril formation [24–27]. Especially lipid membranes are often sensitive to these species [28]. One of the precautionary measures taken in curli-producing bacteria is the synthesis of the chaperones CsgC and CsgE for handling the amyloidogenic

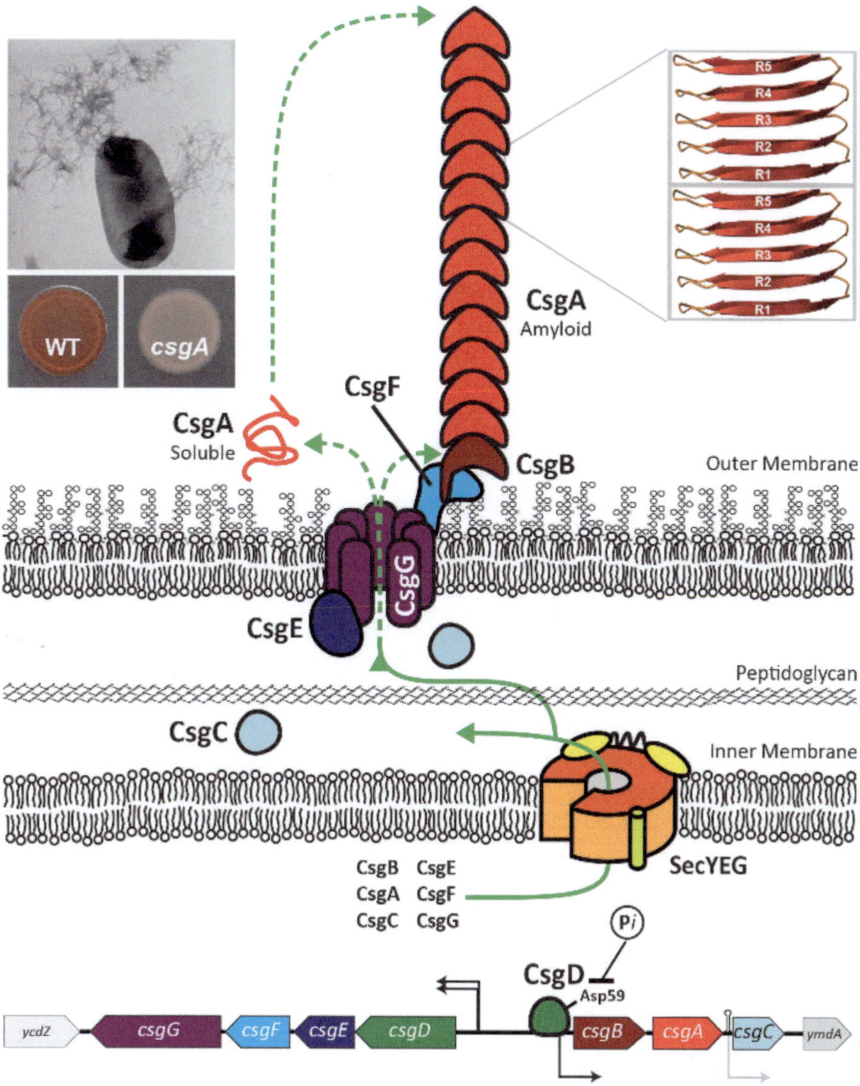

Fig. 4.1 Model for how fibrils are formed in the curli system. (Left upper corner) Curli formation can be detected by growth on Congo Red (CR) agar plates where CR binding to the amyloid structure leads to red colonies unless CsgA is missing. (Main figure) Suggested roles of the curli proteins. Transcription of the *csgDEFG* operon leads to production of CsgD is synthesized which activates transcription of the *csgBAC* operon. The Sec pathway moves CsgA-C and CsgE-F across the inner membrane, after which CsgG transports. CsgA, CsgB and CsgF across the outer membrane; specificity is controlled by CsgE which forms a capping complex on the periplasmic side of CsgG. On the cell surface, CsgF probably anchor CsgB to the cell surface, allowing it to nucleate CsgA fibrillation. CsgC acts as a chaperone to prevent CsgA from forming toxic oligomeric species inside the cell. (Right upper corner) Alignment of the repeats of two CsgA molecules in a β-helical structure. The repeats are connected through short β-turns. (Modified with permission from [18])

proteins, CsgA and CsgB, until they are safely transported to the outside [29, 30]. The three last proteins involved in curli formation are CsgF, CsgD and CsgG which act as a curli assembly factor, a transcriptional regulator and an outer membrane (OM) pore-forming protein, respectively. Transport across the inner membrane (IM) into the periplasm happens through the Sec translocase for which all the Csg proteins, with the exception of CsgD, have a signal sequence [14].

4.2.1 The Partnership of CsgB and CsgA: An Anchor for a Roving Sailor

The two components of curli fibers, **CsgA** and **CsgB**, have been heavily investigated. The two proteins are homologs, share 49% sequence similarity and each contain five imperfect repeats (R1-R5) in addition to an N-terminal signal sequence. In amyloidogenic proteins, repeats have previously been shown to facilitate fibril formation [31, 32]. The repeats in CsgA follow a conserved $SX_5QXGXGNXAX_3G$ motif while the repeats R1-R4 in CsgB follow a related $AX_3QXX2NXAX_3N$ motif [5]. These conserved Gln and Asn residues in the CsgA repeats help stabilize the amyloid structure [33]. Interestingly, three out of the five repeats, R1/R3/R5 from CsgA and R1/R2/R4 from CsgB, are able to form amyloid fibrils when expressed as isolated peptides *in vitro* [34, 35]. This repeat property might also drive amyloid formation *in vivo*. The R5 in CsgB stand out by having four positively charged residues instead of the motif with conserved Asn, Gln and Gly residues [5]. Removing R5 (amino acids 133–151) from CsgB results in a truncated version, $CsgB_{trunk}$, that is secreted from cells, indicating a role for R5 in tethering CsgB to the cell membrane [5]. Deletion of CsgB R4 also leads to secretion of CsgB, while membrane-localized CsgB mutants lacking either R1, R2 or R3 are still able to accommodate curli formation *in vivo* [35].

E. coli mutants that lack either CsgB (Δ*csgB*-) or CsgA (Δ*csgA*-) are not able to form curli [17, 36], even though purified CsgA, CsgB and $CsgB_{trunk}$ readily forms Congo Red (CR) and/or ThT positive fibrils *in vitro* [5, 34, 37, 38]. One reason is that in the absence of CsgB, fibrillation-competent secreted CsgA diffuses away from the cell [36]. In a process called interbacterial complementation, CsgA produced from Δ*csgB*- mutants is able to assemble into fibrils on nearby CsgB-positive bacteria [6, 36]. These investigations demonstrate the nucleating role of CsgB in CsgA polymerization. Normal nucleation-dependent fibrillation in CsgA has a lag phase, an exponential growth phase and a stationary phase, where the nucleus formation during the lag phase is proposed to be the rate limiting step *in vivo* [39]. There is no evidence of secondary nucleation and fragmentation in CsgA fibrillation [40]. In terms of kinetics, nucleation of CsgA by CsgB means that the 2 h lag phase of CsgA fibrillation observed *in vitro* [34, 37] can be greatly reduced – if not eliminated – and the fibrillation can jump straight to the growth phase. Because of the toxicity of many on-pathway oligomers [24–27], expression of a CsgA nucleator in the curli system elegantly minimizes this potential problem of forming toxic oligomers by accelerating fibril formation.

In addition to WT CsgB protein, $CsgB_{trunk}$ is also able to seed CsgA fibrillation *in vitro*, indicating that R5 in CsgB might not be involved in the *in vivo* nucleation of CsgA [5]. Furthermore, CsgA protein monomers can also be seeded not only with pre-formed CsgA fibrils from the same species but also CsgA from other curli-expressing species [7, 37]. Interestingly, CsgA fibrillates under practically any condition in terms of pH and ionic strength [37], possibly emphasizing its importance in bacterial life and survival. After secretion of CsgA from the bacteria, which depends on CsgG [17, 41], the protein is largely unstructured [34]. However, upon fibrillating in the presence of CsgB, the secondary structure changes into β-sheet structure both *in vitro* and *in vivo* [5, 37]. The structure of *in vitro* formed CsgA fibrils have been investigated thoroughly with transmission electron microscopy (TEM), electron diffraction, X-ray fiber diffraction and solid state NMR [37, 38]. These experimental investigations suggest a cross-β structure of stacked β-helical subunits [38]. The β-helical structure was also supported by a study that used simulations to identify low-energy structural models of CsgA [42].

4.2.2 All in the Fibril Family: Cooperation Within the Curli Operon

Figure 4.1 summarizes our current understanding of how the different curli proteins work together to form amyloid. Here we will go through what is known of the individual components. They reveal an exquisitely controlled nanotechnological amyloid generating machinery.

A prerequisite for *in vivo* CsgA/CsgB expression is a functional **CsgD** which acts as a positive transcriptional regulator of the *csgBAC* operon [17]. The C-terminal of CsgD shows a characteristic helix-turn-helix motif known to enable DNA binding.

CsgC was recently shown to be a potent amyloid inhibitor [29]. The crystal structure of oxidized CsgC (PDB: 2Y2Y) reveals seven β-strands and a disulphide bond between Cys29 and Cys31 on the same β-strand. CsgC inhibits *in vitro* CsgA and $CsgB_{trunk}$ fibrillation, however it is 50 times more effective against CsgA compared to $CsgB_{trunk}$ [29]. CsgC inhibits primary nucleation – and to a smaller degree elongation – of CsgA by inhibiting the formation of some intermediate species on the fibrillation pathway [40]. This delays the nucleation rate by a factor of 2–4 times [40]. Mutational studies have shown that especially positively charged residues on the protein surface are important for the fibrillation inhibition potency of CsgC [40]. The difference in the pI of CsgC/CsgH and CsgA (> 2 pH units) is probably responsible for accelerating the encounter between the proteins. This is supported by a CsgC homolog from *Salmonella typhimurium* that shows less potency in CsgA inhibition and at the same time has a pI much closer to that of CsgA (a difference of only ≈ 0.5 units) [40]. Identifying CsgA fibrils in the periplasm of *csgC-* mutants confirmed this fibrillation-inhibitory role of CsgC *in vivo*.

Transport across the IM happens through the Sec pathway and results in the proteolytic cleavage of the Csg proteins (except CsgD) to remove the signal sequence [14]. Further transport across the outer membrane is dependent on the lipoprotein **CsgG** [5, 41, 43]. CsgG is a 30 kDa lipoprotein that forms a pore in the OM of curli-producing bacteria by clustering into discrete areas [43, 44]. The clustering of CsgG is dependent on expression of CsgA, CsgB, CsgF and CsgE as well as on fibrillation of CsgA [44]. The crystal structure of the CsgG OM pore complex shows that nine CsgG molecules together form a 120 Å diameter complex with nine-fold rotational symmetry, a height of 85 Å and an inner diameter of 40 Å [45]. The transmembrane part of the pore is a β-barrel composed of 36 β-strands (4 strands from each CsgG monomer) [45, 46], which may explain the complex' resistance to denaturation by heat and SDS [44]. The CsgG pore consists of both an extracellular domain and a periplasmic domain separated by a narrow 9 Å diameter central channel [45]. Lining this central pore are three stacked concentric rings formed by the sidechain of residues Tyr51, Asn55 and Phe56 [45, 46]. Particularly Tyr51 and Phe56 stabilize the CsgG nonamer [46]. The periplasmic part is composed of a three-strand β-sheet and three α-helices from each CsgG monomer and from this side, a large cavity of 24,000 $Å^3$ is accessible.

What drives CsgA export? The outer membrane lacks a proton motive force or an ATP supply to drive active transport. Some systems such as the chaperone-usher pilus assembly cleverly use a ratchet-type mechanism where folding of exported proteins prevents backsliding (since the folded state will be less flexible than the unfolded state) [47], however CsgA only folds once it gets out of the membrane and binds to the nucleator protein CsgB or elongates existing CsgA fibrils. Rather, *E. coli* exploits a concentration gradient. Upon entering the CsgG cavity, CsgA is encapsulated by **CsgE** by the formation of a dynamic CsgG:CsgE complex with a 9:9 stoichiometry [45]. As a consequence, a entropy potential is formed across the membrane and this gradient drives the diffusion of CsgA to the extracellular space [40, 45, 48]. Single channel recordings reveal the ability of this cap to silence ion conductance [45]. Without CsgE, CsgG works like an ungated pore which can translocate erythromycin and small periplasmic proteins across the outer membrane [30]. CsgE takes care of gatekeeping of the CsgG pore by recognizing the N-terminal sequence of CsgA and CsgB. Thus, heterologous polypeptides fused to full-length CsgA or the first 41 residues of CsgA are not necessarily blocked from CsgG secretion by CsgE [30, 49, 50]. Therefore, secretion through CsgG is not sequence- or conformation specific and can accommodate export of CsgA proteins fused to large proteins such as the natively unfolded 260-residue ERD10 protein or small disulphide-folded proteins like the scFv protein Nb208; all it requires is a substrate traverse diameter of less than 2 nm to avoid blocking the channel [49]. (It is another matter that export of such hybrid proteins can lead to a mixture of fibrils and more amorphous aggregates). Purified CsgE has been shown to interact directly with CsgA [51] and completely inhibit CsgA fibrillation – like CsgC – when added in a 1:1 CsgE:CsgA ratio [30]. Together these results account for two functions of CsgE:

to ensure some secretion specificity for CsgG and to inhibit self-polymerization of CsgA through the periplasmic space.

Another Csg protein, the assembly factor **CsgF**, also interacts with CsgG and is dependent on the membrane protein for localization on the cell surface. CsgG probably also plays a role in CsgF stability as steady-state levels of CsgF are greatly decreased in *csgG*- mutants [52]. CsgF is most likely involved in the localization and nucleation activity of CsgB because *csgF*- mutants resemble $csgB_{trunk}$- mutants with large amounts of stable CsgB being released to the cell exterior [5, 6, 52]. CsgF also mediates CsgB protease resistance on the cell surface [6, 52]. *csgF*- mutants are not able to assemble CsgA fibrils [52], however, CsgA produced from these mutants is fibrillation-competent and can polymerize on the surface of *csgA*- mutants [6]. Direct interaction between CsgF and CsgB still needs to be established, but CsgF might interact with R4/R5 of CsgB since removal of these repeats also results in CsgB secretion [35]. The interaction between CsgB and CsgF and possible involvement of CsgF in CsgB nucleus formation might also be facilitated by a high percentage of Glu and Asn in CsgF (9.2% and 13.4%, respectively), which is similar to CsgA (8.4% and 12.2%) and CsgB (11.5% and 9.2%) [52].

On the cell surface, CsgB is suggested to quickly adopt an R4/R5-dependent amyloid fold that templates CsgA fibrillation [35]. This important role of CsgB R5 in both nucleation and membrane-anchoring [5] is especially interesting because the isolated R5 peptide does not form fibrils *in vitro* [35]. Not forming amyloid structures in the membrane portion of the cells could be another mechanism to avoid forming membrane-permeabilizing oligomers.

An eighth member of the curli family, **CsgH**, has recently been shown to inhibit amyloid formation [40]. The *csgH* gene has been observed in many *Alphaproteobacteria* where it is always situated next to *csgA/B* genes [21]. CsgH is – like CsgC – composed of seven β-strands and the two proteins share great structural similarity (rmsd = 2.6 Å) despite a sequence identity below 20% [40]. CsgH is able to inhibit CsgA fibrillation in a concentration-dependent manner. Interestingly, both CsgH and CsgC are also able to inhibit the unrelated bacterial amyloidogenic protein FapC [40].

4.2.3 Younger Kid on the Block: Fap Fimbria Are Composed of Mainly FapC

The curli system probably evolved from a common ancestor long time ago [21]. Another FuBA system, the Fap system, appears to be evolutionarily younger [53]. The *fapABCDEF* operon encodes 6 proteins [8] and has been found within several classes of the phylum *Proteobacteria* [53]. Fap fimbria were initially identified and purified from a *Pseudomonas* strain and were characterized as a FuBA due to the *in vivo* binding of ThT and amyloid conformational specific antibodies [3] together with the characteristic cross-β structure and extreme stability of purified Fap fibrils

[8]. Fap fimbria are believed to be involved in adhesion and biofilm formation, since overexpression of Fap results in a highly aggregative phenotype with increased biofilm formation [8, 54]. Fap fibrils greatly increase the stiffness ($\approx$ 20-fold) and the hydrophobicity of the biofilm [55]. Biofilm can, however, still be formed in Fap-negative bacteria. In addition to biofilm formation, Fap fibrils might also play a role in virulence as over-expression of the *fap* operon leads to a more mucoid, alginate-rich phenotype which mimics the biofilm formation seen in cystic fibrosis patients [56].

Despite the high stability of Fap fibrils, Dueholm et al. succeeded in purifying the major fibril component, which was termed **FapC** [8]. In a functional genomics study, performed to identify the function of unknown proteins in *P. aeruginosa*, one of the most attenuated phenotypes towards worm infection was a mutant with a deletion in the *fapC* gene (NCBI locus tag PA1954) [57]. This again indicates that Fap fibrils play an important role in virulence – a function that has also been proposed for the analogous curli fibrils due to their interactions with different host proteins [58–60]. Sequencing of the major amyloid subunit, FapC, has shown that it contains a thrice repeated motif (R1-R3) which – like in the repeats of the curli CsgA and CsgB [5] – have conserved Glu and Asn residues, at least across 18 strains [54]. The R1 and R2 of FapC are 26–27 residues longer than the CsgA repeats while R3 is 11–12 residues longer. Linkers connect the different FapC repeats and can be up to 275 residues long between R2 and R3 [8] This contrasts with CsgA and CsgB which have a tight β-turn composed of 4–5 residues between each repeat [35, 61]. Some amyloids depend on interactions between aromatic residues [62, 63], but the lack of any aromatic sidechains in the FapC repeats emphasizes that this is not the case for these amyloid fibrils. Furthermore, aromatic residues were also found to be dispensable in curli formation [33].

4.2.4 Another Study in Team-Work: The Role of Fap Proteins

The FapC homolog, **FapB**, has similar, but shorter, repeats (17 residues versus FapC's 34 and 49) [8] and less variable linker lengths than FapC [53]. Furthermore, the FapB repeats may follow a strand-turn-strand-like motif [54]. Bioinformatics suggest that FapB is a nucleation factor (like CsgB in the curli system) and biophysical investigations have identified FapB and FapE as minor constituent of the Fap fibrils [54, 56]. In any case, FapB is necessary for Fap biogenesis [55].

The roles of the remaining Fap proteins (A, D, E and F) are not entirely clear and are still under scrutiny. However, there are some clues (Fig. 4.2). As an example, fibrils consisting almost entirely of FapB have been isolated from *fapA*- mutants, showing not only that FapB is indeed amyloidogenic, but also that **FapA** probably has a regulating role in the assembly of Fap fibrils *in vivo* [54]. FapA's function is unclear, however, as the deletion of the *fapA* gene is relatively frequent across different species [53]. A change in fibril composition has also been observed in the

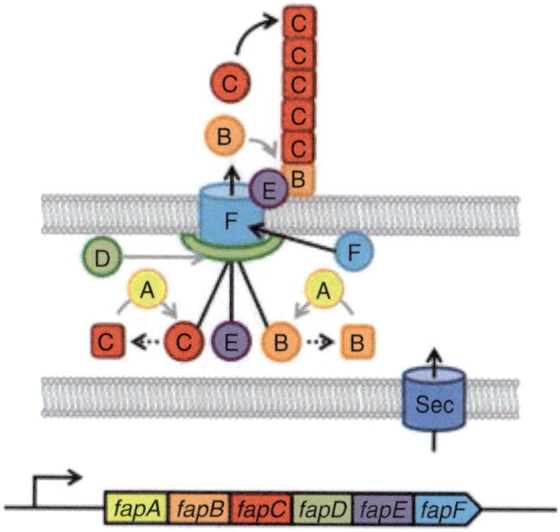

Fig. 4.2 Proposed model of Fap fibril biogenesis. All Fap proteins are encoded by the same operon and contain a secretion signal that enables them to cross the inner membrane through the Sec translocon. FapA is proposed to act as a chaperone for FapB and FapC to inhibit intracellular fibrillation. FapF is a bioinformatically predicted outer membrane β-barrel protein which transports the proposed nucleator FapB and the main fibril component FapC out of the cell. FapE constitutes a minor part of the mature fibrils and might act as an assembly factor. FapD probably has protease activity and might modify some of the Fap proteins before they are secreted from the cell. (Reproduced with permission from [64])

curli system in a *csgC*- mutant [65] and therefore FapA might serve a similar role as a chaperone in Fap formation, just as CsgC does in the curli system [29].

A proteomics study has shown **FapF** to be membrane-associated [66], in good agreement with bioinformatic predictions of FapF as a β-barrel membrane pore protein [54].

The identification of small amounts of **FapE** in mature Fap fibrils suggests that FapE is transported across the OM through FapF and that it interacts with mature fibrils. FapE might interact with FapC through conserved Cys residues in the C-terminal of both proteins and aid in export of FapC [54]. Comparison of sequences of the *fapABCDEF* operon across *Pseudomonas* genera showed the proteins FapD and FapE to be the most conserved, indicating important regulatory roles of these proteins. The regulatory role of FapE could be similar to the role of CsgF in the curli system in organizing the extracellular Fap apparatus. A protein homology recognition tool has recognized **FapD** as a likely cysteine protease [54]. This proteolytic activity might be relevant in processing of some or all of the Fap proteins.

Common to all the Fap proteins is the presence of a Sec secretion signal which enables them to utilize the Sec pathway across the IM [67]. In the curli system, which also uses this IM transport system, only the transcription factor CsgD does not have a secretion sequence [14]. Since Fap proteins are encoded within one operon, no analogue to CsgD is needed in the Fap system.

4.2.5 *Other Bacterial Amyloid Systems*

Besides the curli and Fap systems, numerous other bacterial FuBA systems have been described with various degrees of detail. An overview is presented in Table 4.1. Here we provide a brief description of some of these systems. The continued discovery of new functional amyloids with new and diverse roles within microbial communities, ranging from biofilm over cell communication, cellular warfare, communication and replication control, serves to emphasize that there is not a single unifying role for the cross-β motif – but conversely also illustrates that amyloid can be put to many different uses.

4.2.5.1 TasA: Cell Anchoring and Susceptibility to D-Amino Acids

TasA from the Gram-positive bacteria *Bacillus subtilis* is important for solid-surface and pellicle biofilm formation and stabilization [11] and forms amyloid fibers as seen with CR binding and TEM (including gold-labeled anti-TasA antibody) [73]. TasA extracted from cells readily forms ThT-positive fibrils of 10–15 nm in width. However, TasA fibrils do not share the extreme stability of other well-known functional bacterial amyloids; they dissolve in as little as 10% formic acid [73]. Compared to other amyloids, the TasA fibrils also differs in secondary structure: besides pure β-strands, they also contain α-helices and random coil structure [73]. This is an intriguing extension of the canonical fibril structure. The altered structure and lowered stability may also reflect the fact that monomeric TasA folds to a well-defined structure as monomer: a recently solved crystal structure of TasA reveals a central antiparallel 9-stranded β sandwich flanked by loop regions and 6 α-helices [100]. TasA is encoded by the *tapA-sipW-tasA* operon; together with exopolysaccharide, these three proteins they are the most abundant components of *B.subtilis* biofilm [11, 101]. TasA fibers are anchored to the cell wall and can thus link different cells. Interestingly, TasA does not have any imperfect repeats [102], highlighting a different fibrillation strategy than the multi-repeat structure seen in CsgA and FapC. TasA fibrillation is assisted by TapA which may have nucleation activity; *tapA*- mutants show a significant reduction in TasA fibrillation [74]. The signal peptide SipW helps process and secrete TapA and TasA and has a regulatory effect on TapA and TasA in solid surface biofilm but not pellicle biofilm [74]. TasA and TapA must be produced in the same cell to produce a functional biofilm, as colony assays shows no extracellular complementation [74]. In the ageing biofilm, cells synthesize D-amino acids which upon integration into the peptidoglycan cell wall structure triggers fibril liberation and biofilm degradation [75]. TasA also shows antimicrobial activity, and is a major coat component in spores [76].

Table 4.1 Overview of different functional amyloid systems in bacteria

Name of species	FuBA	Amyloid protein(s)	Operon name(s)	Members of operon(s)	Operon member functions	Role of amyloid	References
Enterobacteriaceae (*E. coli*, *Salmonella* species)	Curli	CsgA	*csgBAC* and *csgDEFG*	CsgA-G	CsgA: major curli subunit CsgB: Nucleator component CsgC: may regulate CsgG through C230 CsgD: transcriptional regulator CsgE: chaperone stabilizing CsgA monomer before release CsgF: chaperone-like, may help surface anchor CsgB CsgG: β-barrel membrane pore	Adhesion to surfaces and host cells, biofilm formation and other community behaviours. Heterogeneous curli fibril formation may facilitate multispecies biofilms	[7, 17, 36, 45, 52, 58, 68, 69]
Salmonella typhimurium and *S. enteritidis*	Tafi	AgfA	*agfBAC* and *agfDEFG*	AgfA-G	These Agf proteins have the same functions as their corresponding Csg-protein	Adhesion to surfaces and host cells, biofilm formation and other community behaviours	[19, 20, 70]
Pseudomonas	Fap	FapC	*fapABCDEF*	FapA-F	FapA: regulatory chaperone? FapB: amyloid nucleator FapC: main fibril monomer FapD: putative C39 peptidase FapE: minor component of amyloid fibers FapF: membrane pore	Involved in aggregation and biofilm formation and able to bind small metabolites	[8, 55, 71, 72]

(continued)

Table 4.1 (continued)

Name of species	FuBA	Amyloid protein(s)	Operon name(s)	Members of operon(s)	Operon member functions	Role of amyloid	References
Bacillus subtilis	TasA	TasA	*tapA-sipW-tasA*	TasA, TapA and SipW	TasA: main fibril monomer TapA: stabilizes TasA and anchors it to cell wall peptidoglycan layer SipW: signal peptidase processing and secretion of TasA and TapA. Also helps incorporate TasA into spores.	Forms hydrophobic colony biofilm and pellicle biofilms at the air-liquid interface and is a component of spores TasA also have antibacterial activity against other bacteria in the same environment	[11, 73–76]
Methanosaeta thermophile PT	MspA	MspA	*mspA* ORF not in operon	–	–	Forms extracellular tubular sheaths	[10]
Xanthomonas	Harpins	HpaG	Encoded by the Hrp PAI which in total encodes 9 *hrp*, 9 *hrc* and 8 *hpa* genes		Most of the Hrp and Hpa protein encoded by the PAI function as type III secretion systems HpaG acts as a type III secreted effector protein	Both proto-fibrils and mature fibrils of HpaG induce hypersensitive responses (HRs) in plants	[77, 78]
Streptomyces coelicolor	Chaplins	ChpD-H	*chpABCDEFGH*	ChpA-H	ChpA-C: possibly tether CphD-H to the cell surface (contain a LAXTG sequence and are sortase substrates) ChpD-H: fibril formation ChpE: may be responsible for fibril assembly of other chaplins	Attachment and aerial formation and growth (lowers surface tension to allow aerial hyphae to grow into the air and forms coats to cover aerial hyphae and spores)	[79–84]

Mycobacterium tuberculosis (maybe also in *bovis* and *avium* which also encode *mtp*)	Pili (cross-β structure remains to be confirmed)	MTP	The *mtp* gene is found in the H37Rv ORF, designated Rv3312A (not in an operon)	–	–	Adhesive properties: inds to laminin. Role in infections? (serum from tuberculosis patients contain anti-MPT antibodies). Involved in biofilm formation and cell-cell communication	[85–87]
Streptococcus mutans	Pili (cross-β structure remains to be confirmed)	P1	P1 is encoded by the *spaP* gene	–	Sortase transpeptidase (encoded by the *srtA* gene) is responsible for tethering P1 to the cell surface	P1 is involved in biofilm formation. Amyloid is detected in dental caries	[88]
Klebsiella pneumoniae RYC492 (homologs are found in *E. coli*)	*K. pneumoniae:* Microcin E492 (Mcc) *E. coli:* Microcin 24	MceA (*E. coli*: mtfS)	MceA is located inside a 3 kb *Cla*I DNA fragment together with its immunity protein encoded by the *mceB* gene	MceA, MceB	MceA: has one transmembrane region; forms cytotoxic oligomers or inactive fibrils *in vivo* MceB: integral inner membrane protein with three predicted transmembrane regions. Confers self-protective immunity to MceA.	Kills Enterobacteriaceae (competitors for same ecological niche) with oligomers that form pores in the cytoplasmic membrane, causing lethal loss of membrane potential	[25, 89–92]
Listeria monocytogenes	Listeriloysin O (LLO) toxin	LLO (encoded by the *hly* gene).	PAI LIPI-1 comprises six genes (*prfA, plcA, hly, mpl, actA, plcB*)		The products of LIPI-1 are required for crucial steps in the intracellular life cycle of the bacteria	LLO oligomers form pores in the membrane of phagolysosomes, enabling the bacterium to escape into the cytosol	[93–95]

(continued)

Table 4.1 (continued)

Name of species	FuBA	Amyloid protein(s)	Operon name(s)	Members of operon(s)	Operon member functions	Role of amyloid	References
Staphylococcus aureus	Phenol soluble modulins, PSMs	PSMs and the delta hemolysin (δ-toxin)	PSMs are encoded by the alpha (*αPSM1–4*) and beta (*βPSM1–2*) operon and the δ-toxins are encoded within the Agr regulatory RNA, RNAIII (*hld*)	αPSM1–4, βPSM1–2 and δ-toxin	A Δ*βPSM* double KO mutant does not form fibrils Transcription of *psm* operons is controlled by the AgrA DNA binding protein. An *agr* quorum sensing mutant does not produce fibrils either	PSMs enhance biofilm integrity and homeostasis while protecting against mechanical and enzymatic attach. Also enable the biofilm to detach and regrow as needed.	[9, 96, 97]
Pseudomonas	RepA	WH1 domain of RepA	Single *repA* gene	–	WH1 forms amyloid fibrils upon binding to DNA *in vitro*.	Inhibition of plasmid DNA replication by inactivating the replication origin	[98, 99]
Gordonia amarae	Unknown	–	–	–	(Neither operon nor protein identified)	Part of cell envelope	[22]

4.2.5.2 MspA: The First Archaeal FuBA

MspA (major sheath protein) from the archaea *Methanosaeta thermophila* PT is the first and so far only example of functional amyloid from Archaea. MS/MS revealed the sheaths to be composed of this single protein [10]. As might be expected, MspA is not evolutionarily related to other known functional amyloids but constitutes a novel amyloid motif that will have to be incorporated into future sequence analyses hunting for functional amyloid [103].

4.2.5.3 Harpins: Green Oligomeric Weapons

Harpins produced by plant pathogenic bacterial species are type III-secreted proteins which cause a hypersensitive response (HR) in the intercellular space of leaves in several plant types. The HR is known to be an early defensive strategy of the plant, to stop further growth of plant pathogens by inducing apoptosis [104]. The harpin protein HpaG of *Xanthomonas axonopodia*, is a fibril forming protein with imperfect repeats, however HR is induced not by fibrils, but rather by non-amyloid spherical oligomers which form at an earlier stage. Such oligomers are known to be the cytotoxic species in Alzheimer's and Parkinson's Disease due to their cell-permeabilizing properties [105], but HpaG is the only example to our knowledge of bacteria using this strategy as a functional tool, despite the protein's amyloid-forming properties.

4.2.5.4 Chaplins: Breaking the Air-Water Interface Barrier

The filamentous bacterium *Streptomyces coelicolor* produces eight hydrophobic proteins (ChpA-H) which are collectively called chaplins. ChpA-C consist of two chaplin domains ($\sim$ 40 hydrophobic residues) and a C-terminal sorting signal while ChpD-H are composed of a single chaplin domain and an N-terminal secretion signal peptide. As a mycelium forming bacterium, *S. coelicolor* forms aerial hyphae which give rise to spores that can spread and form new mycelium [106]. ChpA-H are expressed in the mycelium and in aerial hyphae [107]. They all contain signal sequences for secretion to the cell-wall. When secreted, the chaplins self-assemble at the air/water interface into an amphiphatic amyloid film which reduces surface tension and promotes spore spreading. This resembles the fungal proteins called hydrophobins (see Chap. 8). ChpD-H complement cells extracellularly and accelerate growth [106]. ChpA-C contain a common motif (LAXTG-Membrane spanning domain-Charged C-terminal) for anchoring surface proteins to the peptidoglycan in the cell wall covalently, suggesting a role in ChpD-H incorporation in the cell wall [106]. It has recently been suggested that ChpD-H, associated with cellulose fibrils and supported by ChpA-C, are responsible for fimbriae formation at adhering hyphae, and ChpD-H monomers assemble upon contact with assembled fibrils or hydrophobic surfaces [79].

4.2.5.5 Phenol-Soluble Modulins (PSM): Amyloid or Antimicrobial Agents?

The pathogen *Staphylococcus aureus* inhabits many bodily cavities and can change from benign to pathological within the host. Both states can form biofilm. Functional amyloid in the form of fibrillar structures have been identified in the biofilm. Unusually, the amyloid consists of small (20–40 residues long) peptides called phenol soluble modulins (PSMs) [9]. The PSMs fall into 3 highly conserved classes (α, β and δ) which are conserved and expressed from three different operons [9, 96, 97]. As small peptides, the PSMs represent a very different aggregation motif compared to the imperfect repeats in CsgA and FapC. While monomeric PSMs have antimicrobial activity and can disassemble biofilm in a surfactant-like manner (based on *agr* quorum sensing), fibrillated PSMs stabilize biofilm against mechanical and enzymatic attack [108]. Amyloid formation is stimulated by extracellular DNA (just like the initiator protein RepA involved in plasmid replication control [98]), suggesting that the antimicrobial effect came first and amyloidogenesis evolved later in an opportunitstic fashion.

4.3 Functional Amyloids *in silico*

Being able to predict which part of a protein sequence (if any) forms amyloid is helpful for identifying potential functional amyloid in hitherto uncharted genomes. The following sections will describe the principles behind the bioinformatics methods that have been used to identify aggregation and amyloid "hot spots" within protein sequences, how these methods have been used on functional amyloid sequences, and how we can go beyond these sequence-based methods in our search for as of yet uncharacterized functional amyloids. Functional amyloid can be identified based on sequence, evolved characteristics or structural prediction and simulation, and each of these will be described in more detail.

4.3.1 Predicting Aggregation and Amyloid Propensity of Proteins Based on Sequences

The most widely used methods for detecting aggregation-prone or amyloidogenic sequences were developed mostly with pathological aggregation and amyloid formation in mind [109–120]. These methods have been reviewed several times already [121, 122]. An overview of methods is provided in Table 4.2. In this section, we will discuss different principles on which these methods are based.

Table 4.2 Description of methods used to predict amyloid and aggregation propensity based on protein sequence information

Name[a]	Principles	Year(s)
	Secondary structure propensity	
SecStr [123]	A consensus secondary structure predictor that can be used to identify short sequences that have both alpha-helical and beta-sheet propensities. These short sequences are dubbed "conformational switches" and are thought to contribute to amyloid formation.	1988
CSSP/NetCSSP [120]	Attempts to identify "chameleon" sequences that are prone to switching from alpha-helical to beta-sheet conformations. Emphasizes the influence of tertiary contacts on the determination of secondary structure, and utilizes a feed-forward artificial neural network trained using back-propagation for making predictions.	2005, 2009
SALSA [124]	Uses a sliding window approach to calculate beta-strand contiguity propensity for short amino acid sequences (4–20 residues in length) by comparing beta-strand propensity to alpha-helical and reverse-turn propensities according to the Chou and Fasman scale. The threshold for being considered amyloidogenic was determined by comparing the output of the algorithm for alpha-synuclein, which is amyloidogenic, and beta-synuclein, which is not.	2007
BETASCAN [118]	Attempts to find all contiguous segments in a protein sequence that have propensity to form beta-sheet structures and all pairs of segments that have a propensity to pair in parallel beta conformations, making it well suited to the detection of beta-helical structures found in several functional amyloids. Strand and pairing propensities were estimated based on statistics of a database of protein structures.	2009
	Sequence patterns and physico-chemical properties	
Zyggregator [116, 125]	Aggregation rates are predicted using a weighted sum of terms that takes into account hydrophobicity, the degree to which an alternating pattern of hydrophobic and hydrophilic residues exists in the sequence, electrostatic charge, secondary structure propensity, pH, ionic concentration, and protein concentration.	2004, 2008
Mean packing density [126, 127]	The statistics for packing density of all 20 residue types were first collected from a database of structures and these values were used to calculate a packing density profile using a sliding window of 3–9 residues. If the value of packing density for 5 or more residues in a row exceed an empirically determined threshold, then this region is considered amyloidogenic.	2006, 2007
FoldAmyloid [115]	Statistics on the average packing density and hydrogen bonding propensity of all 20 residue types were collected from a database of protein structures. A score is computed for each peptide of a sequence using the packing density and hydrogen bonding scores of all residues in a sliding window. A peptide is considered amyloidogenic if the average score of all residues within a sliding window are above an empirically determined threshold for packing density or hydrogen bonding.	2010

(continued)

Table 4.2 (continued)

Name[a]	Principles	Year(s)
Waltz [109]	Computation of a sequence score that is an optimized linear combination of scores based on a position-specific-scoring matrix, nineteen physical properties that were determined to be informative regarding amyloid formation, and a structural score that evaluates compatibility with amyloid conformations.	2010
	Evaluation of statistical potentials	
3D profile method [119]	Evaluation of the ROSETTADESIGN energy function on hexapeptides threaded onto multiple templates based on the crystal structure of an amyloid forming peptide	2005
PASTA/PASTA 2.0 [113, 128]	Evaluation of an empirical energy function derived using statistics from a database of globular proteins for all possible pairs of length 7 segments within a sequence in both parallel and anti-parallel orientations	2006, 2014
PRE-AMYL [129]	Evaluation of residue-residue pairwise statistical potential derived from a database of protein structures on hexapeptide sequences that are threaded onto an amyloid fibril core microcrystal structure. An empirically determined threshold is used to decide whether a hexapeptide is amyloidogenic.	2007
GAP [130]	Makes predictions regarding whether a peptide is likely to form amyloid or amorphous beta-sheet aggregates based on the idea that peptides that form amyloid and amorphous aggregates have distinct preferences for residue pairs in "adjacent" (on the same side of the beta-sheet) and "alternate" (on opposite sides of the beta-sheet) positions. Separate statistical potentials were derived using known amyloid forming and amorphous aggregate forming peptides, and the difference between these potentials (evaluated on short peptide sequences) is used as a measure of the preference for the type of aggregate.	2014
ArchCandy [111]	Evaluation of an empirical energy function that takes into account sterics, electrostatics, packing, hydrogen bonding, the effect of prolines, and solvation on an ensemble of beta-arch structures	2015
	Statistical mechanical models	
TANGO [117]	The aggregation propensity of peptides is determined by approximately computing the partition function of a statistical mechanical model wherein segments of the sequence can be in native, beta-turn, alpha-helix, and beta-aggregate states. The propensity of segments to adopt these states are taken from AGADIR for alpha-helical conformations or newly derived statistical potentials for beta conformations.	2004
AmyloidMutants [131]	A statistical mechanical model that attempts to predict amyloid fibril structures from sequence information alone, as well as how fibril structures change upon making mutations to the sequence. A coarse-grained but still detailed structure space is used for enumerating conformations and a residue-residue statistical potential that takes into account the environment of the interaction is used to compute energies for each conformation.	2011

	Experimentally driven methods	
Amyloidogenic Pattern [132]	A hexapeptide pattern for amyloidogenic peptides was derived based on performing *in vitro* aggregation experiments on peptides. The peptides were obtained using saturation mutagenesis of a *de novo* designed amyloid-forming peptide.	2003
AGGRESCAN [114, 133, 134]	Hotspots are identified by computing the aggregation propensity of all sets of 5 contiguous residues and comparing it to the average aggregation propensity for the entire sequence. Relative aggregation propensities for individual residues were obtained by measuring *in vivo* aggregation of 20 mutants of Abeta42.	2006, 2007
	Machine learning-based methods	
Pafig [135]	A support vector machine for differentiating between amyloidogenic and non-amyloidogenic hexapeptides was trained using positive and negative experimentally verified examples and 531 different numerical properties defined for each type of amino acid as input features. An informative subset of these features was selected using support vector machines and genetic algorithms for performing the final encoding of hexapeptides.	2009
David et al. [136]	A naïve Bayes classifiers and weighted decision trees were trained and used to predict the amyloidogenicity of immunoglobulin sequences.	2010
Nair et al. [137]	Physico-chemical properties of amino acids were used as input to train a hybrid machine learning scheme based on genetic algorithms, support vector machines, and artificial neural networks for differentiating between amyloidogenic and non-amyloidogenic peptides.	2011
FISH Amyloid [138]	A machine learning algorithm that emphasizes the recognition of patterns of co-occurrence of amino acid types in short sequences.	2014
APPNN [139]	Features selected using recursive feature selection on a large number of tabulated amino acid properties were used as input to a feed-forward neural network, which was trained using back propagation and a database of peptides with known amyloid forming tendencies. In the end, a relatively small number of features were selected including beta-sheet propensities, isoelectric point, hydrophobic moment, helix termination propensity, and transfer free energies.	2015
	Consensus predictors	
Amylpred/ Amylpred2 [140, 141]	A consensus predictor that relies on predictions from AGGRESCAN, AmyloidMutants, Amyloidogenic Pattern, Average Packing Density, SALSA, PRE-AMYL, NetCSSP, Pafig, SecStr, Tango, and Waltz. Segments where 5 or more of the 11 methods indicate significant amyloidogenicity are considered amyloidogenic.	2013
MetAmyl [110]	A consensus predictor using an optimized logistic regression model with contributions from PASTA, SALSA, AGGRESCAN, PAFIG, FoldAmyloid, TANGO, and WALTZ.	2013

Notes:
[a]Where the method was not named in the original publication but are included in Amylpred2, we adopt the names given by the creators of the Amylpred2 consensus predictor [140]

4.3.1.1 Secondary Structure Propensity and Physico-Chemical Properties of Amino Acids

Recognizing the importance of the beta-sheet secondary structure to the formation of amyloid and other aggregates, several methods for detecting aggregation prone sequences attempt to estimate the secondary structure propensity of amino acid sequences, i.e., their relative preference for α-helical, β-sheet, and random coil configurations. In these methods, amino acid sequences with a strong preference for beta-sheet conformations are considered more likely to end up in amyloid-like aggregates [118, 124]. A popular variant on this theme is to look specifically for sequences that fit well in both α-helical and β-sheet conformations and could therefore be more easily induced to switch from being α-helical to β-sheet by some kind of environmental perturbation [120, 142, 143].

The approach can be expanded to other physico-chemical properties of amino acid sequences [115, 126, 127], including hydrophobicity, charge, hydrogen-bonding capacity, and packing density. Many of these properties have been collected into large tables [144]. A common approach is to average a given set of properties over a short and contiguous sequence using a sliding-window. A part of a protein sequence is then considered to be aggregation/amyloid prone if one or all of the residues within a window exceeds an empirically determined threshold value.

When looking for short, aggregation prone sequences, it is helpful to consider both the physico-chemical properties of amino acids and their distribution along the sequence, i.e., sequence patterns of amino acid types. This could be as simple as looking for alternating hydrophobic-hydrophilic pairs of amino acids [116], which are favored in amphipathic β-sheets, to more detailed models that differentiate between each position in a short sequence using position specific scoring matrices [109]. Waltz [109] is a method based on position specific scoring matrices, which are widely used in homology searches. The total sequence score assigned to a peptide is a weighted sum of a sequence profile based on alignment of known amyloid-forming peptides, 19 physical properties that were determined to be informative regarding amyloid formation, and a pseudo-energy term related to a sequence's preference for amyloid-like structures. Waltz does a particularly good job of distinguishing actual amyloid-forming sequences from generic aggregation prone sequences, which is a task that has proven difficult for methods based solely on physico-chemical properties of amino acids. Waltz is also unusual in that it included a survey of amyloid forming segments in functional amyloids in its original publication [109]. Of the 22 functional amyloids tested, they were able to identify amyloid forming segments in 17 of them.

4.3.1.2 Statistical Potentials

Statistical potentials try to capture the relevant physical chemistry of, e.g., protein folding by first adopting a simplified functional form for the energy and then parameterizing this simplified energy function using statistics collected from a database of experimentally determined structures. A common simplification scheme in the context of proteins is to assume that a large part of the energy can be decomposed into residue-residue pair interactions rather than explicitly accounting for the presence of every atom within the residues. Implicit in the parameterization based on known structures is the assumption that the proximity of pairs of amino acids within protein structures is a reflection of their intrinsic propensity for interaction. Despite these seemingly drastic simplifications and assumptions, statistical potentials have proven useful in many areas of protein science including structure prediction and design.

In practice, peptide sequences are threaded onto experimentally determined amyloid fibril core structures. The statistical potential is then evaluated for a given sequence-structure pair and those sequences with the lowest energies are assumed to be a good fit and therefore amyloid prone [119, 129]. These methods are reminiscent of early attempts at protein structure prediction and fold recognition based on threading [145]. Depending on the structure that the sequences are threaded onto, these methods can potentially take into account inter-strand hydrogen bonding energies, β-sheet stacking energies or even intra-strand hairpin formation [111], but the fact that only a fixed and finite set of structures is considered in these methods will always limit their general purpose use.

4.3.1.3 Statistical Mechanical Models

Theoretical studies have shown that the chain entropy of the disordered part of the protein can be an important factor in favoring particular ensembles of aggregated conformations [117, 121, 146, 147]. It is therefore useful to develop models that allow partial ordering of a sequence (where energetic terms are dominant) but also take into account disorder from the rest of the chain. After (usually approximately) computing the partition function, it is possible to identify the lowest free energy conformational ensembles, and the ordered parts of the sequence in these ensembles are understood to be the parts of the sequence that are driving aggregation.

TANGO is a commonly used aggregation predictor that combines secondary structure propensities and physico-chemical preferences for burial into a statistical mechanical model for β-sheet aggregation [117]. TANGO has shown promise in predicting whether a peptide will aggregate as well as the differences in aggregation propensity when mutations are made to a given sequence, which is particularly relevant when investigating disease-causing mutants. A more recently developed statistical mechanical method attempts to predict not only which parts of the structure are ordered in aggregates but also the structures of fibrils in significant detail and how they change upon mutation [131].

4.3.1.4 Experimentally Driven Methods

The previous methods used data from experiments which were not designed specifically to determine which protein sequences will aggregate or form amyloid. However, there are also methods based directly on experiments purpose-built to discover aggregation prone sequences. One approach involves deriving many variants of a designed amyloid forming peptide through saturation mutagenesis [132]. The secondary structure content of the peptides was measured by circular dichroism spectroscopy and the peptides' tendency to form fibrils was evaluated by electron microscopy after 1 month. The results of these experiments were then used to derive short sequence patterns that reflected which amino acids were allowed in each position of amyloid forming peptides. Consequently, these sequence patterns can then be used to rapidly search for potential amyloid forming sequences in large sequence databases. A possible limitation of methods like these is a bias towards the initial peptide sequence used to derive the short peptides.

Although most aggregation experiments are performed *in vitro*, ultimately we are interested in aggregation propensity of proteins *in vivo*. AGGRESCAN [114] is similar to methods that scan sequences looking for short stretches of highly aggregation prone residues, but is unique in that the aggregation propensities of the 20 amino acid residues were obtained by monitoring aggregation via fluorescence of Abeta-42 variants with single amino acid substitutions *in vivo*.

4.3.1.5 Machine Learning Methods

Methods discussed thus far are all based on underlying physical principles of one kind or another, and can be more easily parameterized on the basis of limited experimental data to provide accurate predictions. Therefore, evaluating the accuracy models based on different physical principles can help us to test our understanding of aggregation. However, in some cases, we simply want to know how specific changes to a sequence will affect aggregation propensity and the detailed mechanisms are of relatively little importance. Any learning problem, including prediction of aggregation and amyloid propensity from sequence information, where sufficient labeled data is available, is potentially susceptible to supervised machine learning techniques [148]. The number of peptides that have been experimentally validated as either fibril forming or non-fibril forming is large and, presumably, still growing [149]. This represents a potentially rich source of training data. Indeed, machine learning and Bayesian statistics based methods such as support vector machines [135], artificial neural networks [137], and decision trees [136] have already been fruitfully applied to the problem of predicting amyloid and aggregation prone peptide sequences. Although these methods perhaps provide less direct physical insight into the problem of amyloid formation than other available methods, given enough training data they could eventually become the most reliable way to engineer the aggregation propensity of an amino acid sequence.

4.3.1.6 Consensus Predictors

Due in part to the proliferation of methods that can be used to predict aggregation and amyloid propensity on the basis of sequence information, consensus predictors have become a popular option that allow non-experts to combine results from different methods and thus hedge their bets when looking for aggregation "hot spots" [110, 140]. The success of consensus methods is likely due to the diversity of principles underlying the constituent methods, many of which we have discussed in this section.

4.3.2 Detecting Amyloid Prone Sequences in Functional Amyloids

Although the methods described in the previous section were not derived and have not been tested primarily with functional amyloid in mind, the structural similarities between functional and pathological amyloid provide a source of hope that these methods can nonetheless be successfully applied in the search for functional amyloid sequences. Several instances of these methods being applied to functional amyloids already exist in the literature.

4.3.2.1 Sequence-Based Methods Can Detect Amyloidogenic Segments in Biofilm-Associated Proteins

By scanning the sequence of a biofilm associated protein in *Staphylococcus epidermidis* using a combination of several sequence-based amyloid detection methods including AGGRESCAN [114], PASTA [113], and TANGO [117], the peptide STVTVTF was chosen for further biophysical characterization [150]. This seven-residue peptide was shown to spontaneously form amyloid fibers *in vitro*. In another study, the amyloidogenic segments of the yeast cell wall protein Bgl2p were predicted using FoldAmyloid, TANGO, AGGRESCAN, PASTA, and Waltz [151]. Peptides containing these sequences were then synthesized and found to form fibrils. In a similar case, sequence similarity to fimbrial proteins and the amyloid detection methods TANGO and AGGRESCAN were used to argue that HfaA might be a functional amyloid involved in the establishment of extracellular structures [152].

4.3.2.2 Searching for Prion-like Domains Can Uncover Previously Unknown Functional Amyloids

The relationship between amyloids and prions remains to be clearly elucidated. The hallmark of a prion is the ability to convert another prion molecule (with the same sequence and perhaps even those with related sequences) from one conformation to another, typically with functional consequences. One of these states is often aggregated or amyloid-like. Like amyloids, prions were first recognized in a pathological context but, also like amyloids, in retrospect it seems obvious that nature would take advantage of these motifs for functional purposes. A statistical model trained on the sequences of proteins that are known to exhibit prion-like behavior in yeast was used to search against many complete genomes including 839 bacterial genomes [153]. The search generated a list of 2200 putative prions (proteins that contain prion-like domains) in bacteria. Of the proteins in bacteria with annotated function, the prion hits were highly overrepresented in certain function classes, including those involved in the establishment of extracellular structures.

4.3.2.3 The Existence of Imperfect Repeats Is Common to Many Functional Amyloids

The sequences of the primary fibril component in the Curli functional amyloid system, CsgA, and its putative fibril-nucleating component, CsgB, are clearly related, with each containing 5 imperfect repeats. However, according to the consensus predictor Amylpred2, CsgB has a significantly higher aggregation propensity [154]. This is consistent with CsgB's putative role as a nucleator of aggregation and the supposition that functional amyloids should have evolved mechanisms for regulating the timing and location of amyloid formation. MspA, the first functional amyloid described in archaea, was also found to be amyloidogenic according to Amylpred2, and its amyloid nature was confirmed using biophysical techniques [10]. Like many functional amyloids, MspA contains imperfect repeats, as do TapA [155] and FapC [102]. On the basis of sequence alone, the existence of repeats containing amyloid motifs may be the single strongest indicator of a likely functional amyloid. The evolved nature of functional amyloids should provide additional clues when searching for functional amyloids in the genome, however, and some possibilities for leveraging those differences will be discussed in the following section.

4.3.3 Identifying Functional Amyloids Based on Their Evolved Characteristics

The fibrils of functional and pathological amyloids share striking structural similarities, and our knowledge of each type of amyloid can help to inform our understanding of the other. However, the evolutionary forces shaping the sequences of functional and pathological amyloids could hardly be more different. Functional amyloids confer selective advantages to their host organisms and typically evolve alongside a set of helper proteins that ensure that the amyloid is properly localized and can form fibrils where and when they are needed. Additionally, although the reasons are not necessarily intuitive, many amyloidogenic segments of functional amyloids are located in imperfect repeats within the full sequence. Each of these facts should be useful when searching for as-of-yet uncharacterized functional amyloids. Given the relatively nascent stage of functional amyloid research, it seems likely that these searches will bear fruit.

4.3.3.1 Searching for Functional Amyloid Homologues in Large Sequence Databases Reveals Functional Amyloid Sequence Diversity, Phylogeny, and Operon Structure

As the price of whole genome sequencing plummets, the number of complete genomes available is rising rapidly, including genomes of bacteria known to express functional amyloid [156]. For families of functional amyloids where at least one sequence is known, searching for related sequences in large sequence databases provides a fuller picture of the sequence diversity within a family as well as a sense of how widespread the family is in the bacterial kingdom, as illustrated in Fig. 4.3 [21, 53]. Related functional amyloid sequences in different organisms show diversity both at the level of individual protein sequences and at the level of operon structure. Most operons contain a core set of proteins, but the order of the coding sequences in the operon may be shuffled or the orientations inverted. The variation

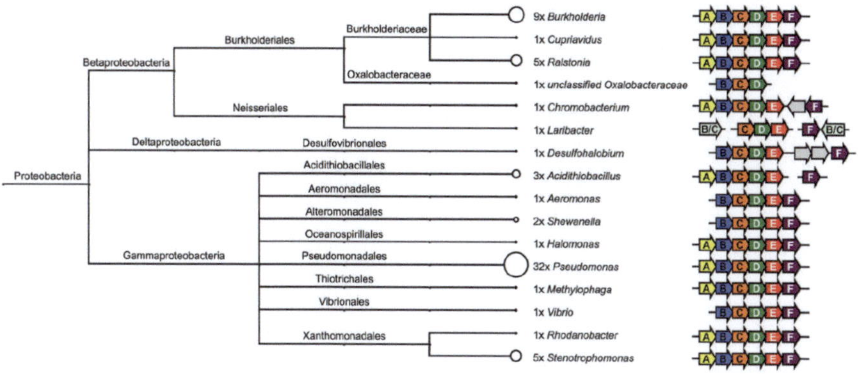

Fig. 4.3 A taxonomic analysis of bacterial strains based on Functional Amyloid in Pseudomonas (Fap) sequences. The Fap operon typically contains 6 proteins: FapA-FapF. The operon structure is illustrated next to each genus. The primary and secondary fibril components are FapC and FapB, respectively. The presence of multiple "helper" proteins within an operon may be a functional requirement and therefore a universal characteristic of functional amyloid operons. If this is the case, bioinformatic algorithms that look for operons containing amyloids and specific types of helper proteins may prove useful in identifying novel functional amyloid systems. (Reproduced with permission from [53])

in operon structure is significant enough that different regulatory mechanisms may be at work, meaning that these functional amyloids have probably been adaptively repurposed during evolution.

The sequence alignments of individual proteins within a functional amyloid operon can be used to build models that are sensitive enough to detect distantly related homologues in newly generated sequencing data, including metagenome sequencing data [21, 53]. We are now only starting to appreciate the importance of metagenomics to human health [157], and studies of environmental microbiomes, such as geothermal vents, have already yielded new putative functional amyloids [158]. A similar approach was used to discover 200 prion candidates in the genome of *S. Cerevisiae*, 24 of which were shown to form amyloid and be passed on indefinitely during cell division, effectively becoming epigenetic units of inheritance [159].

4.3.3.2 Techniques Targeting Evolved Characteristics May Find Unknown Functional Amyloids

But what if we want to use bioinformatics methods to discover functional amyloids with no known examples? The similarity between functional amyloids that are not evolutionarily related may provide some clues [54]. The operons containing the primary fibril components of both Fap and Csg also both contain chaperones, an outer membrane protein, and a minor fibril component with a sequence that is similar to the primary fibril component. Might these proteins comprise a minimal operon for proper functional amyloid formation? The simplest rationale is plausible: chaperones are necessary to keep the amyloid from forming in the cytoplasm; an outer membrane protein must be present to export the fibril components, and a modified version of the primary fibril component can ensure (through interactions with a membrane, for example) that the fibril is formed in the proper location. If analogous proteins accompany other functional amyloids, then bioinformatics searches may be able to leverage this information to identify putative functional amyloid operons even without knowing beforehand the sequences of any of the constituent proteins. This would require algorithms that can recognize generic characteristics of chaperons and outer membrane proteins and look for amyloidogenic segments contained in (imperfect) repeats.

Functional amyloids are frequently used by bacteria in the formation of biofilms [102]. For a species of bacteria that is known to form biofilms but for which no functional amyloid has yet been identified, differential gene expression analysis on cells that are forming biofilms and cells that are not is another promising route to the discovery of novel functional amyloids [160].

4.3.4 *Structure Prediction and Simulations of Functional Amyloids*

When traditional methods of experimental structure determination fail to yield unambiguous structural models, *in silico* methods are often useful for filling in the gaps. When good templates are available, homology modeling regularly results in high quality models of protein structures. Even when no good templates are available, both coarse-grained models and fully atomistic models are closing in on being able to reliably predict structures from sequence information alone. And when it comes to understanding the molecular details that underpin biology, few methods have proven to be as general and powerful as fully atomistic and coarse-grained molecular simulation. The formation of functional amyloids should be no exception. From folding and chaperoning to export and polymerization, the formation of functional amyloids involves many molecular processes that will be difficult to completely resolve using experimental techniques alone. Indeed, some of these processes are already beginning to be elucidated using molecular simulation.

4.3.4.1 Molecular Modeling Techniques Can Propose Structural Models of Functional Amyloids without Experimental Structural Data Using Evolutionary Constraints

The repeat domain of Pmel17, a functional amyloid in mammals, has resisted experimental structure determination. Homology modeling, supplemented by molecular simulation, recently provided strong evidence that the repeat domain forms a β-solenoid structure [161]. If no good template is available, or if modelers want to avoid biases that may be inadvertently introduced in homology modeling, a template-free structure prediction scheme may be preferable or even necessary. Without being supplemented with some kind of additional information, however, most molecular simulation models presently do a poor job of predicting the structure of large proteins with complex topologies. One of the recent and promising advents in molecular simulation is the combination of molecular simulation models with information derived from large sequence databases. In particular, techniques for inferring amino acid pair contacts from the large and growing number of protein sequences have significantly improved our ability to predict structures from sequence information alone. Here functional amyloid has the distinct advantage (compared to pathological counterparts) that their structure is linked to function and therefore under evolutionary constraints. Comparison of related sequences of functional amyloid therefore provides useful constraints, and was recently applied to the primary fibril component of the Curli functional amyloid system, CsgA [42]. Several methods for inferring contacts between amino acids based on sequence alignments were used along with an efficient coarse-grained molecular simulation model to propose structural models of CsgA. According to this model, CsgA adopts a β-helical structure that stacks the imperfect repeats perpendicular to the putative

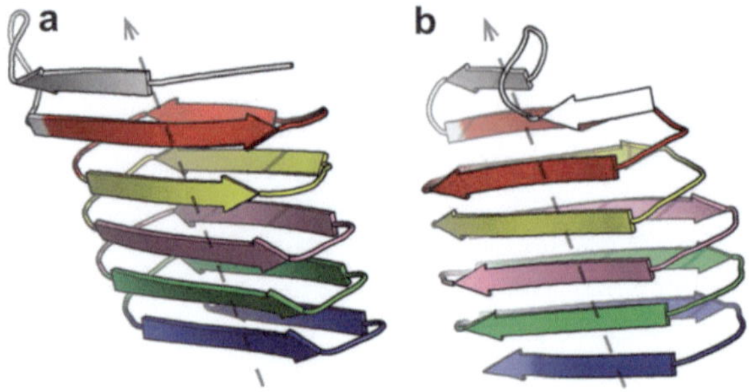

Fig. 4.4 Two predicted structures of the CsgA monomer. The 5 internal repeats are shown in red, yellow, pink, green, and blue. The putative fibril axis is shown as a dashed arrow. The simulations yielded both (**a**) left-handed and (**b**) right-handed beta-helices. The predictions were made on the basis of Monte Carlo sampling of a model that combined an all-atom energy function with a contact potential derived from sequence covariation within CsgA homologues. This approach highlights the synergy between molecular modeling and information-theoretic analyses of large sequence databases. (Reproduced with permission from [42])

fibril axis (Fig. 4.4). Some ambiguities remain, however, as the contacts derived from the sequence information proved insufficient to resolve the handedness of the β-helix and the inter-monomer contacts that contribute to fibril formation. An alternative to inferring docking contacts from sequence information is to use specialized docking protocols that can take advantage of available experimental information [162]. In one study, models of both CsgA and CsgB were generated by homology modeling using AfgA and AfgB structures as templates [154]. The authors then docked two copies of the CsgB model and extended the resulting docking pose to generate a model of a CsgB fiber.

4.3.4.2 Simulation Can Help to Elucidate the Molecular Details of Functional Amyloid Formation

In addition to proposing structural models, molecular simulation can be used to gain insight into molecular mechanisms. When applied to the folding and dimerization of the imperfect repeats of CsgA, atomistic simulations have provided information about the early stages of amyloid formation in Curli [163]. According to this model, the individual repeats of CsgA are largely disordered in isolation and show dimerization propensities that are in line with experiments on the aggregation propensities of the repeats. In another recent study, a complex mechanism involving the conversion of oligomers of the *Aplysia* sea slug translational regulator CPEB from a coiled-coil state to an amyloid state was proposed on the basis of coarse-grained molecular simulations [164]. CPEB is thought to be an important factor in the formation of long-term memories [165]. The conversion from coiled-coil oligomer to amyloid is facilitated by a pulling force that could be provided in the cell via interactions with actin filaments or motor proteins.

4.4 Uses for Functional Amyloid: Brave New Nanomaterials

E. coli's curli operon makes up a fascinating miniaturized export-and-assembly system and holds great potential as a nanotechnological device. The few examples of controlled amyloid production and export available in the literature all make use of this operon, and we will use the following section to describe results obtained so far.

4.4.1 C-DAG as a Screen for Amyloid: How to Hijack a Robust Amyloid Export System

It has already been established that proteins can be exported provided their cross-sectional diameter is not larger than 2 nm; provided the protein is maintained in an unfolded state which can be threaded through the CsgG pore, there are no upper limits on protein size [49]. This has been put to elegant use as a bacterial screening system to identify proteins capable of fibrillation under "amyloid-friendly" conditions in C-DAG (curli-dependent amyloid generator system) [50]. Given that an overexpression of CsgG enables efficient secretion of CsgA [43] (presumably together with smaller amounts of CsgE and CsgF), it is possible to test any given protein by fusing it to the CsgA signal peptide in a strain overexpressing CsgG and lacking CsgA and CsgB (to avoid spurious nucleation on the surface). In this way, expressed protein is separated from proteins in the cytoplasm and is transported in a (relatively) unfolded conformation through CsgG to accumulate to a high local concentration outside the cell. Here it can aggregate at leisure, possibly stimulated by the anionic lipid environment at the cell surface which is known to promote protein aggregation in general [166, 167]. A number of different proteins have been shown to form fibrils under these conditions according to standard assays (Congo Red plates, apple-green birefringence, SDS-resistance to unfolding and seeding ability), even proteins such as the yeast prion Sup35 that normally is strictly dependent on the PIN factor to form amyloid [168]. A subsequent screen of all *E. coli* ORFs using universal primers also identified a novel amyloidogenic protein (FliE from the flagellar basal body). However, the main strength of the system is probably in the screening of amyloid aggregation modulators by plating cells expressing a given protein on solid medium containing Congo Red; a caveat is that screening of whole ORF libraries can generate a large number of false positives as well as various shades of red in the absence of detectable amyloid aggregates [169]. The reason for this is unclear, though may be related to CR's propensity to bind non-amyloid material, as seen for *e.g.* insulin [170], or even facilitate oligomerization of native proteins by binding as a sandwich between two protein molecules [170].

4.4.1.1 Generating New Binding Properties: How to Hitch a Ride on the Amyloid Ladder

Given the above export abilities, it is obvious to combine CsgA with proteins that display other functionalities. The simplest function is to bind, and this was early on very elegantly exemplified in the BIND (biofilm-integrated nanofiber display) system [171]. The approach that works best is to genetically fuse short peptide binding sequences to the C-terminus of CsgA with a long linker (GSGGSG) and then terminate said peptide with another GSGGSG sequence (N-terminal fusions interfere with CsgG recognition). The authors demonstrated the versatility of this system (Fig. 4.5) by demonstrating binding of *E. coli* overexpressing these fusion proteins to inorganic surfaces (graphene, carbon nanotube, glass, ice crystals), metals (steel, gold), minerals (hydroxyapatite, zinc sulfide, magnetite) and proteins. In the latter case, CsgA is fused to the protein SpyTag which covalently links to 15-kDa SpyCatcher protein [172], allowing generic linkage to any protein that is fused to SpyCatcher. The study by Joshi and coworkers is also worth reading for the rigor with which they quantify production of amyloid, using both colorimetry (Congo Red absorption; birefringence was not quantitative, probably due to fiber dispersion), formation of extracellular assemblies (whole-cell filtration ELISA with anti-CsgA antibodies), verification of intact protein (extracellular fractions washed

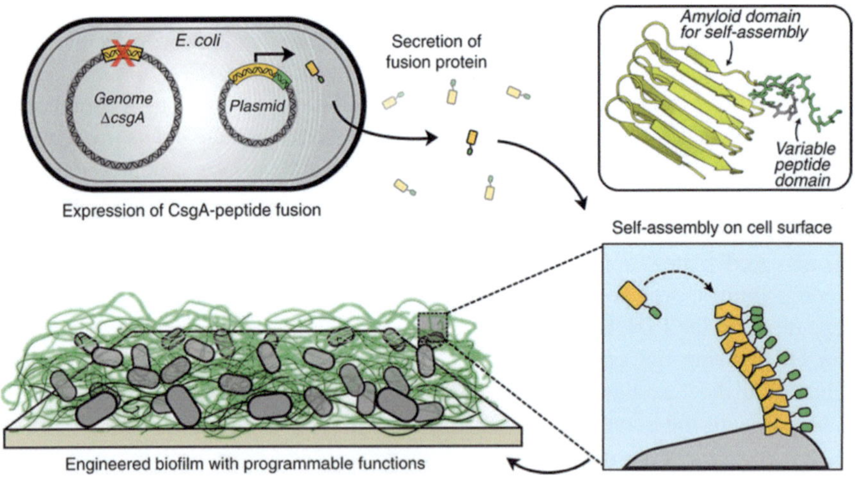

Fig. 4.5 Using fusions of CsgA and peptide binding systems to generate new functionality in biofilm. Top: The endogenous CsgA is knocked out from an *E. coli* strain which from the expression vector produces fusion proteins consisting of CsgA (yellow) and a C-terminally attached peptide domain (green), as shown in the 3D protein model insert where the CsgA sequence was threaded onto an AfgA structure. The linker domain is in grey, followed by the SpyTag domain in green. Peptide structure was predicted with PepFold. Bottom: The expressed protein forms amyloid (curli) fibrils on the surface of the *E. coli* cell and leads to a biofilm with novel function. (Reproduced with permission from [171])

in SDS and the pellet dissolved in HFIP to mobilize for MALDI-TOF analysis) and morphology (SEM, TEM and immunogold labelling with antibodies recognizing the fusion peptide). Another important aspect is to use the right *E. coli* strain. The csgA deletion strain (LSR10) does not produce cellulose, avoiding background effects from CR binding to cellulose [173]. However, for larger-scale production, it is possible to use the K-12 strain PHL628 which does produce cellulose but also has a 3.5-fold increased curli operon expression thanks to a single point mutation in OmpR protein [174], so the whole amyloid producing and secreting machinery is boosted. There are remarkable perspectives in this work at several levels. The system is simple and straightforward and requires no real optimization of expression constructs, yet provides an amyloid matrix that by virtue of its robustness can likely survive cell death to remain behind as a skeletal framework, built from sustainable materials.

4.4.1.2 Amyloid as Underwater Glue: Fusing CsgA to Mussel Foot Proteins

Lu and co-workers have extended this approach by fusing CsgA with mussel foot proteins Mfp3 and Mfp5 from *Mytilus galloprovincialis* as CsgA-Mfp3 and Mfp5-CsgA constructs [175]. Mfp proteins are used as underwater adhesives to silica (rock) surfaces; the proteins are activated by enzymatically converting Tyr residues to the crosslinker 3,4-dihydroxyphenylalanine (DOPA) [176]. By combining these glue proteins with amyloid, it is possible to generate environmentally robust underwater glue that can self-heal through self-assembly and form a large fiber surface area that enhances contact area in the adhesive barnacle plaque [177]. The hybrid CsgA-Mfp proteins lead to fusion fibrils that were three to fivefold wider than pure CsgA fibrils, probably due to intermolecular association between Mfp3 and Mfp5 domains. When hydroxylated, these fibrils had a two to threefold increase in adhesion to silica compared to CsgA. Interestingly, a copolymer made by mixing CsgA-Mfp3 and Mfp5-CsgA gave the greatest effect, showing the strongest underwater adhesion of bio-derived/inspired protein-based underwater adhesives so far.

4.4.1.3 Controlled Combination of Different Amyloid: The Power of Riboregulators

The interface with different organic and inorganic surfaces shown in these previous examples points to a merger of synthetic biology with materials science that will ultimately be able to make large-scale programmable living materials. The trick is to engineer genetic logic gates into this system which could enable us to switch on (and off) different biofilms on demand, using appropriate environmental cues. Such a cue could be provided by the extremely sophisticated riboregulator system employed imaginatively by Lu and co-workers to self-assemble hybrid structures

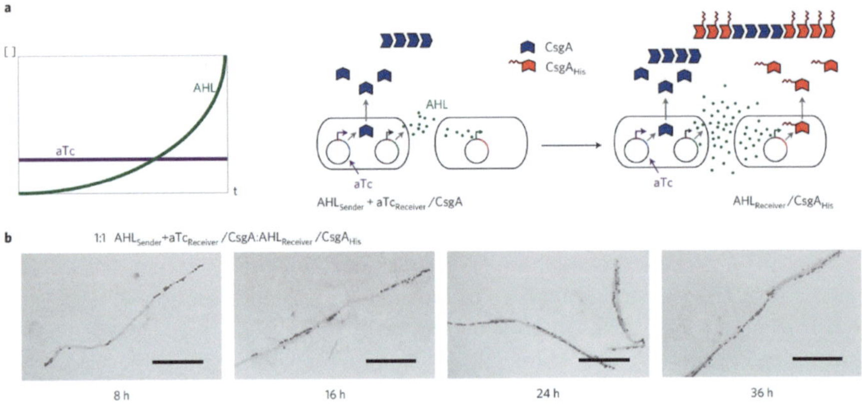

Fig. 4.6 Using riboregulators to make controllable variations in fibril patterns. This leads to synthetic gene circuits where external inducers (aTC or AHL) control protein secretion. (**a**) A constant level of aTc (which drives expression of CsgA) also induces a gradual accumulation of AHL which in term induces expression of $CsgA^{His}$. (**b**) The His-variant can be labelled by gold nanoparticles, and the fraction of fibrils with $CsgA^{His}$ increases with time. (Reproduced with permission from [178])

of CsgA at a variety of different length scales [178]. Riboregulators are RNA molecules that can regulate expression of genes in response to different small molecules such as anhydrotetracycline (aTc) or acyl-homoserine lactone (AHL) [179]. The Lu group used two different *E. coli* strains, regulated by either aTc or AHL, which expressed and secreted CsgA either with or without a His-tail (Fig. 4.6). By growing the two strains together, the production of CsgA and $CsgA^{His}$ could be controlled by sequential pulses or different concentrations of AHL versus aTc, so that the ensuing amyloid fibrils would have different proportions of the two CsgA molecules, depending on the temporal pattern of pulses [178]. This can also be regulated in a cell-to-cell manner if the aTc-inducible strain producing CsgA also produces AHL constitutively, while the $CsgA^{His}$-producing strain is AHL-inducible. As the two strains grow and are induced by aTc, more and more AHL is produced, leading to an increased production (and higher proportion) of $CsgA^{His}$. The absolute proportion will depend on the initial ratio of cell strains. Another elegant variation is growth along two opposing aTc and AHL gradients to get mainly CsgA fibrils in one end and mainly $CsgA^{His}$ fibrils in the other. Finally, the production of an 8-CsgA concatemer with a single His tail together with a single-Csg^{His} can lead to more complex His distribution patterns. Gold nanoparticles containing a Nickel-nitrilo-triacetate group can be deposited on these fibrils in different ways, depending on this $CsgA_8{}^{His}/CsgA^{His}$ pattern, leading to either conducting biofilm (long consecutive gold wires) or shorter gold nanorods. Since the properties of gold rods are very sensitive to their size, it is possible to use the protein distribution pattern to tune rod properties with great versatility. Gold nanoparticles (AuNPs) can even be interfaced and co-assembled with quantum dots (QDs) provided these dots have a spycatcher domain and the CsgA has a SpyTag domain [178]. This is of interest because photon emission properties can be tuned by plasmon-exciton interactions between plasmonic AuNPs and fluorescent QDs [180]. As a more general perspective, the use of such orthogonal affinity tags makes it possible to construct patterned scaffolds for multi-enzyme systems. The Lu approach also illustrates how it is possible to apply useful properties of multicellular communities or organisms into materials fabrication using synthetic circuits.

4.4.2 Controlling Amyloid with Co-Factors: The Case of the Missing Calcium

Apart from these three examples of the use of *bona fide* bacterial amyloid to generate new design properties, there are also other examples of bacterial proteins which self-assemble under unusual circumstances. The **Repeats-in-toxin (RTX) sequence** is a highly conserved nine-peptide Gly- and Asp-rich repeat in the C-terminal Ca^{2+}-binding RTX domain of the *P. aeruginosa* protease ArpA. This domain is disordered in the apo-form, but binding of Ca^{2+} helps fold the RTX domain into compact β-helical structures [181], and thus regulates activity and secretion of these proteins. The Ca^{2+}-binding site provides a useful handle to modulate structure [182]. When heterologously expressed in *E.* coli, the RTX domain can be purified from inclusion bodies as a random coil structure [182]. Addition of Ca^{2+} induces soluble β-sheet structure, while the apoform assembles on the hour-scale to form not just fibrils but also, remarkably, cm^2-large highly regular and uniform sheets with flexible edges. Assembly can be arrested (but not reversed) by addition of Ca^{2+}, while EDTA induces self-assembly. Such scaffolds could incorporate soluble folded proteins as fusion proteins and thus immobilize *e.g.* enzymatic functions in 2 dimensions (with a density regulated by *e.g.* different ratios of fusion proteins and RTX-only proteins). The whole RTX domain is important for this self-assembly; a synthetic β-helical peptide containing 8 copies of the consensus nine-peptide repeats [183] (lacking the terminal capping sequences) did not bind Ca^{2+} or undergo a disorder-to-order transition like the full-length RTX domain. Surprisingly, this construct bound La^{3+} which induced self-assembly and fibrillation, turning the whole metal-handle on its head! Given that the RTX family of exoproteins is associated with virulence of many human pathogens [184], these manipulations are a good example of how to divert the original pathogen virulence functionality to novel purposes.

4.4.3 Inclusion Bodies with Tunable Porosity: Nanopills for Drug Delivery?

Bacterial **protein inclusion bodies** (IBs) are also a source of nanomaterials in their own right. IBs are mechanically stable intracellular inclusions which usually form when bacteria are forced to produce large amounts of recombinant proteins [185]. As with many other protein aggregates, they were previously held in rather low esteem and considered a "protein garbage pile", consisting of completely amorphous aggregates [186]; at best a convenient source of protein for purification and subsequent refolding. However, interest has picked up more recently, not least thanks to the Ventura group which showed that IBs can contain significant amounts of amyloid [187] which survives attack by the aggressive protease Proteinase K [188]. Though aggregated, IB proteins can still retain significant amount of native structure and activity [189], and this makes them interesting as a supply of nanostructured functional material [190]. Indeed, functional protein can be released *in vitro* from IBs under mild washing conditions [191], while IBs taken up and internalized by mammalian cells can work as nanopills to slowly release functional protein for biological effects without losing their overall structure and mechanical integrity [192]. Furthermore, it is possible to modulate the fraction of protein in the amyloid state (as opposed to a functional native state) by using *E. coli* strains lacking either the ATPase ClpA, chaperone DnaK and protease ClpP [193]. Lack of these quality control components (particularly ClpP) increases the amount of proteinase K-resistant components (*i.e.* amyloid) in the IBs. Remarkably, it is only the internal architecture of the IBs that is affected; there is no reduction in IB volume according to DLS, TEM or confocal microscopy, but just increased porosity. Thus IBs should be regarded as a structured sponge where native-like species can fill the gaps in the matrix, allowing co-existence of the same protein as amyloid and native isoforms and slow release of the native state [193]. The relative amounts of the two species can be tuned by the quality control system in a trial-and-error approach. This is a sophisticated expansion upon what is already known about mammalian peptide hormones (as detailed in Chap. 8); these can accumulate in secretory granules in the amyloid state [194] from which they can be slowly released [195], though their small size and consequent lack of a well-defined native fold probably rules out this kind of native-amyloid balance.

4.4.4 Other Amyloid Uses: From Macroscale Films to Bone Replacement and Tissue Engineering

This chapter has concentrated on bacterial amyloid which has intrinsic advantages due to an increasingly well-characterized amyloid production system and great versatility in protein manipulation and production. There are numerous examples of amyloid from non-bacterial sources that can be put to use in many different contexts which space does not allow us to describe in detail. However, some examples are worth highlighting. One approach is to take nonamyloidogenic proteins with natural cross-β structure and turn them into proteins that self-assemble to amyloid fibrils under benign conditions. The β-solenoid proteins (BSPs) have backbones that twist helically in either left- or right-handed direction to form beta-sheets with regular geometric structures (*e.g.* triangles and rectangles) with sides that are 1.5–2 nm long [196]. They avoid aggregation thanks to natural capping features or structural aberrations at the termini. These can be removed to give a seamless interface between N and C-termini of successive monomers, and can be complemented with interfacial salt bridges to stabilize monomer-monomer interactions. When applied to antifreeze BSPs from Spruce Budworm and Ryegrass, the engineered constructs aggregate rapidly and form fibrils; furthermore, kinetics are even compatible with a preformed polymerization nucleus [196]. Slightly more serendipitous approaches include the transcription factor Ultrabithorax from *D. melanogaster* which *in vitro* turns out to self-assemble at the air-water interface (*in vivo* the protein does not encounter air-water interfaces) into nanoscale fibers that further form macroscale films, sheets, ropes and capsules through self-adherence [197]; some of these structures can be lifted from the liquid surface in intact state. The protein aggregates rapidly (within a few hours), even at low (< 0.1 mg/ml) concentrations, and are extremely robust against attempts to solubilize it, though there is no direct evidence that the self-assembly involves amyloid. Finally, it is worth mentioning the milk protein β-lactoglobulin which represents a readily available cheap source of proteins. It does not fibrillate *in vivo*, but a combination of low pH and high temperatures lead to fibrils [198]. These have subsequently been combined with other materials such as hydroxyapatite [199] and silk fibroins [200] to form potential bone replacements or other hybrid materials with a large range of mechanical and physical properties (*e.g.* scaffolds of tunable porosity, akin to the previously mentioned IBs). The ultimate irony is served by Fmoc-modified peptides derived from the archetypal "bad fibril" Aβ; they self-assemble to form nanofibrils and subsequently hydrogels which are thermoreversible, non-toxic and thixotropic and show potential for the attachment and spreading of mammalian cells [201]. The authors elegantly show that gel stiffness can be tuned by modulating peptide concentration and salt concentration to drive differentiation of mesenchymal cells. This supportive property appears to be a general amyloid property, since amyloids formed by lysozyme [202] and transthyretin [203] also show promise for cell attachment and tissue engineering. With all these developments under way, there is little doubt that there will be growing interest in the use of functional amyloid for applications in both materials science and medicine in the future.

4.5 Perspectives

The focus of the present chapter has been on naturally occurring amyloid, which offers great inspiration due to its fascinating biological origin, optimized amyloid design and carefully orchestrated assembly process. It is clear that the amyloid fibril structure shows great potential as a source of structural strength in the construction of many different biomaterials. Nevertheless, applications are not limited to functional amyloid. As alluded to in the previous section, numerous different proteins which do not form amyloid under physiological conditions have been investigated as building materials. Space does not permit a detailed description, but it is worth mentioning that amyloids made from proteins as diverse as CsgA, lysozyme, β-lactoglobulin, Sup35p, gonadotropin-releasing hormone and diphenylalanine peptides have found use as nanowires for electronics [178, 204–206], biosensing [207–209], drug delivery [192, 195, 210], wound healing [211] and even artificial motors [212]. Amyloids are appealing in view of their biocompatibility, self-assembling properties, nanoscale dimensions and overall low cost (particularly when using bulk proteins such as the milk protein β-lactoglobulin [199]). Especially attractive is the use of peptide fibrils as biocompatible nano-scaffolds that form spontaneously when heated. Such peptide fibrils have already shown great promise in promoting cell adhesion, differentiation and migration [202, 213–216] and their use is likely to expand to other cell-based therapies. This may be expanded by functionalizing amyloid nanotubes or amyloid films with various ligands; i.e. enzymes [208], fluorophores [217] or adhesion moieties [218, 219] as sensors or in tissue engineering applications. Predicting how the incorporation of ligands will affect fibrillation and the final fibril structure (whether in film or tube formation) is, however, extremely challenging and even further complicated by the fact that the ligand needs to be accessible and/or active. This is obvious from the previous section about functional amyloid *in silico*, which describe the many factors that need to be considered when trying to design peptides and predict their amyloid propensity.

4.5.1 Challenges in the Development of New Amyloid-Based Biomaterials and -Medicine

In the development of novel amyloid-based biomaterials, the functional (bacterial) amyloids systems hold great promise as these have been optimized by evolution to quickly form the robust fibril structure and at the same time avoid the potentially toxicity of intermediate species like oligomers and protofibrils. Being able to produce large amount of well-characterized fibril material using bacteria as work horses is extremely appealing and should be addressed in future work. It would require that we learn even more about these promising systems – the curli and Fap systems in particular – in terms of fibrillation kinetics and how to shift between amyloid structural phases like the stiff fibrils structure described in this review, liquid crystals and especially hydrogels which show great promise in the field of regenerative medicine and injectable therapeutics [211, 220]. Mutational studies of the imperfect repeats in CsgA/FapC, leading to "perfect repeats", could also be very valuable as a potential amyloid scaffold for biomaterial development. The next challenge will then be to purify these materials – here their exceptional stability towards i.e. boiling SDS [44] should be exploited. Finally, for these materials to be used in drug delivery, bone replacement and at other human interfaces we need to be sure that toxicity aspects are addressed and avoided. This could be the fibril reversibly dissolving into not only monomers but also potentially toxic oligomers [221] or the fibrils fragmenting, resulting in free ends that might act as new growth sites for fibrillation as seen for the prion protein Sup35 [222]. However, by focusing on the mature, stable fibrils, these concerns can be largely eliminated. Furthermore, the toxic oligomer seems to be largely bypassed during fibrillation of functional amyloid as an evolutionary adaptation. However, another source of concern is that the fibrils structure itself might serve as a so-called amyloid enhancing factor when consumed. Aged rats and worms fed with curli-producing *E. coli* displayed deposits of the Parkinson's disease-relevant protein α-synuclein in both gut and brain compared to animals fed with mutant bacteria that do not produce curli [223]. In addition, mice also displayed extensive systemic pathological deposits of amyloid protein A when injected with, or fed with, amyloids extracted from commercially available foie gras [224]. While these results remain to be confirmed in additional studies, they emphasize that the amyloid fold is indeed a double-edged sword – and should be treated as such.

All in all, functional amyloids and proteins reviewed in this chapter pose great potential for manifold applications as novel biomaterials. Nevertheless, for successful applications more knowledge needs to be gathered concerning the stability, purification and toxicity of these systems.

References

1. Sawaya MR, Sambashivan S, Nelson R, Ivanova MI, Sievers SA et al (2007) Atomic structures of amyloid cross-beta spines reveal varied steric zippers. Nature 447:453–457
2. Larsen P, Nielsen JL, Otzen D, Nielsen PH (2008) Amyloid-like adhesins produced by floc-forming and filamentous bacteria in activated sludge. Appl Environ Microbiol 74:1517–1526
3. Larsen P, Nielsen JL, Dueholm MS, Wetzel R, Otzen D et al (2007) Amyloid adhesins are abundant in natural biofilms. Environ Microbiol 9:3077–3090
4. Hung C, Zhou Y, Pinkner JS, Dodson KW, Crowley JR et al (2013) Escherichia coli biofilms have an organized and complex extracellular matrix structure. MBio 4:e00645–e00613
5. Hammer ND, Schmidt JC, Chapman MR (2007) The curli nucleator protein, CsgB, contains an amyloidogenic domain that directs CsgA polymerization. Proc Natl Acad Sci U S A 104:12494–12499
6. Chapman MR, Robinson LS, Pinkner JS, Roth R, Heuser J et al (2002) Role of Escherichia coli curli operons in directing amyloid fiber formation. Science 295:851–855
7. Zhou Y, Smith D, Leong BJ, Brannstrom K, Almqvist F et al (2012) Promiscuous cross-seeding between bacterial amyloids promotes interspecies biofilms. J Biol Chem 287:35092–35103
8. Dueholm MS, Petersen SV, Sonderkaer M, Larsen P, Christiansen G et al (2010) Functional amyloid in pseudomonas. Mol Microbiol 77:1009–1020

9. Schwartz K, Syed AK, Stephenson RE, Rickard AH, Boles BR (2012) Functional amyloids composed of phenol soluble modulins stabilize Staphylococcus aureus biofilms. PLoS Pathog 8:e1002744
10. Dueholm MS, Larsen P, Finster K, Stenvang MR, Christiansen G et al (2015) The tubular sheaths encasing methanosaeta thermophila filaments are functional amyloids. J Biol Chem 290:20590–20600
11. Romero D, Aguilar C, Losick R, Kolter R (2010) Amyloid fibers provide structural integrity to Bacillus subtilis biofilms. Proc Natl Acad Sci U S A 107:2230–2234
12. White AP, Collinson SK, Banser PA, Gibson DL, Paetzel M et al (2001) Structure and characterization of AgfB from Salmonella enteritidis thin aggregative fimbriae. J Mol Biol 311:735–749
13. Gophna U, Barlev M, Seijffers R, Oelschlager TA, Hacker J et al (2001) Curli fibers mediate internalization of Escherichia coli by eukaryotic cells. Infect Immun 69:2659–2665
14. Barnhart MM, Chapman MR (2006) Curli biogenesis and function. Annu Rev Microbiol 60:131–147
15. Austin JW, Sanders G, Kay WW, Collinson SK (1998) Thin aggregative fimbriae enhance Salmonella enteritidis biofilm formation. FEMS Microbiol Lett 162:295–301
16. Zogaj X, Bokranz W, Nimtz M, Romling U (2003) Production of cellulose and Curli fimbriae by members of the family Enterobacteriaceae isolated from the human gastrointestinal tract. Infect Immun 71:4151–4158
17. Hammar M, Arnqvist A, Bian Z, Olsén A, Normark S (1995) Expression of two *csg* operons is required for production of fibronectin- and Congo red-binding curli polymers in *Escherichia coli* K-12. Mol Microbiol 18:661–670
18. Blanco LP, Evans ML, Smith DR, Badtke MP, Chapman MR (2012) Diversity, biogenesis and function of microbial amyloids. Trends Microbiol 20:66–73
19. Romling U, Sierralta WD, Eriksson K, Normark S (1998) Multicellular and aggregative behaviour of Salmonella typhimurium strains is controlled by mutations in the agfD promoter. Mol Microbiol 28:249–264
20. Collinson SK, Emody L, Muller KH, Trust TJ, Kay WW (1991) Purification and characterization of thin, aggregative fimbriae from Salmonella enteritidis. J Bacteriol 173:4773–4781
21. Dueholm MS, Albertsen M, Otzen D, Nielsen PH (2012) Curli functional amyloid systems are phylogenetically widespread and display large diversity in operon and protein structure. PLoS One 7:e51274
22. Jordal PB, Dueholm MS, Larsen P, Petersen SV, Enghild JJ et al (2009) Widespread abundance of functional bacterial amyloid in mycolata and other gram-positive bacteria. Appl Environ Microbiol 75:4101–4110
23. Otzen D (2010) Functional amyloid – turning swords into plowshares. Prion 4:256–264
24. Winner B, Jappelli R, Maji SK, Desplats PA, Boyer L et al (2011) In vivo demonstration that alpha-synuclein oligomers are toxic. Proc Natl Acad Sci U S A 108:4194–4199
25. Shahnawaz M, Soto C (2012) Microcin amyloid fibrils a are reservoir of toxic oligomeric species. J Biol Chem 287:11665–11676
26. He Y, Zheng MM, Ma Y, Han XJ, Ma XQ et al (2012) Soluble oligomers and fibrillar species of amyloid beta-peptide differentially affect cognitive functions and hippocampal inflammatory response. Biochem Biophys Res Commun 429:125–130
27. Kayed R, Head E, Thompson JL, McIntire TM, Milton SC, Cotman CW, Glabe CG (2003) Common structure of soluble amyloid oligomers implies common mechanism of pathogenesis. Science 300:486–489
28. Kayed R, Sokolov Y, Edmonds B, McIntire TM, Milton SC et al (2004) Permeabilization of lipid bilayers is a common conformation-dependent activity of soluble amyloid oligomers in protein misfolding diseases. J Biol Chem 279:46363–46366
29. Evans ML, Chorell E, Taylor JD, Aden J, Gotheson A et al (2015) The bacterial curli system possesses a potent and selective inhibitor of amyloid formation. Mol Cell 57:445–455
30. Nenninger AA, Robinson LS, Hammer ND, Epstein EA, Badtke MP et al (2011) CsgE is a curli secretion specificity factor that prevents amyloid fibre aggregation. Mol Microbiol 81:486–499

31. Wright CF, Teichmann SA, Clarke J, Dobson CM (2005) The importance of sequence diversity in the aggregation and evolution of proteins. Nature 438:878–881
32. Ross ED, Minton A, Wickner RB (2005) Prion domains: sequences, structures and interactions. Nat Cell Biol 7:1039–1044
33. Wang X, Chapman MR (2008) Sequence determinants of bacterial amyloid formation. J Mol Biol 380:570–580
34. Wang X, Smith DR, Jones JW, Chapman MR (2007) In vitro polymerization of a functional Escherichia coli amyloid protein. J Biol Chem 282:3713–3719
35. Hammer ND, McGuffie BA, Zhou Y, Badtke MP, Reinke AA et al (2012) The C-terminal repeating units of CsgB direct bacterial functional amyloid nucleation. J Mol Biol 422:376–389
36. Hammar M, Bian Z, Normark S (1996) Nucleator-dependent intercellular assembly of adhesive curli organelles in Escherichia coli. Proc Natl Acad Sci U S A 93:6562–6566
37. Dueholm MS, Nielsen SB, Hein KL, Nissen P, Chapman M et al (2011) Fibrillation of the major curli subunit CsgA under a wide range of conditions implies a robust design of aggregation. Biochemistry 50:8281–8290
38. Shewmaker F, McGlinchey RP, Thurber KR, McPhie P, Dyda F et al (2009) The functional curli amyloid is not based on in-register parallel beta-sheet structure. J Biol Chem 284:25065–25076
39. Rochet JC, Lansbury PT Jr (2000) Amyloid fibrillogenesis: themes and variations. Curr Opin Struct Biol 10:60–68
40. Taylor JD, Hawthorne WJ, Lo J, Dear A, Jain N et al (2016) Electrostatically-guided inhibition of Curli amyloid nucleation by the CsgC-like family of chaperones. Sci Rep 6:24656
41. Loferer H, Hammar M, Normark S (1997) Availability of the fibre subunit CsgA and the nucleator protein CsgB during assembly of fibronectin-binding curli is limited by the intracellular concentration of the novel lipoprotein CsgG. Mol Microbiol 26:11–23
42. Tian P, Boomsma W, Wang Y, Otzen DE, Jensen MH et al (2015) Structure of a functional amyloid protein subunit computed using sequence variation. J Am Chem Soc 137:22–25
43. Robinson LS, Ashman EM, Hultgren SJ, Chapman MR (2006) Secretion of curli fibre subunits is mediated by the outer membrane-localized CsgG protein. Mol Microbiol 59:870–881
44. Epstein EA, Reizian MA, Chapman MR (2009) Spatial clustering of the curlin secretion lipoprotein requires curli fiber assembly. J Bacteriol 191:608–615
45. Goyal P, Krasteva PV, Van Gerven N, Gubellini F, Van den Broeck I et al (2014) Structural and mechanistic insights into the bacterial amyloid secretion channel CsgG. Nature 516:250–253
46. Cao B, Zhao Y, Kou Y, Ni D, Zhang XC et al (2014) Structure of the nonameric bacterial amyloid secretion channel. Proc Natl Acad Sci U S A 111:E5439–E5444
47. Geibel S, Procko E, Hultgren SJ, Baker D, Waksman G (2013) Structural and energetic basis of folded-protein transport by the FimD usher. Nature 496:243–246
48. Van Gerven N, Klein RD, Hultgren SJ, Remaut H (2015) Bacterial amyloid formation: structural insights into curli biogensis. Trends Microbiol 23:693–706
49. Van Gerven N, Goyal P, Vandenbussche G, De Kerpel M, Jonckheere W et al (2014) Secretion and functional display of fusion proteins through the curli biogenesis pathway. Mol Microbiol 91:1022–1035
50. Sivanathan V, Hochschild A (2012) Generating extracellular amyloid aggregates using E. coli cells. Genes Dev 26:2659–2667
51. Andersson EK, Bengtsson C, Evans ML, Chorell E, Sellstedt M et al (2013) Modulation of curli assembly and pellicle biofilm formation by chemical and protein chaperones. Chem Biol 20:1245–1254
52. Nenninger AA, Robinson LS, Hultgren SJ (2009) Localized and efficient curli nucleation requires the chaperone-like amyloid assembly protein CsgF. Proc Natl Acad Sci U S A 106:900–905

53. Dueholm MS, Otzen D, Nielsen PH (2013) Evolutionary insight into the functional amyloids of the pseudomonads. PLoS One 8:e76630
54. Dueholm MS, Sondergaard MT, Nilsson M, Christiansen G, Stensballe A et al (2013) Expression of Fap amyloids in Pseudomonas aeruginosa, P. fluorescens, and P. putida results in aggregation and increased biofilm formation. Microbiology 2:365–382
55. Zeng G, Vad BS, Dueholm MS, Christiansen G, Nilsson M et al (2015) Functional bacterial amyloid increases Pseudomonas biofilm hydrophobicity and stiffness. Front Microbiol 6:1099
56. Herbst FA, Sondergaard MT, Kjeldal H, Stensballe A, Nielsen PH et al (2015) Major proteomic changes associated with amyloid-induced biofilm formation in Pseudomonas aeruginosa PAO1. J Proteome Res 14:72–81
57. Wiehlmann L, Munder A, Adams T, Juhas M, Kolmar H et al (2007) Functional genomics of Pseudomonas aeruginosa to identify habitat-specific determinants of pathogenicity. Int J Med Microbiol 297:615–623
58. Olsen A, Jonsson A, Normark S (1989) Fibronectin binding mediated by a novel class of surface organelles on Escherichia coli. Nature 338:652–655
59. Olsen A, Arnqvist A, Hammar M, Sukupolvi S, Normark S (1993) The RpoS sigma factor relieves H-NS-mediated transcriptional repression of csgA, the subunit gene of fibronectin-binding curli in Escherichia coli. Mol Microbiol 7:523–536
60. Collinson SK, Doig PC, Doran JL, Clouthier S, Trust TJ et al (1993) Thin, aggregative fimbriae mediate binding of Salmonella enteritidis to fibronectin. J Bacteriol 175:12–18
61. Collinson SK, Parker JM, Hodges RS, Kay WW (1999) Structural predictions of AgfA, the insoluble fimbrial subunit of Salmonella thin aggregative fimbriae. J Mol Biol 290:741–756
62. Gazit E (2002) A possible role for pi-stacking in the self-assembly of amyloid fibrils. FASEB J 16:77–83
63. Azriel R, Gazit E (2001) Analysis of the minimal amyloid-forming fragment of the islet amyloid polypeptide. An experimental support for the key role of the phenylalanine residue in amyloid formation. J Biol Chem 276:34156–34161
64. Dueholm M, Nielsen PH (2016) Amyloids – a neglected child of the slime. In: Flemming H-C, Neu T, Wingender J (eds) The perfect slime. IWA Publishing, London
65. Gibson DL, White AP, Rajotte CM, Kay WW (2007) AgfC and AgfE facilitate extracellular thin aggregative fimbriae synthesis in Salmonella enteritidis. Microbiology 153:1131–1140
66. Manara A, DalCorso G, Baliardini C, Farinati S, Cecconi D et al (2012) Pseudomonas putida response to cadmium: changes in membrane and cytosolic proteomes. J Proteome Res 11:4169–4179
67. Lewenza S, Gardy JL, Brinkman FS, Hancock RE (2005) Genome-wide identification of Pseudomonas aeruginosa exported proteins using a consensus computational strategy combined with a laboratory-based PhoA fusion screen. Genome Res 15:321–329
68. Bian Z, Normark S (1997) Nucleator function of CsgB for the assembly of adhesive surface organelles in Escherichia coli. EMBO J 16:5827–5836
69. Taylor JD, Zhou Y, Salgado PS, Patwardhan A, McGuffie M et al (2011) Atomic resolution insights into curli fiber biogenesis. Structure 19:1307–1316
70. Collinson SK, Clouthier SC, Doran JL, Banser PA, Kay WW (1996) Salmonella enteritidis agfBAC operon encoding thin, aggregative fimbriae. J Bacteriol 178:662–667
71. Dueholm MS, Søndergaard MT, Nilsson M, Christiansen G, Stensballe A, Overgaard MT, Givskov M, Tolker-Nielsen T, Otzen DE, Nielsen PH (2013) Expression of Fap amyloids in *Pseudomonas aeruginosa*, *P. fluorescens*, and *P. putida* results in aggregation and increased biofilm formation. MicrobiologyOpen 2:365–382
72. Seviour T, Hansen SH, Yang L, Yau YH, Wang VB et al (2015) Functional amyloids keep quorum-sensing molecules in check. J Biol Chem 290:6457–6469
73. Chai L, Romero D, Kayatekin C, Akabayov B, Vlamakis H et al (2013) Isolation, characterization, and aggregation of a structured bacterial matrix precursor. J Biol Chem 288:17559–17568

74. Terra R, Stanley-Wall NR, Cao G, Lazazzera BA (2012) Identification of Bacillus subtilis SipW as a bifunctional signal peptidase that controls surface-adhered biofilm formation. J Bacteriol 194:2781–2790
75. Romero D, Vlamakis H, Losick R, Kolter R (2011) An accessory protein required for anchoring and assembly of amyloid fibres in B. subtilis biofilms. Mol Microbiol 80:1155–1168
76. Stover AG, Driks A (1999) Secretion, localization, and antibacterial activity of TasA, a Bacillus subtilis spore-associated protein. J Bacteriol 181:1664–1672
77. Oh J, Kim JG, Jeon E, Yoo CH, Moon JS et al (2007) Amyloidogenesis of type III-dependent harpins from plant pathogenic bacteria. J Biol Chem 282:13601–13609
78. Kim JG, Park BK, Yoo CH, Jeon E, Oh J et al (2003) Characterization of the Xanthomonas axonopodis pv. glycines Hrp Pathogenicity Island. J Bacteriol 185:3155–3166
79. de Jong W, Wosten HA, Dijkhuizen L, Claessen D (2009) Attachment of Streptomyces coelicolor is mediated by amyloidal fimbriae that are anchored to the cell surface via cellulose. Mol Microbiol 73:1128–1140
80. Claessen D, Rink R, de Jong W, Siebring J, de Vreugd P et al (2003) A novel class of secreted hydrophobic proteins is involved in aerial hyphae formation in Streptomyces coelicolor by forming amyloid-like fibrils. Genes Dev 17:1714–1726
81. Sawyer EB, Claessen D, Haas M, Hurgobin B, Gras SL (2011) The assembly of individual chaplin peptides from Streptomyces coelicolor into functional amyloid fibrils. PLoS One 6:e18839
82. Bokhove M, Claessen D, de Jong W, Dijkhuizen L, Boekema EJ et al (2013) Chaplins of Streptomyces coelicolor self-assemble into two distinct functional amyloids. J Struct Biol 184:301–309
83. Elliot MA, Karoonuthaisiri N, Huang J, Bibb MJ, Cohen SN et al (2003) The chaplins: a family of hydrophobic cell-surface proteins involved in aerial mycelium formation in Streptomyces coelicolor. Genes Dev 17:1727–1740
84. Di Berardo C, Capstick DS, Bibb MJ, Findlay KC, Buttner MJ et al (2008) Function and redundancy of the chaplin cell surface proteins in aerial hypha formation, rodlet assembly, and viability in Streptomyces coelicolor. J Bacteriol 190:5879–5889
85. Alteri CJ, Xicohtencatl-Cortes J, Hess S, Caballero-Olin G, Giron JA et al (2007) Mycobacterium tuberculosis produces pili during human infection. Proc Natl Acad Sci U S A 104:5145–5150
86. Ramsugit S, Guma S, Pillay B, Jain P, Larsen MH et al (2013) Pili contribute to biofilm formation in vitro in Mycobacterium tuberculosis. Antonie Van Leeuwenhoek 104:725–735
87. Velayati AA, Farnia P, Masjedi MR (2012) Pili in totally drug resistant Mycobacterium Tuberculosis (TDR-TB). Int J Mycobacteriol 1:57–58
88. Oli MW, Otoo HN, Crowley PJ, Heim KP, Nascimento MM et al (2012) Functional amyloid formation by Streptococcus mutans. Microbiology 158:2903–2916
89. Bieler S, Estrada L, Lagos R, Baeza M, Castilla J et al (2005) Amyloid formation modulates the biological activity of a bacterial protein. J Biol Chem 280:26880–26885
90. Lagos R, Villanueva JE, Monasterio O (1999) Identification and properties of the genes encoding microcin E492 and its immunity protein. J Bacteriol 181:212–217
91. de Lorenzo V (1984) Isolation and characterization of microcin E492 from Klebsiella pneumoniae. Arch Microbiol 139:72–75
92. de Lorenzo V, Martinez JL, Asensio C (1984) Microcin-mediated interactions between Klebsiella pneumoniae and Escherichia coli strains. J Gen Microbiol 130:391–400
93. Gekara NO, Jacobs T, Chakraborty T, Weiss S (2005) The cholesterol-dependent cytolysin listeriolysin O aggregates rafts via oligomerization. Cell Microbiol 7:1345–1356
94. Bavdek A, Kostanjšek R, Antonini V, Lakey JH, Dalla Serra M et al (2012) pH dependence of listeriolysin O aggregation and pore-forming ability. FEBS J 279:126–141
95. Vazquez-Boland JA, Dominguez-Bernal G, Gonzalez-Zorn B, Kreft J, Goebel W (2001) Pathogenicity islands and virulence evolution in Listeria. Microbes Infect 3:571–584

96. Vuong C, Durr M, Carmody AB, Peschel A, Klebanoff SJ et al (2004) Regulated expression of pathogen-associated molecular pattern molecules in Staphylococcus epidermidis: quorum-sensing determines pro-inflammatory capacity and production of phenol-soluble modulins. Cell Microbiol 6:753–759
97. Periasamy S, Joo HS, Duong AC, Bach TH, Tan VY et al (2012) How Staphylococcus aureus biofilms develop their characteristic structure. Proc Natl Acad Sci U S A 109:1281–1286
98. Moreno-Del Alamo M, de la Espina SM, Fernandez-Tresguerres ME, Giraldo R (2015) Pre-amyloid oligomers of the proteotoxic RepA-WH1 prionoid assemble at the bacterial nucleoid. Sci Rep 5:14669
99. Molina-Garcia L, Gasset-Rosa F, Moreno-Del Alamo M, Fernandez-Tresguerres ME, Moreno-Diaz de la Espina S et al (2016) Functional amyloids as inhibitors of plasmid DNA replication. Sci Rep 6:25425
100. Diehl A, Roske Y, Ball L, Chowdhury A, Hiller M et al (2018) Structural changes of TasA in biofilm formation of Bacillus subtilis. Proc Natl Acad Sci U S A 115:3237–3242
101. Nagorska K, Ostrowski A, Hinc K, Holland IB, Obuchowski M (2010) Importance of eps genes from Bacillus subtilis in biofilm formation and swarming. J Appl Genet 51:369–381
102. Dueholm MS, Petersen SV, Sønderkær M, Larsen P, Christiansen G et al (2010) Functional amyloid in *Pseudomonas*. Mol Microbiol 77:1009–1020
103. Dueholm MS, Larsen P, Finster K, Stenvang MR, Christiansen G et al (2015) The tubular sheaths encasing Methanosaeta thermophila are functional amyloids. J Biol Chem 290:20590–20600
104. Oh J, Kim J-G, Jeon E, Yo C-H, Moon JS et al (2007) Amyloidogenesis of type III-dependent harpins from plant pathogenic bacteria. J Biol Chem 282:13601–13609
105. Volles MJ, Lee SJ, Rochet JC, Shtilerman MD, Ding TT et al (2001) Vesicle permeabilization by protofibrillar α-synuclein: implications for the pathogenesis and treatment of Parkinson's disease. Biochemistry 40:7812–7819
106. Claessen D, Rink R, de Jong W, Siebring J, de Vreughd P et al (2003) A novel class of secreted hydrophobic proteins is involved in aerial hyphae formation in *Streptomyces coelicolor* by forming amyloid-like fibrils. Genes Dev 17:1714–1726
107. Elliot MA, Karoonuthaisir N, Huang J, Bibb MJ, Cohen SN et al (2003) The chaplins: a family of hydrophobic cell-surface proteins involved in aerial mycelium formation in *Streptomyces coelicolor*. Genes Dev 17:1727–1740
108. Schwartz K, Ganesan M, Payne DE, Solomon MJ, Boles BR (2015) Extracellular DNA facilitates the formation of functional amyloids in Staphylococcus aureus biofilms. Mol Microbiol 99(1):123–134
109. Maurer-Stroh S, Debulpaep M, Kuemmerer N, de la Paz ML, Martins IC et al (2010) Exploring the sequence determinants of amyloid structure using position-specific scoring matrices (vol 7, pg 237, 2010). Nat Methods 7:855–855
110. Emily M, Talvas A, Delamarche C (2013) MetAmyl: a METa-predictor for AMYLoid proteins. PLoS One 8:e79722
111. Ahmed AB, Znassi N, Chateau MT, Kajava AV (2015) A structure-based approach to predict predisposition to amyloidosis. Alzheimers Dement 11:681–690
112. Stanislawski J, Kotulska M, Unold O (2013) Machine learning methods can replace 3D profile method in classification of amyloidogenic hexapeptides. BMC Bioinformatics 14:21
113. Walsh I, Seno F, Tosatto SCE, Trovato A (2014) PASTA 2.0: an improved server for protein aggregation prediction. Nucleic Acids Res 42:W301–W307
114. Conchillo-Sole O, de Groot NS, Aviles FX, Vendrell J, Daura X et al (2007) AGGRESCAN: a server for the prediction and evaluation of "hot spots" of aggregation in polypeptides. BMC Bioinformatics 8:65
115. Garbuzynskiy SO, Lobanov MY, Galzitskaya OV (2010) FoldAmyloid: a method of prediction of amyloidogenic regions from protein sequence. Bioinformatics 26:326–332
116. Tartaglia GG, Vendruscolo M (2008) The Zyggregator method for predicting protein aggregation propensities. Chem Soc Rev 37:1395–1401

117. Fernandez-Escamilla AM, Rousseau F, Schymkowitz J, Serrano L (2004) Prediction of sequence-dependent and mutational effects on the aggregation of peptides and proteins. Nat Biotechnol 22:1302–1306
118. Bryan AW, Menke M, Cowen LJ, Lindquist SL, Berger B (2009) BETASCAN: probable beta-amyloids identified by pairwise probabilistic analysis. PLoS Comput Biol 5:e1000333
119. Thompson MJ, Sievers SA, Karanicolas J, Ivanova MI, Baker D et al (2006) The 3D profile method for identifying fibril-forming segments of proteins. Proc Natl Acad Sci U S A 103:4074–4078
120. Kim C, Choi J, Lee SJ, Welsh WJ, Yoon S (2009) NetCSSP: web application for predicting chameleon sequences and amyloid fibril formation. Nucleic Acids Res 37:W469–W473
121. Hamodrakas SJ (2011) Protein aggregation and amyloid fibril formation prediction software from primary sequence: towards controlling the formation of bacterial inclusion bodies. FEBS J 278:2428–2435
122. Ahmed AB, Kajava AV (2013) Breaking the amyloidogenicity code: methods to predict amyloids from amino acid sequence. FEBS Lett 587:1089–1095
123. Hamodrakas SJ (1988) A protein secondary structure prediction scheme for the Ibm Pc and compatibles. Comput Appl Biosci 4:473–477
124. Zibaee S, Makin OS, Goedert M, Serpell LC (2007) A simple algorithm locates beta-strands in the amyloid fibril core of alpha-synuclein, a beta, and tau using the amino acid sequence alone. Protein Sci 16:1242–1242
125. Dubay KF, Pawar AP, Chiti F, Zurdo J, Dobson CM et al (2004) Prediction of the absolute aggregation rates of amyloidogenic polypeptide chains. J Mol Biol 341:1317–1326
126. Galzitskaya OV, Garbuzynskiy SO, Lobanov MY (2006) Prediction of amyloidogenic and disordered regions in protein chains. PLoS Comput Biol 2:1639–1648
127. Galzitskaya OV, Garbuzynskiy SO, Lobanov MY (2007) Expected packing density allows prediction of both amyloidogenic and disordered regions in protein chains. J Phys Condens Matter 19:285225
128. Trovato A, Chiti F, Maritan A, Seno F (2006) Insight into the structure of amyloid fibrils from the analysis of globular proteins. PLoS Comput Biol 2:1608–1618
129. Zhang ZQ, Chen H, Lai LH (2007) Identification of amyloid fibril-forming segments based on structure and residue-based statistical potential. Bioinformatics 23:2218–2225
130. Thangakani AM, Kumar S, Nagarajan R, Velmurugan D, Gromiha MM (2014) GAP: towards almost 100 percent prediction for beta-strand-mediated aggregating peptides with distinct morphologies. Bioinformatics 30:1983–1990
131. O'Donnell CW, Waldispuhl J, Lis M, Halfmann R, Devadas S et al (2011) A method for probing the mutational landscape of amyloid structure. Bioinformatics 27:I34–I42
132. de la Paz ML, Serrano L (2004) Sequence determinants of amyloid fibril formation. Proc Natl Acad Sci U S A 101:87–92
133. de Groot NS, Aviles FX, Vendrell J, Ventura S (2006) Mutagenesis of the central hydrophobic cluster in A beta 42 Alzheimer's pepticle – side-chain properties correlate with aggregation propensities. FEBS J 273:658–668
134. de Groot NS, Pallares I, Aviles FX, Vendrell J, Ventura S (2005) Prediction of "hot spots" of aggregation in disease-linked polypeptides. BMC Struct Biol 5:1–15
135. Tian J, Wu NF, Guo J, Fan YL (2009) Prediction of amyloid fibril-forming segments based on a support vector machine. BMC Bioinformatics 10:S45
136. David MPC, Concepcion GP, Padlan EA (2010) Using simple artificial intelligence methods for predicting amyloidogenesis in antibodies. BMC Bioinformatics 11:1–13
137. Nair SSK, Reddy NVS, Hareesha KS (2011) Exploiting heterogeneous features to improve in silico prediction of peptide status – amyloidogenic or non-amyloidogenic. BMC Bioinformatics 12:S21
138. Gasior P, Kotulska M (2014) FISH Amyloid – a new method for finding amyloidogenic segments in proteins based on site specific co-occurence of aminoacids. BMC Bioinformatics 15:1–8

139. Famlia C, Dennison SR, Quintas A, Phoenix DA (2015) Prediction of peptide and protein propensity for amyloid formation. PLoS One 10:e0134679
140. Tsolis AC, Papandreou NC, Iconomidou VA, Hamodrakas SJ (2013) A consensus method for the prediction of 'Aggregation-Prone' peptides in globular proteins. PLoS One 8:e54175
141. Frousios KK, Iconomidou VA, Karletidi CM, Hamodrakas SJ (2009) Amyloidogenic determinants are usually not buried. BMC Struct Biol 9:44
142. Kallberg Y, Gustafsson M, Persson B, Thyberg J, Johansson J (2001) Prediction of amyloid fibril-forming proteins. J Biol Chem 276:12945–12950
143. Yoon S, Welsh WJ (2004) Detecting hidden sequence propensity for amyloid fibril formation. Protein Sci 13:2149–2160
144. Kawashima S, Ogata H, Kanehisa M (1999) AAindex: amino acid index database. Nucleic Acids Res 27:368–369
145. Bowie JU, Eisenberg D (1993) Inverted protein-structure prediction. Curr Opin Struct Biol 3:437–444
146. Zheng WH, Schafer NP, Wolynes PG (2013) Frustration in the energy landscapes of multidomain protein misfolding. Proc Natl Acad Sci U S A 110:1680–1685
147. Zheng WH, Schafer NP, Wolynes PG (2013) Free energy landscapes for initiation and branching of protein aggregation. Proc Natl Acad Sci U S A 110:20515–20520
148. Jain AK, Duin RPW, Mao JC (2000) Statistical pattern recognition: a review. IEEE Trans Pattern Anal Mach Intell 22:4–37
149. Beerten J, Van Durme J, Gallardo R, Capriotti E, Serpell L et al (2015) WALTZ-DB: a benchmark database of amyloidogenic hexapeptides. Bioinformatics 31:1698–1700
150. Lembre P, Vendrely C, Di Martino P (2014) Identification of an Amyloidogenic peptide from the bap protein of Staphylococcus epidermidis. Protein Pept Lett 21:75–79
151. Bezsonov EE, Groenning M, Galzitskaya OV, Gorkovskii AA, Semisotnov GV et al (2013) Amyloidogenic peptides of yeast cell wall glucantransferase Bgl2p as a model for the investigation of its pH-dependent fibril formation. Prion 7:175–184
152. Hardy GG, Allen RC, Toh E, Long M, Brown PJB et al (2010) A localized multimeric anchor attaches the Caulobacter holdfast to the cell pole. Mol Microbiol 76:409–427
153. Iglesias V, de Groot NS, Ventura S (2015) Computational analysis of candidate prion-like proteins in bacteria and their role. Front Microbiol 6:1123
154. Louros NN, Bolas GMP, Tsiolaki PL, Hamodrakas SJ, Iconomidou VA (2016) Intrinsic aggregation propensity of the CsgB nucleator protein is crucial for curli fiber formation. J Struct Biol 195:179–189
155. Romero D, Vlamakis H, Losick R, Kolter R (2014) Functional analysis of the accessory protein TapA in Bacillus subtilis amyloid fiber assembly. J Bacteriol 196:1505–1513
156. Ivanova N, Sikorski J, Jando M, Munk C, Lapidus A et al (2010) Complete genome sequence of Geodermatophilus obscurus type strain (G-20(T)). Stand Genomic Sci 2:158–167
157. Althani AA, Marei HE, Hamdi WS, Nasrallah GK, El Zowalaty ME et al (2016) Human microbiome and its association with health and diseases. J Cell Physiol 231:1688–1694
158. Eloe-Fadrosh EA, Paez-Espino D, Jarett J, Dunfield PF, Hedlund BP et al (2016) Global metagenomic survey reveals a new bacterial candidate phylum in geothermal springs. Nat Commun 7:10476
159. Alberti S, Halfmann R, King O, Kapila A, Lindquist S (2009) A systematic survey identifies prions and illuminates sequence features of prionogenic proteins. Cell 137:146–158
160. Hobley L, Harkins C, MacPhee CE, Stanley-Wall NR (2015) Giving structure to the biofilm matrix: an overview of individual strategies and emerging common themes. FEMS Microbiol Rev 39:649–669
161. Louros NN, Baltoumas FA, Hamodrakas SJ, Iconomidou VA (2016) A beta-solenoid model of the Pmel17 repeat domain: insights to the formation of functional amyloid fibrils. J Comput Aided Mol Des 30:153–164
162. De Vries SJ, van Dijk M, Bonvin AMJJ (2010) The HADDOCK web server for data-driven biomolecular docking. Nat Protoc 5:883–897

163. Tian PF, Lindorff-Larsen K, Boomsma W, Jensen MH, Otzen DE (2016) A Monte Carlo Study of the early steps of functional amyloid formation. PLoS One 11:e0146096
164. Chen MC, Zheng WH, Wolynes PG (2016) Energy landscapes of a mechanical prion and their implications for the molecular mechanism of long-term memory. Proc Natl Acad Sci U S A 113:5006–5011
165. Si K, Lindquist S, Kandel ER (2003) A neuronal isoform of the Aplysia CPEB has prion-like properties. Cell 115:879–891
166. Wang F, Wang X, Yuan CG, Ma J (2010) Generating a prion with bacterially expressed recombinant prion protein. Science 327:1132–1135
167. Knight JD, Miranker AD (2004) Phospholipid catalysis of diabetic amyloid assembly. J Mol Biol 341:1175–1187
168. Derkatch IL, Bradley ME, Zhou P, Chernoff YO, Liebman SW (1997) Genetic and environmental factors affecting the de novo appearance of the [PSI+] prion in Saccharomyces cerevisiae. Genetics 147:507–519
169. Sivanathan V, Hochschild A (2013) A bacterial export system for generating extracellular amyloid aggregates. Nat Protoc 8:1381–1390
170. Khurana R, Uversky VN, Nielsen L, Fink AL (2001) Is Congo Red an amyloid-specific dye? J Biol Chem 276:22715–22721
171. Nguyen PQ, Botyanszki Z, Tay PK, Joshi NS (2014) Programmable biofilm-based materials from engineered curli nanofibres. Nat Commun 5:4945
172. Zakeri B, Fierer JO, Celik E, Chittock EC, Schwarz-Linek U et al (2012) Peptide tag forming a rapid covalent bond to a protein, through engineering a bacterial adhesin. Proc Natl Acad Sci U S A 109:E690–E697
173. Teather RM, Wood PJ (1982) Use of Congo red-polysaccharide interactions in enumeration and characterization of cellulolytic bacteria from the bovine rumen. Appl Environ Microbiol 43:777–780
174. Vidal O, Longin R, Prigent-Combaret C, Dorel C, Hooreman M et al (1998) Isolation of an Escherichia coli K-12 mutant strain able to form biofilms on inert surfaces: involvement of a new ompR allele that increases curli expression. J Bacteriol 180:2442–2449
175. Zhong C, Gurry T, Cheng AA, Downey J, Deng Z et al (2014) Strong underwater adhesives made by self-assembling multi-protein nanofibres. Nat Nanotechnol 9:858–866
176. Brubaker CE, Messersmith PB (2012) The present and future of biologically inspired adhesive interfaces and materials. Langmuir 28:2200–2205
177. Barlow DE, Dickinson GH, Orihuela B, Kulp JL, Rittschof D et al (2010) Characterization of the adhesive plaque of the barnacle *Balanus amphitrite:* amyloid-like nanofirils are a major component. Langmuir 26:6549–6556
178. Chen AY, Deng Z, Billings AN, Seker UO, Lu MY et al (2014) Synthesis and patterning of tunable multiscale materials with engineered cells. Nat Mater 13:515–523
179. Callura JM, Cantor CR, Collins JJ (2012) Genetic switchboard for synthetic biology applications. Proc Natl Acad Sci U S A 109:5850–5855
180. Polman A, Atwater HA (2012) Photonic design principles for ultrahigh efficiency photovoltaics. Nat Mater 11:174–177
181. Zhang L, Conway JF, Thibodeau PH (2012) Calcium-induced folding and stabilization of the Pseudomonas aeruginosa alkaline protease. J Biol Chem 287:4311–4322
182. Zhang L, Franks J, Stolz DB, Conway JF, Thibodeau PH (2014) Inducible polymerization and two-dimensional assembly of the repeats-in-toxin (RTX) domain from the Pseudomonas aeruginosa alkaline protease. Biochemistry 53:6452–6462
183. Lilie H, Haehnel W, Rudolph R, Baumann U (2000) Folding of a synthetic parallel beta-roll protein. FEBS Lett 470:173–177
184. Welch RA, Forestier C, Lobo A, Pellett S, Thomas W Jr et al (1992) The synthesis and function of the Escherichia coli hemolysin and related RTX exotoxins. FEMS Microbiol Immunol 5:29–36
185. Villaverde A, Carrio MM (2003) Protein aggregation in recombinant bacteria: biological role of inclusion bodies. Biotechnol Lett 25:1385–1395

186. Marston FAO (1986) The purification of eukaryotic polypeptides synthesized in *Escherichia coli*. Biochem J 240:1–12
187. Carrió M, González-Montalbán N, Vera A, Villaverde A, Ventura S (2005) Amyloid-like properties of bacterial inclusion bodies. J Mol Biol 347:1025–1037
188. Morell M, Bravo R, Espargaro A, Sisquella X, Aviles FX et al (2008) Inclusion bodies: specificity in their aggregation process and amyloid-like structure. Biochim Biophys Acta 1783:1815–1825
189. Garcia-Fruitos E, Gonzalez-Montalban N, Morell M, Vera A, Ferraz RM et al (2005) Aggregation as bacterial inclusion bodies does not imply inactivation of enzymes and fluorescent proteins. Microb Cell Factories 4:27
190. Mitraki A (2010) Protein aggregation from inclusion bodies to amyloid and biomaterials. Adv Protein Chem Struct Biol 79:89–125
191. Peternel S, Grdadolnik J, Gaberc-Porekar V, Komel R (2008) Engineering inclusion bodies for non denaturing extraction of functional proteins. Microb Cell Factories 7:34
192. Vazquez E, Corchero JL, Burgueno JF, Seras-Franzoso J, Kosoy A et al (2012) Functional inclusion bodies produced in bacteria as naturally occurring nanopills for advanced cell therapies. Adv Mater 24:1742–1747
193. Cano-Garrido O, Rodriguez-Carmona E, Diez-Gil C, Vazquez E, Elizondo E et al (2013) Supramolecular organization of protein-releasing functional amyloids solved in bacterial inclusion bodies. Acta Biomater 9:6134–6142
194. Maji SK, Perrin MH, Sawaya MR, Jessberger S, Vadodaria K et al (2009) Functional amyloids as natural storage of peptide hormones in pituitary secretory granules. Science 325:328–332
195. Maji SK, Schubert D, Rivier C, Lee S, Rivier JE et al (2008) Amyloid as a depot for the formulation of long-acting drugs. PLoS Biol 6:e17
196. Peralta MD, Karsai A, Ngo A, Sierra C, Fong KT et al (2015) Engineering amyloid fibrils from beta-solenoid proteins for biomaterials applications. ACS Nano 9:449–463
197. Greer AM, Huang Z, Oriakhi A, Lu Y, Lou J et al (2009) The Drosophila transcription factor ultrabithorax self-assembles into protein-based biomaterials with multiple morphologies. Biomacromolecules 10:829–837
198. Gosal WS, Clark AH, Ross-Murphy SB (2004) Fibrillar beta-lactoglobulin gels: part 1. Fibril formation and structure. Biomacromolecules 5:2408–2419
199. Li C, Born AK, Schweizer T, Zenobi-Wong M, Cerruti M et al (2014) Amyloid-hydroxyapatite bone biomimetic composites. Adv Mater 26:3207–3212
200. Ling S, Li C, Adamcik J, Shao Z, Chen X et al (2014) Modulating materials by orthogonally oriented beta-strands: composites of amyloid and silk fibroin fibrils. Adv Mater 26:4569–4574
201. Jacob RS, Ghosh D, Singh PK, Basu SK, Jha NN et al (2015) Self healing hydrogels composed of amyloid nano fibrils for cell culture and stem cell differentiation. Biomaterials 54:97–105
202. Reynolds NP, Charnley M, Mezzenga R, Hartley PG (2014) Engineered lysozyme amyloid fibril networks support cellular growth and spreading. Biomacromolecules 15:599–608
203. Reynolds NP, Charnley M, Bongiovanni MN, Hartley PG, Gras SL (2015) Biomimetic topography and chemistry control cell attachment to amyloid fibrils. Biomacromolecules 16:1556–1565
204. Malisauskas M, Meskys R, Morozova-Roche LA (2008) Ultrathin silver nanowires produced by amyloid biotemplating. Biotechnol Prog 24:1166–1170
205. Reches M, Gazit E (2003) Casting metal nanowires within discrete self-assembled peptide nanotubes. Science 300:625–627
206. Scheibel T, Parthasarathy R, Sawicki G, Lin XM, Jaeger H et al (2003) Conducting nanowires built by controlled self-assembly of amyloid fibers and selective metal deposition. Proc Natl Acad Sci U S A 100:4527–4532
207. Knowles TP, Oppenheim TW, Buell AK, Chirgadze DY, Welland ME (2010) Nanostructured films from hierarchical self-assembly of amyloidogenic proteins. Nat Nanotechnol 5:204–207

208. Men D, Guo YC, Zhang ZP, Wei HP, Zhou YF et al (2009) Seeding-induced self-assembling protein nanowires dramatically increase the sensitivity of immunoassays. Nano Lett 9:2246–2250
209. Men D, Zhang ZP, Guo YC, Zhu DH, Bi LJ et al (2010) An auto-biotinylated bifunctional protein nanowire for ultra-sensitive molecular biosensing. Biosens Bioelectron 26:1137–1141
210. Silva RF, Araujo DR, Silva ER, Ando RA, Alves WA (2013) L-diphenylalanine microtubes as a potential drug-delivery system: characterization, release kinetics, and cytotoxicity. Langmuir 29:10205–10212
211. Loo Y, Wong YC, Cai EZ, Ang CH, Raju A et al (2014) Ultrashort peptide nanofibrous hydrogels for the acceleration of healing of burn wounds. Biomaterials 35:4805–4814
212. Ikezoe Y, Washino G, Uemura T, Kitagawa S, Matsui H (2012) Autonomous motors of a metal-organic framework powered by reorganization of self-assembled peptides at interfaces. Nat Mater 11:1081–1085
213. Silva GA, Czeisler C, Niece KL, Beniash E, Harrington DA et al (2004) Selective differentiation of neural progenitor cells by high-epitope density nanofibers. Science 303:1352–1355
214. Kisiday J, Jin M, Kurz B, Hung H, Semino C et al (2002) Self-assembling peptide hydrogel fosters chondrocyte extracellular matrix production and cell division: implications for cartilage tissue repair. Proc Natl Acad Sci U S A 99:9996–10001
215. Holmes TC, de Lacalle S, Su X, Liu G, Rich A et al (2000) Extensive neurite outgrowth and active synapse formation on self-assembling peptide scaffolds. Proc Natl Acad Sci U S A 97:6728–6733
216. Reynolds NP, Styan KE, Easton CD, Li Y, Waddington L et al (2013) Nanotopographic surfaces with defined surface chemistries from amyloid fibril networks can control cell attachment. Biomacromolecules 14:2305–2316
217. Baxa U, Speransky V, Steven AC, Wickner RB (2002) Mechanism of inactivation on prion conversion of the Saccharomyces cerevisiae Ure2 protein. Proc Natl Acad Sci U S A 99:5253–5260
218. Gras SL, Tickler AK, Squires AM, Devlin GL, Horton MA et al (2008) Functionalised amyloid fibrils for roles in cell adhesion. Biomaterials 29:1553–1562
219. Bongiovanni MN, Scanlon DB, Gras SL (2011) Functional fibrils derived from the peptide TTR1-cycloRGDfK that target cell adhesion and spreading. Biomaterials 32:6099–6110
220. Yang Z, Liang G, Wang L, Xu B (2006) Using a kinase/phosphatase switch to regulate a supramolecular hydrogel and forming the supramolecular hydrogel in vivo. J Am Chem Soc 128:3038–3043
221. Koffie RM, Meyer-Luehmann M, Hashimoto T, Adams KW, Mielke ML et al (2009) Oligomeric amyloid beta associates with postsynaptic densities and correlates with excitatory synapse loss near senile plaques. Proc Natl Acad Sci U S A 106:4012–4017
222. Collins SR, Douglass A, Vale RD, Weissman JS (2004) Mechanism of prion propagation: amyloid growth occurs by monomer addition. PLoS Biol 2:e321
223. Chen SG, Stribinskis V, Rane MJ, Demuth DR, Gozal E et al (2016) Exposure to the functional bacterial amyloid protein Curli enhances alpha-Synuclein aggregation in aged Fischer 344 rats and Caenorhabditis elegans. Sci Rep 6:34477
224. Solomon A, Richey T, Murphy CL, Weiss DT, Wall JS et al (2007) Amyloidogenic potential of foie gras. Proc Natl Acad Sci U S A 104:10998–11001

Chapter 5
Fungal Hydrophobins and Their Self-Assembly into Functional Nanomaterials

Victor Lo, Jennifer I-Chun Lai, and Margaret Sunde

Abstract In recent years, much attention has focused on incorporating biological and bio-inspired nanomaterials into various applications that range from functionalising surfaces and enhancing biomolecule binding properties, to coating drugs for improved bioavailability and delivery. Hydrophobin proteins, which can spontaneously assemble into amphipathic layers at hydrophobic:hydrophilic interfaces, are exciting candidates for use as nanomaterials. These unique proteins, which are only expressed by filamentous fungi, have been the focus of increasing interest from the biotechnology industry, as evidenced by the sharply growing number of hydrophobin-associated publications and patents. Here, we explore the contribution of different hydrophobins to supporting fungal growth and development. We describe the key structural elements of hydrophobins and the molecular characteristics that underlie self-assembly of these proteins at interfaces. We outline the multiple roles that hydrophobins can play in supporting aerial growth of filamentous structures, facilitating spore dispersal and preventing an immune response in the infected host. The growing understanding of the hydrophobin protein structure and self-assembly process highlights the potential for hydrophobin proteins to be engineered for use in a variety of novel applications that require biocompatible coatings.

Keywords Hydrophobin · Self-assembly · Functional amyloid · Amphipathic · Surface active · Biomaterial

---

V. Lo · J. I-Chun Lai · M. Sunde (✉)
Discipline of Pharmacology, School of Medical Sciences, Faculty of Medicine and Health, The University of Sydney, Sydney, NSW, Australia

Sydney Nano, NSW, Australia
e-mail: victor.lo@sydney.edu.au; jlai8610@uni.sydney.edu.au; margaret.sunde@sydney.edu.au

S. Perrett et al. (eds.), *Biological and Bio-inspired Nanomaterials*, Advances in Experimental Medicine and Biology 1174,
https://doi.org/10.1007/978-981-13-9791-2_5

5.1 Introduction

Filamentous fungi produce and secrete small, amphipathic proteins called hydrophobins (~70–150 residues) which spontaneously self-assemble into amphipathic protein layers at hydrophobic:hydrophilic interfaces. These hydrophobin layers are functional for fungi because they underpin the ability of filamentous fungi to breach air:water boundaries, where surface tension could otherwise impede growth on a microbial scale [1]. They serve to inhibit wetting of aerial spores, allowing for efficient dispersal and they protect spores from the external environment, as well as facilitating attachment to surfaces necessary for infection of a variety of hosts and germination [2–5].

The ability of the surface-active hydrophobins to self-assemble into robust and biocompatible amphipathic films makes them attractive candidates for applications that involve interfacial modification and activity, including emulsion stabilisation, biosensor development and surface functionalisation. With an understanding of the manner in which the properties and structures of hydrophobins contribute to fungal growth and development, it may be possible to utilise these unique proteins for the design of functional nanomaterials with a range of exciting biotechnological purposes.

5.2 The Discovery of Hydrophobins

The presence of arrays of fibrillar structures on the surface of spores from wind-dispersed fungi was first reported in the 1960s [6, 7]. The function of the rodlet layer to provide a water repellent coating to the spores was discovered by Dempsey and Beever, when they observed that loss of the rodlet coating, which occurred in a gene knockout of *Neurospora crassa*, rendered the spores easily wettable [2, 8]. The gene was named "eas" and the protein it encodes named EAS, for the "easily" wettable phenotype. Attempts to purify the protein from the spores were challenged by the highly insoluble nature of the rodlet film, which resists solubilisation in a variety of denaturing buffers [2, 9]. In 1995, Templeton and colleagues were able to solubilise the rodlets in trifluoroacetic acid and determine that the rodlets were composed of a single protein with the expected sequence for the gene product of eas [10].

The term "hydrophobin" was first used by Wösten and Wessels, when they identified and characterised the products of genes that are highly expressed during the formation of the aerial hyphae and fruiting bodies of *Schizophyllum commune* [11, 12]. More recent studies that include genome sequence information have shown that all filamentous fungi produce hydrophobin proteins [13–15]. Members of the hydrophobin protein family contain eight conserved cysteine residues that are organised in a distinct pattern (X_{2-38}-C-X_{5-9}-C-C-X_{11-44}-C-X_{8-23}-C-X_{5-9}-C-

C-X_{6-18}-C-X_{2-14}), have a relatively high level of hydrophobic residues and share a characteristic hydropathy profile [16].

Many fungi produce multiple hydrophobins, with distinct spatial and temporal distribution, that support development and growth at different stages of the fungal life cycle (Fig. 5.1; Table 5.1). For example, the genome of *S. commune* contains four genes which encode the hydrophobin proteins Sc1, Sc3, Sc4, and Sc6 that contribute to the hydrophobic nature of fungal surfaces [17]. The gene Sc3 is upreg-

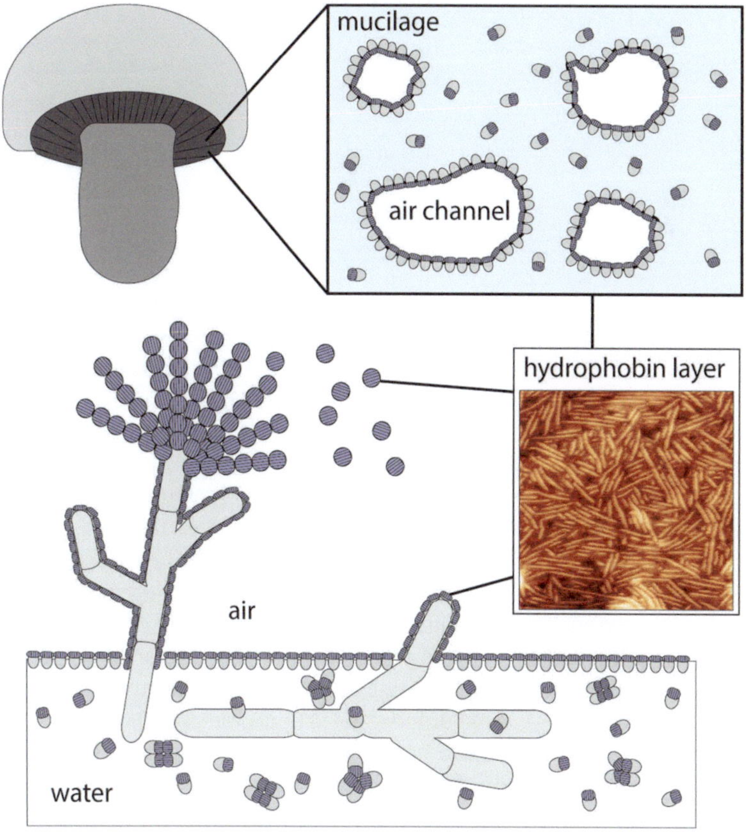

Fig. 5.1 Hydrophobin proteins (shown in purple) play multiple, diverse roles in the fungal life cycle. Hydrophobin proteins are secreted by all filamentous fungi. They are secreted into the aqueous environment surrounding the mycelium and assemble into an amphipathic layer at the air:water interface. This reduces surface tension and facilitates breaching of the air:water interface to allow growth into the air. Hydrophobins form amphipathic protein layers on aerial structures such as hyphae and spores, which prevent wetting, improve dispersal and can also prevent recognition of fungal spores by host immune response. On the surface of spores, the hydrophobin layer consists of fibrillar structures known as rodlets. Hydrophobins are also found as a lining on the surface of gas-exchange structures, for instance in fruiting bodies including mushrooms, to prevent water-induced collapse of the air channels and to maintain gaseous exchange

Table 5.1 Hydrophobin proteins discussed in this chapter

Hydrophobin	Species	Class
EAS	*Neurospora crassa*	I
NC2		II
Sc1, Sc3, Sc4, Sc6	*Schizophyllum commune*	I
HFBI, HFBII	*Trichoderma reesei*	II
RodA, DewA, DewB	*Aspergillus nidulans*	I
DewC, DewD, DewE		Intermediate I/II
RodA, RodB, RodC, RodD, RodE	*Aspergillus fumigatus*	I
HCf1, HCf2, HCf3, HCf4	*Cladosporium fulvum*	I
HCf5, HCf6		II
MPG1	*Magnaporthe oryzae*	I
MHP1		II
Vmh2	*Pleurotus ostreatus*	I
HGFI	*Grifola frondosa*	I
HYD3	*Metarhizium brunneum*	I
HYDPt1	*Pisolithus tinctorius*	I

ulated during the formation of aerial hyphae. This hydrophobin functions to lower the surface tension of the growth medium and it creates a coat around the aerial hyphae, as the fungi breaches through the medium, that prevents the aerial structures from wetting and thus it facilitates dispersal in air. The hydrophobins Sc1, Sc4, and Sc6 support the fruiting bodies, in particular Sc4 forms the hydrophobic film that lines the gas channels of fruiting bodies to prevent water logging. *Aspergillus nidulans* produces six different hydrophobins, RodA, DewA, DewB, DewC, DewD, and DewE, which can all form amphipathic films and contribute to spore hydrophobicity although RodA is the main contributor [18]. The tomato pathogen, *Cladosporium fulvum* also secretes six hydrophobins, two of which are present on the aerial hyphae and the spores, while one is found on young aerial hyphae and another forms a coat around and beneath the fungus, suggesting interactions with the extracellular matrix [19, 20]. In different ways, all hydrophobins support the growth and reproduction of fungi. Some hydrophobins have been shown to be upregulated during nutrient, carbon or nitrogen starvation, while the gene encoding the hydrophobin EAS is activated by light, the circadian clock and nutrient deficiency [4, 21–24].

5.3 Class I and Class II Hydrophobins

Early studies by Wessels and colleagues showed that members of the hydrophobin family could be separated into two distinct classes based on the solubility of the protein layers and the hydropathy profiles of the proteins: class I and class II [5]. As more hydrophobins have been identified and studied in detail, this distinction has generally been maintained and the sequence and structural differences underlying the character of the two main classes have been elucidated [13, 14, 25].

Class I hydrophobin proteins are on average larger than class II hydrophobins and display more variability in the overall length of the protein chain and in the length of the inter-cysteine regions. It is only class I hydrophobins that self-assemble into films that exhibit the distinctive rodlet pattern (Fig. 5.2a). These rodlet films are very robust under physical stress, thermally stable up to 100 °C in detergent

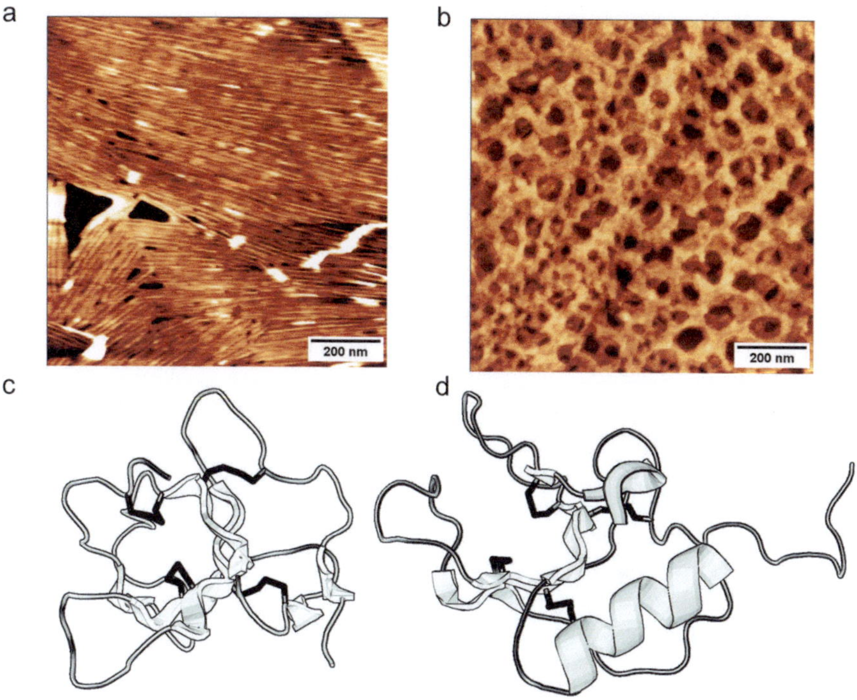

$EAS_{\Delta 15}$ X_{17}CSIDDYKPYCCQSMSGPAGSPGLLNLIPVDLSASLGCVVGVIGSQCGASVKCCKDDVTNTGNSFLIINAANCX_2
NC2 X_{13}CSGLYGTAQCCATDVLGVADLDCANPPATLANATHFESTCAAIGQRARCCVLPILGQDILCX_6

Fig. 5.2 Class I and class II hydrophobins form protein layers with distinct morphologies. (**a**) Atomic force micrograph of the surface morphology of a protein layer formed by self-assembly of the class I hydrophobin $EAS_{\Delta 15}$ when dried onto highly-oriented pyrolytic graphite. The $EAS_{\Delta 15}$ protein, a variant of EAS from *Neurospora crassa,* forms fibrillar rodlets ~10 nm wide and microns long which pack laterally to form a monolayer. (**b**) The class II hydrophobin NC2, from *Neurospora crassa* self-assembles into a spongiform protein layer when dried onto highly-oriented pyrolytic graphite. Although the morphologies of the self-assembled layers formed by EAS and NC2 are very different, the monomeric structures of the two proteins in solution show similar features. Figure reproduced with permission from Ren et al. [34]. Ribbon representations of (**c**) $EAS_{\Delta 15}$ and (**d**) NC2. Both hydrophobin classes display a β-barrel structure, stabilised by the presence of four disulphide bonds, indicated by black sticks. Comparison of the sequences of $EAS_{\Delta 15}$ and NC2 highlights the eight conserved cysteine residues in all hydrophobins (coloured in red) and the longer inter-cysteine segments in the class I hydrophobin compared to the class II hydrophobin

and have been shown to be chemically stable under alkaline wash (3 M NaOH) and acidic wash (3 M HCl) [5, 26]. The rodlets have a diameter of ~10 nm and assemble laterally to form films with a thickness of 2.5–3 nm [26, 27]. They can only be depolymerised and dissociated to monomeric protein by treatment with agents such as trifluoroacetic acid (TFA) or formic acid [12, 28]. Once dissolved into their monomeric form, class I hydrophobins can reassemble into rodlets in the presence of an appropriate interface [29].

The class I rodlets share many amyloid-like characteristics. Investigations of EAS and Sc3 showed that the rodlet form of these hydrophobins interacts with the amyloid-specific dyes, thioflavin-T and Congo red, which are widely used to detect cross-β structures [16, 30]. Further studies with techniques such as circular dichroism and attenuated-total-reflectance Fourier-transform infrared spectroscopy (ATR-FTIR) indicated the presence of β-sheet structure in the rodlet form [16, 27]. X-Ray fibre diffraction has confirmed that the class I hydrophobin rodlets from EAS contain a repeating β-structure, similar to amyloid fibrils [31]. The fibre diffraction pattern generated by rodlets displays a reflection at 4.8 Å that arises from the spacing between the strands in a β-sheet and an orthogonal reflection at 10–12 Å, generated by repetition of the spacing between β-sheets parallel to the fibril long axis.

Strikingly, Kershaw and colleagues were able to show that by substituting genes of different hydrophobins (*sc3* and *sc4* from *S. commune*, *rodA* and *dewA* from *Aspergillus nidulans*, *eas* from *Neurospora crassa*, and *ssgA* from *Metarhizium anisopliae*) into a mutated *Magnaporthe oryzae* strain lacking the gene for the class I hydrophobin MPG1, it was possible to partially supplement the multiple functions of MPG1 [32]. This demonstrates that although there is little sequence homology between class I hydrophobins from different species, the biological functions they perform in different fungi are very similar.

Class II hydrophobins are generally smaller than class I hydrophobins, with up to 63 residues between cysteines one and eight, instead of up to 105 residues between the outer cysteines in class I hydrophobins. Although they do not form rodlets when they self-assemble, class II hydrophobins also spontaneously form amphipathic layers at hydrophobic:hydrophilic interfaces (Fig. 5.2b). These layers are much less robust than class I rodlet-containing layers: they can be disrupted by treatment with alcohols and detergents and the application of mild pressure or temperatures [33, 34]. Under certain conditions some level of macromolecular organisation is apparent in class II hydrophobin monolayers. HFBI and HFBII are known to produce highly orientated crystalline monolayers at the atomic scale under compression in certain experimental conditions, while in NC2 monolayers a porous network structure has been observed (Fig. 5.2b) [34, 35].

5.4 Structures of Class I and Class II Hydrophobins

When the first three-dimensional structure of a hydrophobin was determined, that of HFBII by crystallography, it revealed that the protein core consisted of a β-barrel with four antiparallel β-strands and an external α-helix [36]. Although sequence conservation between hydrophobins is low between species, a similar core β-barrel structure has since been observed in the structures of other hydrophobins, whether determined by crystallography or in solution by nuclear magnetic resonance [31, 34, 37–39] (Fig. 5.2c, d).

A close examination and comparison of all of the known hydrophobin structures reveals that the conserved pattern of disulphide bonds is integral to the common observed hydrophobin structure [13, 40]. Two of the four conserved disulphide bonds lock in the β-barrel structure while the other two disulphide bonds act as tethers, holding relatively separated parts of the sequence in close proximity to the barrel core. This generates a very stable and compact core structure with, in some cases, long loop regions between cysteine residues. In EAS the long loops are flexible while in DewA and MPG1 the long loops contain stable elements of secondary structure [38, 39]. In some of the hydrophobins the β-barrel is relatively complete or "closed" while in others it is open in a half-barrel structure. The class II hydrophobins HFBI and HFBII display very compact structures with loops contributing to the closing of the barrel but in NC2, the structure of which was determined in solution, these loops are open. It is possible that crystallisation conditions may have contributed to the observation of a more closed conformation in HFBI and HFBII.

5.5 The Surface Activity of Hydrophobins

The biophysical basis for the surface activity of the hydrophobins has become apparent from determination of the three-dimensional structures of the proteins. All of the hydrophobins have relatively large exposed hydrophobic areas on the protein surfaces [34]. The monomeric forms of hydrophobin proteins display exposed hydrophobic regions that make up 60–67% of the solvent accessible surface area (Table 5.2). This compares to 37% in the case of the soluble protein ubiquitin. This characteristic of hydrophobins is likely to result in these proteins concentrating and aligning at interfaces and to be responsible for the observed surface activity of the proteins [39, 41] (Fig. 5.3). Such large solvent-exposed hydrophobic areas are unusual in proteins but the hydrophobins have a very stable core structure as a result of the presence of the disulphide bonds, which maintains the protein fold.

Both classes of hydrophobins can lower water surface tension, however while solutions of class I members can take up to several hours to reach the lowest surface tension, class II solutions reach this in several minutes [42]. Hydrophobins, such as Sc3, are able to reduce the surface tension of phosphate buffer to 43 mJ/m^2 with 0.1 mg/mL, while other proteins such as BSA (49 mJ/m^2), IgG (42 mJ/m^2) or lysozyme (40 mJ/m^2), require 5 mg/mL to reach a similar surface energy. The surface tension for purified water (72 mJ/m^2) can be reduced using a variety of different hydrophobins at very low protein concentrations [43]. Sc3 in water reduces the surface tension to 32 mJ/m^2 at 0.1 mg/mL. The class II hydrophobin HFBII can decrease the surface tension to 28 mJ/m^2 with only 0.02 mg/mL [36]. Both classes are active at hydrophobic:hydrophilic interfaces in a way that allows them

Table 5.2 Calculated hydrophobic contribution to the total solvent accessible surface area and the area of the largest single exposed hydrophobic patch. For hydrophobins, the presence of large solvent-exposed hydrophobic patches is likely to underlie the high surface activity of these proteins. Data from Ren et al. [34]

Protein	Total solvent accessible surface area (SASA) that is hydrophobic	Total area of the largest solvent-accessible hydrophobic patch	
		Percentage of total SASA	Area of largest hydrophobic patch (Å^2)
NC2	64%	12%	693
EAS	61%	13%	758
DewA	60%	6%	424
HFBI	62%	20%	783
HFBII	67%	23%	891
HFBI (tetramer)	48%	–	–
HBFII (dimer)	33%	–	–
Ubiquitin	37%	–	–

to, for example, stabilise oil emulsions or coat hydrophobic surfaces to render them wettable. Class I hydrophobins create more stable emulsions than class II representatives and form stronger interactions with hydrophobic surfaces than those formed by Class II.

Some hydrophobins have been shown to oligomerise in solution and this may serve to reduce the exposed hydrophobic surface. For example, Sc3 associates into dimers, with some tetramer and monomer in different buffers but can remain mainly in the monomeric form in pure water [43, 44]. DewA has been shown to populate a dimer in solution but it is likely that this must dissociate to the monomer in order to assemble into rodlets [38].

5.6 Mechanism of Hydrophobin Assembly from Monomer to Amphipathic Monolayer

The mechanism underlying spontaneous assembly of hydrophobins into structured amphipathic films at hydrophobic:hydrophilic interfaces is intriguing. Self-assembly of class I hydrophobins into amphipathic monolayers does not occur in the absence of a distinct interface [41]. When the liquid-air interface is removed from solutions of EAS, DewA or MPG1, no amphipathic film forms [26]. The hydrophobins remain in a monomeric form in solution and can only self-assemble once presented with a liquid-air interface. In a solution of hydrophobins in contact with air, the surface tension at the air:solution interface plays an important role in determining the formation of these hydrophobin films. When the surface tension of solutions of $EAS_{\Delta 15}$ or DewA is reduced by addition of alcohol to

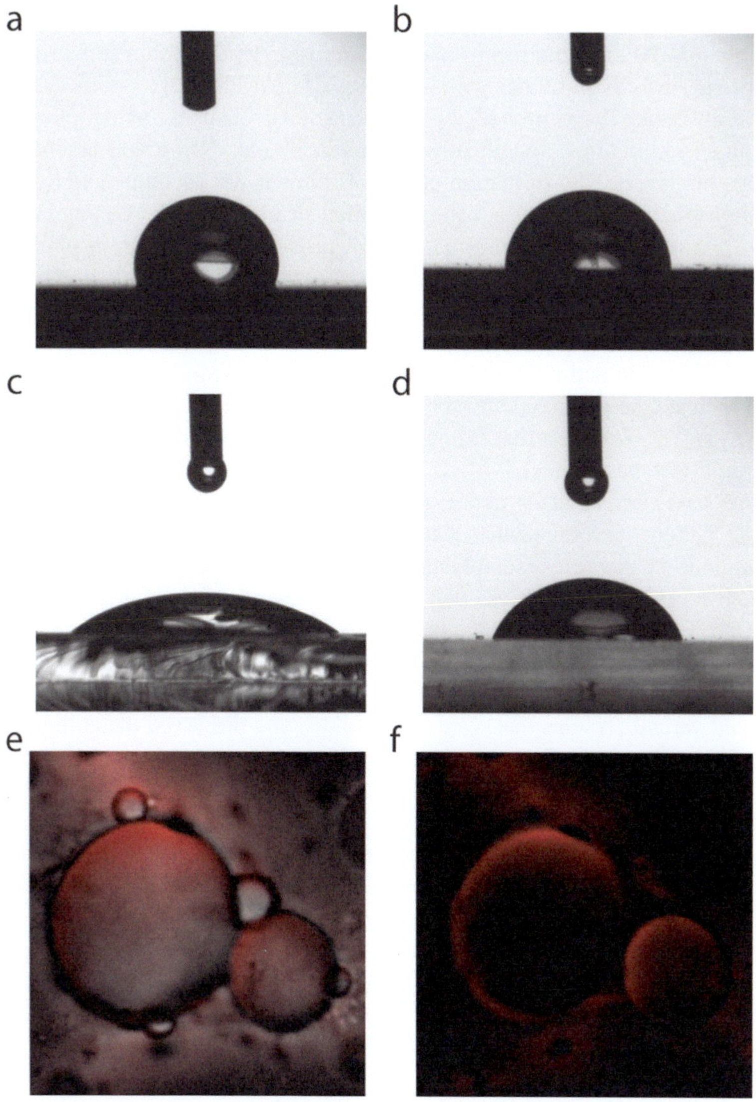

Fig. 5.3 The amphipathic hydrophobin proteins can reduce surface tension and invert the wettability of that surface. When the hydrophobin $EAS_{\Delta 15}$ forms a rodlet coating on the hydrophobic surface, octadecyltrichlorosilane-coated silica, the surface becomes hydrophilic. (**a**) Image showing the high contact angle formed by a water droplet on octadecyltrichlorosilane-coated silica. (**b**) Image showing the reduction in contact angle formed by a water droplet when the surface is coated by $EAS_{\Delta 15}$. When the hydrophobin $EAS_{\Delta 15}$ forms a rodlet coating on a hydrophilic surface, e.g. mica, the surface becomes more hydrophobic. (**c**) Image showing the relatively low contact angle formed by a water droplet on uncoated mica and (**d**) the increase in contact angle formed by a water droplet when the mica surface is pre-coated with the hydrophobin $EAS_{\Delta 15}$. Hydrophobins also assemble at hydrophobic:hydrophilic interfaces between different solutions and can stabilise emulsions of oil in water. (**e**) Fluorescently-tagged hydrophobin $EAS_{\Delta 15}$ forms a protein layer that encapsulates oil droplets in water and stabilises them. (**f**) Fluorescence-only image to show that the hydrophobin layer surrounds the oil

below ~54 mN.m^{-1}, no rodlet formation could be detected [41]. Control of self-assembly may be a natural mechanism for localising hydrophobin self-assembly to appropriate surfaces and times within the fungal life cycle [39].

Several studies have probed the conformational changes taking place in the hydrophobin structures, specifically at interfaces and during self-assembly into amphipathic layers. The formation of class II films is not accompanied by observable conformational change but the formation of class I rodlet-containing layers involves some significant conformational change and the formation of the intermolecular hydrogen bonding common to amyloid fibrils [34]. Sc3 has been observed to populate three different states prior to interaction with a hydrophobic interface: a monomeric state, an α-helix interaction state, and the final β-sheet state which forms the insoluble amphipathic film [45]. Molecular dynamics simulations of the folding of Sc3 indicate the preference for the protein to adsorb to the interface (within pico seconds) and also suggest that formation of the β-sheet secondary structure occurs more rapidly at the interface [46].

Given the nature of the structure and the constraints imposed by the disulphide-bonding pattern, several studies have focussed on the role of the hydrophobin loop regions in the self-assembly process. These are the segments between the third and fourth cysteines (Cys3 and Cys4), between the fourth and fifth cysteines (Cys4 and Cys5) and between the seventh and eighth cysteine residues (Cys7 and Cys8). Truncation of up to 17 residues within the Cys3-Cys4 loop of EAS, or introduction of a foreign highly-charged 8-amino acid sequence into this loop, generated variant proteins that were stably folded and preserved the ability to form native-like rodlets [47]. Contrastingly, when Niu and colleagues created a chimeric protein by substituting the Cys3-Cys4 loop of the class I hydrophobin HGFI from the edible mushroom *Grifola frondosa* with that from the class II hydrophobin HFBI from *Trichoderma reesei*, the resultant protein was not able to form rodlets [48]. This was a substitution of 32 amino acids with the 11-amino acid sequence from the class II protein. In the hydrophobin Sc3 the Cys3-Cys4 loop binds strongly to hydrophobic interfaces, such as colloidal Teflon™, indicating a role in alignment of molecules ready for self-assembly [44]. Molecular dynamics studies have also suggested that the long flexible Cys3-Cys4 loop in EAS may prevent unwanted intermolecular interactions in solution [49].

Studies of single-site mutants of the EAS protein, in which a hydrophobic residue is replaced by a glycine residue in order to increase flexibility and decrease amyloidogenic potential in either the Cys7-Cys8 loop or the Cys4-Cys5 loop have suggested that only the Cys3-Cy4 loop is involved in amyloid hydrogen bonding in EAS rodlets [50]. Furthermore, studies of a chimeric protein where the Cys7-Cys8 loop of the class I hydrophobin EAS was grafted into the non-rodlet forming class II hydrophobin NC2 generated a version of the class II hydrophobin that is able to self-assemble into stable rodlets that are ThT-positive. A molecular model for the assembled EAS rodlet structure has been developed based on the data from these mutational studies and this is consistent with the formation of amphipathic EAS rodlets. Taken together, these results suggest that individual hydrophobins may contain single or multiple amyloidogenic regions and that these may be accommodated in different inter-cysteine segments in different hydrophobins.

5.7 Hydrophobins Have Multiple Functions in the Fungal Life Cycle

The ability to reduce surface tension at air:water interfaces is critical for the production of aerial hyphae to allow the growth and spread of filamentous fungi and hydrophobins, both class I and class II, are deployed for this purpose. Additionally, the hydrophobin coatings are important in mediating fungal:host interactions in ways that reduce detection and killing of fungal spores and improve attachment to host surfaces to facilitate infection.

5.7.1 Hydrophobin Coatings Shield Fungal Structures from Host Immune Recognition

Where fungi express multiple different hydrophobins, at least one is found to form a coat on the surface of aerial spores. The best characterized of these is the hydrophobin RodA protein that is found in a rodlet form coating the exterior of dormant *A. fumigatus* spores. In 2009, Aimanianda et al. provided evidence that host immune cells were unable to activate or mature in the presence of the RodA protein [3]. The study compared the maturation and activation of immune cells such as dendritic cells, alveolar macrophages, and helper T cells in the presence and absence of RodA. Resting spores coated with a RodA layer were incapable of activating the immune cells. However, once the RodA protein was removed, through the process of swelling and germination that results in disappearance of the RodA layer and exposure of the pathogen associated molecular patterns on the fungal cell wall, host immune cells underwent activation and maturation. The presence of hydrophobin RodA on the spore and hyphal surfaces also reduces the stimulation of human platelets, a novel component of the innate immune network [51].

Rohde et al. have utilised scanning electron microscopy to illustrate the shedding of the rodlet layer of hydrophobin RodA upon spore swelling [52]. Dague et al. also observed the phenomenon using atomic force microscopy and further validated the loss of hydrophobicity with loss of hydrophobins upon swelling of the spore [53] (Fig. 5.4). The underlying cell wall components are purely hydrophilic and are recognized by dectin-1 and dectin-2 receptors on the host immune cells [54, 55]. This recognition of the cell wall components results in induction of pro-inflammatory cytokines, chemokines, recruitment of neutrophils, and ultimately killing of the spore [56, 57].

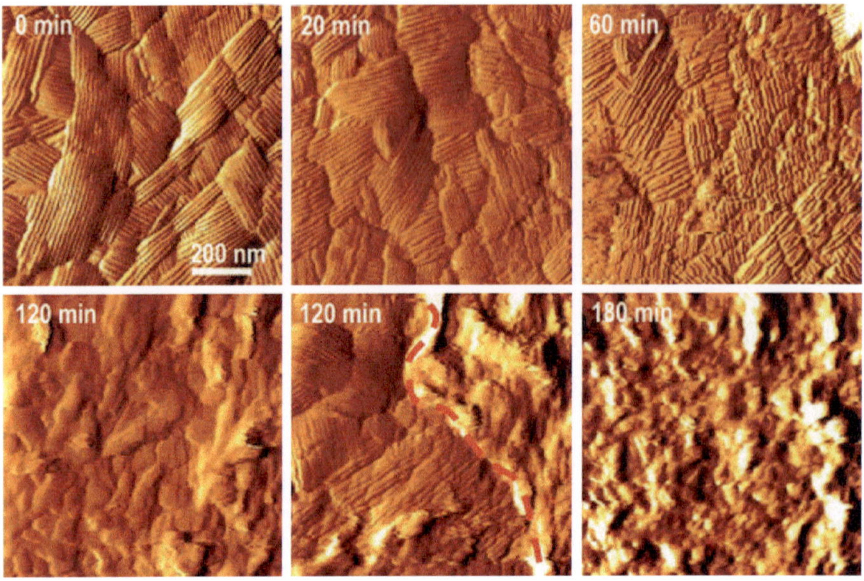

Fig. 5.4 The outer surface of the spores produced by *Aspergillus fumigatus* is coated with a fibrillar rodlet layer composed of the class I hydrophobin RodA. A series of high-resolution atomic force micrographs recorded from the surface of a single *Aspergillus fumigatus* spore during the process of spore germination. The highly structured rodlet layer is visible at 1 and 20 min but changes are apparent at 60 min as the rodlet layer starts to disappear and to be replaced by an amorphous layer. Figure reproduced with permission from Dague et al. [53]

These studies have therefore demonstrated that the RodA protein layer exists as a shield for the underlying fungal cell wall components, including β-glucan and melanin, which would otherwise serve as antigens and cause continuous immune activation and detrimental inflammatory responses. The presence of RodA on the spore surface allows dormant spores to escape detection until conditions become conducive for germination [3, 58].

A. fumigatus is an opportunistic human pathogen and, in immune-compromised individuals, it is able to germinate within the lung, produce invasive hyphae and disseminate throughout the bloodstream, causing the difficult-to-treat invasive aspergillosis. Beauvais et al. have investigated the role of *A. fumigatus* hydrophobins in the static aerial growth conditions that resemble the human lung [59]. This has revealed an extracellular hydrophobic matrix that supports the fungal cells in a fungal biofilm community. The matrix consists of galactomannan, α1,3 glucan, monosaccharide, polyol, melanin and proteins such as major antigens and hydrophobins. Although RodA is absent from the mycelial stage under both static aerial or liquid shaken growth conditions, the class I hydrophobin RodB is found expressed exclusively in the aerial static culture condition resembling growth in the lung. The *A. fumigatus* hydrophobins RodB–RodE continue to be expressed throughout biofilm formation, indicating a potential role for these hydrophobins in formation, maintenance and surface adhesion of the extracellular matrix [60, 61].

5.7.2 *Hydrophobins Facilitate Attachment of Fungi to Host Cells for Colonisation*

The fungus *Magnaporthe oryzae* causes rice blast, a devastating disease that is responsible for loss of up to a third of global annual rice production [62]. The hydrophobin MPG1 is highly expressed during rice blast infections [63]. MPG1 is found on the spore surface of *M. oryzae* spores where it functions to resist wetting and assist adherence to plant tissue [4]. Moreover, MPG1 has been shown to be involved in sensing of a hydrophobic surface, initiation of appressorium differentiation, host penetration and virulence [62]. Knockout or gene silencing studies of MPG1 resulted in reduced appressorium formation and pathogenicity, and the spores of *M. oryzae* lacking MPG1 showed a wettable phenotype [64, 65]. Silencing of the class II hydrophobin from *M. oryzae*, MHP1, does not affect pathogenicity [65].

Pham et al. have studied the interaction between the two hydrophobins, MPG1 and MHP1, and the cutinase cut2 that is associated with growth through the waxy rice leaf cuticle [39]. This study demonstrated that assembly of the class I hydrophobin MPG1 is strictly limited to an interface and the class II hydrophobin MHPI is able to prevent MPG1 assembly by reduction of surface tension in a solution containing both hydrophobins. This suggests that some fungi may produce multiple hydrophobins in order to control the spatial and temporal pattern of rodlet formation. The assembly of the hydrophobins was also shown to recruit cutinase activity to a hydrophobic surface, potentially offering a mechanism for control of the infection interface in rice blast disease.

5.8 Harnessing Hydrophobins for Biotechnological Purposes

Hydrophobins are natural products with properties that are potentially useful for a wide variety of applications. The ability of hydrophobins to self-assemble into biocompatible amphipathic films at liquid-oil, liquid-liquid, air:liquid and liquid-solid interfaces has underpinned studies of their potential use in biotechnology and the food and cosmetic industries [12, 66]. Studies of hydrophobins are reported in multidisciplinary fields, including food technology, pharmacology, chemistry and surface engineering and hydrophobin use in these areas has been heavily patented in recent years [67]. The number of publications and patents associated with hydrophobins shows a striking increase in recent years (Fig. 5.5).

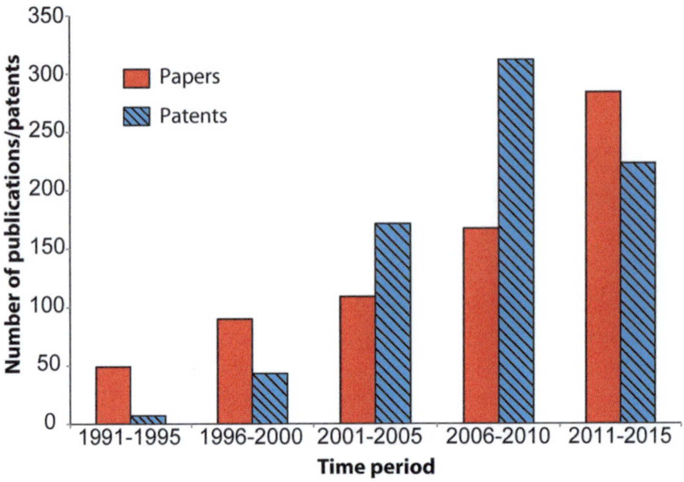

Fig. 5.5 Increase in hydrophobin-related publications and patents over recent years. Number of publications taken from Web of Science and patents from FreePatentsOnline in five-year periods with the search word "hydrophobin"

5.8.1 Hydrophobins Used to Modify or to Functionalise Surfaces

The ability of a hydrophobin layer to stably reverse the wettability of hydrophobic surfaces opens up the possibility of using hydrophobin coatings to improve the utility of hydrophobic materials in aqueous and physiological environments (Fig. 5.6). Immobilisation of active biomolecules on hydrophobic substrates can be achieved through coating the hydrophobic material with hydrophobins and then allowing non-covalent absorption of target proteins to the hydrophobin-coated surface or by covalent attachment (Fig. 5.6a). Non-covalent absorption is dependent on pH and ionic strength, suggesting it is mediated by selective Coulombic charge interactions [68].

An example of hydrophobins being used for protein immobilization is to enable more rapid protein analysis by matrix-assisted laser desorption/ionization (MALDI) mass spectrometry. Pre-coating of the MALDI steel plate with the hydrophobin Vmh2 enables subsequent non-covalent coupling of enzymes including trypsin, V8 protease, PNGaseF, and alkaline phosphatase onto the plate without interfering with the enzymatic activity [69]. The functionalised plates facilitate efficient protein digestion reactions directly on the plate, shortening the process from an 18 h, 37 °C incubation to a 5 min, room temperature incubation. Enzyme immobilization condenses the time-consuming protein digestion steps and allows sampling in small volumes and the possibility of multiple enzymatic digestions on the steel plate.

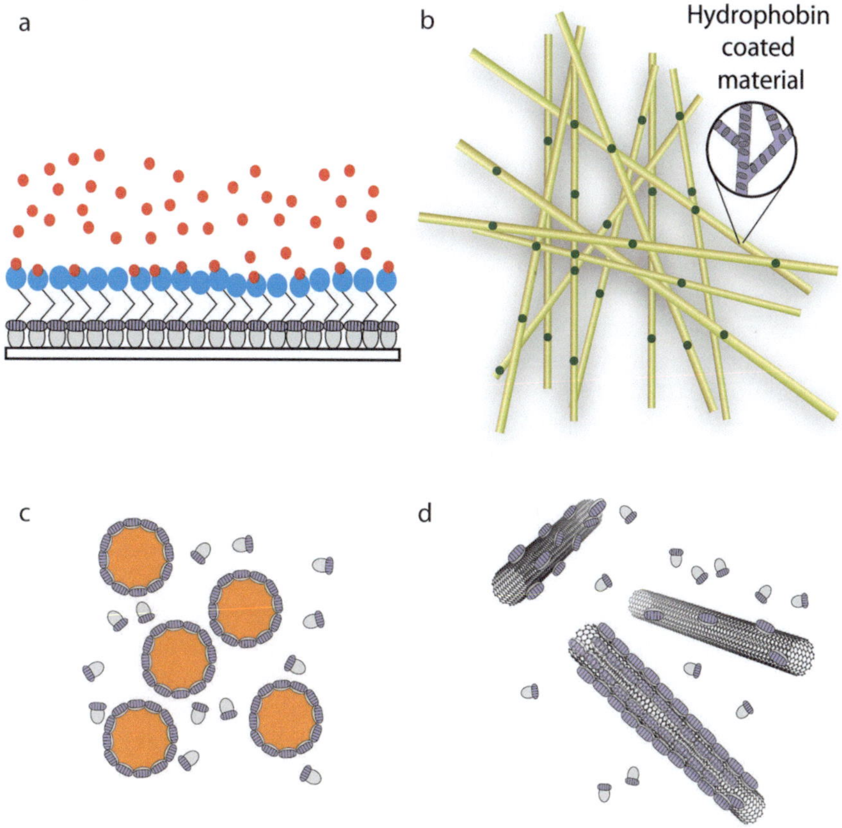

Fig. 5.6 Hydrophobins (shown in purple) show potential for the development of functional nanomaterials. (**a**) As amphipathic biopolymers, hydrophobin coatings may allow immobilisation and maintenance of biological activity, for the functional biomolecules utilised in biosensors. (**b**) Hydrophobin-coatings used as primers on materials can improve the adhesion of nanoparticles that provide desirable characteristics, such as antimicrobial zinc oxide nanoparticles. (**c**) Hydrophobins can be used to encapsulate hydrophobic drugs to enhance bioavailability and increase the half-life of the drug. (**d**) Hydrophobin layers assembled on the surface of hydrophobic carbon nanotubes can increase dispersion of CNTs in aqueous solutions and prevent agglomeration

Hydrophobins have also been applied for biosensor development, where the hydrophobins are used as an intermediate layer to alter surface wettability and support target protein deposition. Creation of a biosensing electrode requires a conductive material, a hydrophilic surface, and sufficient immobilisation of the hydrophilic biomolecule in a non-obstructive manner onto the conductive material while retaining its functional activities. By simply dip-casting a surface using a solution of a low concentration of hydrophobins, it is possible to immobilise proteins, enzymes, quantum dots and nanoparticles [70], unlike other techniques which may require more complicated immobilisation strategies including the use of antibodies [71], amine or thiol terminated silane [72] or UV photo-crosslinking [73]. Zhao et al. have reported the construction of a novel biosensor electrode for evaluation of choline concentration [74]. The inert metal, gold, was coated with an intermediate robust layer of the hydrophobin HFBI protein to enhance surface hydrophilicity and enable immobilisation of the hydrophilic protein choline oxidase in its bioactive form. The same laboratory also used HFBI to create a glucose biosensor with multi-wall carbon nanotubes (MWNTs) [75]. MWNTs are known for their excellent electrical conductivity, however they also represent an extremely insoluble material. A coating of HFBI is able to solubilise MWNTs and at the same time provide a hydrophilic surface that facilitates immobilisation of glucose oxidase. Such an HFBI-MWNT nanocomposite could be adapted for attachment of other enzymes. The hydrophobin Vmh2 from *Pleurotus ostreatus* has been shown to bind glucose directly and has been adapted for measurement of glucose on a silicon surface or on the surface of gold nanoparticles, which removes the need for immobilisation of the enzyme glucose oxidase [76, 77]. Vmh2 has also been shown to effectively immobilise quantum dots, graphene oxide, and functional proteins such as bovine serum albumin and anti-immunoglobulin (IgG) antibodies onto superhydrophilic glass [70]. The superhydrophilic glass (water contact angle of 0°) moderately increased in hydrophobicity when incubated with the hydrophobin (water contact angle of 60°). This degree of water contact angle has been shown to be proficient in causing adhesion of biomolecules to the surface, as shown by the retention of immobilised molecules following harsh washes using detergent-containing solutions.

The hydrophobin HYDPt1 has been patented as a matrix for coating glassy carbon electrodes; it allows immobilisation of small electroactive molecules such as coenzyme Q10 from solution [78]. One of the major advantages of using hydrophobins is the protective barrier the hydrophobin films create around the probe. The layer is very thin and does not affect the capacity current like alkanethiols, and can also protect the probe over a wide pH range.

Hydrophobins have been used as an intermediate primer for textiles, to enhance adhesion of nanoparticles (Fig. 5.6b). Hydrophobins incorporated on materials such as cotton and polyethylene terephthalate improve the hydrophobic properties of both materials [79]. Furthermore, these materials are resistant to common laundry processes and harsh extraction procedures. Studies have observed that hydrophobins improve the adhesiveness of antimicrobial particles. With the addition of a hydrophobin coating as a primer for zinc oxide on cotton/polyester substrate, zinc oxide nanoparticles were up to eight times more resistant to washing. The combined zinc oxide-hydrophobin-coated textiles were 50% more resistant to *Candida albicans* growth and 30% more resistant to mixed mold inoculum cultures, compared to zinc oxide-only coated textiles [80].

5.8.2 Hydrophobins Used to Coat Stents for Anti-Fouling Properties

The biocompatible nature of the hydrophobin layers has stimulated their testing both to improve desirable cell adhesion in cell culture and to prevent unwanted protein deposition on medical implants. The addition of a hydrophobin coating that is not cytotoxic could improve cell adhesion to an implant or culture materials by adjustment of the surface wettability or by specific functionalisation of the hydrophobic surface. Cell adhesion and proliferation rates can be affected by surfaces that are too hydrophobic or too hydrophilic and the optimal surface for cell adhesion has been suggested to have a water contact angle of around 70° [81]. Compared to hydrophobic surfaces such as bare Teflon™, coating of the hydrophobin variant TrSc3 or the hydrophobin Sc4 has been shown to improve the wettability of the surface and assist the adhesion of mouse fibroblast L929 cells [82]. The hydrophobin coating can also be functionalized by fusing the binding sites for integrin receptors such as the Arg-Gly-Asp (RGD) sequence from fibronectin or the laminin globular domain (LG3) with DewA, to facilitate cell adhesion [83]. The coating of titanium surfaces with RGD-DewA and LG3-DewA fusions significantly enhanced adhesion of mesenchymal stem cells, osteoblasts, fibroblasts, and chondrocytes. Importantly, adhesion of the pathogen *Staphylococcus* aureus was not elevated by the treatment of the surface. Weickert and colleagues reported that coating of plastic biliary stents with the hydrophobin H star protein® A reduced adherent material; however, signs of amorphous material and bacteria were still present [84]. A combination of H star protein® A and the anticoagulant heparin further reduced the amount of material adhering to the stent and showed minimal bacterial binding. Hydrophobin coatings therefore show promise for promoting cell adhesion and better tissue integration while minimising biofouling on stents and other implant materials.

5.8.3 Hydrophobins Used to Stabilise Emulsions

The amphipathic structure of hydrophobins makes them attractive candidates for biosurfactants and bioemulsifiers, being a natural alternative to polymer micelles or Janus particles. HFBII has been shown to have relatively high dilatational and shear elastic modulus compared with other proteins, allowing it to create highly stable emulsions and foams [85].

An example of a hydrophobin-stabilised emulsion can be found in the beer industry. Hydrophobins are associated with fungal infections of barley grain and are released into the beer when the malt is crushed into the wort. Hydrophobins that spontaneously interact with calcium oxalate present in the beer can result in the formation of larger oxalate crystals, which release more CO_2 and cause gushing. Deckers et al. have used molecular dynamic simulations to study this phenomenon and suggest that hydrophobins also form an amphipathic film which attaches to the wall of the bottle and can encapsulate CO_2 gas in nanobubbles [86]. When the bottle is depressurised upon opening, the hydrophobin film around the nanobubbles can break, causing release of a high concentration of CO_2 and over-foaming.

This phenomenon of gas encapsulation by hydrophobins in nanobubbles can be adapted for use in food preparation, where stable emulsions are often desired. Hydrophobin self-assembly can generate a natural and biocompatible Pickering emulsion. A Pickering emulsion is formed when the interface of two different phases is stabilised by solid particles, such as hydrophobins, at the hydrophobic:hydrophilic interface. This prevents coalescence of the two different phases, hence stabilising the mixture (Illustrated in Fig. 5.3e, f). Hydrophobins show promise as emulsion stabilisers for foods [67]. The stability of aerated solutions of the hydrophobin HBFII has been tested against other commercially available aerating agents and food emulsifiers, across a range of different pH conditions (pH 3.8 – pH 8.4) [87]. Across all pH values tested, the hydrophobin was able to support aerated foam in water for up to 5 days with no additional thickening agents. With the addition of thickening agents, there was little to no air phase loss for up to 4 months, while in chilled conditions the foam was stable for 2.5 years. Similar results were observed when the hydrophobin HBFII was added to a chocolate milk shake mixture [87, 88]. Hydrophobin-stabilisation of foods may reduce or remove the fat content currently associated with current food emulsions [88]. While hydrophobin proteins are found naturally in a variety of different foods and beverages, such as mushrooms and beer, hydrophobin proteins engineered for use in the food industry will be classified as novel products. This means that any such hydrophobins must be shown to comply with food regulatory affairs prior to distributions or use [89].

5.8.4 Hydrophobins Applied for Improved Drug Delivery

The majority of drugs in development fail due to low solubility, an indicator of low bioavailability. Around 40% of market-approved drugs and about 90% of novel molecules in development are poorly soluble [90]. The self-assembly properties and the amphipathic nature of hydrophobins allows these proteins to be used to encapsulate and solubilise hydrophobic compounds (Fig. 5.6c). Drug encapsulation may reduce systemic toxicity and side effects, increase the solubility and efficacy of therapeutics and enable sustained release of active compounds over an extended period of time. A comparison of two hydrophobic drugs, nifedipine and cyclosporine A, without any additives, with conventional formulations, and with Sc3 hydrophobin-based drug particle suspension demonstrated that the Sc3-based suspension was able to increase bioavailability [91]. Furthermore, the authors found that the drug uptake rate was slower, which resulted in a more constant and prolonged release of the drugs. Another novel nanoparticle formulation using H star Protein® B, a DewA hydrophobin-derivative, to solubilise the hydrophobic anti-cancer agent docetaxel, also demonstrated that the hydrophobin formulation is more bioavailable and biocompatible [92]. The hydrophobic drug beclomethasone dipropionate formulated in water with the addition of the class II hydrophobin HFBII has been found to produce a stable nanoparticle suspension [93].

In addition to being used directly to formulate drugs, hydrophobins may also be applied to improve the properties of other drug carriers. Validated drug delivery candidates such as carbon nanotubes and porous silicon nanoparticles possess a high capacity for loading drugs. However, solubility and rapid elimination issues hinder the possibility for further development. The hydrophobic surface of amphipathic class I hydrophobins strongly adheres to hydrophobic surfaces, and in turn reverses the wettability, resulting in a hydrophilic surface [94]. This property has been used to coat a range of hydrophobic nanomaterials, such as single-walled carbon nanotubes (SWCNTs), graphene sheets and highly oriented pyrolytic graphite using two class I hydrophobins EAS and HYD3 (Fig. 5.6d). In particular, hydrophobin-coated SWCNTs are stable, well-dispersed, and not cytotoxic in aqueous solutions. Surface modification of thermally hydrocarbonised porous silicon nanoparticles by coating with the class II hydrophobin HFBII significantly increases biocompatibility against macrophages and liver cell cultures [95]. The authors also noted significant alteration in liver and spleen accumulation of hydrophobin-coated porous silicon nanoparticles. This is particularly important to reduce rapid elimination of nanoparticles following administration.

5.9 Conclusions

The fungal hydrophobin proteins are multifunctional proteins which are critical for supporting fungal growth and development. The combination of surface activity with the ability to form stable amphipathic layers provides a means for fungi to exploit life at multiple different environmental interfaces. Recent biophysical and biochemical studies of the properties of hydrophobins and their effects on animal hosts have generated an understanding of the self-assembly properties of these unusual proteins and their non-immunostimulatory nature. This growing understanding of the mechanisms that drive the self-assembly of hydrophobins into functional *biomaterials* should expedite the utilisation of hydrophobins for the development of novel functional *nanomaterials*.

Acknowledgements The authors gratefully acknowledge the support of the Australian Research Council in the form of Discovery Grants DP120100756 and DP150104227 to MS, which have supported research into the structure and properties of hydrophobin proteins. VL and JL are supported by Australian Postgraduate Awards. We also thank Dr. Ann Kwan for her contributions to this work.

References

1. Wösten HAB, van Wetter MA, Lugones LG, van der Mei HC, Busscher HJ, Wessels JGH (1999) How a fungus escapes the water to grow into the air. Curr Biol 9:85–88. https://doi.org/10.1016/S0960-9822(99)80019-0
2. Beever RE, Dempsey GP (1978) Function of rodlets on the surface of fungal spores. Nature 272:608–610
3. Aimanianda V, Bayry J, Bozza S, Kniemeyer O, Perruccio K, Elluru SR, Clavaud C, Paris S, Brakhage AA, Kaveri SV, Romani L, Latgé JP (2009) Surface hydrophobin prevents immune recognition of airborne fungal spores. Nature 460:1117–1121. https://doi.org/10.1038/nature08264
4. Talbot NJ, Ebbole DJ, Hamer JE (1993) Identification and characterization of MPG1, a gene involved in pathogenicity from the rice blast fungus Magnaporthe grisea. Plant Cell 5:1575–1590. https://doi.org/10.1105/tpc.5.11.1575
5. Wösten HAB, Schuren FH, Wessels JGH (1994) Interfacial self-assembly of a hydrophobin into an amphipathic protein membrane mediates fungal attachment to hydrophobic surfaces. EMBO J 13:5848–5854
6. Hess WM, Sassen MMA, Remsen CC (1966) Surface structures of frozen-etched Penicillium conidiospores. Naturwissenschaften 53:708
7. Hess WM, Sassen MMA, Remsen CC (1968) Surface characteristics of Penicillium conidia. Mycologia 60:290–303
8. Dempsey GP, Beever RE (1979) Electron microscopy of the rodlet layer of Neurospora crassa conidia. J Bacteriol 140:1050–1062
9. Beever RE, Redgwell RJ, Dempsey GP (1979) Purification and chemical characterization of the rodlet layer of Neurospora crassa conidia. J Bacteriol 140:1063–1070
10. Templeton MD, Greenwood DR, Beever RE (1995) Solubilization of Neurospora crassa rodlet proteins and identification of the predominant protein as the proteolytically processed eas (ccg-2) gene product. Exp Mycol 19:166–169
11. Wessels JGH (1993) Cell wall growth, protein excretion and morphogenesis in fungi. New Phytol 123:397–413
12. Wessels JGH, de Vries OMH, Asgeirsdottir SA, Schuren FHJ (1991) Hydrophobin genes involved in formation of aerial hyphae and fruit bodies in Schizophyllum. Plant Cell 3:793–799
13. Linder MB, Szilvay GR, Nakari-Setälä T, Penttilä ME (2005) Hydrophobins: the protein-amphiphiles of filamentous fungi. FEMS Microbiol Rev 29:877–896. https://doi.org/10.1016/j.femsre.2005.01.004
14. Yang K, Deng Y, Zhang C, Elasri M (2006) Identification of new members of hydrophobin family using primary structure analysis. BMC Bioinforma 7(Suppl 4):S16. https://doi.org/10.1186/1471-2105-7-S4-S16
15. Jensen BG, Andersen MR, Pedersen MH, Frisvad JC, Søndergaard I (2010) Hydrophobins from Aspergillus species cannot be clearly divided into two classes. BMC Res Notes 3:344. https://doi.org/10.1186/1756-0500-3-344
16. Mackay JP, Matthews JM, Winefield RD, Mackay LG, Haverkamp RG, Templeton MD (2001) The hydrophobin EAS is largely unstructured in solution and functions by forming amyloid-like structures. Structure 9:83–91. https://doi.org/10.1016/S0969-2126(00)00559-1
17. Wessels JGH, Asgeirsdottir SA, Birkenkamp KU, de Vries OMH, Lugones LG, Scheer JMJ, Schuren FH, Schuurs TA, van Wetter M-A, Wösten HAB (1995) Genetic regulation of emergent growth in Schizophyllum commune. Can J Bot 73(Suppl. 1):S273–S281
18. Grünbacher A, Throm T, Seidel C, Gutt B, Röhrig J, Strunk T, Vincze P, Walheim S, Schimmel T, Wenzel W, Fischer R (2014) Six hydrophobins are involved in hydrophobin rodlet formation in Aspergillus nidulans and contribute to hydrophobicity of the spore surface. PLoS One 9:e94546. https://doi.org/10.1371/journal.pone.0094546

19. Lacroix H, Whiteford JR, Spanu PD (2008) Localization of Cladosporium fulvum hydrophobins reveals a role for HCf-6 in adhesion. FEMS Microbiol Lett 286:136–144
20. Whiteford JR, Spanu PD (2001) The hydrophobin HCf-1 of Cladosporium fulvum is required for efficient water-mediated dispersal of conidia. Fungal Genet Biol 32:159–168. https://doi.org/10.1006/fgbi.2001.1263
21. Lau G, Hamer JE (1996) Regulatory genes controlling MPG1 expression and pathogenicity in the rice blast fungus Magnaporthe grisea. Plant Cell 8:771–781. https://doi.org/10.1105/tpc.8.5.771
22. Nakari-Setälä T, Aro N, Ilmén M, Muñoz G, Kalkkinen N, Penttilä M (1997) Differential expression of the vegetative and spore-bound hydrophobins of Trichoderma reesei–cloning and characterization of the hfb2 gene. Eur J Biochem 248:415–423
23. Arpaia G, Loros JJ, Dunlap JC, Morelli G, Macino G (1993) The interplay of light and circadian clock: independent dual regulation of clock-controlled gene ccg-2 (eas). Plant Physiol 102:1299–1305
24. Sokolovsky VY, Lauter FR, Muller-Rober B, Ricci M, Schmidhauser TJ, Russo VEA (1992) Nitrogen regulation of blue light-inducible genes in Neurospora crassa. J Gen Microbiol 138:2045–2049
25. Ren Q, Kwan AHY, Sunde M (2013) Two forms and two faces, multiple states and multiple uses: properties and applications of the self-assembling fungal hydrophobins. Biopolymers 100:601–612. https://doi.org/10.1002/bip.22259
26. Lo VC, Ren Q, Pham CLL, Morris VK, Kwan AHY, Sunde M (2014) Fungal hydrophobin proteins produce self-assembling protein films with diverse structure and chemical stability. Nano 4:827–843. https://doi.org/10.3390/nano4030827
27. de Vocht ML, Reviakine I, Ulrich WP, Bergsma-Schutter W, Wösten HAB, Vogel H, Brisson A, Wessels JGH, Robillard GT (2002) Self-assembly of the hydrophobin Sc3 proceeds via two structural intermediates. Protein Sci 11:1199–1205
28. de Vries OMH, Fekkes MP, Wösten HAB, Wessels JGH (1993) Insoluble hydrophobin complexes in the walls of Schizophyllum commune and other filamentous fungi. Arch Microbiol 159:330–335
29. Wösten HAB, de Vries OMH, Wessels JGH (1993) Interfacial self-assembly of a fungal hydrophobin into a hydrophobic rodlet layer. Plant Cell 5:1567–1574
30. Butko P, Buford JP, Goodwin JS, Stroud PA, McCormick CL, Cannon CC (2001) Spectroscopic evidence for amyloid-like interfacial self-assembly of hydrophobin Sc3. Biochem Biophys Res Commun 280:212–215. https://doi.org/10.1006/bbrc.2000.4098
31. Kwan AHY, Winefield RD, Sunde M, Matthews JM, Haverkamp RG, Templeton MD, Mackay JP (2006) Structural basis for rodlet assembly in fungal hydrophobins. Proc Natl Acad Sci USA 103:3621–3626. https://doi.org/10.1073/pnas.0505704103
32. Kershaw MJ, Wakley G, Talbot NJ (1998) Complementation of the mpg1 mutant phenotype in Magnaporthe grisea reveals functional relationships between fungal hydrophobins. EMBO J 17:3838–3849. https://doi.org/10.1093/emboj/17.14.3838
33. Hektor HJ, Scholtmeijer K (2005) Hydrophobins: proteins with potential. Curr Opin Biotechnol 16:434–439. https://doi.org/10.1016/j.copbio.2005.05.004
34. Ren Q, Kwan AHY, Sunde M (2014) Solution structure and interface-driven self-assembly of NC2, a new member of the class II hydrophobin proteins. Proteins 82:990–1003. https://doi.org/10.1002/prot.24473
35. Szilvay GR, Paananen A, Laurikainen K, Vuorimaa E, Lemmetyinen H, Peltonen J, Linder MB (2007) Self-assembled hydrophobin protein films at the air-water interface: structural analysis and molecular engineering. Biochemistry 46:2345–2354. https://doi.org/10.1021/bi602358h
36. Hakanpää J, Paananen A, Askolin S, Nakari-Setälä T, Parkkinen T, Penttilä M, Linder M, Rouvinen J (2004) Atomic resolution structure of the HFBII hydrophobin, a self-assembling amphiphile. J Bio Chem 279:534–539. https://doi.org/10.1074/jbc.M309650200
37. Hakanpää J, Szilvay GR, Kaljunen H, Maksimainen M, Linder MB, Rouvinen J (2006) Two crystal structures of Trichoderma reesei hydrophobin HFBI—the structure of a protein amphiphile with and without detergent interaction. Protein Sci 15:2129–2140. https://doi.org/10.1110/ps.062326706

38. Morris VK, Kwan AHY, Sunde M (2013) Analysis of the structure and conformational states of DewA gives insight into the assembly of the fungal hydrophobins. J Mol Biol 425:244–256. https://doi.org/10.1016/j.jmb.2012.10.021
39. Pham CLL, Rey A, Lo VC, Soulès M, Ren Q, Meisl G, Knowles TPJ, Kwan AHY, Sunde M (2016) Self-assembly of MPG1, a hydrophobin protein from the rice blast fungus that forms functional amyloid coatings, occurs by a surface-driven mechanism. Sci Rep 6:25288. https://doi.org/10.1038/srep25288
40. Sunde M, Kwan AHY, Templeton MD, Beever RE, Mackay JP (2008) Structural analysis of hydrophobins. Micron 39:773–784. https://doi.org/10.1016/j.micron.2007.08.003
41. Morris VK, Ren Q, Macindoe I, Kwan AHY, Byrne N, Sunde M (2011) Recruitment of class I hydrophobins to the air:water interface initiates a multi-step process of functional amyloid formation. J Biol Chem 286:15955–15963
42. Askolin S, Linder MB, Scholtmeijer K, Tenkanen M, Penttilä M, de Vocht ML, Wösten HAB (2006) Interaction and comparison of a class I hydrophobin from Schizophyllum commune and class II hydrophobins from Trichoderma reesei. Biomacromolecules 7:1295–1301. https://doi.org/10.1021/bm050676s
43. van der Vegt W, van der Mei HC, Wösten HAB, Wessels JGH, Busscher HJ (1996) A comparison of the surface activity of the fungal hydrophobin Sc3p with those of other proteins. Biophys Chem 57:253–260
44. Wang X, Graveland-Bikker JF, de Kruif CG, Robillard GT (2004) Oligomerization of hydrophobin Sc3 in solution: from soluble state to self-assembly. Protein Sci 13:810–821. https://doi.org/10.1110/ps.03367304
45. de Vocht ML, Scholtmeijer K, van der Vegte EW, de Vries OMH, Sonveaux N, Wösten HAB, Ruysschaert JM, Hadziloannou G, Wessels JGH, Robillard GT (1998) Structural characterization of the hydrophobin Sc3, as a monomer and after self-assembly at hydrophobic/hydrophilic interfaces. Biophys J 74:2059–2068
46. Zangi R, de Vocht ML, Robillard GT, Mark AE (2002) Molecular dynamics study of the folding of hydrophobin Sc3 at a hydrophilic/hydrophobic interface. Biophys J 83:112–124. https://doi.org/10.1016/S0006-3495(02)75153-9
47. Kwan AHY, Macindoe I, Vukasin PV, Morris VK, Kass I, Gupte R, Mark A, Templeton MD, Mackay JP, Sunde M (2008) The Cys3-Cys4 loop of the hydrophobin EAS is not required for rodlet formation and surface activity. J Mol Biol 382:708–720. https://doi.org/10.1016/j.jmb.2008.07.034
48. Niu B, Gong T, Gao X, Xu H, Qiao M, Li W (2014) The functional role of Cys3-Cys4 loop in hydrophobin HGFI. Amino Acids 46:2615–2625. https://doi.org/10.1007/s00726-014-1805-0
49. Simone A, Kitchen C, Kwan AHY, Sunde M, Dobson C, Frenkel D (2012) Intrinsic disorder modulates protein self-assembly and aggregation. Proc Natl Acad Sci USA 109:6951–6956
50. Macindoe I, Kwan AHY, Ren Q, Morris VK, Yang W, Mackay JP, Sunde M (2012) Self-assembly of functional, amphipathic amyloid monolayers by the fungal hydrophobin EAS. Proc Natl Acad Sci USA 109:E804–E811. https://doi.org/10.1073/pnas.1114052109
51. Rambach G, Blum G, Latgé JP, Fontaine T, Heinekamp T, Hagleitner M, Jeckström H, Weigel G, Würtinger P, Pfaller K, Krappmann S, Löffler J, Lass-Flörl C, Speth C (2015) Identification of Aspergillus fumigatus surface components that mediate interaction of conidia and hyphae with human platelets. J Infect Dis 212:1140–1149. https://doi.org/10.1093/infdis/jiv191
52. Rohde M, Schwienbacher M, Nikolaus T, Heesemann J, Ebel F (2002) Detection of early phase specific surface appendages during germination of Aspergillus fumigatus conidia. FEMS Microbiol Lett 206:99–105
53. Dague E, Alsteens D, Latgé JP, Dufrêne YF (2008) High-resolution cell surface dynamics of germinating Aspergillus fumigatus conidia. Biophys J 94:656–660. https://doi.org/10.1529/biophysj.107.116491
54. Carrion SDJ, Leal SM, Ghannoum MA, Aimanianda V, Latgé JP, Pearlman E (2013) The RodA hydrophobin on Aspergillus fumigatus spores masks dectin-1- and dectin-2-dependent responses and enhances fungal survival in vivo. J Immunol 191:2581–2588. https://doi.org/10.4049/jimmunol.1300748

55. Loures FV, Röhm M, Lee CK, Santos E, Wang JP, Specht CA, Calich VLG, Urban CF, Levitz SM (2015) Recognition of Aspergillus fumigatus hyphae by human plasmacytoid dendritic cells is mediated by dectin-2 and results in formation of extracellular traps. PLoS Pathog 11:e1004643. https://doi.org/10.1371/journal.ppat.1004643
56. Steele C, Rapaka RR, Metz A, Pop SM, Williams DL, Gordon S, Kolls JK, Brown GD (2005) The beta-glucan receptor dectin-1 recognizes specific morphologies of Aspergillus fumigatus. PLoS Pathog 1:e42. https://doi.org/10.1371/journal.ppat.0010042
57. Khan NS, Kasperkovitz PV, Timmons AK, Mansour MK, Tam JM, Seward MW, Reedy JL, Puranam S, Feliu M, Vyas JM (2016) Dectin-1 controls TLR9 trafficking to phagosomes containing β-1,3 glucan. J Immunol 196:2249–2261. https://doi.org/10.4049/jimmunol.1401545
58. Paris S, Debeaupuis JP, Crameri R, Carey M, Charlès F, Prévost MC, Schmitt C, Philippe B, Latgé JP (2003) Conidial hydrophobins of Aspergillus fumigatus. Appl Environ Microbiol 69:1581–1588
59. Beauvais A, Schmidt C, Guadagnini S, Roux P, Perret E, Henry C, Paris S, Mallet A, Prévost MC, Latgé JP (2007) An extracellular matrix glues together the aerial-grown hyphae of Aspergillus fumigatus. Cell Microbiol 9:1588–1600. https://doi.org/10.1111/j.1462-5822.2007.00895.x
60. Bruns S, Seidler M, Albrecht D, Salvenmoser S, Remme N, Hertweck C, Brakhage AA, Kniemeyer O, Müller FMC (2010) Functional genomic profiling of Aspergillus fumigatus biofilm reveals enhanced production of the mycotoxin gliotoxin. Proteomics 10:3097–3107. https://doi.org/10.1002/pmic.201000129
61. Gibbons JG, Beauvais A, Beau R, McGary KL, Latgé JP, Rokas A (2012) Global transcriptome changes underlying colony growth in the opportunistic human pathogen Aspergillus fumigatus. Eukaryot Cell 11:68–78. https://doi.org/10.1128/EC.05102-11
62. Skamnioti P, Gurr SJ (2007) Magnaporthe grisea cutinase2 mediates appressorium differentiation and host penetration and is required for full virulence. Plant Cell 19:2674–2689. https://doi.org/10.1105/tpc.107.051219
63. Matsumura H, Reich S, Ito A, Saitoh H, Kamoun S, Winter P, Kahl G, Reuter M, Kruger DH, Terauchi R (2003) Gene expression analysis of plant host-pathogen interactions by SuperSAGE. Proc Natl Acad Sci USA 100:15718–15723. https://doi.org/10.1073/pnas.2536670100
64. Talbot NJ, Kershaw MJ, Wakley GE, De Vries OMH, Wessels JGH, Hamer JE (1996) MPG1 encodes a fungal hydrophobin involved in surface interactions during infection-related development of Magnaporthe grisea. Plant Cell 8:985–999. https://doi.org/10.1105/tpc.8.6.985
65. Inoue K, Kitaoka H, Park P, Ikeda K (2015) Novel aspects of hydrophobins in wheat isolate of Magnaporthe oryzae: Mpg1, but not Mhp1, is essential for adhesion and pathogenicity. J Gen Plant Pathol 82:18–28. https://doi.org/10.1007/s10327-015-0632-9
66. Khalesi M, Deckers SM, Gebruers K, Vissers L, Verachtert H, Derdelinckx G (2012) Hydrophobins: exceptional proteins for many applicaitons in brewery environment and other bio-industries. Cerevisia 37:3–9
67. Wösten HAB, Scholtmeijer K (2015) Applications of hydrophobins: current state and perspectives. Appl Microbiol Biotechnol 99:1587–1597. https://doi.org/10.1007/s00253-014-6319-x
68. Wang Z, Lienemann M, Qiau M, Linder MB (2010) Mechanisms of protein adhesion on surface films of hydrophobin. Langmuir 26:8491–8496. https://doi.org/10.1021/la101240e
69. Longobardi S, Gravagnuolo AM, Funari R, Della Ventura B, Pane F, Galano E, Amoresano A, Marino G, Giardina P (2015) A simple MALDI plate functionalization by Vmh2 hydrophobin for serial multi-enzymatic protein digestions. Anal Bioanal Chem 407:487–496. https://doi.org/10.1007/s00216-014-8309-3
70. Gravagnuolo AM, Morales-Narváez E, Matos CRS, Longobardi S, Giardina P, Merkoçi A (2015) On-the-spot immobilization of quantum dots, graphene oxide, and proteins via Hydrophobins. Adv Funct Mater 25(38):6084–6092. https://doi.org/10.1002/adfm.201502837
71. Sapsford KE, Medintz IL, Golden JP, Deschamps JR, Uyeda HT, Mattoussi H (2004) Surface-immobilized self-assembled protein-based quantum dot Nanoassemblies. Langmuir 20(18):7720–7728. https://doi.org/10.1021/la049263n

72. Haddada MB, Blanchard J, Casale S, Krafft J-M, Vallée A, Méthivier C, Boujday S (2013) Optimizing the immobilization of gold nanoparticles on functionalized silicon surfaces: amine- vs thiol-terminated silane. Gold Bull 46(4):335–341. https://doi.org/10.1007/s13404-013-0120-y
73. Alves NJ, Kiziltepe T, Bilgicer B (2012) Oriented surface immobilization of antibodies at the conserved nucleotide binding site for enhanced antigen detection. Langmuir 28(25):9640–9648. https://doi.org/10.1021/la301887s
74. Zhao ZX, Wang HC, Qin X, Wang XS, Qiao MQ, Anzai JI, Chen Q (2009) Self-assembled film of hydrophobins on gold surfaces and its application to electrochemical biosensing. Colloids Surf B Biointerfaces 71:102–106. https://doi.org/10.1016/j.colsurfb.2009.01.011
75. Wang X, Wang H, Huang Y, Zhao Z, Qin X, Wang Y, Miao Z, Chen Q, Qiao M (2010) Noncovalently functionalized multi-wall carbon nanotubes in aqueous solution using the hydrophobin HFBI and their electroanalytical application. Biosens Bioelectron 26:1104–1108. https://doi.org/10.1016/j.bios.2010.08.024
76. Rea I, Giardina P, Longobardi S, Porro F, Casuscelli V, Rendina I, De Stefano L (2012) Hydrophobin Vmh2-glucose complexes self-assemble in nanometric biofilms. J R Soc Interface 9:2450–2456. https://doi.org/10.1098/rsif.2012.0217
77. Politi J, De Stefano L, Rea I, Gravagnuolo A, Giardina P, Methivier C, Casale S, Spadavecchia J (2016) One-pot synthesis of a gold nanoparticle-Vmh2 hydrophobin nanobiocomplex for glucose monitoring. Nanotechnology 27:195701. https://doi.org/10.1088/0957-4484/27/19/195701
78. Bilewicz R, Witomski J, Van der Heyden A, Tagu D, Palin B, Rogalska E (2001) Modification of electrodes with self-assembled Hydrophobin layers. J Phys Chem B 105(40):9772–9777. https://doi.org/10.1021/jp0113782
79. Opwis K, Gutmann JS (2011) Surface modification of textile materials with hydrophobins. Text Res J 81:1594–1602
80. Popescu AC, Stan GE, Duta L, Dorcioman G, Iordache O, Dumitrescu I, Pasuk I, Mihailescu IN (2013) Influence of a hydrophobin underlayer on the structuring and antimicrobial properties of ZnO films. J Mater Sci 48:8329–8336
81. Lee JH, Khang G, Lee JW, Lee HB (1998) Interaction of different types of cells on polymer surfaces with wettability gradient. J Colloid Interf Sci 205:323–330. https://doi.org/10.1006/jcis.1998.5688
82. Janssen MI, van Leeuwen MBM, Scholtmeijer K, van Kooten TG, Dijkhuizen L, Wösten HAB (2002) Coating with genetic engineered hydrophobin promotes growth of fibroblasts on a hydrophobic solid. Biomaterials 23:4847–4854
83. Boeuf S, Throm T, Gutt B, Strunk T, Hoffmann M, Seebach E, Mühlberg L, Brocher J, Gotterbarm T, Wenzel W, Fischer R, Richter W (2012) Engineering hydrophobin DewA to generate surfaces that enhance adhesion of human but not bacterial cells. Acta Biomater 8:1037–1047. https://doi.org/10.1016/j.actbio.2011.11.022
84. Weickert U, Wiesend F, Subkowski T, Eickhoff A, Reiss G (2011) Optimizing biliary stent patency by coating with hydrophobin alone or hydrophobin and antibiotics or heparin: an in vitro proof of principle study. Adv Med Sci 56:138–144. https://doi.org/10.2478/v10039-011-0026-y
85. Stanimirova RD, Gurkov TD, Kralchevsky PA, Balashev KT, Stoyanov SD, Pelan EG (2013) Surface pressure and elasticity of hydrophobin HFBII layers on the air-water interface: rheology versus structure detected by AFM imaging. Langmuir 29:6053–6067. https://doi.org/10.1021/la4005104
86. Deckers SM, Venken T, Khalesi M, Gebruers K, Baggerman G, Lorgouilloux Y, Shokribousjein Z, Ilberg V, Schönberger C, Titze J, Verachtert H, Michiels C, Neven H, Delcour J, Martens J, Derdelinckx G, de Maeyer M (2012) Combined modeling and biophysical characterisation of CO2 interaction with class II hydrophobins: new insight into the mechanism underpinning primary gushing. J Am Soc Brew Chem 70:249–256
87. Cox AR, Aldred DL, Russell AB (2009) Exceptional stability of food foams using class II hydrophobin HFBII. Food Hydrocoll 23:366–376. https://doi.org/10.1016/j.foodhyd.2008.03.001

88. Tchuenbou-Magaia FL, Norton IT, Cox PW (2009) Hydrophobins stabilised air-filled emulsions for the food industry. Food Hydrocoll 23:1877–1885. https://doi.org/10.1016/j.foodhyd.2009.03.005
89. Green A, Littlejohn K, Hooley P, Cox P (2013) Formation and stability of food foams and aerated emulsions: hydrophobins as novel functional ingredients. Curr Opin Colloid Interface Sci 18(4):292–301
90. Loftsson T, Brewster ME (2010) Pharmaceutical applications of cyclodextrins: basic science and product development. J Pharm Pharmacol 62:1607–1621. https://doi.org/10.1111/j.2042-7158.2010.01030.x
91. Haas Jimoh Akanbi M, Post E, Meter-Arkema A, Rink R, Robillard GT, Wang X, Wösten HAB, Scholtmeijer K (2010) Use of hydrophobins in formulation of water insoluble drugs for oral administration. Colloids Surf B Biointerfaces 75:526–531. https://doi.org/10.1016/j.colsurfb.2009.09.030
92. Fang G, Tang B, Liu Z, Gou J, Zhang Y, Xu H, Tang X (2014) Novel hydrophobin-coated docetaxel nanoparticles for intravenous delivery: in vitro characteristics and in vivo performance. Eur J Pharm Sci 60:1–9. https://doi.org/10.1016/j.ejps.2014.04.016
93. Valo HK, Laaksonen PH, Peltonen LJ, Linder MB, Hirvonen JT, Laaksonen TJ (2010) Multifunctional hydrophobin: toward functional coatings for drug nanoparticles. ACS Nano 4:1750–1758. https://doi.org/10.1021/nn9017558
94. Yang W, Ren Q, Wu YN, Morris VK, Rey AA, Braet F, Kwan AHY, Sunde M (2013) Surface functionalization of carbon nanomaterials by self-assembling hydrophobin proteins. Biopolymers 99:84–94. https://doi.org/10.1002/bip.22146
95. Sarparanta M, Bimbo LM, Rytkönen J, Mäkilä E, Laaksonen TJ, Laaksonen P, Nyman M, Salonen J, Linder MB, Hirvonen J, Santos HA, Airaksinen AJ (2012) Intravenous delivery of hydrophobin-functionalized porous silicon nanoparticles: stability, plasma protein adsorption and biodistribution. Mol Pharm 9:654–663. https://doi.org/10.1021/mp200611d

Chapter 6
Nanostructured, Self-Assembled Spider Silk Materials for Biomedical Applications

Martin Humenik, Kiran Pawar, and Thomas Scheibel

Abstract The extraordinary mechanical properties of spider silk fibers result from the interplay of composition, structure and self-assembly of spider silk proteins (spidroins). Genetic approaches enabled the biotechnological production of recombinant spidroins which have been employed to unravel the self-assembly and spinning process. Various processing conditions allowed to explore non-natural morphologies including nanofibrils, particles, capsules, hydrogels, films or foams. Recombinant spider silk proteins and materials made thereof can be utilized for biomedical applications, such as drug delivery, tissue engineering or 3D-biomanufacturing.

Keywords Biofabrication · Drug delivery · Fibers · Genetic engineering · Recombinant production · Self-assembly · Spider silk · Tissue engineering

6.1 Introduction

Spider webs and spider silk fibers have been fascinating mankind since ages, exemplarily seen in ancient myths about Arachne, a mortal woman who challenged

---

M. Humenik
Biomaterials, Faculty of Engineering Science, University of Bayreuth, Bayreuth, Germany

K. Pawar
Blusson Spinal Cord Centre, University of British Columbia, Vancouver, Canada

T. Scheibel (✉)
Bayreuth Center for Colloids and Interfaces (BZKG), University of Bayreuth, Bayreuth, Germany

Research Center Bio-Macromolecules (BIOmac), University of Bayreuth, Bayreuth, Germany

Bayreuth Center for Molecular Biosciences (BZMB), University of Bayreuth, Bayreuth, Germany

Bayreuth Center for Material Science (BayMAT), University of Bayreuth, Bayreuth, Germany

Bavarian Polymer Institute (BPI), University of Bayreuth, Bayreuth, Germany
e-mail: thomas.scheibel@bm.uni-bayreuth.de

S. Perrett et al. (eds.), *Biological and Bio-inspired Nanomaterials*, Advances in Experimental Medicine and Biology 1174,
https://doi.org/10.1007/978-981-13-9791-2_6

Athena in weaving, or in stories about SpiderMan, a mutated young man acquiring spider abilities after a bite of a radioactively modified spider. Apart fiction, several specific properties of spider silk webs have been recognized early-on allowing applications, such as wound coatings in ancient Greece and fishing nets of tribes from Papua-New Guinea and Australia [1]. Nowadays, scientific insights into the material properties have been gained which allow to mechanistically understand the assembly of this proteinaceous material, as well as to technically apply it for multiple purposes. Spider silk, like bone, nacre, or wood, reveals a hierarchically ordered structure, which self-assembles from simple building blocks just upon weak (but numerous) non-covalent interactions, whereas its chemical and structural compatibility, probed by evolution from nanoscales to macroscales, results in coherent materials [2].

Here, spider silk composition and fiber structure will be elucidated, including self-assembly principles underlining its hierarchical setup. We will also present bioinspired technologies, which on the one hand helped to understand the self-assembly and spinning process, and on the other hand enabled the application of silk-based materials. Spider-silk based morphologies, such as nanofibrils, particles, capsules, hydrogels, films, and foams, allowed the development of biomedical applications (among others), which are presented in the final section of this book chapter.

6.2 Natural Spider Silk

Currently more than 48.000 spider species, from tiny Peacock to large Goliath bird-eating spiders, are known, which are divided into 109 families [3], occupying every possible niche on Earth and representing one of the most diverse orders of all organisms. All spider species spin silk threads, proteinaceous materials with quite different composition and structure, which are also produced by a broad range of other Arthropods [4, 5]. Many spiders weave silk webs of varying architecture, such as funnel, sheet, bola, cob and orb webs, serving different purposes ranging from an extension of their sensory system to catching prey to protective shells of eggs. Orb webs represent the most fascinating and probably the most complex features of silk architecture. They are produced by the suborder *Araneomorphae* of the *Orbiculariae* clade [6] being typically composed of five different silks with specific mechanical, physical, and chemical properties, which are combined to stop effectively the flying prey by dissipating the kinetic energy of the impact via nonlinear mechanical stress response to minimize possible damages of the web and, at the same time, to keep the pray within the web [7].

Different types of silk proteins are produced by specialized glands in the spider abdomen [8–10]. Orb weaving spiders of *Nephila* or *Araneus* genera show seven individual types (one of which represents a glue), including major and minor ampullate, flagelliform, tubuliform, aciniform, pyriform and aggregate glands [11].

The frame and radii of the orb web are made of proteins produced in major ampullate (MA) glands, and the respective silk fibers are constantly drawn to be also used as a lifeline. Minor ampullate (MI) silk is used as a temporary support spiral and, together with flagelliform silk, it represents the capture spiral. Junctions between MA and flagelliform fibers are reinforced by pyriform silk, which also serves as an attachment cement to anchor the frame on substrates. Aggregate silk is deposited on the capture spiral as a glue. Two silk types are produced to protect offspring in eggs. The cases are made of tough tubuli- /cylindriform silk (both names describe the same type), whereas a soft layer in the egg interior consists of aciniform silk, also used in prey-wrapping [6, 12].

6.2.1 Protein Composition of Major Ampullate Silk

Spider silks typically comprise one or more proteins with various molecular weights. They share a general composition of a repetitive core domain with varying primary sequences among the silk types and small flanking C- and N-terminal domains which are highly conserved between the silk types but also between spider species [11, 13, 14]. Different sequences in the core domain result consequently in different mechanical properties of the respective threads ranging from extensive rubber-like features of the capture spiral to stiff and tough MA silk [15–17]. Aggregate silk, as the only exception in morphology and composition, is deposited as glue droplets and composed of at least two small (< 65 kDa), highly glycosylated proteins and two peptides [18]. In addition to these well described spidroins [6], recent genomic and proteomic approaches have identified a plethora of additional proteins associated with silk production, storage and spinning [19]. There were more than 600 silk-gland specific transcripts (SSTs) identified by Clarke et al. in the case of the black widow spider (*Latrodectus hesperus*) [20]. The majority of the transcripts showed no annotation, but they may include novel silk-associated proteins (SAPs) [8], such as low molecular weight cysteine-rich proteins (CRPs), which were found in the MA dope as well as in the related fiber forming macromolecular complexes [21]. Egg case proteins (ECPs) represent a non-spidroin family identified in black widow tubuliform glands, and respective fibers made of TuSp1 are thought to be cross-linked with ECPs [12].

Details of duct morphologies and spinning of particular spider silks can be found elsewhere [6, 22]. Here, we introduce the general but at the same time most sophisticated features of silk composition, structure and assembly into fibers using the example of MA silk, which represents the best studied silk type. The MA gland is the largest one amongst the spider silk glands providing sufficient amounts of the material for analysis using X-ray diffraction, NMR, or IR spectroscopy [23–26].

The major ampulla is divided in a long tail (zone A) and a sac (zone B and C) continuing into a tapered S-shaped duct connected to an external spigot [27, 28]. In zones A and B proteins are secreted, whereas in zone C they are stored at high concentration (~25–50 wt %) [22]. Major ampullate silk proteins can be classified into spidroin 1 and 2 (MaSp1 and MaSp2) with molecular weights often exceeding 250 kDa [29–31], but low MW variants consisting of a small nonrepetitive core domain do exist [32]. The main difference between MaSp1 and MaSp2 is the proline content, being high in MaSp2 and quite low in MaSp1. In general, *Nephila* and *Latrodectus* MA dopes show only 1–2% of proline, whereas *Araneus* and *Argiope* MA silk contain proline residues in the range of 10–12% [33]. Interestingly, when comparing native dragline silks, they display similar mechanical performance in the dry state, but under humid conditions, they reveal pronounced supercontraction correlating with the increased proline content. Thus, regulating the ratio of MaSp1 and MaSp2 levels enables spiders to alter the material properties of their fibers in response to environmental changes [34, 33].

Core domains of MaSps are prevalently composed of glycine (Gly), alanine (Ala), glutamine (Gln), proline (Pro), serine (Ser) and tyrosine (Tyr) residues [35], which are clustered into small motifs, such as poly-alanine stretches $(Ala)_n$ (n = 4–12), GA, and GGX (X = Y, L and Q) repeats. MaSp2 contains additionally GPGXX repeats (X = Q, G, Y) which alternate with GGX [6, 36, 37]. In case of fully sequenced MaSp1 and MaSp2 of *L. hesperus* [30] and of the wasp spider (*Argiope bruennichi*) [38], as well as proteomically analyzed MaSps of golden silk orb-weaver *Nephila clavipes* [39] it has been shown that distinct sequence motifs can be repeated ~100 times within the core domain. Several phosphorylation sides have been also found in MaSps [39, 40].

In solution i.e., in the sac, the repetitive core of MaSps is intrinsically unstructured. NMR and IR spectroscopy studies have shown no distinct secondary structures but only small contributions of polyproline II (PPII) and α-helices [10, 25, 41–43]. In contrast, terminal domains represent globular, folded structures consisting of bundles of five α-helices [14, 44, 45]. Immunological experiments, using specific antibodies, and mass-spectrometry analysis have shown the presence of folded C- as well as N-terminal domains in spinning dopes as well as in the fibers [19, 27, 46]. C-terminal domains contain highly conserved Cys residues and are assumed to dimerize in a parallel orientation. N-terminal domains are monomeric in the dope and dimerize upon acidification in an antiparallel fashion, as described in the next section. The sequences of the domains are more hydrophobic than the repetitive core, however their native folds expose prevalently hydrophilic residues ensuring the spidroin solubility [30, 44, 45].

6.2.2 Processing of Spider Silk Proteins into Fibers

In the sac, highly concentrated soluble spidroins are stabilized in the presence of NaCl and at neutral pH. Along the spinning duct acidification takes place as well as an exchange of chaotropic Na^+ and Cl^- ions for more kosmotropic K^+ and PO_4^{3-} [42, 47–49]. The structure of the C-terminal domain is sensitive to these changes. Destabilization of the structure, partial unfolding, and exposure of its hydrophobic core have been shown to support aggregation under influence of shear forces [45, 50, 51]. In contrast, the monomeric N-terminal domain dimerizes in an antiparallel fashion, driven by electrostatic interactions upon slight acidification to pH 5–6 [52–60]. The repetitive core domain experiences a transformation of random coil into β-sheet rich structures, mainly induced by the already mentioned kosmotropic ions in the spinning duct [24]. NMR spectroscopy has shown that mainly poly-alanine and GA motifs form the β-sheets, whereas GGX and especially GPGXX motifs form helical or less ordered structures yielding an amorphous matrix. X-ray diffraction (XRD) and solid-state NMR studies revealed an alignment of the stacked β-sheets along the fiber axis, which resemble small crystallites with less than 10 nm edge length. The amorphous region revealed both, isotropic and anisotropic components [26, 61, 62–65]. Structural transformations of spidroins upon the chemical and physical changes in the spinning duct are summarized in Fig. 6.1.

6.2.3 Structure-Mechanics Relationships

Dragline fibers comprise a lipid and glycoprotein shell surrounding the MaSp core composed of fibrillar substructures (Fig. 6.2a) in which the above mentioned β-sheet nanocrystals are embedded in an amorphous matrix [66–68]. Such a setup enables good extensibility (~25%), high strength (~1.2 GPa) and an extraordinary toughness exceeding every known natural or man-made fibrous material [16, 69, 70] (Table 6.1).

Pulling a fiber results in a nonlinear response with three typical regimes (Fig. 6.2b, regimes highlighted in three green shades) as a response to sequential deformation of the individual structures. Homogeneous stretching of partially aligned amorphous regions occurs in the linear elastic phase [71] (Fig. 6.2c, Regime I). Young's (elastic) moduli, representing the stiffness of the material, are in the range of 10 GPa. Upon reaching the yield point, the H-bonds in the amorphous phase rupture followed by unfolding of the protein chains (Fig. 6.2c, Regime II), drop of stiffness and gain of viscoelastic behavior [72]. The β-sheet nanocrystals provide cohesion points transferring the load between peptide chains and enabling full extension of the amorphous regions [72, 73]. Final stiffening is based on fully stretched amorphous chains and a transfer of the load onto the β-sheet crystals [71–73] (Fig. 6.2c, Regime III). The alignment of the β-sheets with the fiber axis, which

Table 6.1 Mechanical performance of major ampullate silk of different orb weavers in tensile-stress tests in comparison to that of high performance man-made fibers[a]

Species	Stiffness (GPa)	Strength (MPa)	Extensibility (%)	Toughness (MJ/m^3)
Argiope trifasciata	5.2	1584	29	163
Araneus diadematus	3.6	1599	33	193
Nephila clavipes	8.4	1725	28	206
Caerostris darwini	11.5	1850	33	271
High-tensile steel	200	1500	0.8	6
Carbon fiber	300	4000	1.3	25
Kevlar	130	3600	2.7	50

[a]Values taken from Refs. [70] and [16]

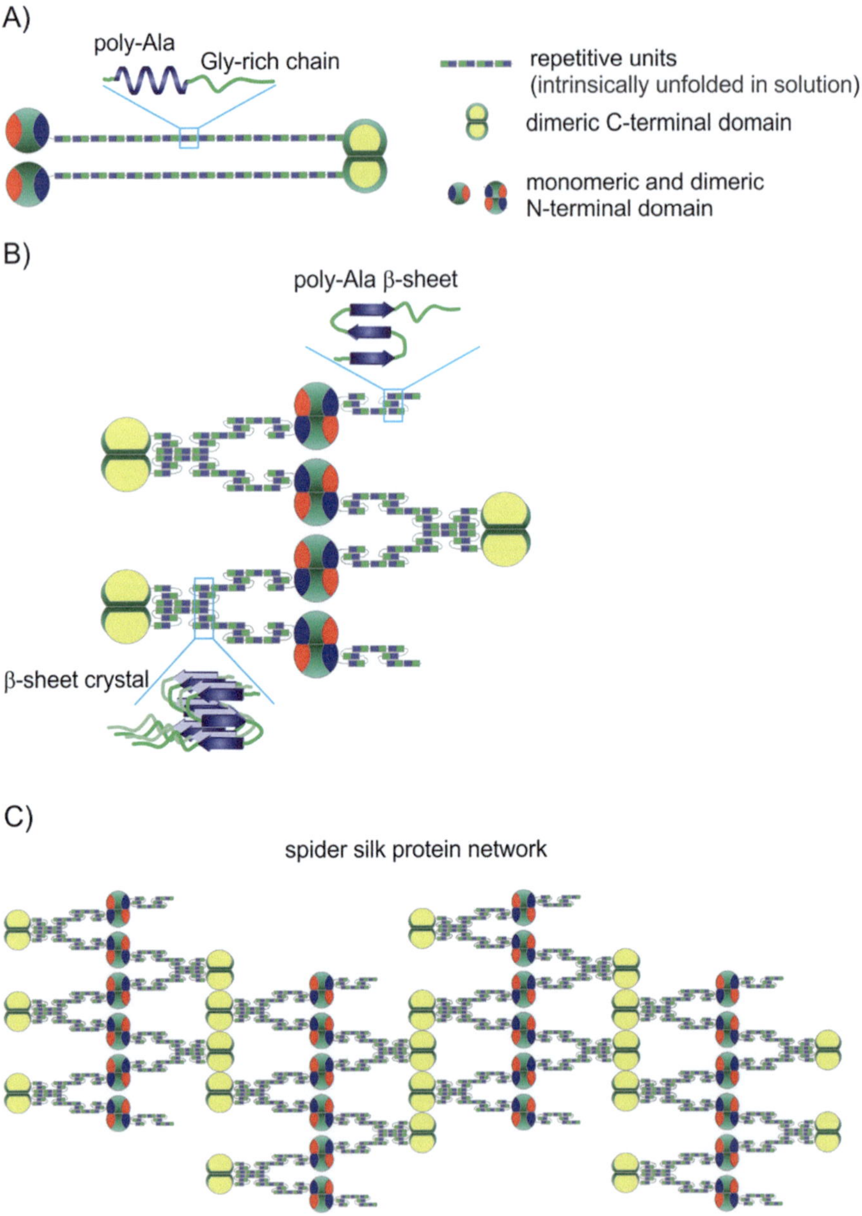

Fig. 6.1 Molecular spider silk network. (**a**) Schematic structure of spidroins with core, dimeric C-terminal and monomeric N-terminal domains. (**b**) Structural transformation of the core domain and assembly occurring upon chemical (pH drop, salting-out) and physical triggers (shear forces) in the spinning duct. (**c**) Schematic representation of a spidroin network cross-linked by dimeric C-terminal and N-terminal domains

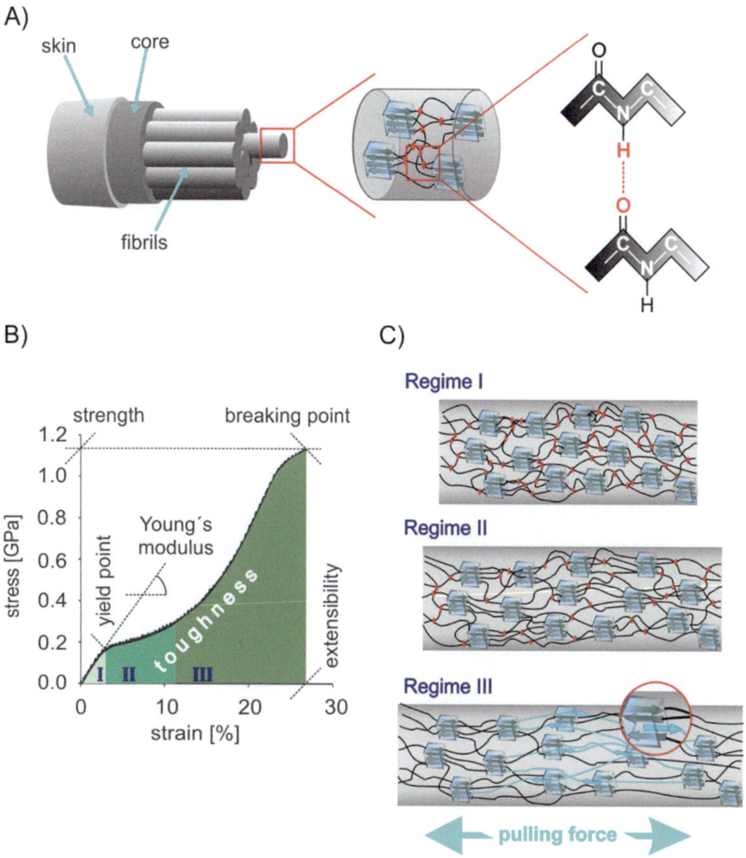

Fig. 6.2 Mechanical performance of dragline fibers. (**a**) Core-shell structure of a fiber built on fibrils comprising amorphous/crystallite domains crosslinked by H-bonds; (**b**) Schematic representation of a typical stress-strain curve with an initial linear region indicating the Young's modulus i.e., the fiber stiffness (Regime I), a visco-elastic region (Regime II), and a Regime III in which stiffening occurs. Maximal values at the break point determine the extensibility and strength of the fiber, whereas the integral of the curve represents the fiber's toughness; (**c**) Structural changes in amorphous matrix and crystallites upon pulling. (Reproduced from Ref. [3], with permission of Creative Commons Attribution License (https://creativecommons.org/licenses/by-nc-sa/3.0/))

is even enhanced at applied stress [74, 75], is crucial for energy dissipation. A stick-slip mechanism has been proposed for multiple and concerted breaking and re-bonding of H-bonds during lateral withdrawal of the chain from the crystal [73] (Fig. 6.2c, Regime III, light turquoise β-sheets). Final pulling of the individual chains out of crystallites causes rapid disintegration of the silk fiber.

6.3 Recombinant Spider Silk Proteins

The extraordinary mechanical properties of spider silk together with its biocompatibility, accompanied by the fact of spider's cannibalistic behavior and the lower quality of the fibers when harvested in captivity, initiated efforts to recombinantly produce spidroins. The recombinant production of spider silk proteins has been reviewed thoroughly in recent years [69, 76, 77]. Here, we will focus on different recombinant spider silk proteins which served as models for understanding self-assembly and spinning processes, and provided a basis for new self-assembled, nanostructured, hierarchical materials including various applications thereof.

6.3.1 Self-Assembly of Artificial Spider Silk Proteins

As described above, different domains of spidroins fulfill distinct roles during storage as well as final fiber assembly. Genetic engineering has allowed the establishment of molecular spidroin "Lego" [78, 75] to study the impact of individual domains on solubility, self-assembly and fiber mechanics [45, 51, 79–82]. Engineered spidroins can be studied using microfluidics to determine the effects of solvent changes and shear forces on the formation of spidroin fibers. Microfluidic chips have been designed to control mixing and flow conditions along with chemical changes of the protein solution [81]. Shear forces together with a simultaneous pH switch are required to initiate fiber assembly. Further, it was shown that ordered assembly of a MaSp2 analogue of *A. diadematus* (ADF3) is possible in the presence of the C-terminal domain under shear stress, resulting in fibrillar structures with aligned β-sheets. In contrast, the absence of this domain resulted in poorly-defined β-sheet aggregates with no alignment [51, 79].

Recombinant mini-spidroins consisting of four repetitive units (4Rep, 1Rep = GSGNSGIQGQGGYGGLGQGGYGQGAGSSAAAAAAAAAAAA) with or without corresponding C-terminal (CT) and/or N-terminal (NT) domains, based on MaSp1 from *Euprosthenops australis,* were used to probe effects of pH and salt on self-assembly. 4Rep and 4RepCT are prone to assemble slowly at neutral as well as acidic pH. In the presence of NT the assembly slows down at neutral conditions. However, acidification induces significantly faster self-assembly of NT4Rep and NT4RepCT into macroscopic fibers. Such effects of NT on spidroin assembly has been confirmed in a microfluidic continuous spinning process [83].

Variants of spider silk core domains have been studied to elucidate the influence of distinct amino acid motifs on assembly. Hydrophobicity plots of spidroins [30, 80, 84, 85] revealed the amphiphilic nature of the spider silk core domain with alternating hydrophobic and hydrophilic blocks, which is clearly important for phase separation during the spinning process [22, 86]. Recombinant spidroins with hydrophobic poly-alanine (A blocks; GAGAAAAAGGAGTS) and hydrophilic glycine-rich regions (B blocks; SQGGYGGLGSQGSGRGGLGGQLTS), derived from MaSp1 of *Nephila clavipes,* showed that sequentially increasing the number of A blocks has a direct impact on the crystallinity index and β-sheet content [87, 88]. Since the sequence variability of spidroin repetitive units on the amino acid level is quite high [84], the prediction of self-assembly behavior of silk proteins is more complex in comparison to that of simple synthetic hydrophilic/hydrophobic block copolymers [89]. Thus, recent approaches have combined biopolymer synthesis and *in silico* modeling to elucidate the role of hydrophobic and hydrophilic components concerning self-assembly and mechanics to allow a rational design of spider silk materials [90–92]. To understand the effect of sequence length on self-assembly propensity, two and twelve AB block-containing proteins were studied experimentally as well as in coarse-grained simulations [90]. The simulation results were consistent with the experimentally observed trends, in which the protein with twelve AB blocks formed larger spherical aggregates in comparison to $(AB)_2$. In this context, it has been demonstrated that the molecular weight of the spider silk core domain significantly affects the ability to form protein networks. In computational models, the chains extended under shear flow, and the size of aggregates increased with increasing numbers of the A block by merging multiple micelles [90, 91]. This also explains the previous observation of a monotonic increase in fiber mechanics with the molecular weight of the recombinant proteins up to 96 repeated units (molecular weight of 285 kDa; the unit: SGRGGLGGQGAGMAAAAAMGGAGQGGYGGLGSQGT) [93]. Thus, increased interconnections between the spidroins in the spinning process suggest the advantage of longer chain length to form more robust fibers [91]. However, the combination of high and low molecular weight spidroins in one spinning dope, which can interact through their terminal domains, empowers additional possibilities to adjust the mechanical properties of spider silk fibers [94].

The influence of the number of repeats has been investigated thoroughly in terms of assembly kinetics using core domains with increasing repeat numbers in the engineered MaSp2 derivative eADF4(Cn). Whereas the β-sheet content in the assemblies remains the same for 2–16 repetitive units, the kinetics of the self-assembly process is increasingly faster with increasing numbers of repetitive sequences. This study also revealed that at least 2 blocks are necessary for self-assembly since the single peptide unit eADF4(C1) (C-module: GSSAAAAAAAASGPGGYGPENQG-PSGPGGYGPGGP) does not self-assemble [95]. Interestingly, eADF4(C1) could be incorporated in growing fibrils upon seeding, highlighting the active role of the respective seed/fibril growth interface for β-sheet transformation of the docking monomeric spidroin (see also Sect. 6.3.2.1).

Many recombinant variants with different MWs, mainly based on MA core domains of *Nephila* and *Araneus* species, have been used to process fibers. Typical approaches included the production of spinning dopes in strong organic solvents, such as hexafluoroisopropanol (HFIP) or formic acid (FA), in which high spidroin concentrations can be reached. However, due to strong protein/solvent interactions no native molecular protein structures could be formed. Consequently, subsequent spinning processes, utilizing extrusion of fibers into coagulation baths or electrospinning approaches based on thin fast drying jets produced in high electric fields, resulted in significantly lower mechanical performance than that of for the natural dragline fiber. Correct protein-protein interactions, which are a prerequisite for spidroin pre-arrangement and self-assembly, are only provided in aqueous spinning dopes. However, the utilization of such aqueous spidroin solutions is often accompanied by inherent low spidroin solubility or premature aggregation. In this context, the presence/absence of the terminal domains was found to be crucial for proper assembly with high impact on the fiber mechanical performance. Comprehensive overviews on artificial spinning approaches have been provided recently [1, 47, 96–98].

The distinct molecular design of spidroins (Fig. 6.3a) has been used to study their self-assembly into mechanically stable fibers. The preparation of biomimetic spinning dopes upon protein phase separation induced a molecular pre-organization allowing wet spinning of fibers with high molecular alignment and fiber toughness (180 MJ/m^3) matching that of natural dragline silk [75, 82, 94] (Fig. 6.3b). However, the high toughness was based on the fact that the engineered fibers were much more extensible and less strong than natural fibers. Further improvements of the system by combination of MaSp2 and MaSp1 recombinant analogues and development of a biomimetic spinning technology will probably yield fibers with more adjustable mechanical features [94].

6.3.2 *Materials Made of Recombinant Spider Silk Proteins*

There are two main principles for preparation of spider silk-based biomaterials. In the first approach, self-assembly of the recombinant spidroins is carried out in an aqueous milieu and generally at ambient temperature. Depending on the particular spidroin and environmental conditions, formation of supramolecular assemblies, such as nanofibrils, particles, capsules or hydrogels, is possible. The second approach exploits spidroin solutions in organic solvents followed by different processing and drying steps. The templates used can define the final material morphology especially for films or foams (Fig. 6.4).

6.3.2.1 Nanofibrils

The formation of nanofibrils was observed in spider silk solutions upon addition of potassium ions *in vitro* [106], in the spinning duct of a spider after dissection [107] as well as upon acidification of diluted spinning dope solutions [42]. Such nanofibril assembly is very slow and, therefore, not a part of the natural spider dragline silk assembly [62, 68, 74, 108], which takes place within milliseconds during the spinning process [55]. Nanofibrils could also be produced from short peptides, such as GAGAAAAGAGA, as well as from the recombinant spider silk pS(4 + 1) derived from *Nephila clavipes* dragline silk which formed β-sheet-rich fibrils in aqueous buffers within several hours [109, 110]. The fibrils showed a length distribution between 60–600 nm and a segmented substructure

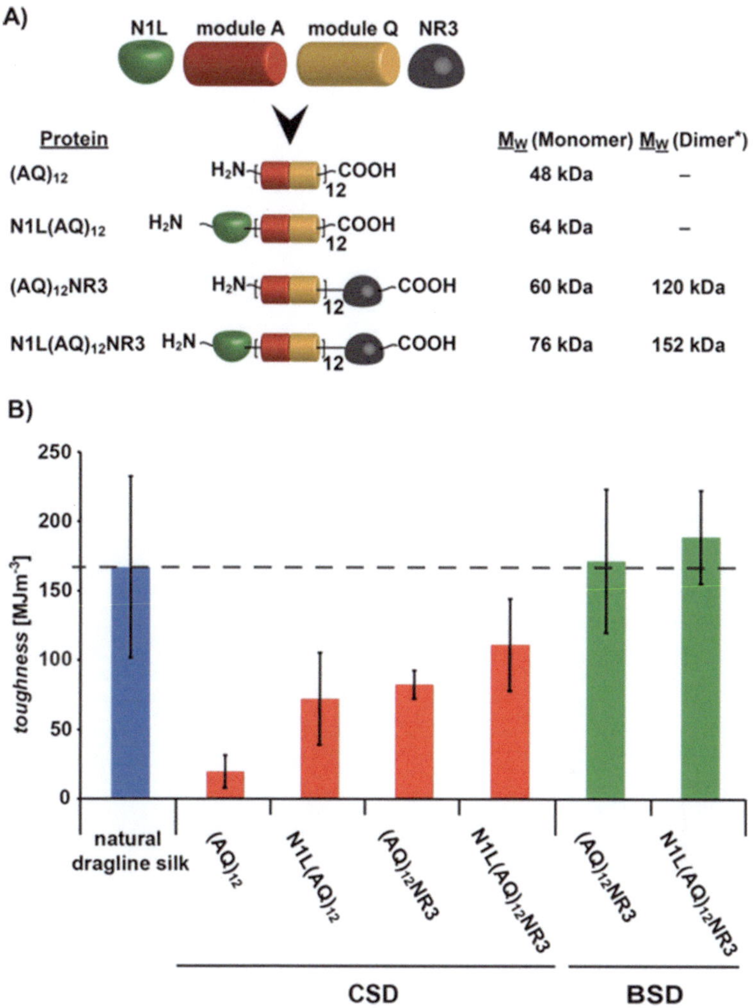

Fig. 6.3 Biotech-approach towards though fibers. (**a**) Molecular spidroin "Lego" of designed recombinant spider silk proteins combining consensus sequence motifs A (GPYGPGASAAAAAAGGYGPGSGQQ) and Q (GPGQQGPGQQGPGQQGPGQQ) as well as the C-terminal domain NR3 (124 amino acids), based on ADF3 of *A. diadematus,* and the N-terminal domain N1L (180 amino acids), based on the aminoterminal non-repetitive domain of MaSp2 of *L. hesperus*. (**b**) Fibers were spun out of classical spinning dopes (CSD) or out of biomimetic spinning dopes (BSD). The toughness of the fibers depended on protein composition and dope preparation. (Reproduced from [82] with permission of Wiley-VCH)

with segment diameters and heights of approximately 35 nm and 3 nm, respectively. The molecules were proposed to have β-sheets in a parallel orientation to the fibril axis [110]. *Araneus diadematus* derived recombinant eADF4(C16) serves as a role-model for fibril self-assembly in aqueous conditions. The fibrils (Fig. 6.4a) formed

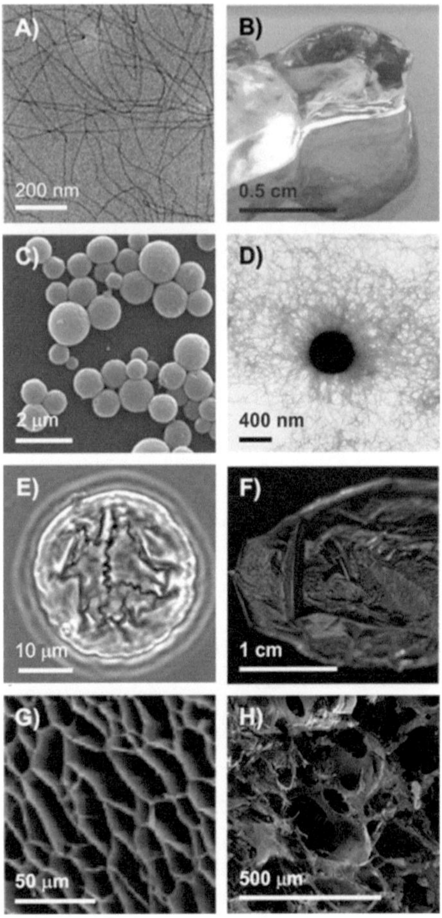

Fig. 6.4 Non-natural morphologies made from recombinant spider silk protein eADF4(C16). (**a**) Nanofibrils assembled at low phosphate concentrations. (**b**) Hydrogels formed due to interactions and entanglements of nanofibrils. (**c**) Particles made at high phosphate concentrations. (**d**) Seeding of nanofibrils from the particle surface. (**e**) Capsules assembled using a water-in-oil emulsion. (**f**) Cast spidroin film. (**g**) Freeze dried hydrogels with porous morphology. (**h**) Stable, β-sheet-rich foam prepared using salt leaching. Images: Cryo-TEM in A; bright field in B, E and F; SEM in C, G and H; TEM in D. (Reproduced as follows: **a** from [99], **b** and **g** from [100], **c** from [101], **d** from [102], **e** from [103], **f** from [104] and **h** from [105], with permission of the American Chemical Society (**a**, **b**, **g** and **h**), Wiley-VCH (**c** and **e**), Elsevier (**d**) and Springer-Verlag (**f**))

at low phosphate concentrations (< 300 mM) and showed a high content of β-sheets (40%). In contrast to the arrangement in natural fibers a cross-β conformation was seen using X-ray diffraction, Thioflavin T and Congo Red binding [95, 111]. Assembly kinetics confirmed protein and phosphate concentration-dependent oligomerization and growth rates [102], which also increased with the number of repeats in the case of eADF4(Cn) variants with n = 2–16 [95]. Fibril morphologies and secondary structures of all variants were, however, indistinguishable. Formation of the fibrils followed a nucleation mechanism with a slow oligomerization and nucleus formation and a fast growth phase induced specifically by K^+ and PO_4^{3-} ions [102, 112]. Importantly, the formation of cross-β fibrils required hours to completion unlike the fast natural fiber assembly process [55]. The fibril formation rather reflects a generic feature of intrinsically unstructured proteins to transform into a thermodynamically stable cross-β sheet structure than being part of the natural fiber assembly process.

Mechanical analysis of cross-β fibrils [113–117] revealed a stiffness in the range of 0.2–14 GPa depending on the underlying protein, which is in the range of the most rigid proteinaceous materials, such as keratin, collagen in tendons, natural dragline spider silk [118] or lacewing silk [119], enabling utilization of the fibrils as scaffolds for self-assembly of nanostructured, hierarchically ordered materials [118, 120–124].

In this context, self-assembly properties of spidroins have been combined with different functionalities such as nucleic acids in DNA-spider silk conjugates [99, 125, 126] or enzymes in genetic fusions with silk moieties [127, 128]. In a chemical approach, eADF4(C16) was modified with short oligonucleotides [126] using azide and alkyne linkers enabling "click" conjugation. The short nucleic acid moieties in the DNA-spider silk hybrids did not interfere with self-assembly of the spidroins nor with the final fibril morphology [99, 126]. However, the presence of the nucleic acid strands on the fibril surface allowed specific functionalization e.g., with complementary DNA-gold nanoparticles (Fig. 6.5a). Temperature gradients applied during assembly of complementary DNA-spider silk hybrids [99] enabled the formation of ordered fibril nano-ribbons and micro-rafts (Fig. 6.5b).

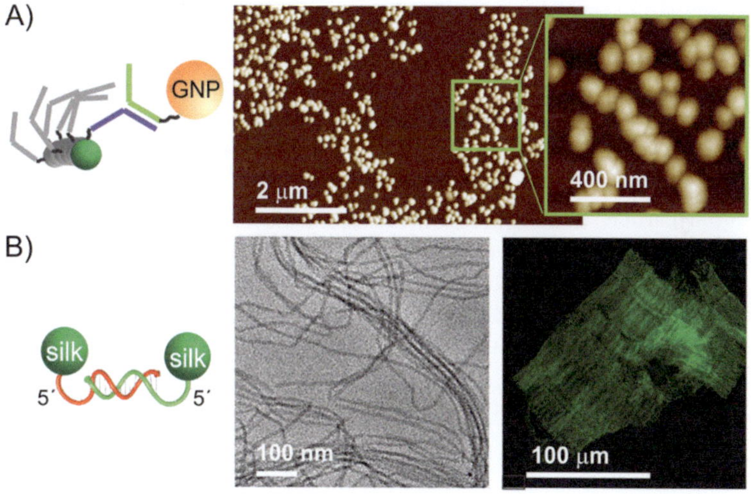

Fig. 6.5 Hierarchical assemblies made of DNA-spidroin hybrids. Specific attachment of DNA-gold nanoparticles onto hybrid fibrils. (**b**) Self-organization of hybrid fibrils into nano-ribbons and micro-rafts upon fibril assembly and annealing at temperatures close to the Tm of DNA moieties of the hybrid. AFM scans in **a**; Cryo-TEM (left) and confocal florescence microscopy (right) in **b**. (**a** reproduced from [126] and **b** reproduced from [99] with permission of the American Chemical Society)

6.3.2.2 Hydrogels

Based on nanofibrils (Sect. 6.3.2.1.), spidroins can further assemble into hydrogels (Fig. 6.4b) at protein concentrations above 1% [100, 129–132], representing a hierarchically ordered, supramolecular physical polymer network [133, 134]. Such hydrogels reveal visco-elastic properties upon shear with maximal shear stresses between 200 and 2000 Pa, depending on protein concentration (3–7%) [100]. Good shape fidelity and comprehensive stress of 140 kPa has been also demonstrated [130, 135]. Spidroin hydrogels typically possess high β-sheet content as well as structural stability upon exposure to shear forces [100, 130]. Since such spidroin hydrogels showed a shear thinning behavior [131, 135], they are suitable to be used as injectable or printable materials (see also Sect. 6.5).

6.3.2.3 Particles

High concentrations of phosphate ions (> 400 mM) can promote "salting-out" of spidroins [136, 137] and induce particle formation (Fig. 6.4c), as demonstrated for eADF4(C16) [101, 138–140] as well as for different recombinant variants of *Nephila clavipes* MaSp1 [90, 141, 142] and MaSp2 [143]. The diameters of the particles can be adjusted by protein and phosphate concentration as well as by mixing intensity [138, 143]. Using ionic liquids as a solvent, nanoparticles of 100–200 nm in diameter can be formed [144]. Similar to cross-β fibrils, all spidroin particles show a high content of β-sheets [101, 143], which render them chemically highly stable [139].

Spidroin particles substantially swell upon hydration (swelling factor of 2.3) inducing a drastic drop in elastic modulus from GPa in the dry state to MPa in the wet state [145], depending on the molecular weight of the respective spidroins. Direct force measurements using a colloidal probe technique revealed a polymer brush-like surface i.e., an ion-permeable particle interface protruding several tenths of nm into the solution [146, 147]. Interestingly, it has been shown that particle surfaces can seed fibril growth (Fig. 6.4d) [102].

6.3.2.4 Capsules

The amphiphilic character of spidroins [85] allows fast structural transformations from intrinsically unstructured into β-sheet rich structures at the interface in "water-in-oil" emulsions e.g., in toluene [103, 148] or silicon oil [149]. As a consequence, mechanically stable micrometer-sized capsules (Fig. 6.4e) could be produced in rapid processes and completed within 30 s. The size of these microcapsules (1–30 μm) can be controlled by adjusting the emulsion droplet size through changes in the emulsion shear rates. The resulting capsule membranes allowed free diffusion of small molecules through the membrane, whereas larger macromolecules were retained within the capsule (MW cut-off of several thousand Dalton) [148, 149].

The formation of bowl-shaped (1–3 μm) or giant compound micelles (<50 μm) with porous membranes was observed in case of spidroins in water/2-propanol mixtures driven by spidroin assembly on entrapped air bubbles i.e., on liquid-air interfaces. The shape, the diameter as well as the β-sheet content increased with the increasing numbers of poly-Ala blocks (n = 1–6) [87].

6.3.2.5 Films

Films made of recombinant spider silk proteins (Fig. 6.4f) can be processed from organic or aqueous solutions driven by drying and template surface effects. Thus, films differ from fibrils, particles or hydrogels which are organized in a "bottom-up" manner.

It has been shown that protein secondary structures of spidroin films cast from HFIP, hexafluoroacetone or aqueous buffers consist mainly of helices and random coil structures [92, 104, 150–154]. Utilization of formic acid as a solvent, however, results in films which are enriched in β-sheet structures [130, 150, 152, 154–156]. The secondary structure content in spidroin films influences their chemical stability, as amorphous films with low β-sheet content are water-soluble, while high β-sheet contents support water-stability. Nevertheless, the α-helical and random coil structures in films processed from HFIP, FA or aqueous solutions can be transformed into β-sheets upon post-treatment with phosphate buffers [104, 150], or alcohols, such as methanol, ethanol or isopropanol [150, 153, 154, 157], or heat [155, 158, 159]. The conformational changes, accompanied by an increased surface roughness [150, 151, 156, 160], mainly occur in the poly-Ala regions as determined by solid state NMR [154, 155]. X-ray diffraction indicates a wide distribution of β-sheet crystal sizes from 20 to 400 Å [151]. Interestingly, it is possible to axially align the β-sheets upon film stretching [154]. β-sheet-rich spidroin films possess enhanced stability against strong protein denaturing agents [150]. Interestingly, non-native amino- or carboxy-terminal peptide motifs, such as cell-binding sequences [161, 162] or antimicrobial peptides [153], showed no influence on film formation or silk structure after casting and post-treatment.

Generally, spidroin films show poor mechanical properties including a maximal mechanical stress of 70 MPa and maximal strain typically below 10% [152, 154] depending on the solvents and additives used. Film stretching and concomitant alignment of the crystallites increased the mechanical stress up to 200 MPa and strain up to 30% [130, 154].

6.3.2.6 Foams and Sponges

Porous foam-like structures were prepared either by foaming of aqueous solutions of recombinant spidroins [130, 163, 164], freezing/thawing or freeze-drying of hydrogels [100, 130] or by salt-leaching [105]. Salt-leached foams displayed a high β-sheet content of 42%, which was induced by the NaCl crystal surface without an additional post-treatment, simplifying the fabrication process [105]. The foams showed interconnected porous morphologies (Fig. 6.4h) with porosities above 90% and pore sizes between 31 and 437 μm, influenced by the size of the salt crystals. The compressive moduli of the salt leached foams were in the range of 0.94–3.24 kPa in a hydrated state, depending on the spidroin concentration employed [105]. Structures obtained by freeze-drying of spidroin hydrogels (lyogels) possess sheet-like morphologies of the pores with diameters ranging from 30 to 200 μm (Fig. 6.4g). Chemically cross-linked hydrogels show highly interconnected sponge-like morphologies after lyophilization, with round pores of 10–20 μm in diameter, independent of the spidroin concentration [100]. Spidroin lyogels have been shown to withstand a stress of up to 300 kPa and strain near 40%, whereas water uptake has been reported at 1600% of the original dry weight [130]. Spidroin sponges produced by hydrogel freezing and thawing show shape-recovery upon several mechanical loadings in the hydrated state as well as upon rehydration of compressed and dried scaffolds [130].

6.4 Biomedical Applications of Spider Silk

Various material properties, such as biocompatibility, low or no inflammatory response, biodegradability in a defined time frame, and specific structural and mechanical features, are required for biomedical applications.

Native spider silk fibers, providing biocompatibility as well as good mechanical properties, were tested already as braided microsurgical sutures [165]. However, drawbacks, as mentioned in the previous section, such as variance in mechanical performance of the fibers and low availability of natural spider silk, hamper the large scale application of this material. In contrast, recombinant spider silk, as described above, allows scalable production as well as processability into various two and three dimensional morphologies with adjustable mechanics. These features, along with their biocompatibility, qualify recombinant spider silk materials for utilization in many different biomedical applications [3, 166–168]. Another obvious benefit of the recombinant silk is a possibility to introduce additional peptide motifs with distinct effects on cell growth, differentiation and migration [140, 161].

Biocompatibility studies of native and recombinant spider silk *in vitro* or *in vivo* indicated that they are surprisingly well tolerated. Natural spider silk has been used to test inflammatory response and degradation in split skin wounds [169]. Egg case silk from spiders was used for subcutaneous implantation in rats to determine the effect of protease treatment on the degree of inflammation and fibrosis [170]. Native spider silk was implanted *in vivo* to test its biofunctionality showing no adverse effect due to inflammatory response [171, 172]. In the latter, dragline spider silk has been used as an artificial support for nerve regeneration. These studies have shown successful regeneration of damaged tissue and the absence of adverse reactions, however, a mild inflammatory reaction has been detected. In another study, porous scaffolds made of spidroins were well tolerated in mice upon subcutaneous implantation. Histological analysis revealed ingrowths of vascularized connective tissue and nerve fibers into the scaffold, but also a mild foreign body response has been reported [173]. Fiber bundles of recombinant spider silk protein implanted subcutaneously in rats showed infiltration of newly formed capillaries as well as ingrowth of fibroblasts [174].

In the following sections, different biomedical applications based on spider silk materials will be discussed in more detail.

6.4.1 Drug Delivery and Deposition

The biocompatibility and degradation rate are two important features of biomaterials in drug delivery applications. The resorption rate of chemically synthesized polymers can often be adjusted by controlling the composition of the polymer [175]. Silk provides all desirable properties required for sustained drug delivery. This rare combination of properties includes aqueous-based processing, biodegradation, biocompatibility, drug stabilization, and robust mechanical properties.

A genetic engineering approach has been used to form block copolymers of spider silk consensus repeats and poly (L-lysine) domains to form ionic complexes that are able to deliver plasmid DNA *in vitro* e.g., into human embryonic kidney (HEK) cells. The DNA complexes were also immobilized on silk films allowing cell transfection. Such silk-based gene carriers could be functionalized with cell membrane penetrating peptides to enhance the transfection efficiency [176]. Further, the spider silk-poly(L-lysine) variants were functionalized with tumor homing peptide (THP), and corresponding nanoscale complexes with DNA have been reported as less cytotoxic and highly target-specific gene carriers [177, 178]. Engineered, positively charged spidroin eADF4(κ16) (where all glutamic acid residues are replaced by lysine residues) showed particle formation indistinguishable to that of the poly-anionic variant eADF4(C16), described above. Particles made of eADF4(κ16) were loaded successfully with high molecular weight substances, such as nucleic acids, with a well-controllable release profile [179]. These examples demonstrated the potential of bioengineered silk proteins as a new family of molecules to be used for nucleic acid delivery.

Submicron particles or even nanoparticles have been designed as short acting delivery vehicles and administered through intramuscular, intravenous, subcutaneous, oral or transdermal routes [180]. Particles with submicron size have unique properties including stability, high surface-to-volume ratio, high carrier-capacity of bioactive molecules and targeted delivery [181, 182]. Microspheres are commonly used as drug carriers for long-acting delivery and usually administered intramuscularly or subcutaneously. Bioengineered spider silks have been mostly explored in terms of particle formation, characterization and loading/release of model drugs. Recently, submicron particles produced from engineered spider silk eADF4(C16) were used to deliver highly water soluble drugs [139]. However, loading with water-insoluble drugs is also feasible [183]. Further, crosslinking of spidroins was shown to affect drug loading as well as release behavior of drug molecules. Hence, negatively charged eADF4(C16) particles show good potential for application in controlled release of positively charged drugs [139, 184]. Particles of variants of eADF4(C16) with cell penetrating peptides and receptor interacting motifs, as well as eADF4(κ16) were tested for cellular uptake, which proceeded mostly via clathrin-mediated endocytosis [144]. Specific cell targeting has been demonstrated on a recombinant variant, derived from the *N. clavipes* MaSp1 sequence, which was functionalized by Her2 binding peptides. The exposed binding domains on the surface of the particles allowed their significantly higher binding to Her2 positive cells, in comparison to control spheres and Her2-negative cell binding, and loaded doxorubicin was released in a pH dependent manner [142]. The recombinant spider silk protein eADF4(C16) was fused with the antigenic peptide from ovalbumin, either without or with a cathepsin cleavable peptide linker. The hybrid particles were taken up by dendritic cells and successfully activated cytotoxic T-cells in vitro. In vivo the antigen-modified particles containing a cathepsin-cleavable linker induced a strong antigen-specific proliferation of cytotoxic T-cells, even in the absence of a vaccine adjuvant, demonstrating the efficacy of this new vaccine strategy using a protein-based all-in-one vaccination system [140].

6.4.2 Tissue Engineering

Tissue engineering is an interdisciplinary field combining principles of engineering and life science towards the development of biological substitutes that restore, maintain, or improve tissue function [185]. Three major issues are important to create new tissue: isolation of cells or cell substitutes; identification of tissue inducing substances and controlled cell placement on or within matrices.. Researchers have attempted to engineer virtually every mammalian tissue and replacement of ectodermally (for example, skin, nervous system), endodermally (liver, pancreas) and mesodermally (cartilage, bone and muscle) derived tissues.

Over the last decade, a variety of polymers and other material have been utilized for tissue engineering applications, and polymeric biomaterials are particularly attractive. The biomaterials used in tissue engineering applications are mainly classified as synthetic and naturally occurring polymers. However, natural materials are preferred due to their intrinsic properties, which can facilitate cell attachment or maintenance of differentiated function. The wide range of materials used for tissue engineering arises from a variety of anatomical locations.

Silk has been used in its native form as a suture material for long time, as mentioned above, and recently it is gaining popularity in tissue engineering. It is worth mentioning that recombinant spidroins bearing artificially engineered cell binding domains (Fig. 6.6), in combination with processing into various morphologies [105, 161, 186], open and expand new possibilities of silk-based materials in cell culture and tissue engineering applications [187].

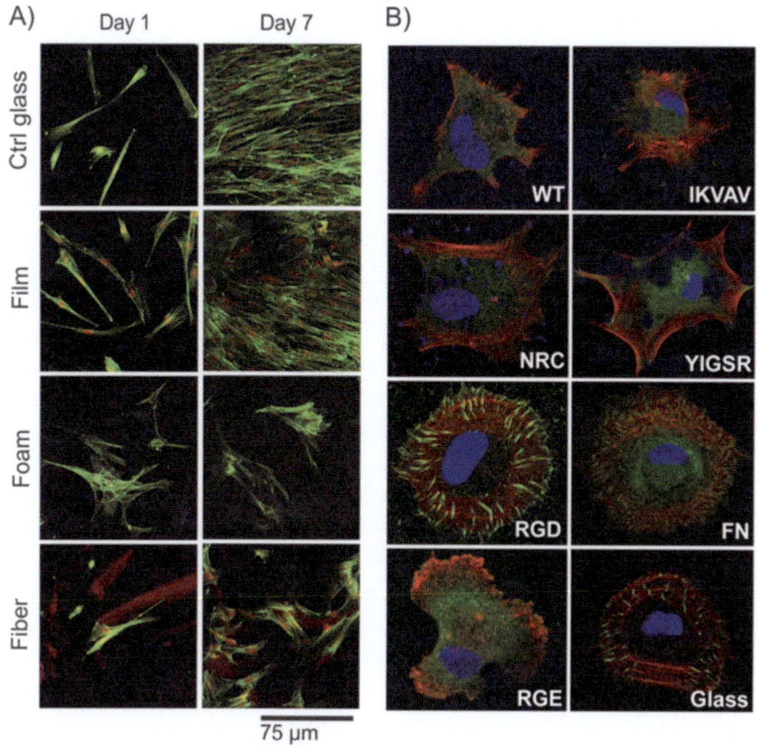

Fig. 6.6 Variants of recombinant spider silk protein (4RepCT) used to fabricate various morphologies for tissue engineering applications. (**a**) Detection of filamentous actin in cells grown on different matrices produced from the recombinant spider silk protein (4RepCT) as indicated. (**b**) Fibroblasts formed stress fibers (F-actin-red) and focal adhesions (Vinculin-green) on films made of variants of 4RepCT within 3 h. Representative micrographs show cells on films made of WT and RGE modified 4RepCT with poorly formed actin filaments and diffuse vinculin staining. Cells cultured on IKVAV and YIGSR modified 4RepCT showed clear stress fibers. Cells on films made of RGD modified 4RepCT showed distinct focal adhesion points and well organized actin stress fibers compared to cells grown on fibronectin (FN as a control). (Adopted and modified from [163] and [162], with permission of Elsevier)

6.4.2.1 Wound Healing Scaffolds

The first tissue engineering approaches used cells on two-dimensional sheets of collagen or collagen-glycosaminoglycan composites to create new skin [188, 189]. Native spider dragline silk has been tested as a suitable matrix for 3D skin cell culture, and both fibroblast and keratinocyte cell lines adhered and proliferated well. A bi-layered co-cultivation in two continuously separated strata was achieved by serum reduction, changing the medium conditions and the cultivation period. Hence, spider silk appears to be a promising biomaterial for the enhancement of skin regeneration [190].

The engineered recombinant spider silk protein eADF4(C16) was modified with the integrin recognition sequence RGD by genetic and chemical approaches. Attachment and proliferation of BALB/3 T3 mouse fibroblasts is significantly improved on films made of RGD-modified silk proteins [161]. In another study, recombinant spider silk was modified with RGD in a structural turn loop, similar to fibronectin (FN), but in the silk-hybrid the loop was stabilized by cysteines (FNcc). Human primary cells cultured on FNcc-silk show increased spreading, attachment and focal adhesion. FNcc-silk supported proliferation and migration of keratinocytes. These results suggest that FNcc-silk can efficiently attract inherent cells for migration into a wound area e.g., from the wound edges from where dermal keratinocytes are usually recruited during wound healing [191]. Thus, these hybrid silk proteins reflect a promising material for future tissue engineering applications.

Highly porous foams made of the recombinant spider silk protein eADF4(C16) and its variant containing the RGD motif were fabricated by salt leaching. Fibroblast cultured on these foams show enhanced adhesion and proliferation as long as RGD is present [105]. The results indicated that spider silk foams with controllable properties can be used as scaffolds for tissue repair and soft tissue engineering e.g., in the case of skin.

6.4.2.2 Bone Tissue Engineering

The burden of musculoskeletal conditions increased at the start of the new millennium [192]. According to published reports by the World health organization in 2011, the cost related to caring for patients with musculoskeletal defects was $796.3 billion in the United States, which is 5.7% of the annual GDP. Many injuries and diseases, such as osteoarthritis, osteogenesis imperfecta, traumatic processes etc., can adversely affect the musculoskeletal system, which decreases quality of life. In Europe and the United States more than 400,000 and 600,000 patients, respectively, receive bone grafts each year [193, 194]. It is a major challenge to repair the bone defects and fractures that occur due to osteoporosis during aging. Bone tissue engineering (TE) is a promising strategy to regenerate bone defects and allow their restoration. Conventional approaches like autogenous bone grafts or allogenic bones show limitations such as limited material supply, donor site morbidity and contour irregularities. Allogenic bone can also be used, but the cell-mediated immune response to transplantation of alloantigens and pathogens can be problematic [195]. Therefore, synthetic and natural polymers have been explored for bone repair, but it has been difficult to create a polymer with optimal strength and degradation properties.

Natural polymers such as silk have been explored in bone tissue engineering. One approach for silk scaffolds in bone tissue engineering is fusion of spider dragline silk from *N. clavipes* with the carboxyl terminal domain of dentin matrix protein 1 (CDMP1). The purified recombinant protein retains native silk-like self-assembly properties and can be processed into films. During incubation in simulated body fluids, processed films show induced growth of hydroxyapatite crystals on silk films along with the fused dentin matrix protein. In this system, spider silk exhibited remarkable mechanical properties while CDMP1 provided controlled nucleation and growth of hydroxyapatite [196]. In another attempt, the R5 peptide derived from silaffin of *Cyloindrotheca fusiformis*, which induces and regulates silica precipitation, was genetically fused to an RGD containing *N. clavipes* spider dragline silk protein. These chimeric silk silica proteins were processed into films and fibers. The R5 peptide induced silica mineralization on the surface [197]. The solution properties of the silk were explored, and optimum conditions for silica deposition were determined [198]. The silk-silica protein films showed osteogenic differentiation of human mesenchymal stem cells (hMSCs) with upregulation of alkaline phosphatase (ALP), bone sialoprotein (BSP) and collagen type I. The deposition of calcium on silk-silica films have further demonstrated enhanced osteogenesis [199].

6.4.2.3 Nerve Tissue Engineering

Peripheral nerve injury contributes to 2.8% of all trauma patients [200]. Annually, 360,000 people suffer from upper paralysis syndrome in the USA alone, whereas in Europe 300,000 people suffer from peripheral nerve injury [201]. The severe consequences of peripheral nerve injury could result in long term disability, functional impairment, and major socio-economic costs. To overcome this problem, attempts using many different approaches have been carried out to repair or to regenerate peripheral nerves. Amongst the approaches tested for nerve regeneration, all strategies focused on guided regeneration of nerve fibers. Current clinical practice of using autologous nerve grafts to replace lost and damaged nervous tissue has drawbacks such as limited availability of nerve grafts as well as loss of sensation at the donor site [202]. Recently, studies have focused on the development of artificial nerve guiding materials as new therapeutic alternatives based on synthetic and natural materials. The production of artificial devices includes a variety of biocompatible non-degradable and degradable materials. Importantly, one major drawback of using non-degradable artificial nerve guidance is the loss of functional recovery in the long term due to progressive myelination and compression of axons within the guidance conduit [203–205].

Natural materials are favored over synthetic materials in nerve regeneration due to their reduced cytotoxicity, enhanced biocompatibility [202], support of cell migration, and lack of toxic effects. Components of the extracellular matrix (ECM) have been the mostly used natural polymers for nerve reconstruction [206]. Spider silk fibers collected from *Nephila* species have been used as guidance material for nerve regeneration. *In vitro* experiments demonstrated remarkable adhesion and proliferation of Schwann cells [172, 171, 207]. Further, recombinant spider silk matrices were evaluated regarding their suitability for *in vitro* cell culture and tissue engineering applications. Neural stem cells (NSCs) cultured on recombinant spider silk matrices differentiated efficiently into neuronal and astrocyte cells but with slightly less efficient oligodendrocyte differentiation in comparison to controls

[164]. The peptides which are responsible for beta sheet formation in *Araneus ventricosus* spider silk were investigated concerning cytotoxic effects on neurons. Silk derived nano-assemblies, nanofilaments and nanofibrils with β-sheet contents ranging from 24% to 40% showed no significant cytotoxic effect on neurons. Amyloid-forming peptides with high β-sheet content, however, showed cytotoxicity [109].

One recent strategy to repair completely transected nerves employs nerve guide conduits (NGCs). Several studies have used dragline spider silk fibers from *Nephila* species placed in isogenic veins and acellularized venules. Such nerve grafts bridged a 20 mm gap injury in the sciatic nerve of rats. Effective regeneration was observed in isogenic grafts as well as grafts containing spider silk. Axons aligned regularly, and the NGCs were highly effective in nerve regeneration. In another study, nerve constructs prepared by decellularized vein grafts filled with spider silk fibers were implanted in 6 cm tibial nerve defects in adult sheep and exhibited axon regeneration and functional recovery through the NGCs. Axons were myelinated indicating migration of Schwann cells into the constructs [171, 172, 207]. The recombinant variant of MaSp1 based on *N. clavipes* supported neural growth as well as axon extension and network connectivity with increased expression levels of the neural cell adhesion molecule (NCAM) on 2D films. The silk modification with a neuron-specific surface binding sequence GRGGL contributes to the biological regulation of neuron growth [97]. From these studies, it can be summarized that silk-based materials enhance axonal regeneration and remyelination with functional recovery in long distance nerve repair.

6.4.2.4 Implant Coating

The coating of implant surfaces with recombinant spider silk was analyzed concerning foreign body reactions [208]. Implants coated with silk and implanted in animals showed low levels of inflammation markers like cytokines. In this context, silicone implants coated with the recombinant spider silk eADF4(C16) show delayed and significantly decreased foreign body reactions and a reduced capsule manifestation [209]. In another study, a coating with recombinantly produced spider silk protein as the outer layer was applied on catheter polymers made of polyurethane, polytetrafluoroethylene and silicone. These silk-coated materials were tested for cell adhesion using various cell lines, such as HaCat, B50, C2C12 and BALB/3 T3. The results indicate low or no adhesion compare to positive tissue culture plates as control [210].

To reduce infection, silk scaffolds or silk-coated implants were functionalized with silver ions known for their antibacterial property. The silver ions were bound through a silver binding peptide hybridized with recombinant spidroins [159]. These reports show that silk coatings have great potential in biomedical applications.

6.5 Biofabrication

Biofabrication can be defined as the simultaneous processing of living cells and biomaterials. It is regarded as a new research field based that bridges life science, physical science and engineering science [211]. The US Defense Advanced Research Projects Agency used the definition ‘Biofabrication- the use of biological materials and mechanism for construction’ to describe methods used to create high resolution 3D structures that mimic biological growth mechanisms [212].

Currently, supply of organs for transplantation in humans is a serious, urgent and emerging medical problem. To overcome this hurdle, biofabrication is a highly desirable new technology to bioengineer living human tissues and organs suitable for implantation. Biofabrication of functional organs could save thousands of lives and it will reduce the cost of healthcare dramatically.

Biofabricated 3D models of human tissues *in vitro* could be superior in comparison to traditional 2D cell culture in terms of replacing animal *in vivo* studies. The important biological problems associated with biofabrication are cell survival during the biofabrication process, tissue self-assembly, vascularization and tissue maturation.

The recent development of 3D printing in the field of tissue engineering has resulted in the development of printed scaffolds loaded with cells for engineering complex tissue structures [213–215]. 3D printers, also referred to as 3D bioprinters, include various elements, such as print head, material cartridge, actuator, nozzle, working area and print stage. 3D bioprinters are most commonly classified based on the mechanism of material deposition e.g., extrusion printing, inkjet printing and laser-assisted bioprinting [213, 214, 216]. The most critical characteristics of 3D bioprinting include cell friendliness, reproducibility, printing of complex geometries as well as physical and chemical gradients. The to-be-printed materials, referred to as bioinks, represent the great challenge in materials engineering due to high demands in biofabrication. Recent studies have demonstrated that hydrogels can show high shape fidelity, cytocompatibility and can be modified to the target tissue, thus hydrogels have a great chance for success in biofabrication [213]. Synthesis of precisely tailored, bioprocessible, functional and biomimetic extracellular matrices and stimuli-sensitive hydrogels is required, and the demand for new biomaterials is continuously growing [135, 217, 218].

Recent studies established the recombinant spidroin eADF4(C16) and a variant with an RGD-motif as bioinks. Hydrogels based on these proteins were evaluated for cell adhesion of BALB/3 T3 mouse fibroblasts, myoblasts, HeLa cells, osteoblasts and keratinocytes. Except for osteoblasts, all other cell types showed low cell adhesion on the non-modified spidroin hydrogels. However, in case of RGD-modification all cell types adhered well to the spidroin scaffolds. The hydrogels were then assessed for their printability. Using robotic dispensing, the hydrogels could be processed with high shape fidelity without the need of additional thickeners, crosslinkers or post-treatment. One important feature needed for cell scaffolds is diffusion of nutrients, oxygen and waste products. It was shown that eADF4(C16) and eADF4(C16)-RGD hydrogels provide good diffusion for model compounds.

Human fibroblasts biofabricated with eADF4(C16) as bioink survived the printing process and achieved 70% viability after 7 days of incubation (Fig. 6.7) [131, 135]. Further adjustments of the rheological properties and/or cell viability were

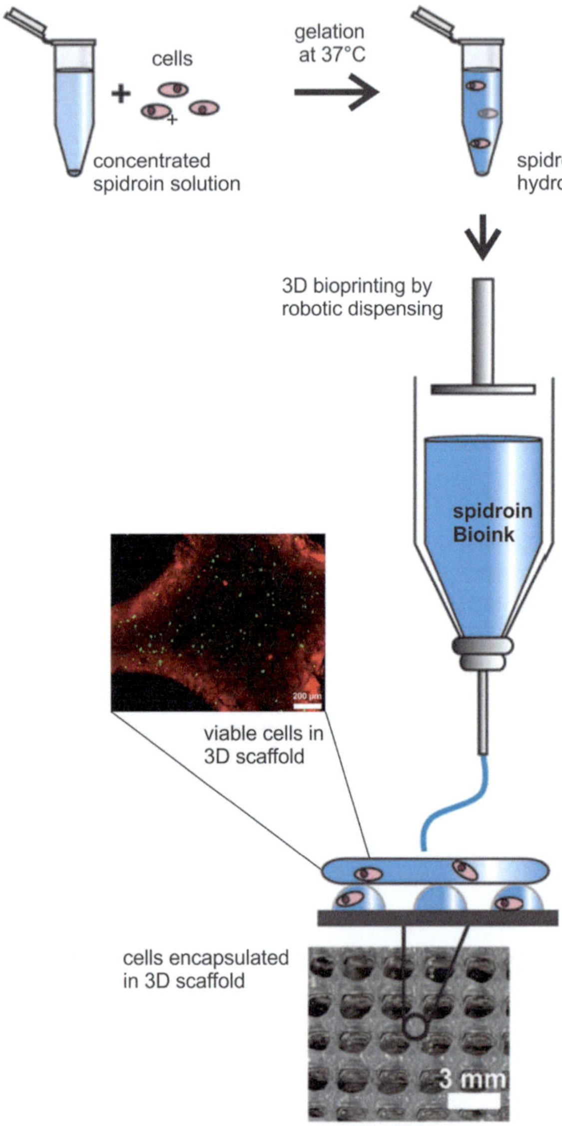

Fig. 6.7 Recombinant spider silk hydrogels as Bioink for 3D printing. Silk gelation in the presence of cells followed by 3D printing using robotic dispensing enables multilayered 3D scaffolds with encapsulated viable cells. (Adapted and modified from [135] with permission of Wiley-VCH)

achieved using a combination of different recombinant spider silk variants derived from different natural blue prints [132] or a combination of recombinant spider silk and collagen in compound hydrogels [131].

6.6 Conclusions

While polymer chemistry is considered a mature disciplinary field, there remains a gap in mimicking and extending the polymer performance of natural systems, since most synthetic polymers do not exhibit hierarchical structures as found in biopolymers. Synthetic or biopolymeric materials allow some degree of control over composition and structure, however, the full potential is often not exploited to maximize their properties. Spider silk is an excellent example of a biopolymer where sequence, structure and processing have been well studied. To enable silk-based medical applications, the biodegradation rate of silk materials has to be engineered to match the re-growth of new tissue, and surface engineering strategies need to be optimized to lower the graft rejection of silk based constructs in clinical applications. A considerable amount of work has been done concerning processing and characterization of silk-based drug delivery vehicles. However, a systematic approach is still needed for formulation development, which includes the selection of viable drugs, as well as manipulation of critical processing parameters to control pharmacokinetics. In this context, genetically engineered spidroins have attracted great attention, since they can be adjusted on demand for the respective use.

Another example for biomedical applications is tissue engineering for reconstruction therapies: cell transplant systems may complement gene therapy approaches, which facilitate transfer of a large population of cells expressing the desired phenotype. In combination with various manufacturing techniques, such as three-dimensional printing, complex tissue structures can be built. Spider silk alone, or composites including spidroins will offer novel options to match complex mechanical and biological functions in hierarchical structures with tissue specific needs.

Acknowledgements This work was supported by Elite Network of Bavaria (ENB) and TRR 225 C01.

References

1. Doblhofer E, Heidebrecht A, Scheibel T (2015) To spin or not to spin: spider silk fibers and more. Appl Microbiol Biotechnol 99(22):9361–9380
2. Meyers MA, McKittrick J, Chen P-Y (2013) Structural biological materials: critical mechanics-materials connections. Science 339(6121):773–779
3. Humenik M, Smith AM, Scheibel T (2011) Recombinant spider silks—biopolymers with potential for future applications. Polymers 3(1):640–661

4. Neuenfeldt M, Scheibel T (2014) Silks from insects: from natural diversity to applications. In: Hoffmann KH (ed) Insect molecular biology and ecology. CRC Press, Boca Raton, pp 376–400
5. Sutherland TD, Young JH, Weisman S, Hayashi CY, Merritt DJ (2010) Insect silk: one name, many materials. Annu Rev Entomol 55(1):171–188
6. Humenik M, Scheibel T, Smith A (2011) Spider silk: understanding the structure–function relationship of a natural fiber. In: Howorka S (ed) Progress in molecular biology and translational science, vol 103. Cambridge, MA, pp 131–185
7. Qin Z, Buehler MJ (2013) Spider silk: webs measure up. Nat Mater 12(3):185–187
8. dos Santos-Pinto JRA, Garcia AMC, Arcuri HA, Esteves FG, Salles HC, Lubec G, Palma MS (2016) Silkomics: insight into the silk spinning process of spiders. J Proteome Res 15(4):1179–1193
9. Vollrath F (2000) Strength and structure of spiders' silks. Rev Mol Biotechnol 74(2):67–83
10. Lefevre T, Boudreault S, Cloutier C, Pezolet M (2011) Diversity of molecular transformations involved in the formation of spider silks. J Mol Biol 405(1):238–253
11. Gatesy J, Hayashi C, Motriuk D, Woods J, Lewis R (2001) Extreme diversity, conservation, and convergence of spider silk fibroin sequences. Science 291(5513):2603–2605
12. Casem ML, Collin MA, Ayoub NA, Hayashi CY (2010) Silk gene transcripts in the developing tubuliform glands of the western black widow, latrodectus hesperus. J Arachnol 38(1):99–103
13. Hu X, Vasanthavada K, Kohler K, McNary S, Moore AMF, Vierra CA (2006) Molecular mechanisms of spider silk. Cell Mol Life Sci 63(17):1986–1999
14. Eisoldt L, Thamm C, Scheibel T (2012) Review the role of terminal domains during storage and assembly of spider silk proteins. Biopolymers 97(6):355–361
15. Hayashi CY, Blackledge TA, Lewis RV (2004) Molecular and mechanical characterization of aciniform silk: uniformity of iterated sequence modules in a novel member of the spider silk fibroin gene family. Mol Biol Evol 21(10):1950–1959
16. Gosline JM, Guerette PA, Ortlepp CS, Savage KN (1999) The mechanical design of spider silks: from fibroin sequence to mechanical function. J Exp Biol 202(23):3295–3303
17. Vollrath F, Porter D, Holland C (2011) There are many more lessons still to be learned from spider silks. Soft Matter 7:9595–9600
18. Agnarsson I, Blackledge TA (2009) Can a spider web be too sticky? Tensile mechanics constrains the evolution of capture spiral stickiness in orb-weaving spiders. J Zool 278(2):134–140
19. Chaw RC, Correa-Garhwal SM, Clarke TH, Ayoub NA, Hayashi CY (2015) Proteomic evidence for components of spider silk synthesis from black widow silk glands and fibers. J Proteome Res 14(10):4223–4231
20. Clarke TH, Garb JE, Hayashi CY, Haney RA, Lancaster AK, Corbett S, Ayoub NA (2014) Multi-tissue transcriptomics of the black widow spider reveals expansions, co-options, and functional processes of the silk gland gene toolkit. BMC Genomics 15(1):1–17
21. Pham T, Chuang T, Lin A, Joo H, Tsai J, Crawford T, Zhao L, Williams C, Hsia Y, Vierra C (2014) Dragline silk: a fiber assembled with low-molecular-weight cysteine-rich proteins. Biomacromolecules 15(11):4073–4081
22. Vollrath F, Knight DP (2001) Liquid crystalline spinning of spider silk. Nature 410(6828):541–548
23. Riekel C, Müller M, Vollrath F (1999) In situ x-ray diffraction during forced silking of spider silk. Macromolecules 32(13):4464–4466
24. Lefevre T, Boudreault S, Cloutier C, Pezolet M (2008) Conformational and orientational transformation of silk proteins in the major ampullate gland of nephila clavipes spiders. Biomacromolecules 9(9):2399–2407
25. Jenkins JE, Holland GP, Yarger JL (2012) High resolution magic angle spinning nmr investigation of silk protein structure within major ampullate glands of orb weaving spiders. Soft Matter 8(6):1947–1954

26. Jenkins JE, Sampath S, Butler E, Kim J, Henning RW, Holland GP, Yarger JL (2013) Characterizing the secondary protein structure of black widow dragline silk using solid-state nmr and x-ray diffraction. Biomacromolecules 14(10):3472–3483
27. Andersson M, Holm L, Ridderstråle Y, Johansson J, Rising A (2013) Morphology and composition of the spider major ampullate gland and dragline silk. Biomacromolecules 14(8):2945–2952
28. Vollrath F, Knight DP (1999) Structure and function of the silk production pathway in the spider nephila edulis. Int J Biol Macromol 24(2–3):243–249
29. Xu M, Lewis RV (1990) Structure of a protein superfiber – spider dragline silk. Proc Natl Acad Sci U S A 87(18):7120–7124
30. Ayoub NA, Garb JE, Tinghitella RM, Collin MA, Hayashi CY (2007) Blueprint for a high-performance biomaterial: full-length spider dragline silk genes. PLoS One 2(6):e514
31. Sponner A, Schlott B, Vollrath F, Unger E, Grosse F, Weisshart K (2005) Characterization of the protein components of nephila clavipes dragline silk. Biochemistry 44(12):4727–4736
32. Han L, Zhang L, Zhao T, Wang Y, Nakagaki M (2013) Analysis of a new type of major ampullate spider silk gene, masp1s. Int J Biol Macromol 56:156–161
33. Liu Y, Sponner A, Porter D, Vollrath F (2008) Proline and processing of spider silks. Biomacromolecules 9(1):116–121
34. Marhabaie M, Leeper TC, Blackledge TA (2014) Protein composition correlates with the mechanical properties of spider (argiope trifasciata) dragline silk. Biomacromolecules 15(1):20–29
35. Andersen SO (1970) Amino acid composition of spider silks. Comp Biochem Physiol 35(3):705–711
36. Rising A, Nimmervoll H, Grip S, Fernandez-Arias A, Storckenfeldt E, Knight DP, Vollrath F, Engstrom W (2005) Spider silk proteins-mechanical property and gene sequence. Zool Sci 22(3):273–281
37. Hinman MB, Jones JA, Lewis RV (2000) Synthetic spider silk: a modular fiber. Trends Biotechnol 18(9):374–379
38. Zhang Y, Zhao A-C, Sima Y-H, Lu C, Xiang Z-H, Nakagaki M (2013) The molecular structures of major ampullate silk proteins of the wasp spider, argiope bruennichi: a second blueprint for synthesizing de novo silk. Comp Biochem Physiol B Biochem Mol Biol 164(3):151–158
39. dos Santos-Pinto JRA, Arcuri HA, Priewalder H, Salles HC, Palma MS, Lubec G (2015) Structural model for the spider silk protein spidroin-1. J Proteome Res 14(9):3859–3870
40. dos Santos-Pinto JRA, Lamprecht G, Chen W-Q, Heo S, Hardy JG, Priewalder H, Scheibel TR, Palma MS, Lubec G (2014) Structure and post-translational modifications of the web silk protein spidroin-1 from nephila spiders. J Proteome 105:174–185
41. Hijirida DH, Do KG, Michal C, Wong S, Zax D, Jelinski LW (1996) 13c nmr of nephila clavipes major ampullate silk gland. Biophys J 71(6):3442–3447
42. Xu D, Guo C, Holland GP (2015) Probing the impact of acidification on spider silk assembly kinetics. Biomacromolecules 16(7):2072–2079
43. Hronska M, van Beek JD, Williamson PTF, Vollrath F, Meier BH (2004) Nmr characterization of native liquid spider dragline silk from nephila edulis. Biomacromolecules 5(3):834–839
44. Askarieh G, Hedhammar M, Nordling K, Saenz A, Casals C, Rising A, Johansson J, Knight SD (2010) Self-assembly of spider silk proteins is controlled by a ph-sensitive relay. Nature 465(7295):236–239
45. Hagn F, Eisoldt L, Hardy JG, Vendrely C, Coles M, Scheibel T, Kessler H (2010) A highly conserved spider silk domain acts as a molecular switch that controls fibre assembly. Nature 465(7295):239–242
46. Sponner A, Unger E, Grosse F, Klaus W (2005) Differential polymerization of the two main protein components of dragline silk during fibre spinning. Nat Mater 4(10):772–775
47. Rising A, Johansson J (2015) Toward spinning artificial spider silk. Nat Chem Biol 11(5):309–315

48. Vollrath F, Knight DP, Hu XW (1998) Silk production in a spider involves acid bath treatment. Proc R Soc London Ser B 265(1398):817–820
49. Knight DP, Vollrath F (2001) Changes in element composition along the spinning duct in a nephila spider. Naturwissenschaften 88(4):179–182
50. Gauthier M, Leclerc J, Lefèvre T, Gagné SM, Auger M (2014) Effect of ph on the structure of the recombinant c-terminal domain of nephila clavipes dragline silk protein. Biomacromolecules 15(12):4447–4454
51. Bauer J, Scheibel T (2017) Conformational stability and interplay of helical n- and c-terminal domains with implications on major ampullate spidroin assembly. Biomacromolecules 18(3):835–845
52. Kronqvist N, Otikovs M, Chmyrov V, Chen G, Andersson M, Nordling K, Landreh M, Sarr M, Jörnvall H, Wennmalm S, Widengren J, Meng Q, Rising A, Otzen D, Knight SD, Jaudzems K, Johansson J (2014) Sequential ph-driven dimerization and stabilization of the n-terminal domain enables rapid spider silk formation. Nat Commun 5:3254
53. Hagn F, Thamm C, Scheibel T, Kessler H (2011) Ph-dependent dimerization and salt-dependent stabilization of the n-terminal domain of spider dragline silk-implications for fiber formation. Angew Chem Int Ed 50(1):310–313
54. Jaudzems K, Askarieh G, Landreh M, Nordling K, Hedhammar M, Jörnvall H, Rising A, Knight SD, Johansson J (2012) Ph dependent dimerization of spider silk n-terminal domain requires relocation of a wedged tryptophan side chain. J Mol Biol 422(4):477–487
55. Schwarze S, Zwettler FU, Johnson CM, Neuweiler H (2013) The n-terminal domains of spider silk proteins assemble ultrafast and protected from charge screening. Nat Commun 4:2815
56. Ries J, Schwarze S, Johnson CM, Neuweiler H (2014) Microsecond folding and domain motions of a spider silk protein structural switch. J Am Chem Soc 136(49):17136–17144
57. Kurut A, Dicko C, Lund M (2015) Dimerization of terminal domains in spiders silk proteins is controlled by electrostatic anisotropy and modulated by hydrophobic patches. ACS Biomater Sci Eng 1(6):363–371
58. Barroso da Silva FL, Pasquali S, Derreumaux P, Dias LG (2016) Electrostatics analysis of the mutational and ph effects of the n-terminal domain self-association of the major ampullate spidroin. Soft Matter 12(25):5600–5612
59. Bauer J, Scheibel T (2017) Dimerization of the conserved n-terminal domain of a spider silk protein controls the self-assembly of the repetitive core domain. Biomacromolecules 18(8):2521–2528
60. Bauer J, Schaal D, Eisoldt L, Schweimer K, Schwarzinger S, Scheibel T (2016) Acidic residues control the dimerization of the n-terminal domain of black widow spiders' major ampullate spidroin 1. Sci Rep 6:34442
61. Holland GP, Creager MS, Jenkins JE, Lewis RV, Yarger JL (2008) Determining secondary structure in spider dragline silk by carbon-carbon correlation solid-state nmr spectroscopy. J Am Chem Soc 130(30):9871–9877
62. van Beek JD, Hess S, Vollrath F, Meier BH (2002) The molecular structure of spider dragline silk: folding and orientation of the protein backbone. Proc Natl Acad Sci U S A 99(16):10266–10271
63. Plaza GR, Perez-Rigueiro J, Riekel C, Perea GB, Agullo-Rueda F, Burghammer M, Guinea GV, Elices M (2012) Relationship between microstructure and mechanical properties in spider silk fibers: identification of two regimes in the microstructural changes. Soft Matter 8(22):6015–6026
64. Simmons AH, Michal CA, Jelinski LW (1996) Molecular orientation and two-component nature of the crystalline fraction of spider dragline silk. Science 271(5245):84–87
65. Jenkins JE, Creager MS, Lewis RV, Holland GP, Yarger JL (2010) Quantitative correlation between the protein primary sequences and secondary structures in spider dragline silks. Biomacromolecules 11(1):192–200
66. Termonia Y (1994) Molecular modeling of spider silk elasticity. Macromolecules 27(25):7378–7381

67. Eisoldt L, Smith A, Scheibel T (2011) Decoding the secrets of spider silk. Mater Today 14(3):80–86
68. Su I, Buehler MJ (2016) Nanomechanics of silk: the fundamentals of a strong, tough and versatile material. Nanotechnology 27(30):302001
69. Heidebrecht A, Scheibel T (2013) Recombinant production of spider silk proteins. Adv Appl Microbiol 82:115–153
70. Agnarsson I, Kuntner M, Blackledge TA (2010) Bioprospecting finds the toughest biological material: extraordinary silk from a giant riverine orb spider. PLoS One 5(9):e11234
71. Lee S-M, Pippel E, Moutanabbir O, Kim J-H, Lee H-J, Knez M (2014) In situ raman spectroscopic study of al-infiltrated spider dragline silk under tensile deformation. ACS Appl Mater Interfaces 6(19):16827–16834
72. Nova A, Keten S, Pugno NM, Redaelli A, Buehler MJ (2010) Molecular and nanostructural mechanisms of deformation, strength and toughness of spider silk fibrils. Nano Lett 10(7):2626–2634
73. Keten S, Xu Z, Ihle B, Buehler MJ (2010) Nanoconfinement controls stiffness, strength and mechanical toughness of beta-sheet crystals in silk. Nat Mater 9(4):359–367
74. Du N, Liu XY, Narayanan J, Li LA, Lim MLM, Li DQ (2006) Design of superior spider silk: from nanostructure to mechanical properties. Biophys J 91(12):4528–4535
75. Anton AM, Heidebrecht A, Mahmood N, Beiner M, Scheibel T, Kremer F (2017) Foundation of the outstanding toughness in biomimetic and natural spider silk. Biomacromolecules 18(12):3954–3962
76. Chung H, Kim TY, Lee SY (2012) Recent advances in production of recombinant spider silk proteins. Curr Opin Biotechnol 23(6):957–964
77. Tokareva O, Michalczechen-Lacerda VA, Rech EL, Kaplan DL (2013) Recombinant DNA production of spider silk proteins. Microb Biotechnol 6(6):651–663
78. Huemmerich D, Helsen CW, Quedzuweit S, Oschmann J, Rudolph R, Scheibel T (2004) Primary structure elements of spider dragline silks and their contribution to protein solubility. Biochemistry 43(42):13604–13612
79. Eisoldt L, Hardy JG, Heim M, Scheibel TR (2010) The role of salt and shear on the storage and assembly of spider silk proteins. J Struct Biol 170(2):413–419
80. Exler JH, Hummerich D, Scheibel T (2007) The amphiphilic properties of spider silks are important for spinning. Angew Chem Int Ed 46(19):3559–3562
81. Rammensee S, Slotta U, Scheibel T, Bausch AR (2008) Assembly mechanism of recombinant spider silk proteins. Proc Natl Acad Sci U S A 105(18):6590–6595
82. Heidebrecht A, Eisoldt L, Diehl J, Schmidt A, Geffers M, Lang G, Scheibel T (2015) Biomimetic fibers made of recombinant spidroins with the same toughness as natural spider silk. Adv Mater 27(13):2189–2194
83. Renberg B, Andersson-Svahn H, Hedhammar M (2014) Mimicking silk spinning in a microchip. Sensors Actuators B Chem 195:404–408
84. Bini E, Knight DP, Kaplan DL (2004) Mapping domain structures in silks from insects and spiders related to protein assembly. J Mol Biol 335(1):27–40
85. Zbilut JP, Scheibel T, Huemmerich D, Webber CL, Colafranceschi M, Giuliani A (2005) Spatial stochastic resonance in protein hydrophobicity. Phys Lett A 346(1–3):33–41
86. Jin HJ, Kaplan DL (2003) Mechanism of silk processing in insects and spiders. Nature 424(6952):1057–1061
87. Rabotyagova OS, Cebe P, Kaplan DL (2009) Self-assembly of genetically engineered spider silk block copolymers. Biomacromolecules 10(2):229–236
88. Rabotyagova OS, Cebe P, Kaplan DL (2010) Role of polyalanine domains in β-sheet formation in spider silk block copolymers. Macromol Biosci 10(1):49–59
89. Borisov O, Zhulina E, Leermakers FM, Müller AE (2011) Self-assembled structures of amphiphilic ionic block copolymers: theory, self-consistent field modeling and experiment. In: Müller AHE, Borisov O (eds) Self organized nanostructures of amphiphilic block copolymers i, Advances in polymer science, vol 241. Springer, Berlin/Heidelberg, pp 57–129

90. Tokareva OS, Lin S, Jacobsen MM, Huang W, Rizzo D, Li D, Simon M, Staii C, Cebe P, Wong JY, Buehler MJ, Kaplan DL (2014) Effect of sequence features on assembly of spider silk block copolymers. J Struct Biol 186(3):412–419
91. Lin S, Ryu S, Tokareva O, Gronau G, Jacobsen MM, Huang W, Rizzo DJ, Li D, Staii C, Pugno NM, Wong JY, Kaplan DL, Buehler MJ (2015) Predictive modelling-based design and experiments for synthesis and spinning of bioinspired silk fibres. Nat Commun 6:6892
92. Krishnaji ST, Bratzel G, Kinahan ME, Kluge JA, Staii C, Wong JY, Buehler MJ, Kaplan DL (2013) Sequence–structure–property relationships of recombinant spider silk proteins: integration of biopolymer design, processing, and modeling. Adv Funct Mater 23(2):241–253
93. Xia X-X, Qian Z-G, Ki CS, Park YH, Kaplan DL, Lee SY (2010) Native-sized recombinant spider silk protein produced in metabolically engineered escherichia coli results in a strong fiber. Proc Natl Acad Sci U S A 107(32):14059–14063
94. Thamm C, Scheibel T (2017) Recombinant production, characterization, and fiber spinning of an engineered short major ampullate spidroin (masp1s). Biomacromolecules 18(4):1365–1372
95. Humenik M, Magdeburg M, Scheibel T (2014) Influence of repeat numbers on self-assembly rates of repetitive recombinant spider silk proteins. J Struct Biol 186:431–437
96. Wohlrab S, Thamm C, Scheibel T (2014) The power of recombinant spider silk proteins. In: Asakura T, Miller T (eds) Biotechnology of silk, Biologically-inspired systems, vol 5. Springer, Dordrecht, pp 179–201
97. An B, Tang-Schomer MD, Huang W, He J, Jones JA, Lewis RV, Kaplan DL (2015) Physical and biological regulation of neuron regenerative growth and network formation on recombinant dragline silks. Biomaterials 48:137–146
98. Fu C, Shao Z, Vollrath F (2009) Animal silks: their structures, properties and artificial production. Chem Commun 43:6515–6529
99. Humenik M, Drechsler M, Scheibel T (2014) Controlled hierarchical assembly of spider silk-DNA chimeras into ribbons and raft-like morphologies. Nano Lett 14(7):3999–4004
100. Schacht K, Scheibel T (2011) Controlled hydrogel formation of a recombinant spider silk protein. Biomacromolecules 12(7):2488–2495
101. Slotta UK, Rammensee S, Gorb S, Scheibel T (2008) An engineered spider silk protein forms microspheres. Angew Chem Int Ed 47(24):4592–4594
102. Humenik M, Smith AM, Arndt S, Scheibel T (2015) Ion and seed dependent fibril assembly of a spidroin core domain. J Struct Biol 191(2):130–138
103. Hermanson KD, Huemmerich D, Scheibel T, Bausch AR (2007) Engineered microcapsules fabricated from reconstituted spider silk. Adv Mater 19(14):1810–1815
104. Huemmerich D, Slotta U, Scheibel T (2006) Processing and modification of films made from recombinant spider silk proteins. Appl Phys A Mater Sci Process 82(2):219–222
105. Schacht K, Vogt J, Scheibel T (2016) Foams made of engineered recombinant spider silk proteins as 3d scaffolds for cell growth. ACS Biomater Sci Eng 2(4):517–525
106. Chen X, Knight DP, Vollrath F (2002) Rheological characterization of nephila spidroin solution. Biomacromolecules 3(4):644–648
107. Kenney JM, Knight D, Wise MJ, Vollrath F (2002) Amyloidogenic nature of spider silk. Eur J Biochem 269(16):4159–4163
108. Du N, Yang Z, Liu XY, Li Y, Xu HY (2011) Structural origin of the strain-hardening of spider silk. Adv Funct Mater 21(4):772–778
109. Numata K, Kaplan DL (2011) Differences in cytotoxicity of beta-sheet peptides originated from silk and amyloid beta. Macromol Biosci 11(1):60–64
110. Oroudjev E, Soares J, Arcidiacono S, Thompson JB, Fossey SA, Hansma HG (2002) Segmented nanofibers of spider dragline silk: atomic force microscopy and single-molecule force spectroscopy. Proc Natl Acad Sci U S A 99(suppl 2):6460–6465
111. Slotta U, Hess S, Spiess K, Stromer T, Serpell L, Scheibel T (2007) Spider silk and amyloid fibrils: a structural comparison. Macromol Biosci 7(2):183–188

112. Humenik M, Smith AM, Arndt S, Scheibel T (2015) Data for ion and seed dependent fibril assembly of a spidroin core domain. Data Brief 4:571–576
113. Kol N, Adler-Abramovich L, Barlam D, Shneck RZ, Gazit E, Rousso I (2005) Self-assembled peptide nanotubes are uniquely rigid bioinspired supramolecular structures. Nano Lett 5(7):1343–1346
114. Smith JF, Knowles TPJ, Dobson CM, MacPhee CE, Welland ME (2006) Characterization of the nanoscale properties of individual amyloid fibrils. Proc Natl Acad Sci U S A 103(43):15806–15811
115. Knowles TP, Fitzpatrick AW, Meehan S, Mott HR, Vendruscolo M, Dobson CM, Welland ME (2007) Role of intermolecular forces in defining material properties of protein nanofibrils. Science 318(5858):1900–1903
116. Adamcik J, Jung J-M, Flakowski J, De Los RP, Dietler G, Mezzenga R (2010) Understanding amyloid aggregation by statistical analysis of atomic force microscopy images. Nat Nanotechnol 5(6):423–428
117. Sachse C, Grigorieff N, Fändrich M (2010) Nanoscale flexibility parameters of alzheimer amyloid fibrils determined by electron cryo-microscopy. Angew Chem Int Ed 49(7):1321–1323
118. Knowles TPJ, Buehler MJ (2011) Nanomechanics of functional and pathological amyloid materials. Nat Nanotechnol 6(8):469–479
119. Lintz ES, Scheibel TR (2013) Dragline, egg stalk and byssus: a comparison of outstanding protein fibers and their potential for developing new materials. Adv Funct Mater 23(36):4467–4482
120. Cherny I, Gazit E (2008) Amyloids: not only pathological agents but also ordered nanomaterials. Angew Chem Int Ed 47(22):4062–4069
121. Knowles TPJ, Oppenheim TW, Buell AK, Chirgadze DY, Welland ME (2010) Nanostructured films from hierarchical self-assembly of amyloidogenic proteins. Nat Nanotechnol 5(3):204–207
122. Reches M, Gazit E (2006) Controlled patterning of aligned self-assembled peptide nanotubes. Nat Nanotechnol 1(3):195–200
123. Aggeli A, Nyrkova IA, Bell M, Harding R, Carrick L, McLeish TCB, Semenov AN, Boden N (2001) Hierarchical self-assembly of chiral rod-like molecules as a model for peptide β-sheet tapes, ribbons, fibrils, and fibers. Proc Natl Acad Sci U S A 98(21):11857–11862
124. Lara C, Adamcik J, Jordens S, Mezzenga R (2011) General self-assembly mechanism converting hydrolyzed globular proteins into giant multistranded amyloid ribbons. Biomacromolecules 12(5):1868–1875
125. Humenik M, Scheibel T (2014) Self-assembly of nucleic acids, silk and hybrid materials thereof. J Phys Condens Matter 26(50):503102
126. Humenik M, Scheibel T (2014) Nanomaterial building blocks based on spider silk–oligonucleotide conjugates. ACS Nano 8(2):1342–1349
127. Humenik M, Mohrand M, Scheibel T (2018) Self-assembly of spider silk-fusion proteins comprising enzymatic and fluorescence activity. Bioconjug Chem 29(4):898–904
128. Jansson R, Courtin CM, Sandgren M, Hedhammar M (2015) Rational design of spider silk materials genetically fused with an enzyme. Adv Funct Mater 25(33):5343–5352
129. Rammensee S, Huemmerich D, Hermanson KD, Scheibel T, Bausch AR (2006) Rheological characterization of hydrogels formed by recombinantly produced spider silk. Appl Phys A Mater Sci Process 82(2):261–264
130. Jones JA, Harris TI, Tucker CL, Berg KR, Christy SY, Day BA, Gaztambide DA, Needham NJC, Ruben AL, Oliveira PF, Decker RE, Lewis RV (2015) More than just fibers: an aqueous method for the production of innovative recombinant spider silk protein materials. Biomacromolecules 16(4):1418–1425
131. DeSimone E, Schacht K, Pellert A, Scheibel T (2017) Recombinant spider silk-based bioinks. Biofabrication 9(4):044104
132. Thamm C, DeSimone E, Scheibel T (2017) Characterization of hydrogels made of a novel spider silk protein emasp1s and evaluation for 3d printing. Macromol Biosci 17(11):1700141

133. Jungst T, Smolan W, Schacht K, Scheibel T, Groll J (2016) Strategies and molecular design criteria for 3d printable hydrogels. Chem Rev 116(3):1496–1539
134. Seiffert S, Sprakel J (2012) Physical chemistry of supramolecular polymer networks. Chem Soc Rev 41(2):909–930
135. Schacht K, Jüngst T, Schweinlin M, Ewald A, Groll J, Scheibel T (2015) Biofabrication of cell-loaded 3d spider silk constructs. Angew Chem Int Ed 54(9):2816–2820
136. Zhang Y, Cremer PS (2006) Interactions between macromolecules and ions: the hofmeister series. Curr Opin Chem Biol 10(6):658–663
137. Heim M, Keerl D, Scheibel T (2009) Spider silk: from soluble protein to extraordinary fiber. Angew Chem Int Ed 48(20):3584–3596
138. Lammel A, Schwab M, Slotta U, Winter G, Scheibel T (2008) Processing conditions for the formation of spider silk microspheres. ChemSusChem 1(5):413–416
139. Blüm C, Scheibel T (2012) Control of drug loading and release properties of spider silk sub-microparticles. BioNanoScience 2(2):67–74
140. Lucke M, Mottas I, Herbst T, Hotz C, Römer L, Schierling M, Herold HM, Slotta U, Spinetti T, Scheibel T, Winter G, Bourquin C, Engert J (2018) Engineered hybrid spider silk particles as delivery system for peptide vaccines. Biomaterials 172:105–115
141. Krishnaji ST, Huang W, Cebe P, Kaplan DL (2014) Influence of solution parameters on phase diagram of recombinant spider silk-like block copolymers. Macromol Chem Phys 215(12):1230–1238
142. Florczak A, Mackiewicz A, Dams-Kozlowska H (2014) Functionalized spider silk spheres as drug carriers for targeted cancer therapy. Biomacromolecules 15(8):2971–2981
143. Jastrzebska K, Felcyn E, Kozak M, Szybowicz M, Buchwald T, Pietralik Z, Jesionowski T, Mackiewicz A, Dams-Kozlowska H (2016) The method of purifying bioengineered spider silk determines the silk sphere properties. Sci Rep 6:28106
144. Elsner MB, Herold HM, Müller-Herrmann S, Bargel H, Scheibel T (2015) Enhanced cellular uptake of engineered spider silk particles. Biomater Sci 3(3):543–551
145. Neubauer MP, Blüm C, Agostini E, Engert J, Scheibel T, Fery A (2013) Micromechanical characterization of spider silk particles. Biomater Sci 1(11):1160–1165
146. Helfricht N, Klug M, Mark A, Kuznetsov V, Blüm C, Scheibel T, Papastavrou G (2013) Surface properties of spider silk particles in solution. Biomater Sci 1(11):1166–1171
147. Helfricht N, Doblhofer E, Duval JFL, Scheibel T, Papastavrou G (2016) Colloidal properties of recombinant spider silk protein particles. J Phys Chem C 120(32):18015–18027
148. Hermanson KD, Harasim MB, Scheibel T, Bausch AR (2007) Permeability of silk microcapsules made by the interfacial adsorption of protein. Phys Chem Chem Phys 9(48):6442–6446
149. Blüm C, Nichtl A, Scheibel T (2014) Spider silk capsules as protective reaction containers for enzymes. Adv Funct Mater 24(6):763–768
150. Slotta U, Tammer M, Kremer F, Koelsch P, Scheibel T (2006) Structural analysis of spider silk films. Supramol Chem 18(5):465–471
151. Metwalli E, Slotta U, Darko C, Roth SV, Scheibel T, Papadakis CM (2007) Structural changes of thin films from recombinant spider silk proteins upon post-treatment. Appl Phys A Mater Sci Process 89(3):655–661
152. Spiess K, Wohlrab S, Scheibel T (2010) Structural characterization and functionalization of engineered spider silk films. Soft Matter 6(17):4168–4174
153. Gomes SC, Leonor IB, Mano JF, Reis RL, Kaplan DL (2011) Antimicrobial functionalized genetically engineered spider silk. Biomaterials 32(18):4255–4266
154. Tucker CL, Jones JA, Bringhurst HN, Copeland CG, Addison JB, Weber WS, Mou Q, Yarger JL, Lewis RV (2014) Mechanical and physical properties of recombinant spider silk films using organic and aqueous solvents. Biomacromolecules 15(8):3158–3170
155. Spiess K, Ene R, Keenan CD, Senker J, Kremer F, Scheibel T (2011) Impact of initial solvent on thermal stability and mechanical properties of recombinant spider silk films. J Mater Chem 21(35):13594–13604

156. Young SL, Gupta M, Hanske C, Fery A, Scheibel T, Tsukruk VV (2012) Utilizing conformational changes for patterning thin films of recombinant spider silk proteins. Biomacromolecules 13(10):3189–3199
157. Wohlrab S, Spie ST (2012) Varying surface hydrophobicities of coatings made of recombinant spider silk proteins. J Mater Chem 22(41):22050–22054
158. Huang W, Krishnaji S, Tokareva OR, Kaplan D, Cebe P (2014) Influence of water on protein transitions: morphology and secondary structure. Macromolecules 47(22):8107–8114
159. Currie HA, Deschaume O, Naik RR, Perry CC, Kaplan DL (2011) Genetically engineered chimeric silk-silver binding proteins. Adv Funct Mater 21(15):2889–2895
160. Junghans F, Morawietz M, Conrad U, Scheibel T, Heilmann A, Spohn U (2006) Preparation and mechanical properties of layers made of recombinant spider silk proteins and silk from silk worm. Appl Phys A Mater Sci Process 82(2):253–260
161. Wohlrab S, Müller S, Schmidt A, Neubauer S, Kessler H, Leal-Egaña A, Scheibel T (2012) Cell adhesion and proliferation on rgd-modified recombinant spider silk proteins. Biomaterials 33(28):6650–6659
162. Widhe M, Johansson U, Hillerdahl C-O, Hedhammar M (2013) Recombinant spider silk with cell binding motifs for specific adherence of cells. Biomaterials 34(33):8223–8234
163. Widhe M, Bysell H, Nystedt S, Schenning I, Malmsten M, Johansson J, Rising A, Hedhammar M (2010) Recombinant spider silk as matrices for cell culture. Biomaterials 31(36):9575–9585
164. Lewicka M, Hermanson O, Rising AU (2012) Recombinant spider silk matrices for neural stem cell cultures. Biomaterials 33(31):7712–7717
165. Ratner BD, Hoffman AS, Schoen FJ, Lemons JE (eds) (2004) Biomaterials science: an introduction to materials in medicine, 2nd edn. Academic, Kidlington
166. Hardy JG, Scheibel TR (2009) Silk-inspired polymers and proteins. Biochem Soc Trans 37(Pt 4):677–681
167. Leal-Egana A, Scheibel T (2010) Silk-based materials for biomedical applications. Biotechnol Appl Biochem 55(3):155–167
168. Aigner TB, DeSimone E, Scheibel T (2018) Biomedical applications of recombinant silk-based materials. Adv Mater 30(19):1704636
169. Vollrath F, Barth P, Basedow A, Engstrom W, List H (2002) Local tolerance to spider silks and protein polymers in vivo. In Vivo 16(4):229–234
170. Gellynck K, Verdonk PC, Van Nimmen E, Almqvist KF, Gheysens T, Schoukens G, Van Langenhove L, Kiekens P, Mertens J, Verbruggen G (2008) Silkworm and spider silk scaffolds for chondrocyte support. J Mater Sci Mater Med 19(11):3399–3409
171. Allmeling C, Jokuszies A, Reimers K, Kall S, Choi CY, Brandes G, Kasper C, Scheper T, Guggenheim M, Vogt PM (2008) Spider silk fibres in artificial nerve constructs promote peripheral nerve regeneration. Cell Prolif 41(3):408–420
172. Radtke C, Allmeling C, Waldmann KH, Reimers K, Thies K, Schenk HC, Hillmer A, Guggenheim M, Brandes G, Vogt PM (2011) Spider silk constructs enhance axonal regeneration and remyelination in long nerve defects in sheep. PLoS One 6(2):e16990
173. Moisenovich MM, Pustovalova OL, Arhipova AY, Vasiljeva TV, Sokolova OS, Bogush VG, Debabov VG, Sevastianov VI, Kirpichnikov MP, Agapov II (2011) In vitro and in vivo biocompatibility studies of a recombinant analogue of spidroin 1 scaffolds. J Biomed Mater Res A 96(1):125–131
174. Camilla Fredriksson MH, Feinstein R, Nordling K, Kratz G, Johansson J, Rising FHA (2009) Tissue response to subcutaneously implanted recombinant spider silk: an in vivo study. Materials 2(4):1908–1922
175. Langer R (1990) New methods of drug delivery. Science 249(4976):1527–1533
176. Numata K, Kaplan DL (2010) Silk-based gene carriers with cell membrane destabilizing peptides. Biomacromolecules 11(11):3189–3195
177. Numata K, Reagan MR, Goldstein RH, Rosenblatt M, Kaplan DL (2011) Spider silk-based gene carriers for tumor cell-specific delivery. Bioconjug Chem 22(8):1605–1610

178. Numata K, Mieszawska-Czajkowska AJ, Kvenvold LA, Kaplan DL (2012) Silk-based nanocomplexes with tumor-homing peptides for tumor-specific gene delivery. Macromol Biosci 12(1):75–82
179. Doblhofer E, Scheibel T (2015) Engineering of recombinant spider silk proteins allows defined uptake and release of substances. J Pharm Sci 104(3):988–994
180. Wang X, Yucel T, Lu Q, Hu X, Kaplan DL (2010) Silk nanospheres and microspheres from silk/pva blend films for drug delivery. Biomaterials 31(6):1025–1035
181. Patil Y, Panyam J (2009) Polymeric nanoparticles for sirna delivery and gene silencing. Int J Pharm 367(1–2):195–203
182. Sushmita Sundar JK, Kundu SC (2010) Biopolymeric nanoparticles. Sci Technol Adv Mater 11(1):1–13
183. Liebmann B, Huemmerich D, Scheibel T, Fehr M (2008) Formulation of poorly water-soluble substances using self-assembling spider silk protein. Colloids Surf A Physicochem Eng Asp 331(1–2):126–132
184. Lammel A, Schwab M, Hofer M, Winter G, Scheibel T (2011) Recombinant spider silk particles as drug delivery vehicles. Biomaterials 32(8):2233–2240
185. Langer R, Vacanti JP (1993) Tissue engineering. Science 260(5110):920–926
186. Schacht K, Scheibel T (2014) Processing of recombinant spider silk proteins into tailor-made materials for biomaterials applications. Curr Opin Biotechnol 29:62–69
187. Petzold J, Aigner TB, Touska F, Zimmermann K, Scheibel T, Engel FB (2017) Surface features of recombinant spider silk protein eadf4(κ16)-made materials are well-suited for cardiac tissue engineering. Adv Funct Mater 27(36):1701427
188. Bell E, Ehrlich HP, Buttle DJ, Nakatsuji T (1981) Living tissue formed in vitro and accepted as skin-equivalent tissue of full thickness. Science 211(4486):1052–1054
189. Burke JF, Yannas IV, Quinby WC Jr, Bondoc CC, Jung WK (1981) Successful use of a physiologically acceptable artificial skin in the treatment of extensive burn injury. Ann Surg 194(4):413–428
190. Wendt H, Hillmer A, Reimers K, Kuhbier JW, Schafer-Nolte F, Allmeling C, Kasper C, Vogt PM (2011) Artificial skin-culturing of different skin cell lines for generating an artificial skin substitute on cross-weaved spider silk fibres. PLoS One 6(7):e21833
191. Widhe M, Shalaly ND, Hedhammar M (2016) A fibronectin mimetic motif improves integrin mediated cell biding to recombinant spider silk matrices. Biomaterials 74:256–266
192. Millennium WSGotBoMCatSotN (2003) The burden of musculoskeletal conditions at the start of the new millennium. World Health Organ Tech Rep Ser 919:1–218
193. Hing KA (2004) Bone repair in the twenty-first century: biology, chemistry or engineering? Philos Transact A Math Phys Eng Sci 362(1825):2821–2850
194. Parikh SN (2002) Bone graft substitutes in modern orthopedics. Orthopedics 25(11):1301–1309. quiz 1310-1301
195. Mark DE, Hollinger JO, Hastings C Jr, Chen G, Marden LJ, Reddi AH (1990) Repair of calvarial nonunions by osteogenin, a bone-inductive protein. Plast Reconstr Surg 86(4):623–630. discussion 631-622
196. Huang J, Wong C, George A, Kaplan DL (2007) The effect of genetically engineered spider silk-dentin matrix protein 1 chimeric protein on hydroxyapatite nucleation. Biomaterials 28(14):2358–2367
197. Wong Po Foo C, Patwardhan SV, Belton DJ, Kitchel B, Anastasiades D, Huang J, Naik RR, Perry CC, Kaplan DL (2006) Novel nanocomposites from spider silk-silica fusion (chimeric) proteins. Proc Natl Acad Sci U S A 103(25):9428–9433
198. Belton DJ, Mieszawska AJ, Currie HA, Kaplan DL, Perry CC (2012) Silk-silica composites from genetically engineered chimeric proteins: materials properties correlate with silica condensation rate and colloidal stability of the proteins in aqueous solution. Langmuir 28(9):4373–4381
199. Mieszawska AJ, Nadkarni LD, Perry CC, Kaplan DL (2010) Nanoscale control of silica particle formation via silk-silica fusion proteins for bone regeneration. Chem Mater 22(20):5780–5785

200. Chen MB, Zhang F, Lineaweaver WC (2006) Luminal fillers in nerve conduits for peripheral nerve repair. Ann Plast Surg 57(4):462–471
201. Ciardelli G, Chiono V (2006) Materials for peripheral nerve regeneration. Macromol Biosci 6(1):13–26
202. Evans GR, Brandt K, Katz S, Chauvin P, Otto L, Bogle M, Wang B, Meszlenyi RK, Lu L, Mikos AG, Patrick CW Jr (2002) Bioactive poly(l-lactic acid) conduits seeded with schwann cells for peripheral nerve regeneration. Biomaterials 23(3):841–848
203. Lietz M, Dreesmann L, Hoss M, Oberhoffner S, Schlosshauer B (2006) Neuro tissue engineering of glial nerve guides and the impact of different cell types. Biomaterials 27(8):1425–1436
204. Lundborg G, Rosen B, Abrahamson SO, Dahlin L, Danielsen N (1994) Tubular repair of the median nerve in the human forearm. preliminary findings. J Hand Surg (Br) 19(3):273–276
205. Stanec S, Stanec Z (1998) Reconstruction of upper-extremity peripheral-nerve injuries with eptfe conduits. J Reconstr Microsurg 14(4):227–232
206. Letourneau PC, Condic ML, Snow DM (1994) Interactions of developing neurons with the extracellular matrix. J Neurosci 14(3 Pt 1):915–928
207. Allmeling C, Jokuszies A, Reimers K, Kall S, Vogt PM (2006) Use of spider silk fibres as an innovative material in a biocompatible artificial nerve conduit. J Cell Mol Med 10(3):770–777
208. Borkner CB, Elsner MB, Scheibel T (2014) Coatings and films made of silk proteins. ACS Appl Mater Interfaces 6(18):15611–15625
209. Zeplin PH, Berninger AK, Maksimovikj NC, van Gelder P, Scheibel T, Walles H (2014) Improving the biocompatibility of silicone implants using spider silk coatings: immunohistochemical analysis of capsule formation. Handchir Mikrochir Plast Chir 46(6):336–341
210. Borkner CB, Wohlrab S, Möller E, Lang G, Scheibel T (2017) Surface modification of polymeric biomaterials using recombinant spider silk proteins. ACS Biomater Sci Eng 3(5):767–775
211. Mironov V, Trusk T, Kasyanov V, Little S, Swaja R, Markwald R (2009) Biofabrication: a 21st century manufacturing paradigm. Biofabrication 1(2):022001
212. Groll J, Boland T, Blunk T, Burdick JA, Cho DW, Dalton PD, Derby B, Forgacs G, Li Q, Mironov VA, Moroni L, Nakamura M, Shu W, Takeuchi S, Vozzi G, Woodfield TB, Xu T, Yoo JJ, Malda J (2016) Biofabrication: reappraising the definition of an evolving field. Biofabrication 8(1):013001
213. Malda J, Visser J, Melchels FP, Jungst T, Hennink WE, Dhert WJ, Groll J, Hutmacher DW (2013) 25th anniversary article: engineering hydrogels for biofabrication. Adv Mater 25(36):5011–5028
214. Murphy SV, Atala A (2014) 3d bioprinting of tissues and organs. Nat Biotechnol 32(8):773–785
215. Skardal A, Atala A (2015) Biomaterials for integration with 3-d bioprinting. Ann Biomed Eng 43(3):730–746
216. Catros S, Fricain JC, Guillotin B, Pippenger B, Bareille R, Remy M, Lebraud E, Desbat B, Amedee J, Guillemot F (2011) Laser-assisted bioprinting for creating on-demand patterns of human osteoprogenitor cells and nano-hydroxyapatite. Biofabrication 3(2):025001
217. Chimene D, Lennox KK, Kaunas RR, Gaharwar AK (2016) Advanced bioinks for 3d printing: a materials science perspective. Ann Biomed Eng 44(6):2090–2102
218. DeSimone E, Schacht K, Jungst T, Groll J, Scheibel T (2015) Biofabrication of 3d constructs: fabrication technologies and spider silk proteins as bioinks. Pure Appl Chem 87(8):737–749

Chapter 7
Protein Microgels from Amyloid Fibril Networks

Lianne W. Y. Roode, Ulyana Shimanovich, Si Wu, Sarah Perrett, and Tuomas P. J. Knowles

Abstract Nanofibrillar forms of amyloidogenic proteins were initially discovered in the context of protein misfolding and disease but have more recently been found at the origin of key biological functionality in many naturally occurring functional materials, such as adhesives and biofilm coatings. Their physiological roles in nature reflect their great strength and stability, which has led to the exploration of their use as the basis of artificial protein-based functional materials. Particularly for biomedical applications, they represent attractive building blocks for the development of, for instance, drug carrier agents due to their inherent biocompatibility and biodegradability. Furthermore, the propensity of proteins to self-assemble into amyloid fibrils can be exploited under microconfinement, afforded by droplet microfluidic techniques. This approach allows the generation of multi-scale functional microgels that can host biological additives and can be designed to incorporate additional functionality, such as to aid targeted drug delivery.

L. W. Y. Roode
Centre for Misfolding Diseases, Department of Chemistry, University of Cambridge, Cambridge, UK

U. Shimanovich (✉)
Department of Materials and Interfaces, Weizmann Institute of Science, Rehovot, Israel
e-mail: ulyana.shimanovich@weizmann.ac.il

S. Wu · S. Perrett (✉)
National Laboratory of Biomacromolecules, CAS Center for Excellence in Biomacromolecules, Institute of Biophysics, Chinese Academy of Sciences, Beijing, China

University of the Chinese Academy of Sciences, Beijing, China
e-mail: sarah.perrett@cantab.net

T. P. J. Knowles (✉)
Centre for Misfolding Diseases, Department of Chemistry, University of Cambridge, Cambridge, UK

Cavendish Laboratory, University of Cambridge, Cambridge, UK
e-mail: tpjk2@cam.ac.uk

S. Perrett et al. (eds.), *Biological and Bio-inspired Nanomaterials*, Advances in Experimental Medicine and Biology 1174,
https://doi.org/10.1007/978-981-13-9791-2_7

Keywords Self-assembled amyloid fibrils · Protein microgels · Droplet microfluidics · Drug carrier agents · Functional materials

7.1 Nature of Amyloid Proteins

7.1.1 Introduction

Amyloid fibrils have become a subject of rapidly increasing research interest across a wide range of scientific disciplines mainly due to their association with a growing number of debilitating medical disorders, ranging from neurodegenerative disorders (e.g. Alzheimer's and Parkinson's diseases) to systemic amyloidoses [1–5]. A number of human diseases associated with amyloid formation are listed in Table 7.1. The origin of such pathological conditions was found to be associated with misfolding of normally soluble, functional peptides and proteins, giving rise to intractable aggregates and their subsequent conversion into mature fibrils, of which the archetypal example are amyloid fibrils. They are highly ordered, thread-like aggregates and their formation is associated with both a loss of function of the proteins involved as well as with the generation of toxic intermediate species that are formed in the process of self-assembly [6–9].

Although the amyloid form was initially identified within the context of disease, an increasing number of proteins that are not associated with any known pathogenic disorder have been found to form amyloid-like fibrils. Furthermore, increasing evidence has pointed towards amyloid formation being a generic property of proteins and polypeptide chains, with many polymorphic forms having been identified [6, 8, 10]. More recently, amyloid or amyloid-like fibrillar structures (which are defined as functionally beneficial fibrillar constructs structurally similar to amyloids) have emerged as key functional components in biological materials found in organisms ranging from bacteria to mammals [11–16]. These amyloid materials are exploited for use in diverse contexts, ranging from adhesives [17–20], catalytic scaffolds [11], bacterial coatings and biofilm formation [13, 21], and mediating epigenetic-information storage and transfer [22–26] (Table 7.2).

The discovery of functional amyloid structures in such a diverse set of roles in nature has transformed the understanding of the evolutionary requirements and design features that favour functional protein assembly versus aberrant association. Although the exact molecular level mechanisms that differentiate toxic amyloid formation and beneficial amyloid species are not fully understood yet, increasing evidence suggests that rather than the mature amyloid fibrils themselves, small soluble intermediate species, namely oligomers, are many responsible for the cellular toxicity associated with aberrant protein aggregation [10, 27–33]. In cases where amyloid structures are used to generate functional materials it seems that nature, therefore, has mechanisms in place to control the self-assembly process, ensuring that it goes to completion and that it does not give rise to the accumulation of potentially highly toxic intermediates [12, 15, 21, 34].

Table 7.1 Human diseases associated with amyloid formation from protein misfolding and aggregation

Aggregating protein or peptide	Polypeptide length (number of residues)	Structure	Associated disease
Neurodegenerative diseases			
Amyloid-β peptide	37–43	Intrinsically disordered	Alzheimer's disease
α-Synuclein	140	Intrinsically disordered	Parkinson's disease
Prion proteins or its fragments	230	Intrinsically disordered and α-helical	Spongiform encephalopathies
Huntingtin fragments	Variable	Mostly intrinsically disordered	Huntington's disease
Superoxide dismutase 1	153	β-sheet and Ig-like	Amyotrophic lateral sclerosis
Transthyretin mutants	127	β-sheet	Familial amyloidotic polyneuropathy
Systemic amyloidoses			
Wild-type transthyretin	127	β-sheet	Senile systemic amyloidosis
Immunoglobin (Ig) light chains or its fragments	~90	β-sheet and Ig-like	Amyloid light chain (AL) amyloidosis
Serum amyloid A1 protein fragments	76–104	α-helical and unknown fold	Amyloid A (AA) amyloidosis
β_2-microglobulin	99	β-sheet and Ig-like	Haemodialysis-related amyloidosis
Lysozyme mutants	130	α-helical and β-sheet	Lysozyme amyloidosis
Localized amyloidoses			
Apolipoprotein A-1 fragments	80–93	Intrinsically disordered	Apolipoprotein A-1 amyloidosis
Amylin	37	Intrinsically disordered	Type II diabetes
Insulin	21 and 30	α-helical and insulin-like	Injection-localized amyloidosis

Reproduced with permission from Macmillan Publishers Ltd.: Nature Reviews Molecular Cell Biology Ref. [9], copyright 2014

The highly organized nature of amyloid fibrils confers unique properties upon them, typically including a high level of kinetic and thermodynamic stability, making amyloid materials very robust and mechanically stable [9, 35, 36]. Their unique mechanical properties, together with the inherent ability of proteins to self-assemble into fibrillar structures under certain conditions [37–39], open up the possibility of using amyloid fibrils as a powerful platform for the design of new functional multi-scale materials [40].

Table 7.2 Functional amyloid-like fibrils and their corresponding biological roles

Organism	Functional amyloid	Major substituent	Structure	Function
Bacteria				
Escherichia coli, Salmonella spp.	Curli	CsgA (15 kDa)	β-sheet rich fibrils at the cell surface	Biofilm formation; host cell adhesion and invasion; inflammatory response activation
Streptomyces coelicolor	Chaplins	ChpA (20 kDa)	β-sheet rich fibrils at soil-air interfaces	Development of aerial structures and spores
Methanosaeta thermophile (archaea)	Cell wall sheaths	MspA (60 kDa)	β-solenoid with disulfide-linked subunits	Encasement of bacterium; protection from heat; diffusion of metabolites
Fungi				
Most fungi	Hydro-phobins	SC3 (12 kDa)	β-barrel monomers form continuous β-sheet amyloid at air-water interface	Host adhesion; aerial growth spore formation
Animals				
Arthropods	Silk	MaSp1, MaSp2 (∼3 kDa)	β-sheet crystalline regions surrounded by amorphous matrix	Shelter; food entrapment; egg protection
Spiders and caterpillars				
Insects and fish	Chorions	Several	Conserved central domain aids β-sheet assembly	Egg production
Homo sapiens	Pmel-17 fibers	Pmel-17 (100 kDa)	Luminal fragment rapidly forms cross-β sheet fibrils	Polymerization of intermediates in melanin synthesis
Plants				
Hervea brasiliensis	Rubber elongation factor	REF (15 kDa)	High β-sheet content transmembrane structure	Biosynthesis of natural rubber

Adapted from Ref. [16], with permission from John Wiley and Sons

Amyloid fibril networks are particularly advantageous as the basis of microgels for biomedical applications, such as drug delivery [41, 42], due to their inherent biocompatibility and their ability to self-assemble under mild conditions in aqueous solution [43, 44]. The approach to synthesize microgels from self-associating species represents itself as a very attractive strategy due to its simplicity in application of molecular self-assembly. By combining the inherent tendency of protein molecules to form fibrils with micron scale structuring that can be achieved through droplet microfluidics [45], control over the microgel size and morphology can be attained during the formation process [46, 47]. This controlled assembly provides a system that is highly effective for applications of tuneable, functional materials.

7.1.2 Detection of Amyloid Structures

The term 'amyloid', originally meaning 'starch like', is a legacy of its first observation more than 150 years ago in systemic amyloidosis, as the deposits observed in patient's tissues and organs could be readily stained with iodine, which is commonly used to detect starch [48, 49]. The repetitive sequence of protein units that forms an organized structure includes regular binding pockets for various dyes by which amyloid fibrils are routinely characterized. Amyloid in human tissue is traditionally visualized by pathologists using the sulfonated azo dye Congo Red, which lends a distinct green birefringence to amyloid deposits under cross-polarized filters [50]. Similarly, *in vitro* studies generally make use of the fluorescent dye Thioflavin T (ThT), giving rise to increased quantum yield and red shifted fluorescence of ThT upon binding to amyloid fibrils [51, 52].

Although these dyes are readily available and easy to handle, they do not have absolute specificity for amyloid structures, and are thus best used against isolated proteins [53, 54]. *For in vivo* studies, it is more appropriate to use conformationally specific monoclonal antibodies, which were originally developed for amyloid-β fibrils but were shown to bind to a multitude of amyloid fibrils tested [55, 56]. In addition, Kaminski and co-workers reported an intrinsic fluorescence signature of aggregating polypeptides, including the amyloidogenic human peptides amyloid-β, lysozyme and tau. The intrinsic fluorescence appears to result from electronic levels that become available when the polypeptide chain folds into a cross-β sheet structure reminiscent of what has been reported to take place in crystals. This allows for label-free quantification of protein aggregation *in vitro* [57]. Following these findings, the same research group developed a Förster resonance energy transfer (FRET) sensor that exploits the observation that amyloid assemblies can act as energy acceptors for fluorescently labelled proteins, thereby permitting the aggregation kinetics of amyloidogenic proteins to be quantified in a non-invasive manner suitable for both *in vitro* and *in vivo* studies [58, 59].

7.1.3 Structure of Amyloid Fibrils

Amyloid fibrils have been observed for a variety of proteins and peptides [8, 60]. Despite the various origins, the fibrils seem to be remarkably similar, having a fibrillar morphology at the nanoscale and generic quaternary protein structure on the molecular scale [61, 62]. A multitude of studies, including solid-state NMR spectroscopy [63–68], single crystal X-ray diffraction analysis [69–71], atomic force microscopy [72], electron microscopy [73], and cryo-electron microscopy [74–77], have given insight into the atomic level architecture of amyloid fibrils. These studies confirm common structural features, notably a high content of hydrogen bond-rich β-sheets and the frequent presence of repetitive hydrophobic or polar interactions along the fibrillar axis [8]. The fibrils are observed to consist

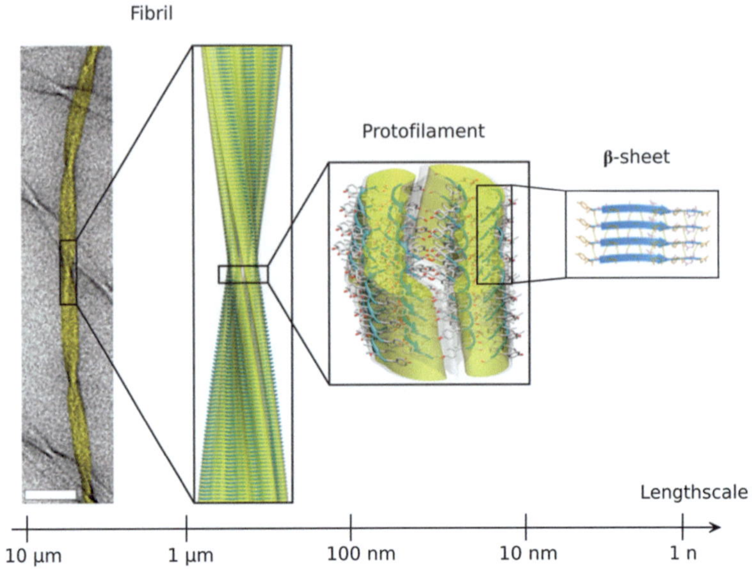

Fig. 7.1 The hierarchical structure of amyloid fibrils. The fibril structure is shown on the left in an image taken using TEM (scale bar = 50 nm); MAS NMR atomic-resolution structure of the fibril is fitted into the cryo-EM reconstruction. Within the protofilament, β-sheets are shown in a ribbon representation, where oxygen, carbon, and nitrogen atoms are shown in red, gray, and blue, respectively. On the far right the β-sheet fragment shows β-strands aligned perpendicular to the fibril axis. (Reproduced from Ref. [74] with permission)

of orderly repeats of protein molecules assembled in a hierarchical manner. β-Strands (separated by ~4 Å) are aligned perpendicularly to the fibril axis, resulting in a dense hydrogen bonded network composed of continuous hydrogen bonded β-sheets (typical intersheet distance ~10–12 Å) along the length of the fibril. These structures are the fundamental components of protofilaments which then twist around each other to form mature fibrils, which under electron and atomic force microscopy appear as unbranched filamentous structures of only a few nanometers in diameter but exceeding micrometers in length (Fig. 7.1) [74, 78].

The high persistence length of amyloid fibrils is primarily due to the strong hydrogen bonded network along the polypeptide backbone within the fibril core [40]. Given that the backbone is common to all polypeptides and main chain hydrogen bonds are accessible to all proteins, the view that the core structure of amyloid fibrils is primarily stabilized by the hydrogen bonded network, supports the finding that a variety of polypeptide chains of very different amino acid sequence can give rise to amyloid fibrils.

7.1.4 Self-Assembly and Polymorphism of Amyloid Fibrils

Although all amyloid systems have a common fibrillar architecture, structural variations in different amyloid fibrils have been observed. Fibrils may vary in β-strand length, number of β-sheets per protofilament, and conformation (parallel/anti-parallel) as a result of the manner in which different sets of side chains interact and are incorporated into the fibrillar architecture [79, 80]. Unlike the native protein state, a single arrangement of a given polypeptide chain in the amyloid core structure is unlikely to provide unique stability relative to all other arrangements. Indeed, the basic amyloid cross-β structure can be adopted through a multitude of possible stacking geometries of β-sheets, giving rise to a large number of polymorphic forms [8, 40, 74].

In addition to structural variation in fibrils from different polypeptides, polymorphic forms have been observed for fibrils from the same polypeptide sequence but formed under varying conditions [82–85]. This highlights differences in the details of the way in which individual molecules are incorporated into the fibrils, depending on the assembly environment. While the essential elements of the amyloid fold are mainly determined by inherent physico-chemical properties of the polypeptide chain, as indicated by studies correlating relative aggregation rates with features such as the hydrophobicity and charge [7, 86], detailed morphological variation arises due to secondary structure and long-range interactions. As such, solvent conditions such as temperature, pH, ionic strength, and protein concentration as well as mechanical influences such as agitation influence the assembly process [39, 78, 81, 87, 88].

Studies showed that, while under physiological conditions the formation of amyloid fibrils appeared to be restricted to a subset of proteins, under mildly denaturing conditions *in vitro* proteins can be induced to form fibrillar aggregates [6, 10, 37, 89]. These *in vitro* studies, together with the recognition of functional amyloidogenic proteins in bacteria and human [12, 13, 90], led to the idea that the formation of amyloid structures is not a rare phenomenon associated with a small number of diseases only, but is rather a generic property of many, if not all, polypeptide chains [8, 10, 91].

The multitude of amyloid fibril forming proteins observed, together with mathematical analysis of the mechanism of amyloid formation, suggests that the amyloid form may be determined by kinetic rather than thermodynamic factors. There is increasing evidence that the native protein state may not always represent the absolute free energy minima in protein folding processes but rather that the amyloid state is thermodynamically more stable, though kinetically inaccessible under some conditions [92, 93]. The amyloid state reflects a well-defined structural form of the protein that is an alternative to the native state and, unlike the native state, its essential architecture is not encoded by the amino acid sequence, although the propensity of formation does vary with sequence [8, 9, 94].

A high kinetic barrier that is associated with amyloid structures makes the formation of the initial amyloid nucleus a slow process. Nevertheless, once formed amyloid fibrils are capable of self-replication through secondary processes where additional protein monomers are incorporated into the fibril structure [95–97]. Pre-formed fibrillar species, referred to as seeds, can overcome sequence- or environment-based structural preferences by acting as a template for further fibril growth; protein monomers added to pre-formed fibril-ends adopt a cross-β conformation to match that of the protein molecules already present in the aggregate [98, 99]. This process results in rapid fibril formation, analogous to crystallization phenomena [100], where the seeds alleviate the high kinetic barrier associated with amyloid structures; thus the fibril morphology and structure is propagated [101–103].

7.1.5 Mechanical Properties of Amyloid Fibrils

The highly-ordered structure of amyloid fibrils is intimately reflected by its properties. The comprehensive hydrogen bonding network of the β-sheet core provides strong intermolecular backbone-backbone interactions between polypeptide chains within the fibrils, resulting in great fibril stability [35, 70]. This is exemplified by amyloid fibrils being very stable, even under extreme conditions, such as high concentrations of (chemical) denaturants, extreme pH environments and high temperatures [16]. Indeed, amyloid materials possess a remarkably high Young's modulus and tensile strength [36, 104–106].

In discussing the physical nature of amyloid fibrils, it is informative to compare them with other types of one-dimensional organic macromolecular and colloidal systems [40]. Proteins show generic polymeric and colloidal features, and many non-biological and synthetic polymers exhibit analogous condensed phases, such as filamentous and particulate gel states [107]. Indeed, some amyloid deposits formed *in vivo* and fibrils generated *in vitro* have been found to contain spherulites and other higher-order assemblies [108]. These structures are also observed in preparations of synthetic polymers, such as polyethylene, and this observation supports the idea that amyloid fibrils have features similar to those of classical polymers [8].

However, in contrast to classical polymeric systems, the persistence length of amyloid fibrils shows remarkable independence of the environmental conditions. The persistence length in amyloid fibrils primarily originates from the hydrogen bonding network in the core, which is less affected by solution conditions than the electrostatic repulsion which is responsible for the persistence length of conventional charged polymers [40]. Furthermore, amyloid fibrils possess the properties of chirality and polarity along their main fibrillar axis, which opens the possibility of manipulations via weak electric fields [109, 110]. The chiral, polar and charged nature, together with the intrinsic rigidity, of amyloid fibrils provides complex physical behaviour in one, two, and three dimensions.

Remarkably, the elastic moduli of amyloid fibrils are amongst the highest recorded for protein-based materials in nature [111, 112]. With stiffness values of E up to 10–20 GPa they are comparable to those of collagen, keratin and silk [36, 113–115] (Fig. 7.2). Considering the maximal density of intermolecular hydrogen bonds that is achievable in proteins, this provides a limit for the material performance

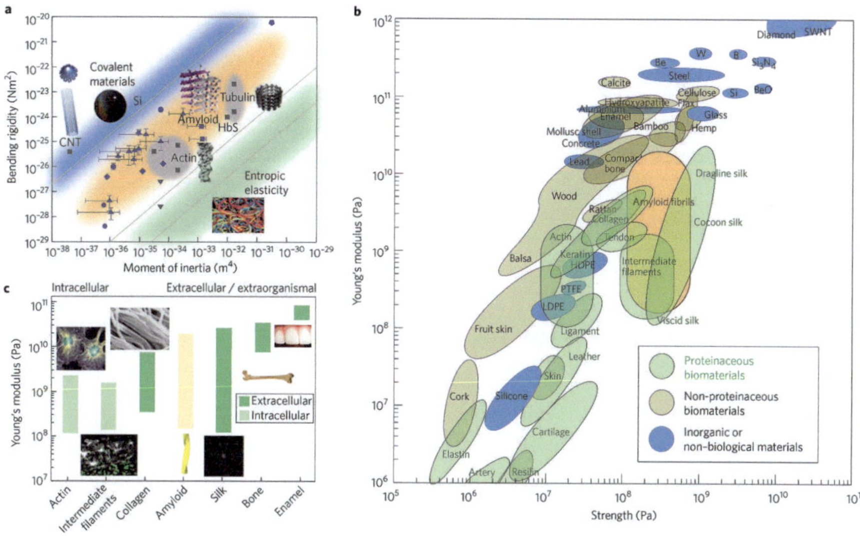

Fig. 7.2 Mechanical properties of amyloid fibrils in comparison to biological and inorganic or non-biological materials. (**a**) Bending rigidity versus moment of inertia for covalent materials (blue region), strong non-covalent interactions (such as hydrogen bonds, orange region) and weak non-covalent interactions (green region). Blue points are amyloid fibrils (different symbols denote data from different studies) and grey symbols are other materials [97, 105, 120–125]. Grey inverted triangles show the values for the response measured in the perpendicular direction to the fibril axis [126]. The upward triangles show data for one- and two-filament forms of bovine insulin, B-chain of bovin insulin, hen-egg-white lysozyme, bovine β-lactoglobulin, Alzheimer's amyloid β-peptide residues 1–40, GNNQQNY fragment of the yeast prion sup35, and human transthyretin residues 105–115 (all experimental) [121]. The pentagons show data for diphenylalanine (experimental) [120], octagons for insulin (experimental) [122], hexagons for ac-[RARADADA]2-am self-assembling peptide (simulation) [124], stars for Alzheimer's amyloid β-peptide residues 1–40 (simulation) [105], and lozenges show data for β-lactoglobulin (experimental) [106]. The circles show data for the N-terminal domain of the hydrogenase maturation factor HypF (experimental) [125], and squares show data for Alzheimer's amyloid β-residues 1–40 (experimental) [123]. (**b**) Young's modulus (which corresponds) to stiffness) versus strength [112, 127] for a range of different materials. Covalent and metallic bonding results in the stiffest and strongest materials, with diamond and single-wall carbon nanotubes (SWNTs) being the best performers. Silks are the strongest and stiffest protein materials, followed by amyloid and collagen; and significantly more rigid materials (for example, bone) contain minerals. Amyloid fibrils are shown in orange to distinguish them from other materials. (**c**) Range of values of Young's modulus for seven different classes of biological materials inside and outside the cell. The stiffest materials (such as collagen, bone, enamel and silk) are found outside the cell. (Reproduced with permission from Macmillan Publishers Ltd.: Nature Nanotechnology Ref. [36], copyright 2014)

that can be obtained from proteinaceous structures [97, 116]. Above this limit, materials contain covalent or metallic interactions which provide significantly higher energy density than hydrogen bonds or other weak non-covalent interactions [117]. Therefore nature, through evolution, has optimized proteinaceous materials to be close to the theoretical performance limits.

It is apparent, however, that the rigidity of intracellular materials, such as cytoskeletal filaments [118], does not reach the maximal values achievable for biological materials. This reflects the requirement of intracellular protein assembly to be readily reversible under physiological conditions to ensure cellular homeostasis [119]. Indeed, the materials exhibiting the highest moduli (cellulose, bone, keratin, silk) seem to be incompatible with intracellular structure and, instead, are primarily found extracellularly in roles where controlled rapid breakdown is not required. This phenomenon alludes to nature's regulation of the formation of such rigid protein materials in functional roles [36].

7.2 Amyloid Proteins for the Development of Functional Microgels

7.2.1 Emerging Applications of Artificial Amyloid Protein-Based Materials and Microgels

The intricate relationship between protein structure and function, which in nature has been tuned by evolutionary pressures over many millennia, can serve as a fruitful source of inspiration for the design of functional materials. The discovery of various biologically active amyloid structure-based materials contrasts markedly with the original picture of amyloid fibrils as inherently pathological structures and it has motivated the exploration of these fibrils as functional materials. Many of the same characteristics that underpin the natural use of amyloid structures, such as the fibril's extraordinary stability and accurate self-assembly, make them attractive as the basis of artificial functional materials [78, 128].

Within the last decade, artificial fibrillar, proteinaceous materials have been developed with applications including sensors [129–131], opto-electronics [132], biomimetic functional materials [133, 134], cell scaffolds [135–138], sustained drug release [139, 140], and food applications [141–145]. Furthermore, composite materials have been developed [146–149], with active properties making them capable of for example capturing carbon dioxide [150, 151], removal of radioactive waste and heavy metal ions from polluted water [152], and continuous flow catalysis [153]. Moreover, the interfacial properties of amyloid fibrils such as adhesiveness – generated via aggregation-induced development of positive surface charge – offered a good biocompatible platform towards biomineralization [154], antimicrobial coatings [155], wound healing [156], protein crystallization [157], and water-based photo/electro-lithography resist [158].

In the context of biomedical applications, where biocompatibility and a low immune-response are of utmost importance, the use of peptides and proteins in the development of new functional materials offers significant advantages over purely synthetic systems due to the biocompatibility and biodegradability of many naturally abundant proteins. Furthermore, owing to the highly variable chemical and physical properties of the wide range of peptides and proteins from which amyloid fibrils can be formed, such amyloid assemblies allow great potential for versatility in terms of material properties. This feature, in combination with the assembly processes being highly sensitive to environmental conditions, offers an opportunity to finely tune the structural features, and thus the properties, of materials based on amyloid fibril networks.

In nature, sophisticated material functionality is commonly achieved through the spatial control of protein structure and localization on both the nano- and micro-scale. Using proteins as building blocks for functional materials offers nano-scale ordering through the proteins' propensity to self-assemble into amyloid fibrils; micro-patterning can be achieved by protein assembly under micro-confinement using techniques such as microfluidics. Particularly, micro-scale systems are explored in the form of microgels and their potential for biomedical applications [41, 159]. Microgels are micrometer-sized colloidal particles composed of and stabilized by a three-dimensional cross-linked network of highly polymerized molecules, e.g. protein networks in which swelling of flexible particles occurs due to the incorporation of solvent molecules within its gel structure [160–162]. Amyloid fibril assembly for microgel formation has been conducted by achieving nucleation and subsequent growth of amyloid assemblies through tuning temperature and chemical conditions [37, 163, 164]. An additional route to manipulate amyloid fibril assembly is through laser trapping by which Yuyama et al. prepared a single spherical assembly of amyloid fibrils, demonstrating the method as a facile route to prepare protein assemblies and microgels [165].

Microgels are increasingly explored for their use in a variety of applications, including microsensors [166], catalysis [167], enzyme immobilization [168] and drug delivery [169–172]. Such applications are based on the ability of microgels to host nano- or mesoscopic additives as a result of their porous architecture, thereby serving as microscopic capsules. With the major component of microgels being the swelling agent, typically water, they are particularly useful for maintaining bioactivity and hosting sensitive biological payloads, including proteins, nucleic acids, and living cells [173–175].

7.2.1.1 Amyloid Microgels as Drug Carrier Agents

Due to the ability of microgels to encapsulate active species, microgels have emerged as important tools for developing clinically useful drug delivery systems. Currently, many drugs cannot be administered using conventional release methods due to undesired pharmacological properties, such as unfavourable solubility and toxicology, and/or inappropriate interactions with cellular components and other chemical species during drug delivery [176, 177]. Amyloid-based microgels, where the gel network comprises amyloid fibrils, have been found to be highly suitable to address such challenges, not only for their loading capacity [178–180], but also for their ability to incorporate additional functionality.

In addition to protein fibril microgels being versatile, biocompatible, biodegradable and of low immunogenicity [181–183], the dynamic nature of fibril assembly opens up the opportunity to modulate microgel behaviour. The composition and physical features of self-assembled fibrils can be modified during microgel formation and, thus, drug release may be tuned for example by controlling pore size, which influences the rate of release [184, 185]. Furthermore, the amyloid fibrils in the microgel network are sensitive to environmental conditions, such as changes in pH and salt conditions, giving rise to stimuli-responsive microgels [174, 186]. Moreover, modification of the self-assembling peptides or proteins with specific ligands can provide additional functionality, making them amenable to chemical surface functionalization for targeted drug delivery [135, 187, 188]. As such, they can be designed to transport therapeutic agents across biochemical barriers, so as to release them in specifically targeted tissues whilst sparing healthy tissues.

7.2.2 Microgel and Microcapsule Formation

7.2.2.1 Microgel Formation Techniques

The fabrication of microgels requires control of compartmentalization on the micro- and/or nano-scale. Methods such as spray drying [189, 190] and electrospray [191] have been used to achieve this objective. One of the most widely adopted techniques is based on microemulsion technology where two immiscible fluids, commonly an oil phase and an aqueous phase, are emulsified through agitation or sonication in bulk [192–195]. (Bio)polymer precursor molecules, such as proteins, forming the microgel network are typically dissolved in the aqueous phase and compartmentalized as dispersed microdroplets within the continuous oil phase. This microenviroment serves as a template to direct further structuring and localization of the precursor species as they assemble into network structures during gelation. The gelled network provides the microgel with the stability needed for it to be self-standing compared to the initial microdroplet form.

In addition to the stability of the microgels and microcapsules, their properties can be finely tuned by controlling their size. The size of the microgels is a key parameter that affects their ability to interface with biological systems and the rate at which encapsulated species are released. The afore-mentioned (bulk) emulsification methods, however, are limited in this respect as they do not allow precise control of the dimensions and internal structure of the microgels. This is unfavourable if the product particles are to serve as drug delivery capsules for which control over size, volume and content is essential. The use of microfluidic methods, rather than bulk emulsification techniques, provides greater control due to the characteristic behaviour of fluids in the micro-level regime [196, 197]. Microfluidics offers fluidic control in a gentle manner with minimal shear stress, realising a protein friendly micro-environment (i.e. mild, aqueous conditions) and precise micron scale structuring, affording control over size, shape, morphology and internal structure of the microparticle [198, 199].

Droplet microfluidics is a miniaturized technique that enables manipulation of ultralow interfacial tensions at the surface of emulsion droplets, resulting in a highly monodisperse emulsion [45, 200]. Several types of microfluidic droplet devices exist, including microfluidic devices assembled from glass microcapillaries [201, 202] and devices fabricated using soft lithography [203]. The latter approach allows for greater flexibility in channel geometries with greater precision than that of pulling glass capillaries. The concept is to fabricate microfluidic devices from to-scale drawings that are printed on a transparent mask and then transferred to a photoactive resin. Following a photolithography process step, the resulting positive relief serves as a master from which a microfluidic device is moulded [204]. An elastomer, such as polydimethylsiloxane, is poured over it, hardened, peeled off and bound to a substrate, producing a device with micrometer-sized channels as illustrated in Fig. 7.3 [205].

The principle of droplet formation in such channels is the use of flow-focusing geometries [203, 207]. A precursor stream of the fluid phase to be dispersed is intersected by an immiscible carrier fluid, which acts as a continuous phase, resulting in the periodic breakup of the stream as induced by the flow-focusing of the two phases at the junction. Droplets are formed due to a balance of interfacial tension and the shear of the continuous phase. The process is influenced by parameters, such as the viscosities, polarities, and flow rates of the fluids as well as the dimensions of the microchannels, giving powerful control over droplet dimensions [208]. In addition, microfluidic techniques enable the formation of structurally more complex microgel structures. Multiple-layered droplets (double or higher-order emulsions, i.e. droplets either containing smaller droplets or having multiple layers of shells) can be produced by consecutive droplet-making. Such higher-order structures serve as templates for multiphase and core-shell capsules rather than dense, uniform gels [209–212]. Furthermore, this principle of droplet formation presents a highly efficient manner of encapsulating active additives during the formation process, simply by flowing the additives in with the precursor solution (Fig. 7.4).

Until recently, most of the structural materials explored for generating microgels have involved polymerization or cross-linking of synthetic molecules. In many of these synthetic systems, this approach requires non-biocompatible conditions or reagents, such as extreme pH, high temperature, exposure to high doses of UV radiation or the use of chemical cross-linkers such as glutaraldehyde, formaldehyde or diacid chloride [41, 213, 214]. These cross-linking strategies can cause degradation of proteins and render the microgels non-biocompatible and potentially toxic due to undesired reactivity of the toxic chemical cross-linkers.

However, the requirement of biocompatibility for biomedical applications, such as drug delivery agents [215], has led to an increase in research activity focussing on

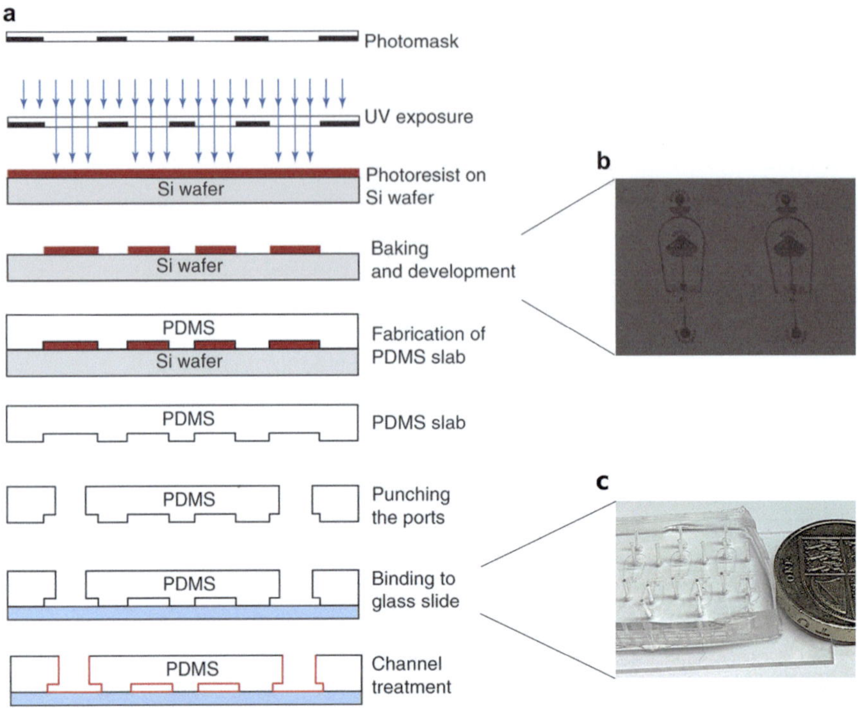

Fig. 7.3 (**a**) Schematic of the soft lithography process steps to prepare (**b**) a master containing the design structure from which (**c**) PDMS-moulded microfluidic devices are fabricated. (The schematic in panel (**a**) is reprinted by permission from Macmillan Publishers Ltd.: Nature Protocols Ref. [206], copyright 2013)

the use of biocompatible species. Microgels composed of biological molecules can mimic biological environments more readily than those formed from synthetic compounds. There are a few examples where naturally occurring biological molecules have been utilized for microgel formation, including agarose [175, 216], chitosan [217, 218], and alginate [219, 220].

Nevertheless, biological molecules with self-assembling properties, such as proteins and peptides, serve to be more attractive as the need for additional cross-linking or other chemical modifications is omitted and a wide variety of material properties may be achieved [221, 222]. Particularly, amyloidogenic proteins are advantageous due to their inherent ability to self-assemble, often under mild conditions in aqueous solution, into strong fibrillar networks. Amyloid hydrogels have been formed from a variety of amyloidogenic proteins, including elastin [223], α-synuclein [224], lysozyme [138, 225], and β-lactoglobulin [226–230], presenting them as opportune microgel building blocks with intrinsic cross-linking and, therefore, biocompatibility and minimal toxicity.

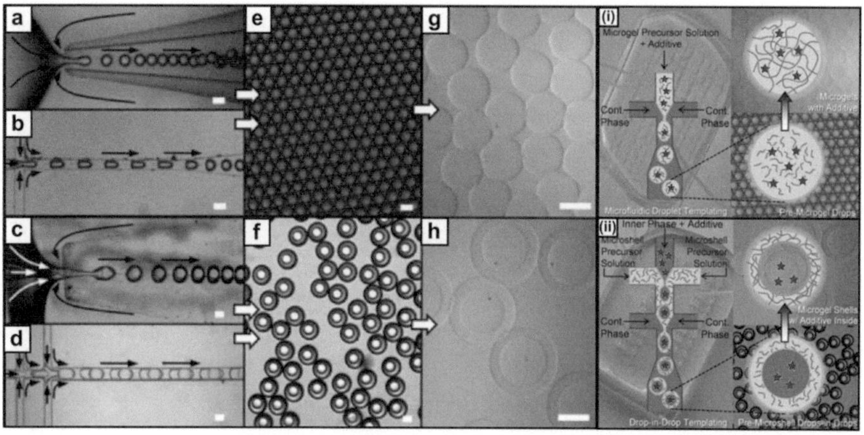

Fig. 7.4 Droplet-based microfluidic templating of microgel particles: (**a** and **c**) Glass microcapillary and (**b** and **d**) PDMS elastomer microfluidic devices producing (**e**) single emulsion droplets and (**f**) double emulsion droplets. (**g**) Monodisperse bulk microgels resulting from the single-emulsion templates in (**e**). (**h**) Monodisperse microgel shells resulting from the double-emulsion templates in (**f**). The panels (*i*, *ii*) to the right show schematics representing the gelation of the microgel precursor solution that contains additives to form bulk microgel capsules from a single emulsification (*i*), entrapping the additives within the meshes of their constituent polymer network, and double emulsion templates, yielding core-shell microgel capsules (*ii*) that can host additives within their hollow, solvent-swollen cores, whilst the polymer network forms the stabilizing surrounding shell. All scale bars = 50 mm. Arrows indicate the direction of flow. (Adapted from Ref. [47], with permission from John Wiley and Sons)

Furthermore, the robust nature of amyloid fibrils results in thermodynamic and kinetic stability of the resulting protein structure, giving rise to stable biocompatible microgels. Indeed, protein-based particles of micrometer-scale sizes, as required for many applications, can be obtained, while keeping proteins fully functional and accessible [231].

7.2.2.2 Structural Changes Accompanying the Formation of Protein Microgels and Protein Microgel Stability

A microemulsion formed from two immiscible phases acts as a template for the formation of capsules around the interface of the phases. Due to the amphipathic nature of proteins, they can localize at the water-oil interface of emulsion droplets. Upon binding to the interface, the protein is able to adjust its globular structure by orientating the hydrophobic residues towards the organic oil phase and the hydrophilic residues towards the aqueous phase, which results in a new energetically favourable 3D structure, where longer amphiphilic peptides yield more stable and smaller protein aqueous spheres [232]. Such a conformational change was observed in the case of, for example, silk fibrils for which an increased β-sheet content was found in silk fibroin microspheres, which is reminiscent of the densely hydrogen bonded amyloid fibril structures [233].

In the case of self-assembled protein-based microgels, stabilization of the gel sphere is driven through intrinsic interactions between protein molecules, providing a native support system. Insights into such interactions and structural changes in the protein following assembly have been gained by Fourier transform infra-red spectroscopy and molecular dynamic simulations that showed conformational changes in the protein secondary and tertiary structure [232]. More recently, structural differences between monomeric, aggregated and aggregating amyloidogenic lysozyme protein were resolved at nano-scale resolution by infra-red spectra from microdoplet contents, showing increased β-sheet structure for aggregating protein [234].

The most prevalent contributor to protein microgel stability has been thought to be the formation of disulfide bonds between cysteine residues of protein molecules [235]. However, it has been shown that the replacement of cysteine residues involved in disulfide bonding by hydrophobic residues can result in stable gels, which suggests that stabilization may also be achieved through hydrophobic interactions [232]. More evidence of a dense hydrogen bonded network was reported for protein bovine serum albumin (BSA)-silk fibroin capsules [236]. Furthermore, Shimanovich et al. demonstrated nanospheres made of nucleic acids, DNA and RNA, where the stabilization of the spheres was proposed to be due to hydrogen bonding and hydrophobic and electrostatic interactions stabilized by counter ions (commonly Ca^{2+}, Na^{+} and Mg^{2+}) [237, 238]. Interestingly, the phenomenon of formation of stabilising hydrogen bonding networks makes the incorporation of different proteins in a single microgel possible.

Furthermore, in recent work by Li et al. it was found that a series of proteins, including lysozyme, insulin, BSA and α-lactalbumin, could undergo superfast amyloid transitions with the treatment of tris(2-carboxyethyl)phosphine (TCEP), which is a highly efficient disulphide bond reducing agent [239]. The reaction mechanism is a novel model towards amyloid assembly in which the integration of three important building blocks in typical globular proteins is crucial for such a superfast protein amyloid-like assembly. They include the segment required for high fibrillation propensity, abundant α-helix structures and intramolecular disulphide bonds to lock the α-helix. With the reduction of the disulphide bonds from the TCEP treatment, the α-helix rapidly unlocked from the protein chain, and the resulting unfolded monomer underwent a fast transition to β-sheet-rich amyloid assemblies.

Thus, rather than relying on extrinsic cross-linking agents, protein-based microgel and microcapsules can be stabilized by native interactions in the form of covalent inter-molecular disulfide bonds, or non-covalent hydrophobic and hydrophilic interactions, and hydrogen bonding networks, which are strong enough to hold entire 3D protein spheres intact. Indeed, Shimanovich et al. demonstrated that microgel structures were significantly stabilized by lysozyme protein amyloid nanofibers relative to microparticles without fibrillar content where the latter were found to decompose rapidly [240]. Therefore, the conversion of monomeric proteins into robust amyloid structures through self-assembly is a convenient approach to produce microgels stabilized by fibril networks.

7.2.3 *Case Study: The Development of Protein Microgels and Gel Shells from Amyloid Fibril Networks as Drug Carrier Agents*

The potential of amyloid protein microgels as drug delivery vehicles has been explored by Shimanovich et al. [42, 241], who demonstrated control over the microgel morphology and subsequent controlled release of encapsulated drug-like small molecules from the microgels, which were shown to be non-toxic to human cells and indicative of enhanced pharmacological action. The nanofibril microgels were synthesized using the droplet microfluidic approach, where lysozyme protein formed the basis building block. A concentrated aqueous lysozyme solution was dispersed into droplets in an immiscible oil phase and conversion of the soluble lysozyme protein into nanofibril gel networks was induced through incubation of the microemulsion at elevated temperatures (65 °C) (Fig. 7.5). By inverting the aqueous (containing the protein gel precursor) and oil phases, water-in-oil and oil-

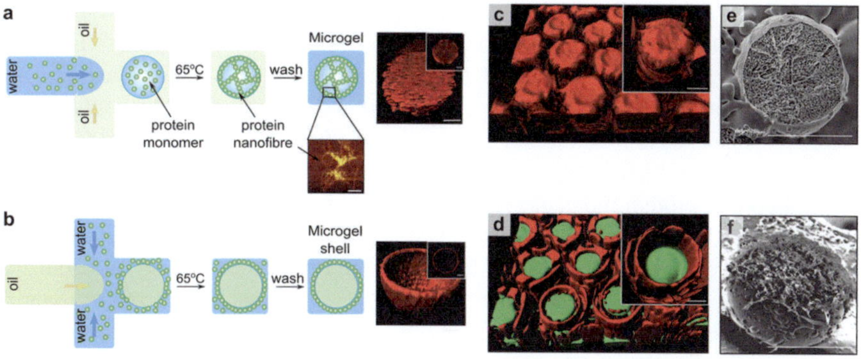

Fig. 7.5 Schematic representation of protein microgel synthesis: (**a**) Water-in-oil microgel. An atomic force microscopy image of the lysozyme protein nanofibrils is shown in the inset of the scheme. Scale bar = 400 nm. (**b**) Oil-in-water microgel shell. The corresponding 3D confocal images of microgel and microgel shell particles stained with Nile Red are shown on the right-hand side of each scheme. (**c**) 3D reconstructions of the confocal images for lysozyme water-in-oil and (**d**) oil-in-water microgel particles stained with Nile Red. The enlarged images of a single microgel and a microgel shell capsule are shown as an insert in the right corner of each image, scale bar = 5 μm. Red emission (excitation at 594 nm/emission at 617 nm) is observed for the aqueous protein component while green emission (excitation at 488 nm/emission at 519 nm) is detected for the oil environment. Scale bars = 5 μm. (**e**) Cryo-scanning electron microscopy images of lysozyme water-in-oil and (**f**) oil-in-water microgel and microgel shell. Scale bars = 20 μm. (Reprinted with permission from Ref. [42]. Copyright 2015 American Chemical Society)

in-water microdroplets were formed, resulting in core microgels and hollow gel shells following protein fibrillation, respectively. The protein fibrils had assembled throughout the water-in-oil droplets to give dense microgels, whereas in the case of the oil-in-water droplets the proteins were localized at the oil/water interface and subsequently formed hollow gel shells or microcapsules (Fig. 7.5).

The gel particles could be formed with sizes ranging from 60 μm down to 2 μm in diameter by tuning the channel width of the microfluidic device and the relative flow rates of the oil and aqueous phases. Furthermore, control over the microgel internal structure and morphology was achieved by varying the ratio of soluble lysozyme monomer and seed concentrations as well as the incubation time (at 65 °C) and pH of the precursor solution (Fig. 7.6). While the initial monomeric lysozyme concentration showed little effect on the microgel morphology, greater seed concentrations gave rise to an increased concentration of amyloid nanofibrils in the final microgels, and, therefore, increase in particle density; an increase in the pH of the precursor solution resulted in a decrease in density of protein nanofibrillar content.

The potential of the lysozyme microgels as drug carriers was demonstrated by the encapsulation of four types of compounds (the dyes ThT and Remazol Brilliant Blue R (RBBR), and the drugs tetracycline and penicillin V) of different hydrophilicity and affinity to proteins. The release rates of such drug-like small molecules were mainly determined by an interaction with amyloid fibrils. For example, slower release rates were observed for the molecules with higher affinity to protein nanofibrils (ThT and tetracycline) compared to release rates of the encapsulates with lower affinity (RBBR and Penicillin V). The release mechanism followed a multistep process. During the first stage, unbound small molecules near the microgel interface are released into solution after which, at a slower time scale, dissolution of the microgels leads to liberation of the remaining trapped molecules (Fig. 7.7a–d).

Similar to the release mechanism of encapsulated small drug-like molecules, the component protein molecules of the microgel itself were found to be released progressively from the microgel particles, indicative of microgel biodegradability. Transfer of the microgels from their formation environment at pH 2.0 to deionized water showed an initial fast phase release of 30–50% of the protein content over a time scale of less than an hour. Following this release, a slow phase, occuring over days to weeks, resulted in the complete dissolution of the gel particles. The ratio of the protein mass released during the fast phase relative to the slow phase could be controlled by varying the density of the gel network and the fraction of free to fibrillar protein. Therefore, the ability to tune the density of the microgels can enable control over drug release rates and supports slow release over extended periods of time.

Furthermore, Shimanovich et al. demonstrated that protein nanofibril self-assembly can continue after the initial microgel particle formation as a result of the dynamic self-assembling nature of amyloid fibrils. Unlike conventional synthetic polymer-based microgels, self-assembling nanofibril microgels allow for

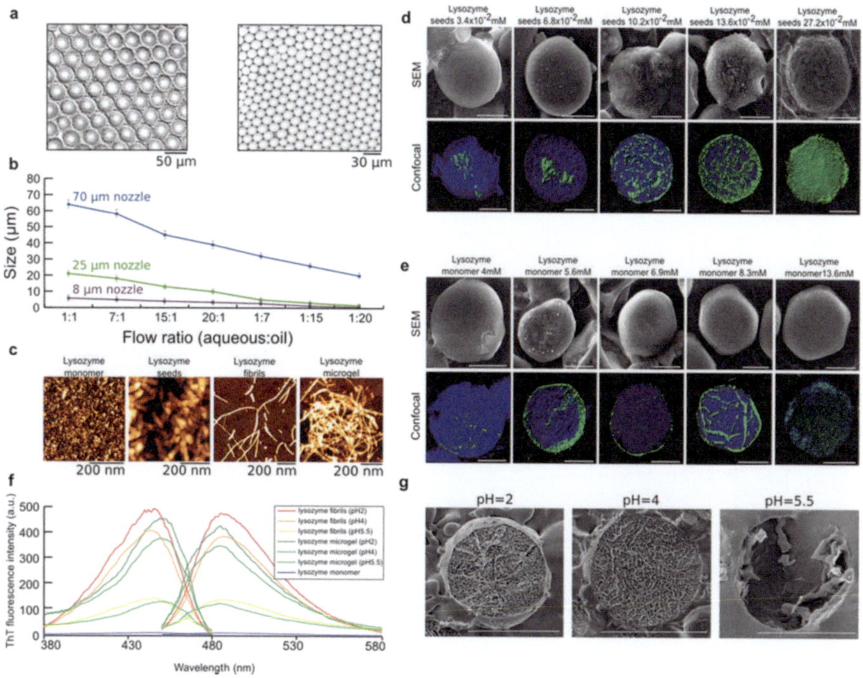

Fig. 7.6 Modulating microgel size and morphology: (**a**) Images from bright field light microscopy of lysozyme microgels of different sizes (from left to right: 49 and 23 μm). (**b**) Graph showing the change of lysozyme capsule diameter as a function of change in microfluidic channel width and the aqueous solution:oil ratio. (**c**) Atomic force microscopy images of lysozyme monomers, seeds, fibrils and microgel. (**d**) Scanning electron microscopy (SEM) images (top) and confocal images (bottom) of lysozyme capsules synthesized with a concentration of soluble protein of 4.08 mM and increasing seed concentrations (left to right). The concentrations are indicated on the top of each image. (**e**) SEM and confocal images of the ThT stained capsule with increasing lysozyme monomer concentration. The blue emission (excitation 350 nm/emission 438 nm) is detected for lysozyme protein monomers, green fluorescent emission (excitation 450 nm/emission 482 nm) detected for lysozyme nanofibrils. Scale bars = 10 μm. (**f**) Excitation and emission spectra showing ThT fluorescence intensity change upon binding to nanofibrillar content of lysozyme protein formed at pH 2, 4 and 5.5 and nanofibrillar content of microgel particles formed at pH 2, 4 and 5.5. (**g**) Cryo-SEM images of microgel particles formed at pH 2, 4 and 5.5. (Reprinted with permission from Ref. [42]. Copyright 2015 American Chemical Society)

the density of the fibril network in their core to be altered in response to exposure to monomeric precursor proteins even after the gel has been formed (Fig. 7.7e). This may further allow the tailoring of chemical stability or reactivity of the nanofibril microgels, giving rise to dynamic, active materials with network densities that can be modulated *in situ*.

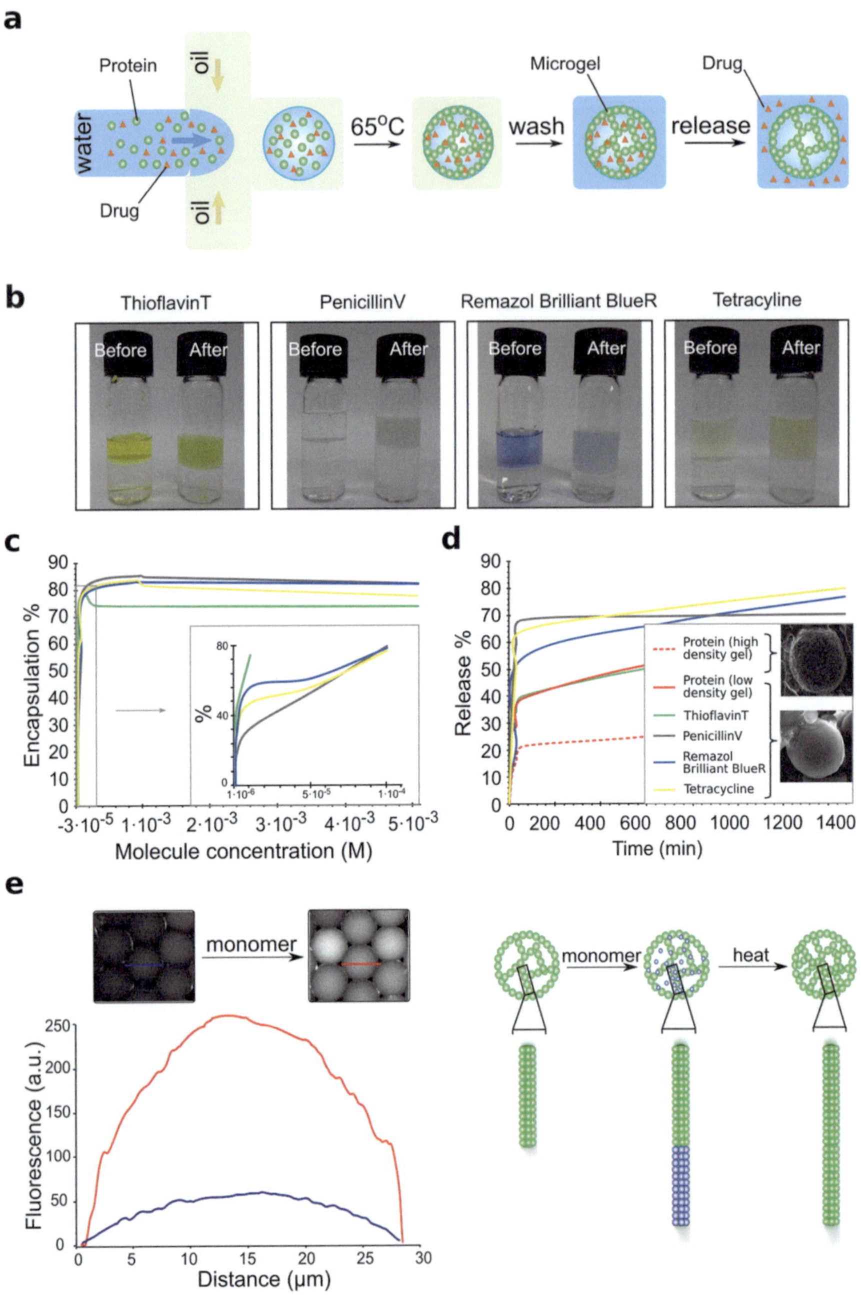

Fig. 7.7 Encapsulation and release of drug-like small molecules thioflavin T, penicillin V, Remazol Brilliant Blue R and tetracycline: (**a**) Schematic representation of protein microgel synthesis and small molecule encapsulation and release from the lysozyme microgels. (**b**) Images of the precursor solutions containing a drug ("before") and the drug loaded microgels ("after")

7.2.4 *Multiphase Protein Microgels – Phase Separation Phenomenon in Microgels*

For microgels to truly emerge and become applicable as drug delivery systems, the crucial challenge to overcome lies within the engineering of the precursor components and the assembly pathway to generate capsules with the desired properties. Significant structural complexity can be accessed by controlling the internal structure and composition of the microgels and by combining different materials within the same microgel particle. Approaches to enable the microgel's overall micron scale morphology to progress beyond spatially uniform gels have emerged, particularly in the form of microfluidic techniques where one of the most attractive features, in addition to precise droplet size control, is the ability to manipulate emulsion composition by consecutive droplet making.

Recently, Knowles and co-workers explored the synthesis of hierarchically ordered, multiphase microgels composed of lysozyme protein [240, 242]. The utilization of aqueous liquid/liquid phase separation afforded control of the micro-scale localization of the proteinaceous component under microconfinement. In combination with droplet microfluidics, this strategy enabled the formation of several types of micro-scale multiphase protein microgels, where water-in-oil microdroplets were generated that serve as a microenvironment within which aqueous phase separation of a polyethylene glycol (PEG)-dextran system was explored.

Aqueous two-phase systems, such as a PEG and dextran mixture in aqueous solution, separate in a two-phase system under certain conditions, such as high polymer concentration and elevated temperatures, where the thermodynamically favoured state is not the homogeneous solution, but rather a two-phase system consisting of a PEG rich and a polysaccharide rich phase [243–245]. The inner water/water and outer water/oil interfaces of the emulsion droplets serve as templates for the formation of microgels and capsules that are stabilized by amyloid nanofibril networks through the localization of the protein at the interface and subsequent gelation (Fig. 7.8). Due to differences in relative affinity of the lysozyme protein to the PEG or dextran phase, selective partitioning of the protein can be achieved, and therefore multiphase microgels can be generated.

Several types of multiphase protein microgels were generated. In the first approach a single emulsion of PEG-dextran in a continuous oil phase was generated

Fig. 7.7 (continued) incorporation. (**c**) Encapsulation efficiency studies. (**d**) Release kinetics as a function of time. The inset images on the right panel show a dense microgel particle (top) and a low density particle (bottom). (**e**) Left panel: Fluorescence intensity profiles for the lysozyme gels before (blue) and after (red) incubation with protein monomer solution with images of the corresponding gels shown above the graph. Right panel: Schematic representation of lysozyme monomer incorporation into microgel fibrils. (Reprinted with permission from Ref. [42]. Copyright 2015 American Chemical Society)

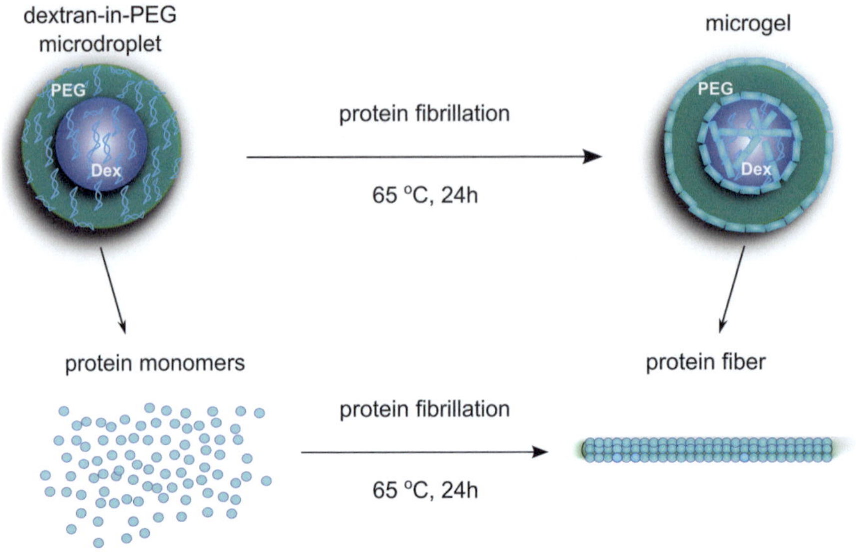

Fig. 7.8 Schematic representation of monomeric lysozyme protein conversion into amyloidogenic nanofibers inside multiphase dextran-PEG microdroplets, forming aqueous phase separated microgels. (Reproduced from Ref. [240] by permission of John Wiley & Sons Ltd.)

using a single junction microfluidic device, after which the PEG-dextran phase was induced to phase separate by subsequent heating of the microemulsion to 65 °C (Fig. 7.9a). Alternatively, spatial localization of the protein was established directly by using a double junction microfluidic device in which the PEG and dextran phases were mixed on chip in the first droplet junction prior to their formation into microdroplets at the second junction. Two-phase aqueous droplets with either a dextran-core in PEG-shell (Fig. 7.9b) or PEG-core in dextran-shell (Fig. 7.9c) were produced by inverting the PEG and dextran phases in the first junction. The resulting PEG/dextran interface within the droplets serves as a template for protein localization, and since lysozyme has a greater affinity towards the dextran phase [246], the protein content has a tendency to migrate towards the dextran phase rather than the PEG phase.

Conversion of the monomeric lysozyme into fibril networks was achieved by incubation at 65 °C for 24 h. This approach, exploiting the tendency of the protein to undergo self-assembly into nanoscale amyloid fibrils, afforded core-shell structures stabilized by fibril networks. Dextran-core-PEG-shell microgels were shown to be fully intact after storage in the continuous oil phase for 6 weeks. By contrast, the PEG-core-dextran-shell structures were found to not be sufficiently stabilized by protein gelation and dissociated within a few minutes after microgel formation.

Using ThT fluorescence, localization of the protein fibrils was observed and, in agreement with greater affinity of lysozyme towards dextran than PEG, fibrils were mainly present within the hydrophilic dextran phase, and dextran-core-

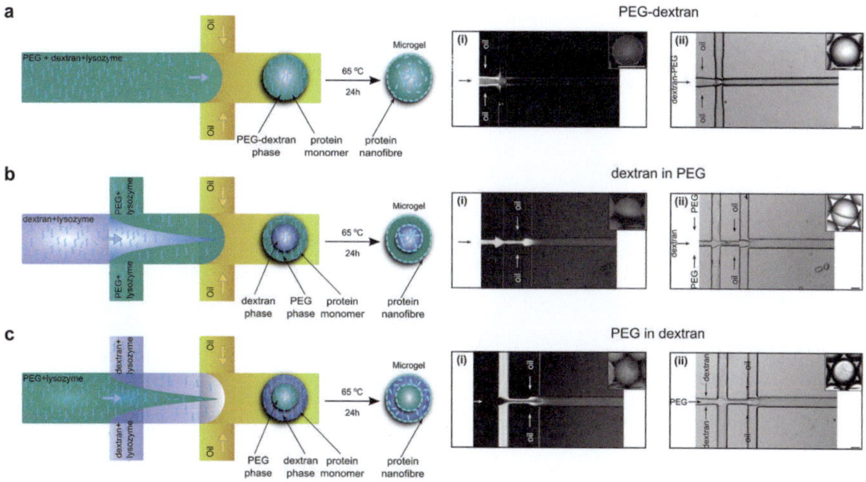

Fig. 7.9 Multiphase protein microgels: Schematic representation of multiphase protein microgel formation via (**a**) single junction microfluidic droplet device and (**b**) and (**c**) double junction microfluidic droplet devices for synthesis of dextran-core-PEG-shell and PEG-core-dextran-shell lysozyme microgels, respectively. The corresponding *(i)* fluorescent microscopy and *(ii)* light microscopy images of PEG dextran lysozyme protein microgel droplet formation are shown on the right-hand side of each scheme. Scale bars = 20 μm. Localization of the PEG and dextran phases is visualized using FITC-labeled dextran as shown in (*i*). (Reproduced from Ref. [240] by permission of John Wiley & Sons Ltd.)

PEG-shell structures are stabilized by protein aggregates localized at the PEG-oil interface (Fig. 7.10c). This work shows that the characteristics of aqueous two-phase systems can be exploited in combination with protein self-assembly for controlling aggregation and structuring of fibrillar proteins, providing a convenient route towards the synthesis of functional microgels.

7.2.5 Microgels from All-Aqueous Emulsions Stabilized by Amyloid Nanofibrils

All-aqueous emulsions as a template for gel capsule formation were investigated by Song et al. who demonstrated the stabilization of all-aqueous microdroplets through amyloid nanofibril networks, giving rise to colloidosome-like two-dimensional networks of lysosome nanofibrils, which were termed fibrillosomes [191]. Due to their water-based nature, all-aqueous systems are of superior biocompatibility and are particularly advantageous for storage and processing of biomacromolecules as non-aqueous interfaces, including water/oil interfaces, can compromise the stability and activity of biomolecules, such as by lipid oxidation or protein denaturation [247, 248].

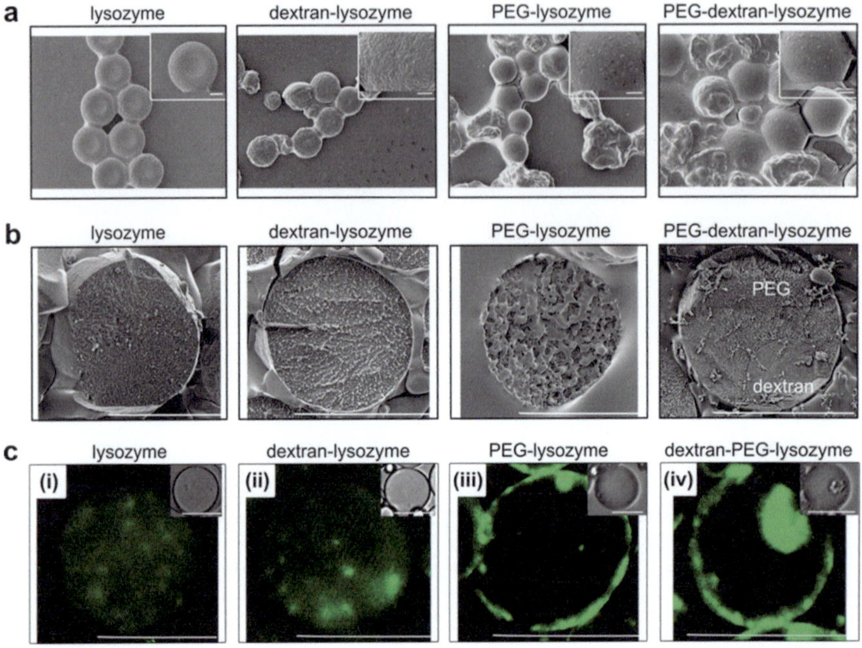

Fig. 7.10 (**a**) Scanning electron microscopy (SEM) and (**b**) cryo-SEM images of lysozyme, dextran-lysozyme, PEG-lysozyme, and dextran-core-PEG-shell lysozyme microgel particles. Scale bars = 20 μm. The polydispersity of the microgels presented in (**a**) is due to the partial deformation of the capsules under the electron beam during SEM analysis. The enlarged images of the microgel surface are shown in the right corner of each SEM image. Scale bars = 5 μm. (**c**) Confocal microscopy images of ThT stained lysozyme fibrils with fluorescent signal emitted. Scale bars = 20 μm. (Reproduced from Ref. [240] by permission of John Wiley & Sons Ltd.)

However, the characteristic ultralow interfacial tension of all-aqueous interfaces (typically in the range from 10^7 to 10^4 Nm^{-1}) represent the primary constraint that limits emulsion stability [249]. Although it is known that the adsorption of nanoparticles can provide emulsion stability, at such ultra-low interfacial tensions, colloidosomal stability (which results from the lowering of free energy at the interface as surface-active species adsorb and desorb) is strongly diminished [250, 251]. Song et al. hypothesized that all-aqueous interface stabilization can be achieved by the combination of the strong interface affinity and the high aspect ratio of protein fibrils that results in a high surface coverage compared with spherical colloids.

Based on the spontaneous liquid-liquid phase separation of the high molecular weight polymers PEG and dextran, water-in-water (W/W) emulsions have been generated in aqueous suspensions containing lysozyme protein. Using ThT labelling, it was shown that mature protein amyloid fibrils accumulate at the W/W interface of the emulsion droplets, forming a monolayer that prevents the droplets from coalescing (Fig. 7.11).

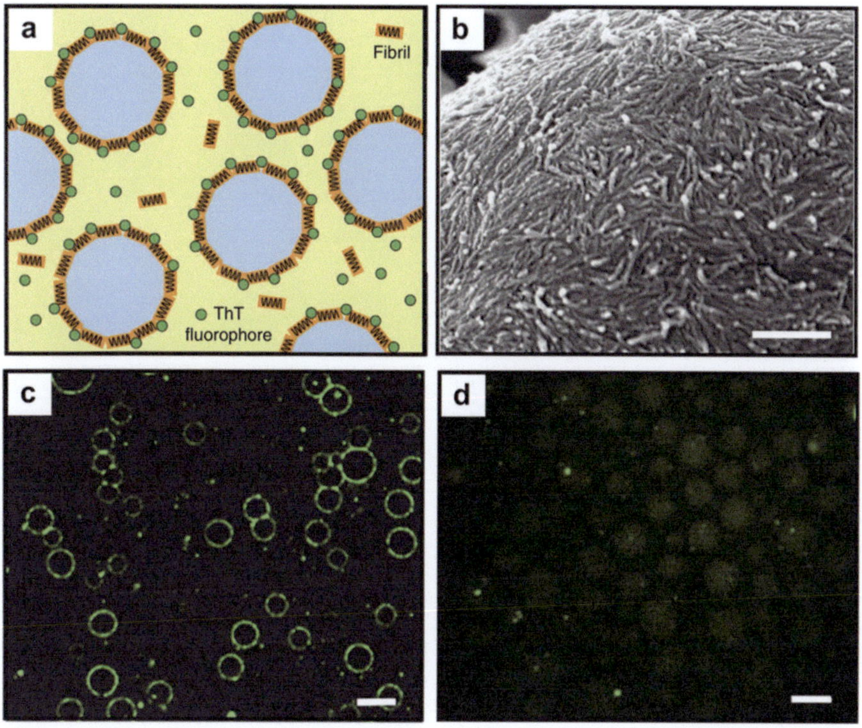

Fig. 7.11 (**a**) Schematic of a monolayer of fibrils, with ThT labelling incorporated, adsorbed at the interface of W/W emulsion droplets. (**b**) SEM image showing lysozyme fbrils deposited as a monolayer at the emulsion interface. Scale bar = 500 nm. (**c**) Fluorescence microscopy image of ThT-labeled lysozyme fibrils accumulated at the interface of W/W emulsion droplets. Scale bar = 20 μm. (**d**) In the absence of fibrils, weak ThT fluorescence is observed in the dextran-rich droplet phase. Scale bar = 20 μm. (Reprinted and adapted from Ref. [191]. This work is licensed under a Creative Commons Attribution 4.0 International License)

Further stabilization was achieved through the formation of 2D multi-layers of fibrils. In this approach, a monolayer of nanofibrils was used to seed further fibril deposition at the W/W interface. Additional lysozyme monomers were converted into mature fibrils upon incubation at favourable aggregating conditions (60 °C, pH 2) and control of the fibril layer thickness could be achieved by varying the added monomer concentration. The stability of the multi-layered fibril emulsions was shown to be enhanced compared to emulsions stabilized by a fibril monolayer, as demonstrated by the small change of the droplet size over a 30-day period (Fig. 7.12).

Furthermore, covalently cross-linking of non-adsorbed fibrils to the fibril monolayer at the W/W interface yielded even more mechanically robust capsules, which were shown to be stable in the complete absence of an interface after replacement of the continuous phase with the same solution as inside the droplet. These

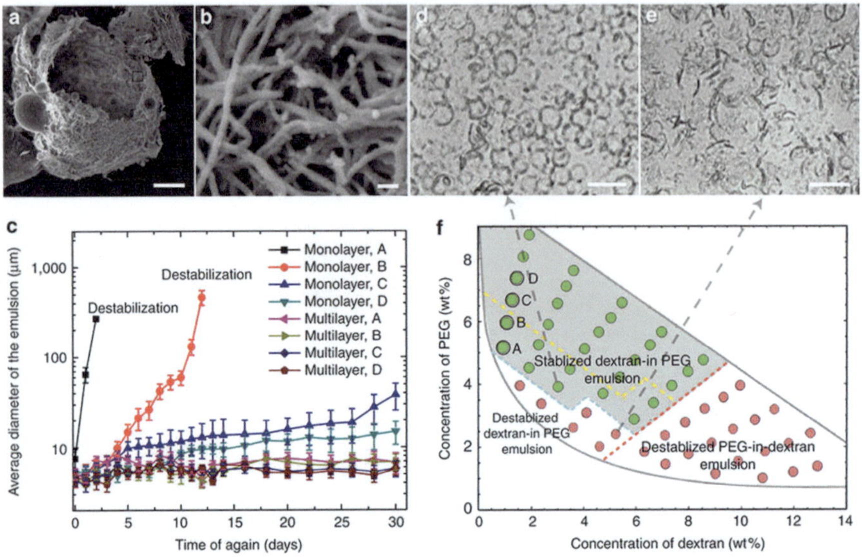

Fig. 7.12 (**a** and **b**) SEM images of lysozyme capsules composed of multilayers of fibrils. W/W emulsions coated with fibril monolayers are used as the seeding templates, and the formation of multilayer fibrils is induced by the fibrillization of monomeric lysozyme. Scale bars = 2 μm (**a**); and 50 nm (**b**). (**c**) A comparative study of the size of emulsion droplets stabilized by a monolayer and a multilayer of fibrils as a function of aging time. Error bars represent the deviation (s.d.) of the droplet diameters. The different compositions of the tested emulsions (A–D) are shown in **f**. (**d** and **e**) Representative microscopic images of the multilayer fibril stabilized and destabilized dextran-in-PEG emulsions are shown. Scale bars = 20 μm. (**f**) Compared with fibril monolayers, fibril multilayers enhance the stability of W/W emulsions over 30 days, as suggested by the enlarged stability region in the phase diagram (grey area) corresponding to emulsions stabilized by the fibril multilayer. Emulsions stabilized by a fibril monolayer remained stable for 30 days only in the area above the yellow dashed lines. (Reprinted from Ref. [191]. This work is licensed under a Creative Commons Attribution 4.0 International License)

self-standing, so-called fibrillosomes showed properties, such as stretch ability and semi-permeability, that make them excellent candidates for the encapsulation and delivery of bioactive species (Fig. 7.13).

7.2.6 *Functionalized Proteinaceous Microgels*

In addition to the stability afforded by the fibrillar protein network, using proteins as building blocks for the synthesis of microgels enables incorporation of further functionality. A convenient approach for incorporating new functionality into the microgel is by using an additional protein with the required properties, either through co-assembly or through tandem construction of fused proteins. In this manner, microgels have been constructed composed of mixed proteins [179], coated

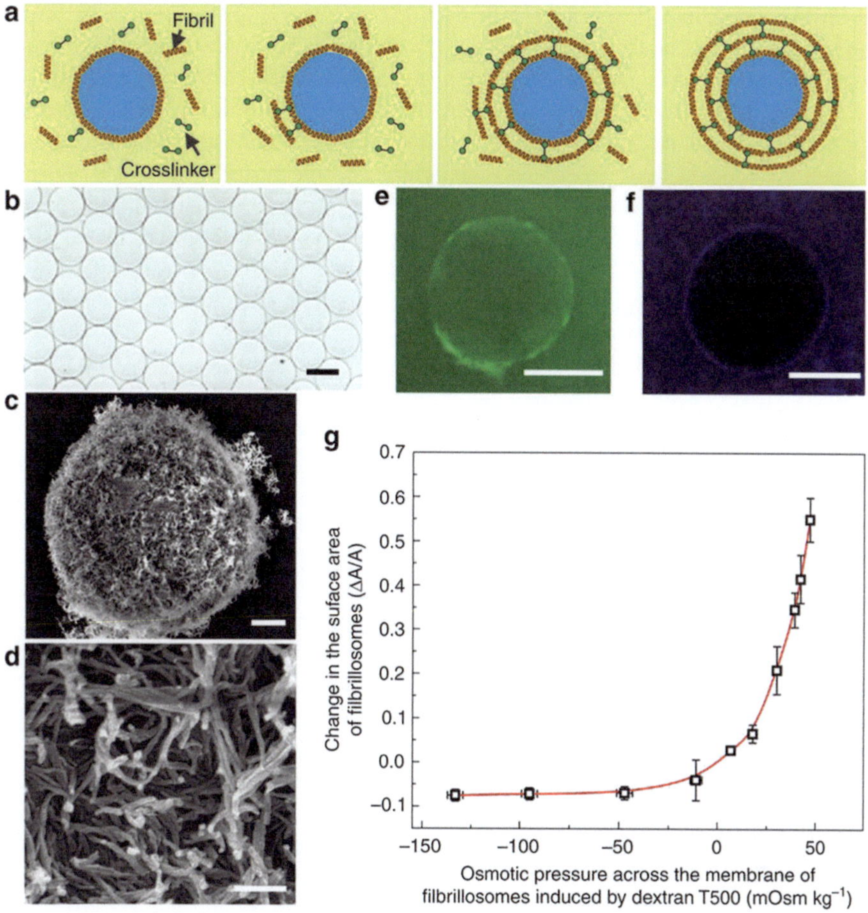

Fig. 7.13 Multi-layered microgel shells templated from W/W emulsion: (**a**) Schematic representation of the formation of protein fibrillosomes by crosslinking fibril-coated microdroplets. (**b**) Optical microscope images of monodisperse fibrillosomes in the complete absence of an interface, obtained after replacing the continuous phase with the same liquid as inside the fibrillosomes. Scale bar = 100 μm. (**c** and **d**) SEM images of fibrillosomes with their walls consisting of amyloid fibrils. Scale bars = 2 μm (**c**) and 200 nm (**d**). (**e**). FITC-dextran macromolecules with hydrodynamic diameters of around 30 nm can penetrate through the membrane of fibrillosomes. Scale bar = 200 μm. (**f**) Fluorescent nanoparticles with diameters of 50 nm fail to penetrate the fibrillosomes. Scale bar = 200 μm. (**g**) The fibrillosomes exhibit excellent elasticity and robustness upon osmotic swelling and shrinking. The concentration of dextran in the continuous phase was varied from 4 to 25 wt%, while the dextran concentration inside the fibrillosomes was kept at 15 wt%, generating osmotic pressure across the membrane. Error bars represent the s.d. (Reprinted from Ref. [191]. This work is licensed under a Creative Commons Attribution 4.0 International License)

with biocompatible polymers [180], and protein spheres conjugated with target ligands, such as folate, for targeted drug delivery [252]. Moreover, Rahimipour and co-workers presented surface-conjugated protein microspheres that capture amyloid-β and inhibit its aggregation and toxicity [253]. In this case, the globular protein BSA (in its non-aggregating form) was the main structural protein of which the capsule shell was composed and the capsule surface was functionalized with the KLVFF peptide [254]. The KLVFF peptide is a known inhibitor of the amyloidogenic Aβ-40 peptide aggregation phenomenon that is associated with Alzheimer's disease. The BSA-KLVFF conjugated protein capsules show high affinity and selectivity for Aβ, sequestering the protein and, thereby, reducing its toxicity [254].

Functionalization of amyloid fibrils *via* gene fusion has been explored by Perrett and co-workers who demonstrated the immobilization of active enzymes, including alkaline phosphatase (AP) and horseradish peroxidase (HRP) on amyloid fibrils formed by the prion domain of the yeast prion protein Ure2 [255]. The gene fusion process gives rise to a polypeptide that maintains the ability to self-assemble as well as to display the desired activity of the fused entity. In following work, Zhou et al. demonstrated the formation of enzymatically active microgels from the Ure2-AP chimeric protein construct. The microgels showed robust material properties and their porous architecture allowed for diffusion of reactants and products in and out of the microgels, illustrating the potential for enzyme immobilization and for biological flow-chemistry or bioactivity assays [231].

The physicochemical properties of the microgel can be further tailored to make them responsive to environmental cues and thus afford targeted drug delivery and modulation of release [256, 257]. For instance, engineering protein capsules with an external hydrophobic layer can introduce specificity to a target site, aid transdermal transport, encode a release switch mechanism or enhance the biological effect of the active species delivered through the capsules. As an example, RNA loaded BSA microspheres (where BSA protein appears in its native fold) were coated with metal (Fe_2O_3), uncharged polyvinylalcohol (PVA) or positively charged polyethyleneimine (PEI) biocompatible synthetic polymers that buried the negative charge of the BSA. This eliminated the adhesion to the cell membrane, which was observed in the case of uncoated capsules, and allowed efficient delivery of RNA into human osteosarcoma U2OS cancer cells and *Trypanosoma* parasites. In addition, the Fe_2O_3 metal coating could allow enhanced contrast in electron microscopy for the spatial localization of the Fe_2O_3 metal coated capsules inside the parasites and mammalian cancer cells [180]. Furthermore, the transport of viruses and metal nanoparticles in living cells was shown to be enhanced through transfection mediated by non-toxic amyloid fibrils [258–261].

In a further example of tuning of microgel physicochemical properties, Levin et al. synthesized hybrid organic/inorganic microcapsules [262]. Lysozyme composites were functionalized with carboxyl-modified Fe_3O_4 nanoparticles (NPs), which have been used extensively in biological applications due to their biocompatibility, small dimension (<30 nm), ease of characterization, and the rich diversity of surface chemistry modifications that can be exploited for biomedical applications

[263, 264]. Lysozyme and Fe_3O_4 NPs exhibit a strong interaction that may be attributed to their electrostatic interactions and there is evidence that NPs may influence amyloid aggregation kinetics [265, 266]. This effect was further explored for lysozyme self-assembly under the microconfinement of microdroplets in Levin's work. It was shown that, in the presence of carboxyl-modified Fe_3O_4 NPs, lysozyme fibrillation was inhibited in unseeded conditions, while under seeded conditions, fibrillation occurred that allowed for extensive decoration of the fibrillary microcapsules. It was suggested that the lysozyme monomers may have a greater affinity to the carboxyl-modified Fe_3O_4 NPs, which would inhibit further protein aggregation and, thus, resulting in less stable microgel structures. In addition, the fibril morphology within the microgels was shown to vary upon protein fibrillation in the presence of the NPs, indicative of their structure-modifying and aggregation-modulating role on protein self-assembly. These findings suggest a novel approach for tuning protein-based microgel formation, allowing control over the morphology and characteristics of the end product [262].

7.3 Conclusions

Although amyloid fibril structures were initially recognized in association with disease, the increase in research focused on their functional roles in nature and their potential as the basis of artificial materials with complex functionality present these as excellent building blocks for the development of functional materials with applications, for example, as drug delivery agents. In addition to their biocompatibility, the self-assembling ability of proteins can be exploited, in combination with microfluidic techniques, to produce multi-scale functional microgels capable of hosting biologically active compounds. The microgel formation process avoids the use of harsh reagents or conditions and enables precise control over capsule size, stability and morphology. The ability to tailor such features is advantageous in the development of biomedical applications, such as drug delivery, as drug release can be controlled. Moreover, the use of proteins and peptides as functional material building blocks allows incorporation of additional functionality through protein modification and/or coupling with other functional molecules, enhancing the versatility of this class of materials as well as leading to responsive microgels, enabling targeted drug delivery. Thus, amyloid fibril microgels represent attractive designer materials for biomedical applications, including drug delivery.

References

1. Hardy J, Selkoe DJ (2002) The amyloid hypothesis of Alzheimer's disease: progress and problems on the road to therapeutics. Science 297:353–356
2. Haass C, Selkoe DJ (2007) Soluble protein oligomers in neurodegeneration: lessons from the Alzheimer's amyloid β-peptide. Nat Rev Mol Cell Biol 8:101–112

3. Ballard C, Gauthier S, Corbett A, Brayne C, Aarsland D, Jones E (2011) Alzheimer's disease. Lancet Lond Engl 377:1019–1031
4. Querfurth HW, LaFerla FM (2010) Mechanisms of disease Alzheimer's disease. N Engl J Med 362:329–344
5. Niedowicz DM, Nelson PT, Murphy MP (2011) Alzheimer's disease: pathological mechanisms and recent insights. Curr Neuropharmacol 9:674–684
6. Dobson CM (1999) Protein misfolding, evolution and disease. Trends Biochem Sci 24:329–332
7. Dobson CM (2003) Protein folding and misfolding. Nature 426:884–890
8. Chiti F, Dobson CM (2006) Protein misfolding, functional amyloid, and human disease. Annu Rev Biochem 75:333–366
9. Knowles TPJ, Vendruscolo M, Dobson CM (2014) The amyloid state and its association with protein misfolding diseases. Nat Rev Mol Cell Biol 15:384–396
10. Stefani M, Dobson CM (2003) Protein aggregation and aggregate toxicity: new insights into protein folding, misfolding diseases and biological evolution. J Mol Med 81:678–699
11. Fowler DM, Koulov AV, Alory-Jost C, Marks MS, Balch WE, Kelly JW (2006) Functional amyloid formation within mammalian tissue. PLoS Biol 4(1):100–107. https://doi.org/10.1371/journal.pbio.0040006
12. Fowler DM, Koulov AV, Balch WE, Kelly JW (2007) Functional amyloid – from bacteria to humans. Trends Biochem Sci 32:217–224
13. Otzen D, Nielsen PH (2008) We find them here, we find them there: functional bacterial amyloid. Cell Mol Life Sci 65:910–927
14. Kelly JW, Balch WE (2003) Amyloid as a natural product. J Cell Biol 161:461–462
15. Maji SK, Perrin MH, Sawaya MR et al (2009) Functional amyloids as natural storage of peptide hormones in pituitary secretory granules. Science 325:328–332
16. Bleem A, Daggett V (2017) Structural and functional diversity among amyloid proteins: agents of disease, building blocks of biology, and implications for molecular engineering. Biotechnol Bioeng 114:7–20
17. Mostaert AS, Higgins MJ, Fukuma T, Rindi F, Jarvis SP (2006) Nanoscale mechanical characterisation of amyloid fibrils discovered in a natural adhesive. J Biol Phys 32:393–401
18. Zhong C, Gurry T, Cheng AA, Downey J, Deng Z, Stultz CM, Lu TK (2014) Strong underwater adhesives made by self-assembling multi-protein nanofibres. Nat Nanotechnol 9:858–866
19. Li C, Qin R, Liu R, Miao S, Yang P (2018) Functional amyloid materials at surfaces/interfaces. Biomater Sci 6:462–472
20. Wang D, Ha Y, Gu J, Li Q, Zhang L, Yang P (2016) 2D protein supramolecular nanofilm with exceptionally large area and emergent functions. Adv Mater 28:7414–7423
21. Chapman MR, Robinson LS, Pinkner JS, Roth R, Heuser J, Hammar M, Normark S, Hultgren SJ (2002) Role of Escherichia coli curli operons in directing amyloid fiber formation. Science 295:851–855
22. Shorter J, Lindquist S (2005) Prions as adaptive conduits of memory and inheritance. Nat Rev Genet 6:435–450
23. Krishnan R, Lindquist SL (2005) Structural insights into a yeast prion illuminate nucleation and strain diversity. Nature 435:765–772
24. Lindquist SL, Henikoff S (2002) Self-perpetuating structural states in biology, disease, and genetics. Proc Natl Acad Sci U S A 99:16377
25. Tanaka M, Collins SR, Toyama BH, Weissman JS (2006) The physical basis of how prion conformations determine strain phenotypes. Nature 442:585–589
26. DePace AH, Weissman JS (2002) Origins and kinetic consequences of diversity in Sup35 yeast prion fibers. Nat Struct Mol Biol 9:389–396
27. Lashuel HA, Hartley D, Petre BM, Walz T, Lansbury PT (2002) Neurodegenerative disease: amyloid pores from pathogenic mutations. Nature 418:291–291

28. Bucciantini M, Giannoni E, Chiti F, Baroni F, Formigli L, Zurdo J, Taddei N, Ramponi G, Dobson CM, Stefani M (2002) Inherent toxicity of aggregates implies a common mechanism for protein misfolding diseases. Nature 416:507–511
29. Walsh DM, Klyubin I, Fadeeva JV, Cullen WK, Anwyl R, Wolfe MS, Rowan MJ, Selkoe DJ (2002) Naturally secreted oligomers of amyloid β protein potently inhibit hippocampal long-term potentiation in vivo. Nature 416:535–539
30. Caughey B, Peter T, Lansbury J (2003) Protofibirls, pores, fibrils, and neurodegeneration: separating the responsible protein aggregates from the innocent bystanders. Annu Rev Neurosci 26:267–298
31. Koffie RM, Meyer-Luehmann M, Hashimoto T et al (2009) Oligomeric amyloid β associates with postsynaptic densities and correlates with excitatory synapse loss near senile plaques. Proc Natl Acad Sci U S A 106:4012–4017
32. Tomic JL, Pensalfini A, Head E, Glabe CG (2009) Soluble fibrillar oligomer levels are elevated in Alzheimer's disease brain and correlate with cognitive dysfunction. Neurobiol Dis 35:352–358
33. Shankar GM, Bloodgood BL, Townsend M, Walsh DM, Selkoe DJ, Sabatini BL (2007) Natural oligomers of the Alzheimer amyloid-β protein induce reversible synapse loss by modulating an NMDA-type glutamate receptor-dependent signaling pathway. J Neurosci 27:2866–2875
34. Hammer ND, Schmidt JC, Chapman MR (2007) The curli nucleator protein, CsgB, contains an amyloidogenic domain that directs CsgA polymerization. Proc Natl Acad Sci U S A 104:12494–12499
35. Knowles TP, Fitzpatrick AW, Meehan S, Mott HR, Vendruscolo M, Dobson CM, Welland ME (2007) Role of intermolecular forces in defining material properties of protein nanofibrils. Science 318:1900–1903
36. Knowles TPJ, Buehler MJ (2011) Nanomechanics of functional and pathological amyloid materials. Nat Nanotechnol 6:469–479
37. Chiti F, Webster P, Taddei N, Clark A, Stefani M, Ramponi G, Dobson CM (1999) Designing conditions for in vitro formation of amyloid protofilaments and fibrils. Proc Natl Acad Sci U S A 96:3590–3594
38. Chiti F, Dobson CM (2009) Amyloid formation by globular proteins under native conditions. Nat Chem Biol 5:15–22
39. Pawar AP, DuBay KF, Zurdo J, Chiti F, Vendruscolo M, Dobson CM (2005) Prediction of "aggregation-prone" and "aggregation-susceptible" regions in proteins associated with neurodegenerative diseases. J Mol Biol 350:379–392
40. Knowles TPJ, Mezzenga R (2016) Amyloid fibrils as building blocks for natural and artificial functional materials. Adv Mater 28:6546–6561
41. Shimanovich U, Bernardes GJ, Knowles TP, Cavaco-Paulo A (2014) Protein micro- and nano-capsules for biomedical applications. Chem Soc Rev 43:1361–1371
42. Shimanovich U, Efimov I, Mason TO et al (2015) Protein microgels from amyloid fibril networks. ACS Nano 9:43–51
43. Cao A, Hu D, Lai L (2004) Formation of amyloid fibrils from fully reduced hen egg white lysozyme. Protein Sci Publ Protein Soc 13:319–324
44. Kelly JW (2002) Towards an understanding of amyloidogenesis. Nat Struct Mol Biol 9:323–325
45. Teh S-Y, Lin R, Hung L-H, Lee AP (2008) Droplet microfluidics. Lab Chip 8:198–220
46. Zhang H, Tumarkin E, Sullan RMA, Walker GC, Kumacheva E (2007) Exploring microfluidic routes to microgels of biological polymers. Macromol Rapid Commun 28:527–538
47. Seiffert S (2013) Microgel capsules tailored by droplet-based microfluidics. ChemPhysChem 14:295–304
48. Sipe JD, Cohen AS (2000) Review: history of the amyloid fibril. J Struct Biol 130:88–98
49. Buxbaum JN, Linke RP (2012) A molecular history of the amyloidoses. J Mol Biol 421:142–159

50. Nilsson MR (2004) Techniques to study amyloid fibril formation in vitro. Methods 34:151–160
51. Groenning M (2009) Binding mode of Thioflavin T and other molecular probes in the context of amyloid fibrils—current status. J Chem Biol 3:1–18
52. Hawe A, Sutter M, Jiskoot W (2008) Extrinsic fluorescent dyes as tools for protein characterization. Pharm Res 25:1487–1499
53. Khurana R, Uversky VN, Nielsen L, Fink AL (2001) Is Congo Red an amyloid-specific dye? J Biol Chem 276:22715–22721
54. Hudson SA, Ecroyd H, Kee TW, Carver JA (2009) The thioflavin T fluorescence assay for amyloid fibril detection can be biased by the presence of exogenous compounds. FEBS J 276:5960–5972
55. O'Nuallain B, Wetzel R (2002) Conformational Abs recognizing a generic amyloid fibril epitope. Proc Natl Acad Sci 99:1485–1490
56. Larsen P, Nielsen JL, Dueholm MS, Wetzel R, Otzen D, Nielsen PH (2007) Amyloid adhesins are abundant in natural biofilms. Environ Microbiol 9:3077–3090
57. Chan FTS, Kaminski Schierle GS, Kumita JR, Bertoncini CW, Dobson CM, Kaminski CF (2013) Protein amyloids develop an intrinsic fluorescence signature during aggregation. Analyst 138:2156–2162
58. Kaminski Schierle GS, Bertoncini CW, Chan FTS et al (2011) A FRET sensor for non-invasive imaging of amyloid formation in vivo. Chemphyschem Eur J Chem Phys Phys Chem 12:673–680
59. Chen W, Young LJ, Lu M, Zaccone A, Ströhl F, Yu N, Kaminski Schierle GS, Kaminski CF (2017) Fluorescence self-quenching from reporter dyes informs on the structural properties of amyloid clusters formed in vitro and in cells. Nano Lett 17:143–149
60. Fändrich M (2007) On the structural definition of amyloid fibrils and other polypeptide aggregates. Cell Mol Life Sci 64:2066–2078
61. Greenwald J, Riek R (2010) Biology of amyloid: structure, function, and regulation. Structure 18:1244–1260
62. Eisenberg D, Jucker M (2012) The amyloid state of proteins in human diseases. Cell 148:1188–1203
63. Sawaya MR, Sambashivan S, Nelson R et al (2007) Atomic structures of amyloid cross-β spines reveal varied steric zippers. Nature 447:453–457
64. Petkova AT, Ishii Y, Balbach JJ, Antzutkin ON, Leapman RD, Delaglio F, Tycko R (2002) A structural model for Alzheimer's β-amyloid fibrils based on experimental constraints from solid state NMR. Proc Natl Acad Sci 99:16742–16747
65. Wasmer C, Lange A, Van Melckebeke H, Siemer AB, Riek R, Meier BH (2008) Amyloid fibrils of the HET-s(218-289) prion form a beta solenoid with a triangular hydrophobic core. Science 319:1523–1526
66. Lührs T, Ritter C, Adrian M, Riek-Loher D, Bohrmann B, Döbeli H, Schubert D, Riek R (2005) 3D structure of Alzheimer's amyloid-β(1–42) fibrils. Proc Natl Acad Sci U S A 102:17342–17347
67. Ritter C, Maddelein M-L, Siemer AB, Lührs T, Ernst M, Meier BH, Saupe SJ, Riek R (2005) Correlation of structural elements and infectivity of the HET-s prion. Nature 435:844–848
68. Tycko R (2011) Solid state NMR studies of amyloid fibril structure. Annu Rev Phys Chem 62:279–299
69. Sunde M, Serpell LC, Bartlam M, Fraser PE, Pepys MB, Blake CCF (1997) Common core structure of amyloid fibrils by synchrotron X-ray diffraction. J Mol Biol 273:729–739
70. Makin OS, Atkins E, Sikorski P, Johansson J, Serpell LC (2005) Molecular basis for amyloid fibril formation and stability. Proc Natl Acad Sci U S A 102:315–320
71. Nelson R, Sawaya MR, Balbirnie M, Madsen AØ, Riekel C, Grothe R, Eisenberg D (2005) Structure of the cross-β spine of amyloid-like fibrils. Nature 435:773–778
72. Adamcik J, Mezzenga R (2012) Study of amyloid fibrils via atomic force microscopy. Curr Opin Colloid Interface Sci 17:369–376

73. Sachse C, Fändrich M, Grigorieff N (2008) Paired β-sheet structure of an Aβ(1-40) amyloid fibril revealed by electron microscopy. Proc Natl Acad Sci 105:7462–7466
74. Fitzpatrick AWP, Debelouchina GT, Bayro MJ et al (2013) Atomic structure and hierarchical assembly of a cross-β amyloid fibril. Proc Natl Acad Sci U S A 110:5468–5473
75. Jiménez JL, Guijarro JI, Orlova E, Zurdo J, Dobson CM, Sunde M, Saibil HR (1999) Cryo-electron microscopy structure of an SH3 amyloid fibril and model of the molecular packing. EMBO J 18:815–821
76. Jiménez JL, Nettleton EJ, Bouchard M, Robinson CV, Dobson CM, Saibil HR (2002) The protofilament structure of insulin amyloid fibrils. Proc Natl Acad Sci 99:9196–9201
77. Fitzpatrick AWP, Falcon B, He S, Murzin AG, Murshudov G, Garringer HJ, Crowther RA, Ghetti B, Goedert M, Scheres SHW (2017) Cryo-EM structures of tau filaments from Alzheimer's disease. Nature 547:185–190
78. Wei G, Su Z, Reynolds NP, Arosio P, Hamley IW, Gazit E, Mezzenga R (2017) Self-assembling peptide and protein amyloids: from structure to tailored function in nanotechnology. Chem Soc Rev 46(15):4661–4708. https://doi.org/10.1039/C6CS00542J
79. Pedersen JS, Andersen CB, Otzen DE (2010) Amyloid structure – one but not the same: the many levels of fibrillar polymorphism. FEBS J 277:4591–4601
80. VandenAkker CC, Schleeger M, Bruinen AL, Deckert-Gaudig T, Velikov KP, Heeren RMA, Deckert V, Bonn M, Koenderink GH (2016) Multimodal spectroscopic study of amyloid fibril polymorphism. J Phys Chem B 120:8809–8817
81. Pedersen JS, Otzen DE (2008) Amyloid—a state in many guises: survival of the fittest fibril fold. Protein Sci Publ Protein Soc 17:2–10
82. Fändrich M, Meinhardt J, Grigorieff N (2009) Structural polymorphism of Alzheimer Aβ and other amyloid fibrils. Prion 3:89–93
83. Auer S (2015) Nucleation of polymorphic amyloid fibrils. Biophys J 108:1176–1186
84. Pedersen JS, Dikov D, Flink JL, Hjuler HA, Christiansen G, Otzen DE (2006) The changing face of glucagon fibrillation: structural polymorphism and conformational imprinting. J Mol Biol 355:501–523
85. Petkova AT, Leapman RD, Guo Z, Yau W-M, Mattson MP, Tycko R (2005) Self-propagating, molecular-level polymorphism in alzheimer's ß-amyloid fibrils. Science 307:262–265
86. Dobson CM, Šali A, Karplus M (1998) Protein folding: a perspective from theory and experiment. Angew Chem Int Ed 37:868–893
87. DuBay KF, Pawar AP, Chiti F, Zurdo J, Dobson CM, Vendruscolo M (2004) Prediction of the absolute aggregation rates of amyloidogenic polypeptide chains. J Mol Biol 341:1317–1326
88. Phan-Xuan T, Durand D, Nicolai T, Donato L, Schmitt C, Bovetto L (2011) On the crucial importance of the pH for the formation and self-stabilization of protein microgels and strands. Langmuir 27:15092–15101
89. Uversky VN, Fink AL (2004) Conformational constraints for amyloid fibrillation: the importance of being unfolded. Biochim Biophys Acta BBA 1698:131–153
90. Sawyer EB, Claessen D, Gras SL, Perrett S (2012) Exploiting amyloid: how and why bacteria use cross-β fibrils. Biochem Soc Trans 40:728–734
91. Auer S, Meersman F, Dobson CM, Vendruscolo M (2008) A generic mechanism of emergence of amyloid protofilaments from disordered oligomeric aggregates. PLoS Comput Biol 4:e1000222
92. Baldwin AJ, Knowles TPJ, Tartaglia GG et al (2011) Metastability of native proteins and the phenomenon of amyloid formation. J Am Chem Soc 133:14160–14163
93. Gazit E (2002) The "correctly folded" state of proteins: is it a metastable state? Angew Chem Int Ed 41:257–259
94. Dobson CM (2001) The structural basis of protein folding and its links with human disease. Philos Trans R Soc Lond Ser B 356:133–145
95. Cohen SIA, Vendruscolo M, Dobson CM, Knowles TPJ (2012) From macroscopic measurements to microscopic mechanisms of protein aggregation. J Mol Biol 421:160–171

96. Cohen SIA, Linse S, Luheshi LM, Hellstrand E, White DA, Rajah L, Otzen DE, Vendruscolo M, Dobson CM, Knowles TPJ (2013) Proliferation of amyloid-β42 aggregates occurs through a secondary nucleation mechanism. Proc Natl Acad Sci 110:9758–9763
97. Knowles TPJ, Waudby CA, Devlin GL, Cohen SIA, Aguzzi A, Vendruscolo M, Terentjev EM, Welland ME, Dobson CM (2009) An analytical solution to the kinetics of breakable filament assembly. Science 326:1533–1537
98. Jarrett JT, Lansbury PT (1993) Seeding "one-dimensional crystallization" of amyloid: a pathogenic mechanism in Alzheimer's disease and scrapie? Cell 73:1055–1058
99. Lorenzen N, Cohen SIA, Nielsen SB, Herling TW, Christiansen G, Dobson CM, Knowles TPJ, Otzen D (2012) Role of elongation and secondary pathways in S6 amyloid fibril growth. Biophys J 102:2167–2175
100. ten Wolde PR, Frenkel D (1997) Enhancement of protein crystal nucleation by critical density fluctuations. Science 277:1975–1978
101. Bousset L, Thomson NH, Radford SE, Melki R (2002) The yeast prion Ure2p retains its native α-helical conformation upon assembly into protein fibrils in vitro. EMBO J 21:2903–2911
102. Jucker M, Walker LC (2011) Pathogenic protein seeding in Alzheimer's disease and other neurodegenerative disorders. Ann Neurol 70:532–540
103. Jones EM, Surewicz WK (2005) Fibril conformation as the basis of species- and strain-dependent seeding specificity of mammalian prion amyloids. Cell 121:63–72
104. Smith JF, Knowles TPJ, Dobson CM, MacPhee CE, Welland ME (2006) Characterization of the nanoscale properties of individual amyloid fibrils. Proc Natl Acad Sci U S A 103:15806–15811
105. Paparcone R, Keten S, Buehler MJ (2010) Atomistic simulation of nanomechanical properties of Alzheimer's Aβ(1–40) amyloid fibrils under compressive and tensile loading. J Biomech 43:1196–1201
106. Adamcik J, Lara C, Usov I, Jeong JS, Ruggeri FS, Dietler G, Lashuel HA, Hamley IW, Mezzenga R (2012) Measurement of intrinsic properties of amyloid fibrils by the peak force QNM method. Nanoscale 4:4426–4429
107. Anderson VJ, Lekkerkerker HNW (2002) Insights into phase transition kinetics from colloid science. Nature 416:811–815
108. Krebs MRH, MacPhee CE, Miller AF, Dunlop IE, Dobson CM, Donald AM (2004) The formation of spherulites by amyloid fibrils of bovine insulin. Proc Natl Acad Sci U S A 101:14420–14424
109. Rogers SS, Venema P, van der Ploeg JPM, van der Linden E, Sagis LMC, Donald AM (2006) Investigating the permanent electric dipole moment of β-lactoglobulin fibrils, using transient electric birefringence. Biopolymers 82:241–252
110. Dzwolak W, Loksztejn A, Galinska-Rakoczy A, Adachi R, Goto Y, Rupnicki L (2007) Conformational indeterminism in protein misfolding: chiral amplification on amyloidogenic pathway of insulin. J Am Chem Soc 129:7517–7522
111. Ashby MF, Gibson LJ, Wegst U, Olive R (1995) The mechanical properties of natural materials. Proc R Soc Lond Math Phys Eng Sci 450:123–140
112. Wegst UGK, Ashby MF (2004) The mechanical efficiency of natural materials. Philos Mag 84:2167–2186
113. Shen ZL, Dodge MR, Kahn H, Ballarini R, Eppell SJ (2008) Stress-strain experiments on individual collagen fibrils. Biophys J 95:3956–3963
114. Yang L, van der Werf KO, Koopman BFJM, Subramaniam V, Bennink ML, Dijkstra PJ, Feijen J (2007) Micromechanical bending of single collagen fibrils using atomic force microscopy. J Biomed Mater Res A 82A:160–168
115. Slotta U, Hess S, Spieß K, Stromer T, Serpell L, Scheibel T (2007) Spider silk and amyloid fibrils: a structural comparison. Macromol Biosci 7:183–188
116. Keten S, Xu Z, Ihle B, Buehler MJ (2010) Nanoconfinement controls stiffness, strength and mechanical toughness of β-sheet crystals in silk. Nat Mater 9:359–367
117. Fratzl P, Weinkamer R (2007) Nature's hierarchical materials. Prog Mater Sci 52:1263–1334

118. Kreplak L, Bär H, Leterrier JF, Herrmann H, Aebi U (2005) Exploring the mechanical behavior of single intermediate filaments. J Mol Biol 354:569–577
119. Vendruscolo M, Knowles TPJ, Dobson CM (2011) Protein solubility and protein homeostasis: a generic view of protein misfolding disorders. Cold Spring Harb Perspect Biol. https://doi.org/10.1101/cshperspect.a010454
120. Kol N, Adler-Abramovich L, Barlam D, Shneck RZ, Gazit E, Rousso I (2005) Self-assembled peptide nanotubes are uniquely rigid bioinspired supramolecular structures. Nano Lett 5:1343–1346
121. Adamcik J, Jung J-M, Flakowski J, De Los Rios P, Dietler G, Mezzenga R (2010) Understanding amyloid aggregation by statistical analysis of atomic force microscopy images. Nat Nanotechnol 5:423–428
122. Meersman F, Cabrera RQ, McMillan PF, Dmitriev V (2009) Compressibility of insulin amyloid fibrils determined by X-ray diffraction in a diamond anvil cell. High Press Res 29:665–670
123. Sachse C, Grigorieff N, Fändrich M (2010) Nanoscale flexibility parameters of Alzheimer amyloid fibrils determined by electron cryo-microscopy. Angew Chem Int Ed Engl 49:1321–1323
124. Park J, Kahng B, Kamm RD, Hwang W (2006) Atomistic simulation approach to a continuum description of self-assembled β-sheet filaments. Biophys J 90:2510–2524
125. Relini A, Torrassa S, Ferrando R, Rolandi R, Campioni S, Chiti F, Gliozzi A (2010) Detection of populations of amyloid-like protofibrils with different physical properties. Biophys J 98:1277–1284
126. Guo S, Akhremitchev BB (2006) Packing density and structural heterogeneity of insulin amyloid fibrils measured by AFM nanoindentation. Biomacromolecules 7:1630–1636
127. Ashby MF, Gibson LJ, Wegst U, Olive R (1995) The mechanical properties of natural materials. Proc Math Phys Sci 450:123–140
128. Mankar S, Anoop A, Sen S, Maji SK (2011) Nanomaterials: amyloids reflect their brighter side. Nano Rev 2:1–12. https://doi.org/10.3402/nano.v2i0.6032
129. Sasso L, Suei S, Domigan L, Healy J, Nock V, MAK W, Gerrard JA (2014) Versatile multi-functionalization of protein nanofibrils for biosensor applications. Nanoscale 6:1629–1634
130. Hauser CAE, Maurer-Stroh S, Martins IC (2014) Amyloid-based nanosensors and nanodevices. Chem Soc Rev 43:5326–5345
131. Yang JE, Park JS, Cho E, Jung S, Paik SR (2015) Robust polydiacetylene-based colorimetric sensing material developed with amyloid fibrils of α-synuclein. Langmuir 31:1802–1810
132. Hamedi M, Herland A, Karlsson RH, Inganäs O (2008) Electrochemical devices made from conducting nanowire networks self-assembled from amyloid fibrils and alkoxysulfonate PEDOT. Nano Lett 8:1736–1740
133. Li C, Born A-K, Schweizer T, Zenobi-Wong M, Cerruti M, Mezzenga R (2014) Amyloid-hydroxyapatite bone biomimetic composites. Adv Mater 26:3207–3212
134. Meier C, Welland ME (2011) Wet-spinning of amyloid protein nanofibers into multifunctional high-performance biofibers. Biomacromolecules 12:3453–3459
135. Jacob RS, Ghosh D, Singh PK et al (2015) Self healing hydrogels composed of amyloid nano fibrils for cell culture and stem cell differentiation. Biomaterials 54:97–105
136. Reynolds NP, Charnley M, Mezzenga R, Hartley PG (2014) Engineered lysozyme amyloid fibril networks support cellular growth and spreading. Biomacromolecules 15:599–608
137. Gras SL, Tickler AK, Squires AM, Devlin GL, Horton MA, Dobson CM, MacPhee CE (2008) Functionalised amyloid fibrils for roles in cell adhesion. Biomaterials 29:1553–1562
138. Yan H, Nykanen A, Ruokolainen J, Farrar D, Gough JE, Saiani A, Miller AF (2008) Thermo-reversible protein fibrillar hydrogels as cell scaffolds. Faraday Discuss 139:71–84; discussion 105–128, 419–420
139. Maji SK, Schubert D, Rivier C, Lee S, Rivier JE, Riek R (2008) Amyloid as a depot for the formulation of long-acting drugs. PLoS Biol 6:e17

140. Silva RF, Araújo DR, Silva ER, Ando RA, Alves WA (2013) L-diphenylalanine microtubes as a potential drug-delivery system: characterization, release kinetics, and cytotoxicity. Langmuir 29:10205–10212
141. Akkermans C, Van der Goot AJ, Venema P, Gruppen H, Vereijken JM, Van der Linden E, Boom RM (2007) Micrometer-sized fibrillar protein aggregates from soy glycinin and soy protein isolate. J Agric Food Chem 55:9877–9882
142. Akkermans C, Venema P, van der Goot AJ, Gruppen H, Bakx EJ, Boom RM, van der Linden E (2008) Peptides are building blocks of heat-induced fibrillar protein aggregates of β-lactoglobulin formed at pH 2. Biomacromolecules 9:1474–1479
143. Bateman L, Ye A, Singh H (2010) In vitro digestion of β-lactoglobulin fibrils formed by heat treatment at low pH. J Agric Food Chem 58:9800–9808
144. Bateman L, Ye A, Singh H (2011) Re-formation of fibrils from hydrolysates of β-lactoglobulin fibrils during in vitro gastric digestion. J Agric Food Chem 59:9605–9611
145. Graveland-Bikker JF, de Kruif CG (2006) Unique milk protein based nanotubes: food and nanotechnology meet. Trends Food Sci Technol 17:196–203
146. Rao SP, Meade SJ, Healy JP, Sutton KH, Larsen NG, Staiger MP, Gerrard JA (2012) Amyloid fibrils as functionalizable components of nanocomposite materials. Biotechnol Prog 28:248–256
147. Li C, Bolisetty S, Mezzenga R (2013) Hybrid nanocomposites of gold single-crystal platelets and amyloid fibrils with Tunable fluorescence, conductivity, and sensing properties. Adv Mater 25:3694–3700
148. Li C, Adamcik J, Mezzenga R (2012) Biodegradable nanocomposites of amyloid fibrils and graphene with shape-memory and enzyme-sensing properties. Nat Nanotechnol 7:421–427
149. Mi R, Liu Y, Chen X, Shao Z (2016) Structure and properties of various hybrids fabricated by silk nanofibrils and nanohydroxyapatite. Nanoscale 8:20096–20102
150. Li D, Furukawa H, Deng H, Liu C, Yaghi OM, Eisenberg DS (2014) Designed amyloid fibers as materials for selective carbon dioxide capture. Proc Natl Acad Sci U S A 111:191–196
151. Li D, Jones EM, Sawaya MR et al (2014) Structure-based design of functional amyloid materials. J Am Chem Soc 136:18044–18051
152. Bolisetty S, Mezzenga R (2016) Amyloid–carbon hybrid membranes for universal water purification. Nat Nanotechnol 11:365–371
153. Bolisetty S, Arcari M, Adamcik J, Mezzenga R (2015) Hybrid amyloid membranes for continuous flow catalysis. Langmuir 31:13867–13873
154. Ha Y, Yang J, Tao F, Wu Q, Song Y, Wang H, Zhang X, Yang P (2018) Phase-transited lysozyme as a universal route to bioactive hydroxyapatite crystalline film. Adv Funct Mater 28:1704476
155. Gu J, Su Y, Liu P, Li P, Yang P (2017) An environmentally benign antimicrobial coating based on a protein supramolecular assembly. ACS Appl Mater Interfaces 9:198–210
156. Zhao J, Qu Y, Chen H, Xu R, Yu Q, Yang P (2018) Self-assembled proteinaceous wound dressings attenuate secondary trauma and improve wound healing in vivo. J Mater Chem B 6:4645–4655
157. Gao A, Wu Q, Wang D, Ha Y, Chen Z, Yang P (2016) A superhydrophobic surface templated by protein self-assembly and emerging application toward protein crystallization. Adv Mater 28:579–587
158. Jiang B, Yang J, Li C, Zhang L, Zhang X, Yang P (2017) Water-based photo- and electron-beam lithography using egg white as a resist. Adv Mater Interfaces 4:1601223
159. Saunders BR, Laajam N, Daly E, Teow S, Hu X, Stepto R (2009) Microgels: from responsive polymer colloids to biomaterials. Adv Colloid Interf Sci 147:251–262
160. Das M, Zhang H, Kumacheva E (2006) Microgels: old materials with new applications. Annu Rev Mater Res 36:117–142
161. Seiffert S (2013) Small but smart: sensitive microgel capsules. Angew Chem Int Ed 52:11462–11468

162. Fernández-Barbero A, Suárez IJ, Sierra-Martín B, Fernández-Nieves A, de las Nieves FJ, Marquez M, Rubio-Retama J, López-Cabarcos E (2009) Gels and microgels for nanotechnological applications. Adv Colloid Interf Sci 147:88–108
163. Maes D, Vorontsova MA, Potenza MA, Sanvito T, Sleutel M, Giglio M, Vekilov PG (2015) Do protein crystals nucleate within dense liquid clusters? Acta Crystallogr Sect F Struct Biol Commun 71:815–822
164. Chatani E, Imamura H, Yamamoto N, Kato M (2014) Stepwise organization of the β-structure identifies key regions essential for the propagation and cytotoxicity of insulin amyloid fibrils. J Biol Chem 289:10399–10410
165. Yuyama K, Ueda M, Nagao S, Hirota S, Sugiyama T, Masuhara H (2017) A single spherical assembly of protein amyloid fibrils formed by laser trapping. Angew Chem Int Ed 56:6739–6743
166. Hu Z, Chen Y, Wang C, Zheng Y, Li Y (1998) Polymer gels with engineered environmentally responsive surface patterns. Nature 393:149–152
167. Lu Y, Mei Y, Ballauff M, Drechsler M (2006) Thermosensitive core−shell particles as carrier systems for metallic nanoparticles. J Phys Chem B 110:3930–3937
168. Schachschal S, Adler H-J, Pich A, Wetzel S, Matura A, van Pee K-H (2011) Encapsulation of enzymes in microgels by polymerization/cross-linking in aqueous droplets. Colloid Polym Sci 289:693–698
169. Vinogradov SV (2006) Colloidal microgels in drug delivery applications. Curr Pharm Des 12:4703–4712
170. Oh JK, Drumright R, Siegwart DJ, Matyjaszewski K (2008) The development of microgels/nanogels for drug delivery applications. Prog Polym Sci Oxf 33:448–477
171. Lopez VC, Hadgraft J, Snowden MJ (2005) The use of colloidal microgels as a (trans)dermal drug delivery system. Int J Pharm 292:137–147
172. Hoare TR, Kohane DS (2008) Hydrogels in drug delivery: progress and challenges. Polymer 49:1993–2007
173. Fettis MM, Wei Y, Restuccia A, Kurian JJ, Wallet SM, Hudalla GA (2016) Microgels with tunable affinity-controlled protein release via desolvation of self-assembled peptide nanofibers. J Mater Chem B 4:3054–3064
174. Du X, Zhou J, Shi J, Xu B (2015) Supramolecular hydrogelators and hydrogels: from soft matter to molecular biomaterials. Chem Rev 115:13165–13307
175. Velasco D, Tumarkin E, Kumacheva E (2012) Microfluidic encapsulation of cells in polymer microgels. Small 8:1633–1642
176. Lipinski CA (2000) Drug-like properties and the causes of poor solubility and poor permeability. J Pharmacol Toxicol Methods 44:235–249
177. Stella VJ, Nti-Addae KW (2007) Prodrug strategies to overcome poor water solubility. Adv Drug Deliv Rev 59:677–694
178. Shimanovich U, Tkacz ID, Eliaz D, Cavaco-Paulo A, Michaeli S, Gedanken A (2011) Encapsulation of RNA molecules in BSA microspheres and internalization into Trypanosoma Brucei parasites and human U2OS cancer cells. Adv Funct Mater 21:3659–3666
179. Angel (Shimanovich) U, Matas D, Michaeli S, Cavaco-Paulo A, Gedanken A (2010) Microspheres of mixed proteins. Chem Eur J 16:2108–2114
180. Shimanovich U, Eliaz D, Zigdon S, Volkov V, Aizer A, Cavaco-Paulo A, Michaeli S, Shav-Tal Y, Gedanken A (2012) Proteinaceous microspheres for targeted RNA delivery prepared by an ultrasonic emulsification method. J Mater Chem B 1:82–90
181. Ma X, Sun X, Hargrove D, Chen J, Song D, Dong Q, Lu X, Fan T-H, Fu Y, Lei Y (2016) A biocompatible and biodegradable protein hydrogel with green and red autofluorescence: preparation, characterization and *In Vivo* biodegradation tracking and modeling. Sci Rep 6:19370
182. Sarkar A, Murray B, Holmes M, Ettelaie R, Abdalla A, Yang X (2016) In vitro digestion of Pickering emulsions stabilized by soft whey protein microgel particles: influence of thermal treatment. Soft Matter 12:3558–3569

183. O'Neill GJ, Jacquier JC, Mukhopadhya A, Egan T, O'Sullivan M, Sweeney T, O'Riordan ED (2015) In vitro and in vivo evaluation of whey protein hydrogels for oral delivery of riboflavin. J Funct Foods 19:512–521
184. Branco MC, Pochan DJ, Wagner NJ, Schneider JP (2009) Macromolecular diffusion and release from self-assembled β-hairpin peptide hydrogels. Biomaterials 30:1339–1347
185. Koutsopoulos S, Unsworth LD, Nagai Y, Zhang S (2009) Controlled release of functional proteins through designer self-assembling peptide nanofiber hydrogel scaffold. Proc Natl Acad Sci 106:4623–4628
186. Ramachandran S, Tseng Y, Yu YB (2005) Repeated rapid shear-responsiveness of peptide hydrogels with tunable shear modulus. Biomacromolecules 6:1316–1321
187. Stendahl JC, Wang LJ, Chow LW, Kaufman DB, Stupp SI (2008) Growth factor delivery from self-assembling nanofibers to facilitate islet transplantation. Transplantation 86:478–481
188. Baldwin AJ, Bader R, Christodoulou J, MacPhee CE, Dobson CM, Barker PD (2006) Cytochrome display on amyloid fibrils. J Am Chem Soc 128:2162–2163
189. Maschke A, Becker C, Eyrich D, Kiermaier J, Blunk T, Göpferich A (2007) Development of a spray congealing process for the preparation of insulin-loaded lipid microparticles and characterization thereof. Eur J Pharm Biopharm 65:175–187
190. Jiang G, Thanoo BC, DeLuca PP (2002) Effect of osmotic pressure in the solvent extraction phase on BSA release profile from PLGA microspheres. Pharm Dev Technol 7:391–399
191. Song Y, Shimanovich U, Michaels TCT, Ma Q, Li J, Knowles TPJ, Shum HC (2016) Fabrication of fibrillosomes from droplets stabilized by protein nanofibrils at all-aqueous interfaces. Nat Commun 7:ncomms12934
192. Erni P, Fischer P, Windhab EJ (2005) Deformation of single emulsion drops covered with a viscoelastic adsorbed protein layer in simple shear flow. Appl Phys Lett 87:244104
193. Gedanken A (2008) Preparation and properties of proteinaceous microspheres made sonochemically. Chem Eur J 14:3840–3853
194. Silva R, Ferreira H, Cavaco-Paulo A (2011) Sonoproduction of liposomes and protein particles as templates for delivery purposes. Biomacromolecules 12:3353–3368
195. Xu H, Zeiger BW, Suslick KS (2013) Sonochemical synthesis of nanomaterials. Chem Soc Rev 42:2555–2567
196. Whitesides GM (2006) The origins and the future of microfluidics. Nature 442:368–373
197. Squires TM, Quake SR (2005) Microfluidics: fluid physics at the nanoliter scale. Rev Mod Phys 77:977–1026
198. Wang J-T, Wang J, Han J-J (2011) Fabrication of advanced particles and particle-based materials assisted by droplet-based microfluidics. Small 7:1728–1754
199. Zhang H, Tumarkin E, Peerani R, Nie Z, Sullan RMA, Walker GC, Kumacheva E (2006) Microfluidic production of biopolymer microcapsules with controlled morphology. J Am Chem Soc 128:12205–12210
200. Tumarkin E, Kumacheva E (2009) Microfluidic generation of microgels from synthetic and natural polymers. Chem Soc Rev 38:2161–2168
201. Utada AS, Chu L-Y, Fernandez-Nieves A, Link DR, Holtze C, Weitz DA (2007) Dripping, jetting, drops, and wetting: the magic of microfluidics. MRS Bull 32:702–708
202. Shum HC, Kim J-W, Weitz DA (2008) Microfluidic fabrication of monodisperse biocompatible and biodegradable polymersomes with controlled permeability. J Am Chem Soc 130:9543–9549
203. Ward T, Faivre M, Abkarian M, Stone HA (2005) Microfluidic flow focusing: drop size and scaling in pressure versus flow-rate-driven pumping. Electrophoresis 26:3716–3724
204. Xia Y, Whitesides GM (1998) Soft lithography. Angew Chem Int Ed 37:550–575
205. McDonald JC, Duffy DC, Anderson JR, Chiu DT, Wu H, Schueller OJA, Whitesides GM (2000) Fabrication of microfluidic systems in poly(dimethylsiloxane). Electrophoresis 21:27–40
206. Mazutis L, Gilbert J, Ung WL, Weitz DA, Griffiths AD, Heyman JA (2013) Single-cell analysis and sorting using droplet-based microfluidics. Nat Protoc 8:870–891

207. Martín-Banderas L, Flores-Mosquera M, Riesco-Chueca P, Rodríguez-Gil A, Cebolla Á, Chávez S, Gañán-Calvo AM (2005) Flow focusing: a versatile technology to produce size-controlled and specific-morphology microparticles. Small 1:688–692
208. Baroud CN, Gallaire F, Dangla R (2010) Dynamics of microfluidic droplets. Lab Chip 10:2032–2045
209. Abate AR, Weitz DA (2009) High-order multiple emulsions formed in poly(dimethylsiloxane) microfluidics. Small 5:2030–2032
210. Kim S-H, Shum HC, Kim JW, Cho J-C, Weitz DA (2011) Multiple polymersomes for programmed release of multiple components. J Am Chem Soc 133:15165–15171
211. Seo M, Paquet C, Nie Z, Xu S, Kumacheva E (2007) Microfluidic consecutive flow-focusing droplet generators. Soft Matter 3:986–992
212. Keating CD (2012) Aqueous phase separation as a possible route to compartmentalization of biological molecules. Acc Chem Res 45:2114–2124
213. Plamper FA, Richtering W (2017) Functional microgels and microgel systems. Acc Chem Res 50:131–140
214. Thorne JB, Vine GJ, Snowden MJ (2011) Microgel applications and commercial considerations. Colloid Polym Sci 289:625
215. Wang Y-X, Robertson JL, Spillman WB, Claus RO (2004) Effects of the chemical structure and the surface properties of polymeric biomaterials on their biocompatibility. Pharm Res 21:1362–1373
216. Kumachev A, Greener J, Tumarkin E, Eiser E, Zandstra PW, Kumacheva E (2011) High-throughput generation of hydrogel microbeads with varying elasticity for cell encapsulation. Biomaterials 32:1477–1483
217. Agnihotri SA, Mallikarjuna NN, Aminabhavi TM (2004) Recent advances on chitosan-based micro- and nanoparticles in drug delivery. J Control Release 100:5–28
218. Mitra A, Dey B (2011) Chitosan microspheres in novel drug delivery systems. Indian J Pharm Sci 73:355–366
219. Augst AD, Kong HJ, Mooney DJ (2006) Alginate hydrogels as biomaterials. Macromol Biosci 6:623–633
220. Tan W-H, Takeuchi S (2007) Monodisperse alginate hydrogel microbeads for cell encapsulation. Adv Mater 19:2696–2701
221. Gazit E (2007) Self-assembled peptide nanostructures: the design of molecular building blocks and their technological utilization. Chem Soc Rev 36:1263–1269
222. Bai S, Pappas C, Debnath S, Frederix PWJM, Leckie J, Fleming S, Ulijn RV (2014) Stable emulsions formed by self-assembly of interfacial networks of dipeptide derivatives. ACS Nano 8:7005–7013
223. Flamia R, Salvi AM, D'Alessio L, Castle JE, Tamburro AM (2007) Transformation of amyloid-like fibers, formed from an elastin-based biopolymer, into a hydrogel: an X-ray photoelectron spectroscopy and atomic force microscopy study. Biomacromolecules 8:128–138
224. Bhak G, Lee S, Park JW, Cho S, Paik SR (2010) Amyloid hydrogel derived from curly protein fibrils of α-synuclein. Biomaterials 31:5986–5995
225. Mains J, Lamprou D, McIntosh L, Oswald IH, Urquhart A (2013) Beta-adrenoceptor antagonists affect amyloid nanostructure; amyloid hydrogels as drug delivery vehicles. Chem Commun 49:5082–5084
226. Gosal WS, Clark AH, Ross-Murphy SB (2004) Fibrillar β-lactoglobulin gels: Part 2. Dynamic mechanical characterization of heat-set systems. Biomacromolecules 5:2420–2429
227. Bolisetty S, Harnau L, Jung J, Mezzenga R (2012) Gelation, phase behavior, and dynamics of β-lactoglobulin amyloid fibrils at varying concentrations and ionic strengths. Biomacromolecules 13:3241–3252
228. Jung J-M, Mezzenga R (2010) Liquid crystalline phase behavior of protein Fibers in water: experiments versus theory. Langmuir 26:504–514
229. Nyström G, Fong W-K, Mezzenga R (2017) Ice-templated and cross-linked amyloid fibril aerogel scaffolds for cell growth. Biomacromolecules 18:2858–2865

230. Langton M, Hermansson A-M (1992) Fine-stranded and particulate gels of β-lactoglobulin and whey protein at varying pH. Food Hydrocoll 5:523–539
231. Zhou X-M, Shimanovich U, Herling TW, Wu S, Dobson CM, Knowles TPJ, Perrett S (2015) Enzymatically active microgels from self-assembling protein nanofibrils for microflow chemistry. ACS Nano 9:5772–5781
232. Silva R, Ferreira H, Azoia NG, Shimanovich U, Freddi G, Gedanken A, Cavaco-Paulo A (2012) Insights on the mechanism of formation of protein microspheres in a biphasic system. Mol Pharm 9:3079–3088
233. Shimanovich U, Ruggeri FS, Genst ED et al (2017) Silk micrococoons for protein stabilisation and molecular encapsulation. Nat Commun 8:ncomms15902
234. Müller T, Simone Ruggeri F, Kulik J, Shimanovich U, Mason TO, Knowles TPJ, Dietler G (2014) Nanoscale spatially resolved infrared spectra from single microdroplets. Lab Chip 14:1315–1319
235. Avivi S, Gedanken A (2002) S–S bonds are not required for the sonochemical formation of proteinaceous microspheres: the case of streptavidin. Biochem J 366:705–707
236. Subia B, Kundu SC (2013) Drug loading and release on tumor cells using silk fibroin–albumin nanoparticles as carriers. Nanotechnology 24:035103
237. Shimanovich U, Volkov V, Eliaz D, Aizer A, Michaeli S, Gedanken A (2011) Stabilizing RNA by the sonochemical formation of RNA nanospheres. Small 7:1068–1074
238. Shimanovich U, Eliaz D, Aizer A, Vayman I, Michaeli S, Shav-Tal Y, Gedanken A (2011) Sonochemical synthesis of DNA nanospheres. Chembiochem 12:1678–1681
239. Li C, Xu L, Zuo YY, Yang P (2018) Tuning protein assembly pathways through superfast amyloid-like aggregation. Biomater Sci 6:836–841
240. Shimanovich U, Song Y, Brujic J, Shum HC, Knowles TPJ (2015) Multiphase protein microgels. Macromol Biosci 15:501–508
241. Knowles T, Shimanovich U, Dobson C, Weitz D (2016) Protein Capsules
242. Volpatti LR, Shimanovich U, Ruggeri FS, Bolisetty S, Müller T, Mason TO, Michaels TCT, Mezzenga R, Dietler G, Knowles TPJ (2016) Micro- and nanoscale hierarchical structure of core–shell protein microgels. J Mater Chem B 4:7989–7999
243. Peters TJ (1987) Partition of cell particles and macromolecules: separation and purification of biomolecules, cell organelles, membranes and cells in aqueous polymer two phase systems and their use in biochemical analysis and biotechnology. Cell Biochem Funct 5:233–234
244. Kroner KH, Hustedt H, Granda S, Kula M-R, Introduction by T Alan Hatton (2009) Technical aspects of separation using aqueous two-phase systems in enzyme isolation processes. Biotechnol Bioeng 104:217–239
245. Diamond AD, Hsu JT (1990) Protein partitioning in PEG/dextran aqueous two-phase systems. AICHE J 36:1017–1024
246. Fele L, Fermeglia M (1996) Partition coefficients of proteins in poly(ethylene glycol) + dextran + water at 298 K. J Chem Eng Data 41:331–334
247. Osborn HT, Akoh CC (2004) Effect of emulsifier type, droplet size, and oil concentration on lipid oxidation in structured lipid-based oil-in-water emulsions. Food Chem 84:451–456
248. Sah H (1999) Stabilization of proteins against methylene chloride/water interface-induced denaturation and aggregation. J Control Release 58:143–151
249. Liu Y, Lipowsky R, Dimova R (2012) Concentration dependence of the interfacial tension for aqueous two-phase polymer solutions of dextran and polyethylene glycol. Langmuir 28:3831–3839
250. Balakrishnan G, Nicolai T, Benyahia L, Durand D (2012) Particles trapped at the droplet interface in water-in-water emulsions. Langmuir 28:5921–5926
251. Nguyen BT, Nicolai T, Benyahia L (2013) Stabilization of water-in-water emulsions by addition of protein particles. Langmuir 29:10658–10664
252. Rollett A, Reiter T, Nogueira P, Cardinale M, Loureiro A, Gomes A, Cavaco-Paulo A, Moreira A, Carmo AM, Guebitz GM (2012) Folic acid-functionalized human serum albumin nanocapsules for targeted drug delivery to chronically activated macrophages. Int J Pharm 427:460–466

253. Richman M, Wilk S, Skirtenko N, Perelman A, Rahimipour S (2011) Surface-modified protein microspheres capture amyloid-β and inhibit its aggregation and toxicity. Chem Weinh Bergstr Ger 17:11171–11177
254. Krysmann MJ, Castelletto V, Kelarakis A, Hamley IW, Hule RA, Pochan DJ (2008) Self-assembly and hydrogelation of an amyloid peptide fragment. Biochemistry 47:4597–4605
255. Zhou X-M, Entwistle A, Zhang H, Jackson AP, Mason TO, Shimanovich U, Knowles TPJ, Smith AT, Sawyer EB, Perrett S (2014) Self-assembly of amyloid fibrils that display active enzymes. ChemCatChem 6:1961–1968
256. Kim J-W, Fernández-Nieves A, Dan N, Utada AS, Marquez M, Weitz DA (2007) Colloidal assembly route for responsive colloidosomes with tunable permeability. Nano Lett 7:2876–2880
257. Li M-H, Keller P (2009) Stimuli-responsive polymer vesicles. Soft Matter 5:927–937
258. Yolamanova M, Meier C, Shaytan AK et al (2013) Peptide nanofibrils boost retroviral gene transfer and provide a rapid means for concentrating viruses. Nat Nanotechnol 8:130–136
259. Dai B, Li D, Xi W et al (2015) Tunable assembly of amyloid-forming peptides into nanosheets as a retrovirus carrier. Proc Natl Acad Sci 112:2996–3001
260. Bolisetty S, Boddupalli CS, Handschin S, Chaitanya K, Adamcik J, Saito Y, Manz MG, Mezzenga R (2014) Amyloid fibrils enhance transport of metal nanoparticles in living cells and induced cytotoxicity. Biomacromolecules 15:2793–2799
261. Shen Y, Posavec L, Bolisetty S et al (2017) Amyloid fibril systems reduce, stabilize and deliver bioavailable nanosized iron. Nat Nanotechnol 12:642–647
262. Levin A, Mason TO, Knowles TPJ, Shimanovich U (2017) Self-assembled protein fibril-metal oxide nanocomposites. Isr J Chem 57(7):724–728
263. Sandhu A, Handa H, Abe M (2010) Synthesis and applications of magnetic nanoparticles for biorecognition and point of care medical diagnostics. Nanotechnology 21:442001
264. Wiogo HTR, Lim M, Bulmus V, Yun J, Amal R (2011) Stabilization of magnetic iron oxide nanoparticles in biological media by eetal bovine serum (FBS). Langmuir 27:843–850
265. Jordens S, Rühs PA, Sieber C, Isa L, Fischer P, Mezzenga R (2014) Bridging the gap between the nanostructural organization and macroscopic interfacial rheology of amyloid fibrils at liquid interfaces. Langmuir 30:10090–10097
266. Linse S, Cabaleiro-Lago C, Xue W-F, Lynch I, Lindman S, Thulin E, Radford SE, Dawson KA (2007) Nucleation of protein fibrillation by nanoparticles. Proc Natl Acad Sci U S A 104:8691–8696

Chapter 8
Protein Nanofibrils as Storage Forms of Peptide Drugs and Hormones

Reeba Susan Jacob, A. Anoop, and Samir K. Maji

Abstract Amyloids are highly organized cross β-sheet protein nanofibrils that are associated with both diseases and functions. Thermodynamically amyloids are stable structures as they represent the lowest free energy state that proteins can attain. However, recent studies suggest that amyloid fibrils can be dissociated by a change in environmental parameters such as pH and ionic strength. This reversibility of amyloids can not only be associated with disease, but function as well. In disease-associated amyloids, fibrils can act as reservoirs of cytotoxic oligomers. Recently, in higher organisms such as mammals, hormones were found to be stored in amyloid-like state, where these were reported to act as a reservoir of functional monomers. These hormone amyloids can dissociate to monomers upon release from the secretory granules, and subsequently bind to their respective receptors and perform their functions. In this book chapter, we describe in detail how these protein nanofibrils represent the densest possible peptide packing and are suitable for long-term storage. Thus, mimicking the feature of amyloids to release functional monomers, it is possible to formulate amyloid-based peptide/protein drugs, which can be used for sustained release.

Keywords Amyloid · Functions · Fibrils · Secretory granules · Hormones

8.1 Introduction

Proteins are the workhorses of the cellular machinery and perform a wide array of functions such as transporting molecules, catalyzing reactions, signal transduction and DNA replication [1]. However, for most proteins to function properly, they must be folded correctly into a three-dimensional native conformation [1, 2], which is

R. S. Jacob · A. Anoop · S. K. Maji (✉)
Department of Biosciences and Bioengineering, Indian Institute of Technology Bombay, Mumbai, Maharashtra, India
e-mail: samirmaji@iitb.ac.in

S. Perrett et al. (eds.), *Biological and Bio-inspired Nanomaterials*, Advances in Experimental Medicine and Biology 1174,
https://doi.org/10.1007/978-981-13-9791-2_8

often considered as their lowest energy conformation. Sometimes the biological function of proteins also depends on their self-assembly into functional polymers [3, 4]. In such cases, proteins are not functional when they are in their monomeric folded states. For example, protein polymers such as actin and tubulin are involved in maintaining cell shape and other vital cellular functions [3, 4]. Occasionally proteins also become aberrantly folded or misfolded to form aggregated structures called amyloids, which are mainly associated with human diseases. Amyloids are highly ordered fibrillar protein aggregates (Fig. 8.1) that are generally associated with more than 40 human diseases including Parkinson's, Alzheimer's and Kuru [5]. These higher order protein aggregates are made up of cross-β-sheet motifs where β-strands run perpendicular to the fibril axis and β-sheets are parallel to the fibril axis [6, 7] (Fig. 8.1). Amyloids appear as unbranched fibrils, mostly of indefinite length and of 2–10 nm in diameter when visualized under a transmission electron microscope (TEM) [6, 7]. Amyloid fibrils are shown to bind small molecule dyes such as Thioflavin T (ThT) [8] and Congo red (CR) [9]. When bound to amyloids, CR most often shows a yellow-green birefringence under cross-polarized light [9]. Studying the molecular details of amyloid fibrils is challenging due to their low solubility and non-crystalline nature. The cross-β-sheet structure of amyloids

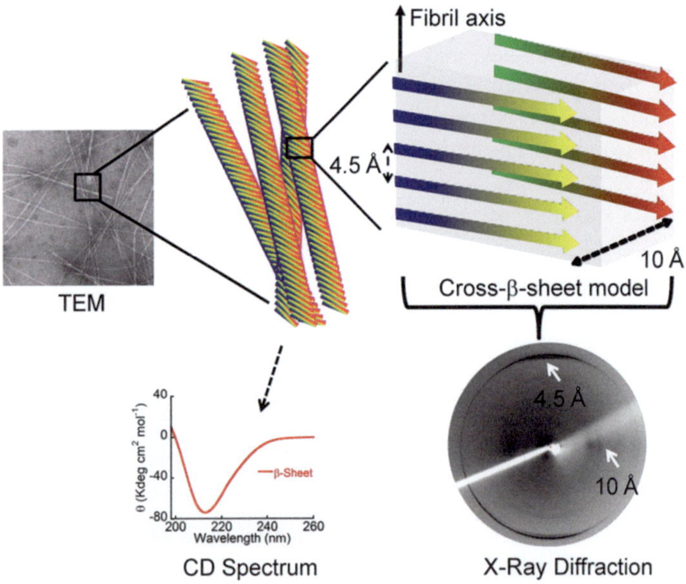

Fig. 8.1 Characterization of amyloid fibrils. Amyloid fibrils are protein/peptide aggregates, which show fibrillar morphology under the transmission electron microscope (TEM) and are made up of cross-β-sheet motifs where β-strands run perpendicular to fibril axis and the β-sheets are parallel to the fibril axis. The β-sheet conformation of amyloid fibrils can be measured by circular dichroism (CD) spectroscopy and the X-ray diffraction pattern of amyloid fibrils that shows two characteristic reflections at 4.5 Å and 10 Å, corresponding to the inter-strand and the inter-sheet distance of the β-sheets, respectively

provides them with very high stability as well as resistance to harsh conditions such as high temperature, wide ranges of pH and proteolytic degradation [10–12].

During aggregation/amyloid formation, both natively structured and unstructured proteins/peptides undergo misfolding into partially folded intermediates, which eventually form soluble oligomers and mature fibrils [13]. Recent studies have shown that soluble oligomers on the pathway of amyloid formation are more toxic and responsible for cell death occurring in human diseases [14–17]. The aggregation of protein into amyloid fibrils generally follows nucleation-dependent polymerization, where three distinct phases are involved. These include the (i) lag phase, where protein monomers associate to form aggregation competent nuclei, but the aggregate mass remains below the threshold for detection (ii) growth phase where monomer addition to nuclei leads to a rapid increase in fibril mass and finally the (iii) stationary phase, where fibril formation is completed and the fibrils remain in equilibrium with monomers [18–20]. The nucleation-dependent polymerization phenomenon in amyloid aggregation is further supported by the fact that addition of preformed fibril seeds (made of the same protein) decreases the lag time for fibril formation (self-seeding) [21–23]. Amyloid formation has thus been considered as an irreversible process, where the fibril is believed to be a thermodynamically controlled product [24, 25]. This signifies that once amyloid is formed, its dissociation into monomeric protein/peptide is a seemingly difficult process. Contrasting this assertion, several research groups have shown that amyloids could be readily 'dissolved' to release their constituent protein/peptides when exposed to different pH or simply by dilution [10, 26–28]. Moreover, in the amyloid aggregation pathway, mature amyloid fibrils remain in equilibrium with monomers, where the monomeric protein/peptide may constantly exchange with its fibrillar counterpart in solution (a process known as 'molecular recycling') [29]. This dynamic nature of amyloid was nicely demonstrated by Carulla et al. using amyloid fibrils of the bovine phosphatidylinositol-3-kinase SH-3 domain [29]. When SH3 domain fibrils were incubated in deuterated buffer for a long time, a substantial amount of amyloid fibrils were found to be deuterated supporting that monomers in amyloid fibrils are getting constantly exchanged with the deuterated monomers in solution [29]. In addition to the 'natural' phenomenon of molecular recycling in fibrils, various external factors that can bring about a shift in the equilibrium between the fibrillar and soluble (monomeric/oligomeric) states include (i) protein/peptide concentration, (ii) ionic strength of the buffer, (iii) pH of the environment and (iv) the presence of other molecules. The pH-induced fibril dissociation study by Tipping et al. showed that β2 microglobulin amyloid fibrils disassemble into toxic oligomers at pH 6.4, whereas the same fibrils dissociate to native non-toxic monomers at physiological pH 7.4 [30]. Interestingly, a similar observation was found in another *in vitro* study, where incubation of α-synuclein fibrils resulted in disaggregation of the fibrils into neurotoxic β-sheet structured oligomers, which on further incubation formed monomers [31]. Overall, the above studies indicate that amyloids can act as a 'reservoir' for either monomeric or oligomeric states of proteins.

8.2 Functional Amyloids

In the last decade, various reports showed that amyloids cannot only cause human diseases but are also able to perform native functions in the host organism. These non-disease associated amyloids, involved in important biological functions are collectively termed 'functional amyloids' [32]. One of the first examples of functional amyloids came from the discovery of the curli fibrils produced by extracellularly secreted proteins of the bacterium *E. coli* [33]. These fibrils were suggested to aid in colonization and biofilm formation of the bacteria. The biofilm is useful for the survival of bacteria in extreme environmental conditions such as drastic pH changes, high salt concentration and chemical toxins [34, 35]. Amyloids owing to their ability to withstand harsh conditions could likely contribute to these characteristics of the biofilm, being a part of the same. Several other bacteria-associated proteins are also reported to polymerize into amyloid fibrils. Examples of these include fimbriae of *Salmonella* spp. [34], harpins of *X. campestris* and *P. syringae* [36], pili from *M. tuberculosis* [37], chaplins from *S. coelicolor* [38, 39]. Similarly, the amyloid fibrils of the adhesin protein were shown to facilitate yeast cells in colonization as well as attachment to the host [40]. Furthermore, hydrophobin amyloids, which provide support for hyphae/spore formation in fungi [38, 41] and prions of yeast and *Podospora anserina* that aid in the better survival of the host during stress conditions are among other examples of fungal functional amyloids [42–45]. Functional amyloids are also widespread in lower animals such as insects, mollusks etc. For example, amyloids of chorion protein can be found in the eggshell of the silk worm, where they protect the growing embryo from environmental hazards [46]. The first mammalian functional amyloid was discovered by Kelly and co-workers, where they showed that amyloid fibrils of Pmel17 inside the melanosome facilitate the melanin polymerization [47]. Recently, it was suggested that pituitary hormones could also form an amyloid-like structure *in vitro* and *in vivo*, which is important for hormone storage in secretory granules and its release [26]. In line with these findings, designing amyloid-based functional material is shown to be promising for their potential application in nano-biotechnology [48, 49]. The following sections in this chapter highlight protein nanofibrils as the densest possible peptide-packing structure suitable for long-term storage. More importantly, it is emphasized here that by utilizing the feature of amyloids to 'release functional monomers', it is possible to formulate amyloid-based peptide/protein drugs, which can be used for sustained release.

8.3 Amyloids as a Depot for Protein/Peptide Storage and Release

Cells usually synthesize a plethora of proteins for their own activity as well as for secretion and to act as chemical messages for other cells, an activity which overall leads to a coordinated functioning of tissues and organs. However, deciding which protein to secrete, when to store and release is a highly regulated and complex process in cells. By default, proteins that are to be secreted immediately after their synthesis in ER are transported to the extracellular space through vesicles that bud from the trans-Golgi complex [50–53]. Many cells such as liver and muscle cells use this constitutive route for protein secretion, where the vesicle contents are released without any external stimulus [54, 55] (Fig. 8.2). However, the secretion of neuronal/endocrine peptides, growth factors and hormones is tightly regulated to maintain physiological homeostasis. Such molecules are required to be stored for long periods in a highly concentrated form until the cells receive signals to release their contents into the extracellular space [52, 53]. This route of protein secretion is called "regulated secretion" and is used by several cells including endocrine, neuroendocrine and mast cells [56–58]. Some eukaryotic cells display both pathways of protein secretion [54]. In such cells, both constitutively released proteins and regulatory secretory proteins need to be sorted into their respective secretory pathways [59]. Although the exact machinery of sorting and signaling of different protein/peptides to different protein secretory pathway is not clearly understood yet, recent evidence suggests that the formation of amyloid-like structures could be the mechanism for selective aggregation and sorting of hormones and neuropeptides into regulatory secretory pathway [26]. Therefore, in

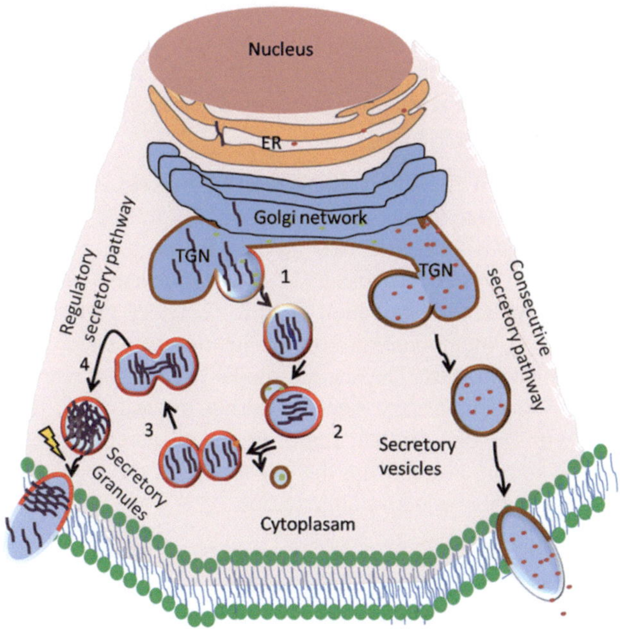

Fig. 8.2 Secretory granule biogenesis. This scheme depicts various stages involved in secretory granule biogenesis, which includes (*1*) formation of immature secretory granules (ISG), (*2*) removal of soluble protein by budding (*3*) homotypic fusion of ISG, (*4*) remodeling of the membrane of ISG to form MSG

order to correctly understand the mechanisms behind sorting of protein/peptides for regulated secretion, it is important to know the events in the regulatory secretory pathway, which are briefly discussed in the following section.

Regulated Secretory Pathway of Protein/Peptide Secretion in Cells Peptide hormones and neurotransmitters, after their synthesis as large precursor proteins in the endoplasmic reticulum, are transported to the Golgi apparatus [60]. At the trans-Golgi network, these precursors and other secretory granule proteins are packed into immature secretory granules (ISGs) [52, 61]. While the transport of proteins to the constitutive secretory pathway involves a passive, bulk flow mechanism, the proteins destined for the regulated secretory pathways apparently undergo various maturation and processing steps to reach mature secretory granules (MSGs) [54] (Fig. 8.2). Immediately after the formation of ISG, acidification of ISGs occurs by a gradual and regulated increase in active H^+ pump density [62]. This acidification step is not only required for the aggregation of proteins inside secretory granules (SG) for protein condensation, but also for activation of proteolytic enzymes that convert prohormones to hormones [60, 63]. Moreover, the condensation promotes higher order intermolecular association of proteins and increases the efficiency of storage of hormones in secretory granules. Insulin, growth hormone (GH) and prolactin (PRL) are some of the hormones reported to undergo condensation to attain higher order protein aggregates for their retention and storage in MSG [64–66]. These MSGs are stored within the cell until they are stimulated by secretagogues to release their content into the extracellular space through membrane fusion [52, 53] (Fig. 8.2). Although there are multiple steps in the formation of MSGs, the cell type and the regulatory secretory proteins [67] determine the extent and influence of each of these steps.

Sorting of Proteins into Secretory Granules by Protein Aggregation Protein aggregation is considered to be an excellent method for sorting regulatory secretory pathway proteins to SGs [68–70]. Two models for protein sorting in secretory granule formation have been proposed [52, 53]. The first model proposed was the "sorting for entry" model, in which prior to secretory granule formation, the selection of the cargo, membrane and the exclusion of non-secretory proteins in the trans-Golgi network occur. The second model "sorting by retention" suggests that secretory granule contents are selected during maturation of the granules by retention along with the concomitant removal of non-secretory proteins [53, 61]. Irrespective of the sorting mechanism, the key feature of the regulated secretion of proteins appears to be their selective aggregation.

Aggregation of secretory proteins in SG could be facilitated by various factors, which include acidification of the cellular compartments from trans-Golgi to SG, proteolytic processing of prohormones, the presence of high concentrations of divalent ions (e.g. calcium, zinc) and biogenic amines [65, 70, 71]. Many proteins are reported to aggregate at acidic pH and in the secretory pathway where the pH decreases from roughly 7.2 in the ER to ~6.0 in the *trans*-Golgi lumen and finally to 5.5 in the secretory granules [72–74]. Hence it was hypothesized that acidic pH of *trans*-Golgi can induce aggregation and condensation of the proteins in the regulatory pathway to form SG [52]. Furthermore, for aggregation to occur, the local concentration of the secretory protein should be increased [52]. However, if the concentration of the secretory protein is less than optimally required for aggregation, some helper molecules and/or proteins (e.g. sulfated proteoglycans, metal ions, glycoproteins and glycosaminoglycans) may assist in the formation of aggregates [75, 76]. These molecules/proteins help in intermolecular interactions and improve the efficiency packaging of regulated secretory proteins into SGs.

Storage State of Secretory Protein/Peptide Hormones In most endocrine cells, the secretion of peptide/protein hormones occurs through the regulatory secretory pathway, wherein the hormones are stored as "condensed" or "insoluble" aggregates intracellularly in membrane-enclosed electron-dense core SGs and released upon extracellular signals [54, 55, 65, 70]. The amount of these hormones released by SG varies from full content (mast cells) [58] to the minimal amount (endocrine) [56] depending upon the type of granule storage. The proteins inside most SGs include a dense-core, which essentially forms the releasable content and other proteins required for granule acidification, transport, targeting and fusion [52]. Although hormones and other regulated secretory proteins have been known to form aggregated structures inside SGs, the exact nature of these aggregates was not clear. Several studies suggest that the aggregates of SG are not amorphous, but possess a stable and distinct molecular organization, which might be crystalline [77–79] or composed of highly ordered aggregates [53, 80, 81]. These aggregates are also stable against increased temperature, mild detergents, enzymatic degradation and within a large pH range [66, 82]. Based on the fact that peptide/protein hormones in SG are not amorphous, but rather possess definite structure, Maji et al. proposed that aggregates of protein/peptides in secretory granule could be stored in an amyloid state [26]. To support this hypothesis, ~40 different protein/peptide hormones were studied for aggregation *in vitro* at pH 5.5 (secretory granule relevant pH). Using various biophysical techniques, the study showed amyloid formation by only a few hormones *in vitro* [26]. Interestingly, when these hormones were allowed to aggregate in the presence of the glycosaminoglycan (GAG) heparin, most of the hormones formed amyloid fibrils *in vitro* [26]. However, certain hormones such as prolactin (PRL) and adrenocorticotropic hormone (ACTH) did not form amyloid fibrils even in the presence of heparin [26]. It was hypothesized that these hormones might require some specific environmental conditions or helper molecules to form amyloid fibrils inside secretory granules. Interestingly, it was found that PRL formed amyloid along with its granule-specific GAG, chondroitin sulfate A (CSA), while ACTH formed amyloid along with its co-storage partner β-endorphin (β-end) in the presence of heparin. This study suggests that not only granule relevant conditions such as low pH or presence of GAGs can induce the formation of amyloid structures, but that each protein/peptide hormone might also require a specific condition for aggregation and amyloid formation in the context of their respective secretory granule formation.

The amyloid-like storage state of peptide/protein hormones in secretory granules can explain most of the processes of sorting, granule biogenesis, storage, and release. It is hypothesized that the high concentration of hormones in the trans-Golgi might initiate hormone aggregation and amyloid formation [26]. Since the conditions for amyloid formation for each hormone may vary, only specific hormones will be sorted into a specific secretory granule. For example, the amyloid formation of PRL requires CSA, whereas growth hormone (GH), a hormone structurally related to PRL, requires zinc (Zn(II)) ions [26, 83]. This suggests that irrespective of their structural similarity and location of storage, both PRL and GH are packed individually into the secretory granule along with their respective helper molecules. Since hormone aggregation starts (either in the presence or absence of helper molecules) in the trans-Golgi [84], amyloid formation might act as a sorting signal for selection of peptide/protein hormones to secretory granules. This process not only ensures the highest possible concentration through dense packing of SG, but also facilitates the exclusion of non-aggregating proteins from SG. Moreover, the strong ability of amyloid fibrils to interact with membranes [85] might further facilitate the granule maturation from trans-Golgi. Since the SG content has to be stored for extended periods of time, the integrity of the granular content has to be protected against degradation by proteases. For this purpose, the inherent stability of amyloid due to their cross-β-sheet rich structures will protect the protein/peptide content and can ensure a sustained release of functional monomer from the fibril ends [27]. Thus, the proposed amyloid-like storage of proteins/peptides is probably conserved across tissues and species (Fig. 8.3).

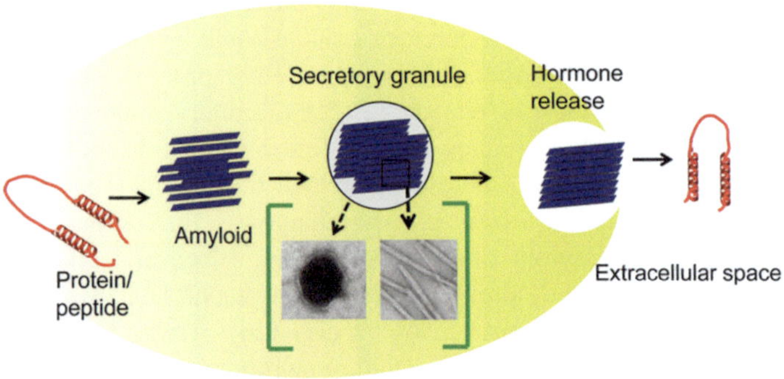

Fig. 8.3 Amyloid as a natural storage state of protein/peptide hormones in secretory granules. Protein/peptide hormones aggregate selectively into amyloid fibrils in the presence or absence of helper molecules and subsequently are packaged into secretory granules in neuroendocrine cells. Upon external stimulus the secretory granules release their contents into the extracellular space. The change in pH, ionic strength and dilution induces the dissolution of the amyloid fibrils and the monomeric forms of hormones are then released from the fibril ends. The monomeric proteins then refold either spontaneously or with the help of an unknown chaperone to attain their functional states

Hormone Amyloids Release Functional Monomer For amyloids to be the 'native state' of hormone storage, they are required to be reversible and be able to release functional monomers. As described in the previous section, amyloids fibrils generally exist in equilibrium with monomers in solution and the components of fibrils are constantly recycled by monomers in solution [29]. Hence, a decrease in monomer concentration might result in fibrils releasing more monomers to maintain the equilibrium. In line with this assumption, it was shown previously that amyloid fibrils of Substance P upon 1000-fold dilution could release monomers and completely dissolve within 48 h [86]. Similarly, in another study, amyloid fibrils of various GnRH analogs were also able to release monomers at physiological conditions [27]. Additionally, Maji et al. showed that many hormone amyloids were able to release monomers upon dilution by dialysis [26] and monomers released from amyloids were indeed in their native conformations and were functional. For example, during amyloid formation, corticotrophin-releasing hormone (hCRF) undergoes a secondary structural transition from α-helix to a β-sheet rich structure. When the amyloid fibrils of hCRF were dialyzed in 10 mM Tris-HCl pH 7.4, the fibrils were able to release α-helix-rich monomeric peptides. These released monomers were functional and could bind to hCRF receptors expressed by CHO cells with similar efficiency as those of monomeric CRF that had not undergone aggregation, and were able to activate cyclic adenosine monophosphates (cAMP) [26]. In line with these observations, the authors reasoned that after fusion of SG to the cell membrane, the release of granule content could be further triggered by a drastic change in pH within SGs (pH ~5.5) to that in the extracellular space (pH ~7.4). This difference in pH combined with dilution might initiate the release of the aggregated hormones from granules. Moreover, this study also showed that monomer release of hormone amyloids at pH 7.4 was slightly faster than the release at pH 6, indicating that the amyloid aggregates of hormones dissociate faster at neutral (or close to physiological) pH [26]. The release of monomer by increasing the pH may not be unique to functional hormone amyloids, as β2-microglobulin amyloid fibrils formed at pH 2 were also reported to release monomers at physiological pH 7.4 [30].

Another recent study by Riek and coworkers on β-end hormone amyloid reported a contrasting observation with respect to fibril disaggregation at pH > 6.0. In this study, β-end was found to form amyloid fibrils faster at pH 6.0 (and above), rather than pH 5.5, but the fibrils released monomers faster at pH 5.5. Here the authors suggest that at lower pH (5.5), the overall positive charge of the β-end increases to a higher value, which results in the slower aggregation rates of this peptide [87]. This study suggested that the pH of the solution as well as the overall charge of the protein/peptide (at a particular pH) could together influence the rate of amyloid fibril formation and fibril disaggregation.

Primary Structure Controlling the Amyloid Formation and Its Release The primary structure of a protein might play an important role in controlling protein secretion from secretory granules. Dannies and co-workers showed that a GH mutant, R183H, can be packaged into SGs in cells [88, 89], however, the secretory granules of this mutant hormone were not able to release monomeric hormone from these granules, which may cause GH deficiency syndromes [88–90] in children. The study suggests that even a single mutation in the protein sequence can interfere with the release of hormone from secretory granules. We hypothesized that primary/secondary structural modification due to mutation(s), posttranslational modifications and other modifications influencing the protein structure could eventually affect the storage (within SGs) and secretion (extracellular release) of the protein. In this aspect, recently our group showed how disulfide bonds might play a role in the amyloid formation and subsequent release of the monomeric peptide from amyloids, using a small cyclic peptide hormone somatostatin-14 as a model system [91].

Somatostatin-14 is a 14-residue peptide with a single disulfide bond between Cys 3 and Cys 14, which helps this hormone to attain a cyclic structure [91]. In this study, it was found that the disulfide-reduced 'non-cyclic' somatostatin (ncSST) aggregates faster into amyloid fibril than the cyclic SST in presence of heparin. This could be due to the higher accessibility of aggregation-prone regions in the linearized ncSST compared to cyclic SST. Molecular dynamics (MD) simulations of aggregation of SST and ncSST showed that most of the residues in ncSST as well as SST participated in H-bonding in the aggregated state of the peptide. But an interesting difference is that the ncSST displayed an organized H-bond network along the length of the peptide, while SST had the inter H-bond donors and acceptors scattered throughout the peptide. This difference in H-bond pattern observed at an atomistic level likely contributed to the experimentally observed variation in structure and morphology of ncSST and SST fibrils. The fibrils of ncSST showed a classical β-sheet signature using circular dichroims (CD) spectroscopy, while the SST fibrils have a mixed secondary structure [91]. Interestingly, the study showed that amyloid fibrils formed from ncSST are slower in releasing the monomers than SST, suggesting different monomer packing in both types of fibrils. The study further suggests that the presence of the disulfide bond limits the conformational flexibility of SST and enables it to form easily reversible amyloids [91]. Thus the correct disulfide bond formation and hence native structure of proteins/peptides may be necessary to regulate the aggregation and amyloid formation for secretory granule storage and its subsequent release.

Although a structural transition and β-sheet formation are required for amyloid formation of protein/peptide hormones, the peptide segment(s) involved in the formation of the cross-β-sheet structure may also play an important role in disaggregation (for monomer release). For small peptides, the formation of correct structure after monomer release might be almost instantaneous. However, for larger protein hormones, if a major segment of the protein is involved in the amyloid formation, the release as well as subsequent folding into the native and functional conformation, might be more difficult. Alternatively, if only a small segment of the large protein is involved in the formation of cross-β-sheet rich structure, rapid refolding to the native state could be achieved. Furthermore, the stability of the resultant fibrils might also control the release of the active protein hormone. In protein hormones such as GH and PRL, the major secondary structure characteristics of the protein are not altered after fibril formation when measured using CD spectroscopy [26, 83]. Both PRL and GH are highly helical proteins and after amyloid formation, these proteins show also a mostly helical conformation, with a slight decrease in helicity [26, 83]. One of the possible explanations for this observation is that only a small segment of these proteins might be involved in amyloid formation, while most of the other segments remain in their native conformation. This will ensure not only facile monomer release but also near-instantaneous folding into the native state for binding to the receptor and other functions. Several other proteins/peptides were also previously reported to show the formation of amyloid-like fibrils without significant secondary structural changes [26, 92–94]. Therefore, we propose that for an amyloid to be an *in vivo* storage state inside the secretory granules, the structural change for amyloid formation may not be aberrant, but rather highly regulated at least for large proteins, so that substantial amount of native structure could be maintained within the amyloid state inside secretory granules, facilitating functional monomer release. Moreover, the monomer release and folding could be also assisted by chaperones or chaperone-like molecules.

Role of Helper Molecules and Buffer Composition on Amyloid Formation and Release Aggregation-inducing helper molecules (e.g. GAGs, divalent ions) and other biomolecules coexisting within the SG are known to facilitate hormone aggregation and amyloid formation. Several *in vitro* aggregation studies of protein/peptide hormones have shown that GAGs such as heparin and chondroitin sulfates induce aggregation [26, 91, 95, 96], wherein the protein and GAGs are used at particular ratios in solution. Although the exact mechanism of amyloid formation by protein/peptides in the presence of heparin is currently unknown, it is proposed that heparin's linear repeat and the electrostatic interactions between protein and GAGs can aid in protein aggregation and amyloid formation. Additionally, a recent study by Christensen et al. on prolactin aggregation suggested that the pH of the surrounding environment could affect the overall charge on the protein hormone, which in turn facilitates the interaction with the helper molecule (GAGs), leading to its aggregation *in vitro* [96]. This could be an important mechanism of hormone condensation inside SGs *in vivo*. In addition to the role in amyloid formation, the GAGs could also influence the release of monomeric hormones from aggregates. For example, a recent study by Nespovitaya et al. showed that the β-end amyloid fibrils formed in the absence of heparin released monomers in less than one hour, whereas the β-end fibrils formed in the presence of heparin released monomers in a sustained manner up to 24 h [87]. Similarly, a previous study from our group on the aggregation of glucagon-like peptides 1 and 2 (GLP1 and GLP2), also emphasized the role of heparin in modulating monomer release from fibrils [95]. In this study, a slower monomer release of GLP1 fibrils was observed, when compared to GLP2 fibrils, both types of fibrils having been formed in the presence of heparin [95]. Interestingly it was shown that heparin participated in the fibril formation of GLP1, but not in GLP2. This difference was found to be due to the presence of a basic-non basic-basic (B-X-B) motif present in GLP1 (and not in GLP2), to which heparin could bind to induce fibrillation, and also further aid fibril stabilization. Taken together, these studies suggest that GAGs can significantly influence the fibril architecture, and as a consequence govern the rate of monomer release from amyloid fibrils.

Similar to GAGs, metal ions can also act as helper molecules in protein aggregation and amyloid formation in the SGs. A recent study from our group showed that amyloid fibrils formed in the presence of Zn(II) can release monomeric GH in native helical conformation upon dialysis of GH fibrils against 10 mM Tris pH 7.4 [83]. Further, this study also showed that GH is co-stored with Zn(II) ions *in vivo*, suggesting the role of Zn(II) in both *in vivo* and *in vitro* amyloid formation of GH. Another study likewise reported the aggregation of prolactin in the presence of heparin, chondroitin sulfate A and Zn(II) ions [96]. These studies suggest that hormones require a specific helper molecule (or a combination of molecules) not only for controlling the formation of individual granules and the content of the granules but also for the subsequent release of the contents.

In addition to the helper molecules and ions, the buffer constituents of the solution are also known to be important for amyloid formation. An extensive study by Nespovitaya et al. on β-end fibrillation under various physiologically relevant environmental conditions, reported that helper molecules, pH and buffer constituents could differentially affect the amyloid formation of the peptide [87]. In this study, the authors showed that β-end was able to form amyloid by itself in multivalent ion buffers such as phosphate, sulfate, and citrate but not in ammonium, acetate and Tris. However, the addition of heparin to β-end in ammonium acetate buffer resulted in its amyloid formation [87]. Nespovitaya et al. also found that the amyloid fibrils formed in multivalent buffers and in the presence of heparin differed in both structure and morphology. This study and previous works by Maji et al. [26, 27] overall suggest that the change of the immediate environment that the secretory proteins encounter along the secretory pathway in terms of pH, GAGs, metal ions, concentration of protein and even ion constituents modulate the amyloid aggregation and subsequent disassembly of hormones.

In line with this evidence, it can be envisioned how the amyloid structure enables the densest peptide-packing possible and is suitable for long-term storage. Moreover, the common structural signature of cross-β-sheets can by itself act as sorting signal for packing of hormone/neuropeptide aggregates into secretory granules. A change in environmental conditions can dissociate these protein nanofibrils and could release functional monomers, thus enabling amyloid fibrils to act as depot both *in vitro* and *in vivo*.

8.4 Amyloid as Long-Acting Depot Formulations

Developing novel biocompatible scaffolds for drug delivery applications is an emerging area in modern science. Small molecular and protein/peptide drugs have short half-life and low bioavailability *in vivo*. Such drugs are often delivered through frequent injections or subcutaneous administrations and infusions to prevent them from degradation [97, 98]. To overcome the delivery related problems, new drug formulations have been created that continuously release and maintain an active drug concentration over an extended period. In this regard, administrating protein drugs like insulin and TGF-β3 in self-assembled crystalline form has shown controlled delivery of these drugs [99, 100]. However, protein crystals tend to be fragile and can disintegrate easily, requiring sophisticated methods to preserve and store them. Hence an alternative peptide/protein self-assembly system needs to be employed, that can deliver active and functional peptide/protein drugs in a sustained manner for an extended period. Since a growing amount of studies have already established that not all amyloids are toxic and can serve as functional protein entity, it is possible to imagine that amyloid fibrils can act as a storage depot of various protein/peptide hormones. In order to explore this possibility, a study utilized the short-acting as well as long-acting analogs of gonadotropin-releasing hormone (GnRH) [27]. When these GnRH analogs were incubated *in vitro*, all long-acting analogs were found to form fibrillar aggregates that bind to amyloid specific dyes ThT and CR. In contrast, only a few short-acting GnRH analogs formed amyloid fibrils *in vitro*. Furthermore, when the monomer releasing capabilities of these amyloid fibrils were studied, the long-acting analogs showed sustained release of monomeric hormone from amyloids, while the fibrils of short-acting analogs either released monomer immediately upon dilution (similar to non-aggregated peptide) or released very slowly such that effective drug concentration cannot be reached [27]. This study clearly indicates that packing and stability of each peptide fibril might dictate the monomer release, which could possibly be determined by the differences in their amino acid composition.

Many earlier studies have reported that amyloid fibrils can be stabilized in the presence of serum amyloid protein P and GAGs [101, 102]. If the protein/peptide drugs are administered in the form of amyloid fibrils, a more sustained and controlled release can be expected, due to the combined effect of the highly ordered structure of the amyloids and the GAG-mediated stabilization of the fibrils. The GnRH analogs mentioned earlier were shown to form amyloid fibrils in the presence as well as absence of heparin [27]. Further, when the release assay was performed with fibrils formed in the presence or absence of heparin, fibrils formed in the presence of heparin displayed a slower release pattern, compared to GnRH analog fibrils formed in the absence of heparin. This observation was found to apply for other peptide hormones as well. For instance, two independent studies (as mentioned in the previous section) showed that the amyloid fibrils of β-endorphin or GLP1 hormone formed in the presence of heparin [87, 95], displayed a slower monomer release profile compared to the respective hormone amyloid formed in the absence of heparin. These studies suggest that GAGs present in body could also influence the formation, stability as well as controlled release capability of administered peptide fibrils [27, 101–104].

***In Vivo* Release of Monomers from Hormone Amyloids** For amyloids to act as drug reservoirs, they should release functional monomeric proteins *in vivo*. While a large range of studies has shown the reversible nature of amyloids *in vitro*, only a handful of studies exist, which demonstrate that amyloid can release functional monomers *in vivo*. In one such study, long-acting GnRH analog (L14) monomers in D-Mannitol were injected subcutaneously into rats, which resulted in *in vivo* amyloid formation at the injection site [27]. GnRH analogs are generally used as suppressing agent of luteinizing hormone (LH) in prostate tumors [105]. The *in vivo* amyloid formation of L14 was confirmed by CR staining and birefringence of the tissue section on the injection site. Thus, the long-acting capability of L14 could be due to its amyloid formation *in vivo* and release of functional monomers from this amyloid depot. This study thereby suggests that the amyloid formation by amyloidogenic peptide hormones could be further facilitated by *in vivo* conditions, which may also play a role in the controlled release of the peptide drug [27]. Moreover, to directly demonstrate that amyloid formation could influence the duration of action of GnRH analogs, a short-acting GnRH analog (S2) was chosen, which did not form amyloid even after 8 days of incubation but formed amyloid after long incubation (Fig. 8.4a). Unlike the L14 analog, the introduction of monomeric GnRH analog S2 did not form amyloid *in vivo* at the site of injection [27]. However, after the subcutaneous injection of aged S2 fibrils in rat, the localization of S2 fibrils was confirmed by CR staining and birefringence at the injected site [27]. When aged S2 fibrils and S2 monomer solutions were subcutaneously injected in a male rat (castrated 10 days prior the experiment), both aged fibrillar and monomeric S2 were able to suppress (measured by radioimmunoassay) LH in the blood of the animals after 24 h of administration. After 48 h of administration, the fibrillar GnRH analogs showed more LH suppressive activity compared to monomeric GnRH analogs. It is interesting to note that, although at 72 h, the monomeric S2 almost lost its LH suppressing activity, the fibrillar S2 was still able to suppress LH significantly at this time (Fig. 8.4b) [27]. This particular study directly shows the ability of amyloid fibrils to act as drug depot *in vivo* and further suggests that amyloid formation of protein/peptide drugs can prolong their activity for sustainable drug delivery applications (Fig. 8.4c) [27].

Toxicity and Amplification of Toxic Protein Aggregates by Hormone Amyloids
Since amyloids are originally associated with cell death in various neurodegenerative diseases such as Alzheimer's and Parkinson's, it is a potential concern that the hormone amyloids may exhibit toxicity upon formation *in vivo* or administration of hormone amyloids as drug depots. However, recently it was suggested that amyloid fibrils, the end-products of the aggregation pathway, are the less toxic species in protein aggregation reactions and protein/peptide oligomers (pathway intermediates) are more toxic species, responsible for cell death [14–17]. Moreover, it was also suggested that amyloid could be a nontoxic reservoir of toxic protein/peptide oligomers [14, 106, 107]. Various studies during the last decades showed that amyloid fibrils are not only less toxic than smaller aggregated species, but also that many amyloids have evolved in nature to perform native functions of the host organism [5, 32] (discussed earlier under Sect. 8.2). Moreover, the toxicity of amyloid fibrils could depend on the protein/peptide sequences that

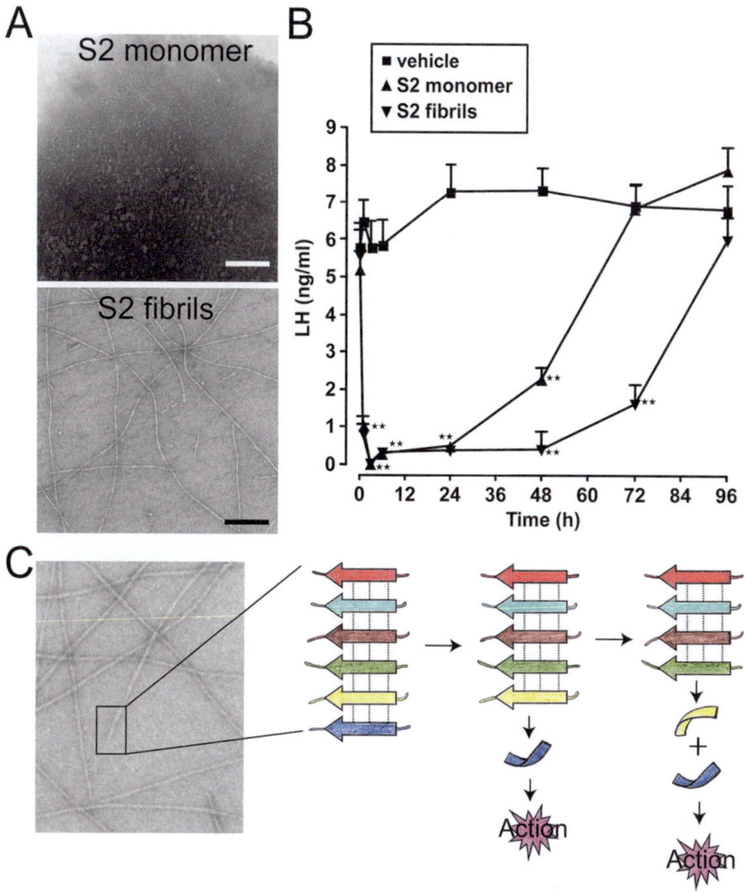

Fig. 8.4 Amyloid fibrils as depots for GnRH analogs. (**a**) TEM of short-acting GnRH analog S2 incubated for 8 days (top) and 30 days (bottom). Scale bar indicates 200 nm. (**b**) Subcutaneous administration of amyloid fibrils of GnRH analog S2 prolonged the inhibition of LH release compared to S2 in monomeric form. (**c**) Schematic representation of the mode of action of amyloid forms of GnRH analogs. The stable amyloid fibrils formed prior (see TEM image) or shortly after subcutaneous injection and gradually release monomeric functional analogs at their ends. (Reprinted from Ref. [27], with kind permission from PLOS Biology)

constitute the amyloid. This is further confirmed by the observation that while amyloid fibrils of the Aβ peptide are moderately toxic, most of the amyloid fibrils of GnRH analogs are nontoxic in nature [27]. Similar to GnRH analogs, many hormone amyloids associated with SG formation also were reported to be nontoxic [26]. However, for a possible formulation of peptide/protein fibrils as drug depot, the toxicity of the fibrils should be tested both *in vitro* as well as *in vivo*. Another concern of using amyloid depots for sustained drug release is that amyloid fibrils from drug depots could seed their own monomeric protein/peptide present in the body and amplify the amyloid form. This could be detrimental to the proper functioning of the normal hormone. However, Maji and colleagues observed that mature fibrils of long-acting GnRH (L14) could not template aggregation of wild type GnRH, suggesting that amyloid fibrils of GnRH analogs might be incapable to seed and amplify GnRH amyloid formation [27]. Similarly, it is also possible that the amyloid depot of the particular hormone (or analogs) can 'cross-seed' other protein/peptides associated with diseases and induce their aggregation, thereby increasing the load of cytotoxic protein/peptide aggregates in the body. However, to cross-seed, both the seed and host protein(s) require high sequence similarity as well as amyloid forming capability [23, 108, 109]. This could be a probable mechanism by which nature prevents the disease-associated amyloids from cross-seeding the various other proteins existing in the body. Nevertheless, one has to test the possibility of amyloid drug depot to cross-seed other proteins, which are potentially amyloidogenic and are associated with diseases. In the study by Maji et al. to demonstrate the safety of GnRH amyloid against cross-seeding of disease associated protein aggregation, a small amount (5%) of mature fibrils of L20 GnRH analog was incubated with the wild-type monomers of α-Synuclein (α-Syn), a protein associated with Parkinson's disease [110–114]. The study showed that although α-Syn spontaneously aggregated and formed amyloid fibrils, the fibril seeds of long-acting hormone L20 did not cross-seed the aggregation and amyloid formation of α-Syn [27]. This study indicates that rationally designing protein analogs that would not seed wild-type proteins (to form toxic aggregates) can also be used as a strategy for developing amyloid formulation of peptide/protein drugs. Moreover, under physiological conditions, different proteins are often compartmentalized, thus minimizing the possibility of encounter and interaction between protein amyloid as depot with other disease-associated proteins such as α-Syn or Aβ.

The reversibility of amyloid fibrils can also be employed for various other functional purposes; for example, in the controlled release of peptide antibiotics. In this regard, Pertinhez et al. have recently shown that the aggregates of an antibody-derived decapeptide KP (killer peptide) can provide targeted delivery as well as possesses slow release properties [115]. The study showed that disulfide-bonded dimers of the synthetic peptide KP (AKVTMTCSAS) are the active form displaying microbicidal activity and these dimers can self-assemble to form higher order aggregates with fibrillar morphology and β-sheet structure over time [115]. The dilution or incubation of these KP aggregates at 35 °C resulted in the soluble form. Moreover, similar to other amyloidogenic peptides, the KP also rapidly self-assembles in the presence of GAGs such as 1, 3 β-glucan, and the aggregates formed in presence of this GAG have slow-releasing capability.

In addition to amyloid fibrils, self-assembled protein/peptide oligomers can also be used as drug depots. A recent study by Gupta et al. suggested that self-assembled oligomers of insulin can be used as a drug depot, after showing that oligomers formed during the aggregation of insulin were capable of releasing functional insulin monomers [116]. In patients with type 1 diabetes, insulin production is generally insufficient to control the blood glucose levels. Insulin injections are administered in these patients for a glycemic control; however, multiple dosages are required, as the half-life of the injected peptide is short [117]. In their study, Gupta et al. characterized three intermediate states in insulin amyloid formation and designated them as supramolecular insulin assembly (SIA) I, II and III. The SIA-II intermediate formed at 12–16 h of incubation of insulin at pH 7.0 was found to release monomeric insulin at a rate that is needed for maintaining the basal level of insulin in blood [116]. The intermediates formed at pH 2.0 released monomers at a slow rate and the amyloid fibrils formed at pH 7.0 or 2.0, did not release any monomers. This observation suggests that the release capability of peptide/protein either from amyloid or oligomeric assemblies depends on the peptide sequences and the conditions used for making the assemblies.

Amyloid-Based Biomaterials as Drug Delivery Vehicles Amyloid fibrils as supramolecular assemblies display charged and hydrophobic surfaces, which could bind both small molecules and proteins [118–121]. The unique surface properties of amyloid fibril networks can be utilized for entrapping drugs, enzymes and proteins, thus amyloid fibrils could be used for developing drug delivery vehicles. Further, amyloid-based scaffolds or hydrogels can be rationally designed such that they can release the encapsulated drugs/protein at the site of action. In this context, Zhang and coworkers have used hydrogels from self-assembling RADA16 peptides as drug delivery vehicles of small molecules and proteins such as trypsin inhibitor, lysozyme, immunoglobulin G (IgG) and bovine serum albumin (BSA) [122, 123]. Moreover, the flanking regions of amyloid-forming peptide sequences can be modified appropriately such that the designed fibril matrix can deliver its payload at target sites. For example, Mazza et al. modified an amphiphilic dalargin (an opioid receptor agonist) derivative with a palmitoyl moiety, which self-assembled into β-sheet fibrils after sonication in water. The intravenous injection of the modified dalargin nanofibrils allowed it to cross the blood-brain barrier and deliver active dalargin into the brain [124].

Since many small molecule drugs are hydrophobic in nature, an increasing number of studies are attempting to encapsulate small-molecule drugs in hydrogels derived from amyloid fibrils for their successful delivery to target sites. For example, a recent study successfully demonstrated the use of protein nanofibril microgels composed of amyloid fibrils for encapsulation and sustained release of antibiotics [125]. In this work, the nanofibril microgels were produced using a water-in-oil method and were tested for their efficacy to act as carriers of small molecules such as dyes and antibiotics. Using this system, it was observed that the drug molecules penicillin and tetracycline were encapsulated in the nanofibril microgels and could be released locally, resulting in increased anti-microbial action. Moreover, these microgels were proven to be non-cytotoxic to human cell lines, and thus are biocompatible [125]. Furthermore, the biocompatibility of drugs can be further enhanced when delivered within an amyloid matrix. For example, Li et al. immobilized carbon nanotubes (NT) within a β-lactoglobulin gel matrix, which made the NT biocompatible. The study further showed that NT can be released from the protein shell by enzymatic digestion and these NT can be used for killing cancer cells by heating the NT by near infrared light indicating possible applications in cancer therapy [126]. Recently, it was also shown that hybrid materials composed of amyloid fibrils-iron colloidal dispersion could be used as oral iron delivery systems, which increases the bioavailability of iron *in vivo* [127].

Intriguingly, it is not always the amyloid fibrils that can modulate the biodistribution or availability of the encapsulated drug/small molecules. The encapsulated drugs or small molecules can also modulate the properties of the gel matrix by interacting with the amyloid fibrils. In a recent study, lysozyme was allowed to form amyloid in the presence of the β1 receptor antagonist drugs, propanolol, timolol and atenolol, and it was found that the drugs modulated the fibril architecture, leading to altered release profiles of the drugs [128]. These studies suggest that the interaction between the various payload and amyloid fibrils can be utilized to fine tune the architecture of amyloid-based drug delivery vehicles to obtain desirable release profiles. Furthermore, it has been shown that the amyloid fibrils can be fabricated into different matrices for various biotechnological applications such as drug delivery. For example, β2-microglobulin amyloid fibrils were fabricated into a nanoporous and high mechanical strength protein matrix, which is suggested to be useful in tissue engineering and drug delivery applications [129]. In another example, hydrogels were prepared from α-synuclein curly-amyloid fibrils and were used for enzyme entrapment. The enzyme trapped inside the amyloid hydrogel was able to withstand heat treatment and maintain enzymatic activity, which suggests a plausible use in therapeutic delivery [130]. Moreover, the entrapping of enzymes and other small molecules inside nano-fibrillar scaffolds of amyloids could improve their stability and activity. For example, organophosphate hydrolase (OPH), an enzyme important for bioremediation and chemical detoxification was covalently cross-linked using glutaraldehyde on bovine insulin fibrils. This enhanced the stability of the encapsulated enzyme at higher temperatures compared to its free counterpart [131]. Such enzyme encapsulated amyloid hydrogels need not always release the enzyme molecule. Since in hydrogels, free diffusion of molecules occurs within and also with the surrounding environment, the substrate molecules can diffuse inside the hydrogel within the proximity of the enzyme molecules and be converted into product, which can then diffuse out of the hydrogel. Alternatively, to enhance the stability of enzymes, amyloid-forming peptide segments can be engineered onto medically relevant enzymes such that these engineered enzymes can form amyloid fibrils retaining the enzymatic activity [92]. These engineered enzymes will be stable and may perform substrate conversion for longer duration than soluble enzymes.

We, therefore suggest that amyloid-based technology could be used for drug delivery in two different ways (Fig. 8.5). In one case, the amyloid fibril form of the protein/peptide drug could be administered directly, as suggested by previous studies [27], which will allow the fibrils to release monomers *in vivo* in a controlled manner. The other possibility is to construct an amyloid-based scaffold where small molecules or drugs can be encapsulated. The latter method can also be used for multiple drug delivery purposes as well as this would provide protein/peptide drug after dissociation and also will deliver the drug molecules of interest which are

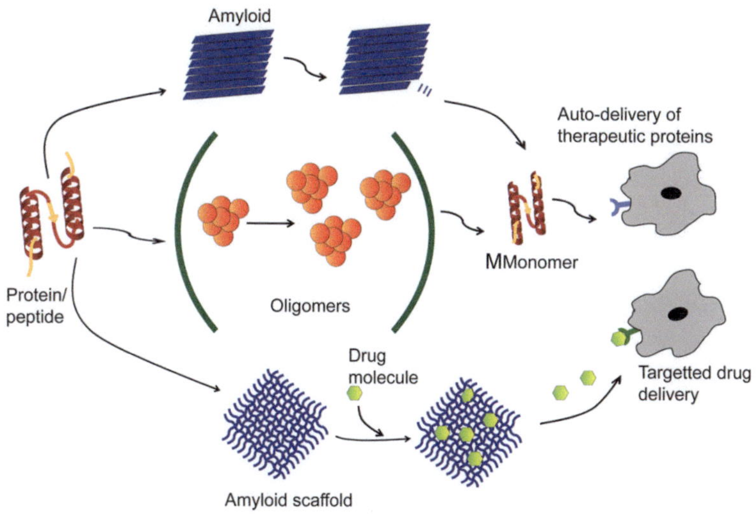

Fig. 8.5 Amyloids as drug delivery systems. Schematic representation of different types of drug delivery using amyloid fibrils. Hormones and other therapeutic peptides can self-assemble to form amyloid fibrils, which can release functional and monomeric peptide/protein from fibril ends after administration (top). Similarly, nontoxic oligomers of protein/peptides can also be used as drug depots (centre). Amyloid-based scaffolds can also be used as drug carriers. The drug molecules (yellow) entrapped within amyloid networks can be released in a controlled manner after administration (bottom). The released drug peptide can bind to specific receptors on the cell surface to perform its intended action

embedded within the amyloid network. The amyloid networks could immobilize drugs, small molecules, protein/peptides including enzymes and function as drug delivery carriers (Fig. 8.5). The molecules that are entrapped inside amyloid hydrogels will be protected from heat and enzymatic degradation [27, 130]. Furthermore, biological macromolecules (e.g. amyloid P and glycosaminoglycans) present inside the host organism may bind to amyloid-carriers and increase their *in vivo* stability as well as reducing the potential immune response against the formulation [101–103]. In addition, such formulations are more easily administered by subcutaneous injection. Therefore, amyloid-based drug delivery systems are devoid of complicated manufacturing procedures and may ensure controlled release of the peptide/drug for an extended time period without recurring drug administration or surgery for implanting depots [27].

8.5 Conclusion

Amyloids are highly ordered protein/peptide aggregates that are implicated in both disease and functions in host organisms [5, 32]. Despite the toxicity of some amyloids, many recent studies however, suggest native biological functions such as protein/peptide storage inside secretory granules of mammals [26]. The higher order structural organization of protein/peptide in a cross-β-sheet rich motif and stability against various harsh physical/chemical conditions such as extreme pHs, temperature and proteases, makes amyloid attractive for versatile nanobiotechnological applications such as drug delivery and tissue engineering [132–134]. Since as few as two amino acid residues are sufficient for amyloid-like assembly [135], tagging short amyloid sequences onto functional proteins can be exploited for protein formulation similar to the case of GnRH analogs [27]. Moreover, many small peptide-based amyloids can also be utilized for encapsulating different drugs/proteins for their possible delivery. We expect to witness many more applications of designed amyloids for diverse biomedical applications in coming years.

Acknowledgments The authors wish to acknowledge DBT (BT/PR9797/NNT/28/774/2014) Government of India and Wadhwani Research Centre for Bioengineering (WRCB) for financial support.

References

1. Jeremy M, Berg JLT, Stryer L (2002) Biochemistry, 5th edn. W H Freeman, New York
2. Dobson CM (2004) Principles of protein folding, misfolding and aggregation. Semin Cell Dev Biol 15(1):3–16
3. Kirschner M, Mitchison T (1986) Beyond self-assembly: from microtubules to morphogenesis. Cell 45(3):329–342
4. Fletcher DA, Mullins RD (2010) Cell mechanics and the cytoskeleton. Nature 463(7280):485–492
5. Chiti F, Dobson CM (2006) Protein misfolding, functional amyloid, and human disease. Annu Rev Biochem 75:333–366
6. Sunde M, Blake C (1997) The structure of amyloid fibrils by electron microscopy and X-ray diffraction. Adv Protein Chem 50:123–159
7. Sunde M, Serpell LC, Bartlam M, Fraser PE, Pepys MB, Blake CC (1997) Common core structure of amyloid fibrils by synchrotron X-ray diffraction. J Mol Biol 273(3):729–739
8. LeVine H III (1993) Thioflavine T interaction with synthetic Alzheimer's disease β-amyloid peptides: detection of amyloid aggregation in solution. Protein Sci 2:404–410
9. Westermark GT, Johnson KH, Westermark P (1999) Staining methods for identification of amyloid in tissue. Methods Enzymol 309:3–25
10. Meersman F, Dobson CM (2006) Probing the pressure-temperature stability of amyloid fibrils provides new insights into their molecular properties. Biochim Biophys Acta 1764(3):452–460
11. Mesquida P, Riener CK, MacPhee CE, McKendry RA (2007) Morphology and mechanical stability of amyloid-like peptide fibrils. J Mater Sci Mater Med 18(7):1325–1331
12. Zurdo J, Guijarro JI, Dobson CM (2001) Preparation and characterization of purified amyloid fibrils. J Am Chem Soc 123(33):8141–8142
13. Uversky VN, Fink AL (2004) Conformational constraints for amyloid fibrillation: the importance of being unfolded. Biochim Biophys Acta 1698(2):131–153
14. Haass C, Selkoe DJ (2007) Soluble protein oligomers in neurodegeneration: lessons from the Alzheimer's amyloid beta-peptide. Nat Rev Mol Cell Biol 8(2):101–112

15. Kayed R, Head E, Thompson JL, McIntire TM, Milton SC, Cotman CW, Glabe CG (2003) Common structure of soluble amyloid oligomers implies common mechanism of pathogenesis. Science 300(5618):486–489
16. Bucciantini M, Giannoni E, Chiti F, Baroni F, Formigli L, Zurdo J, Taddei N, Ramponi G, Dobson CM, Stefani M (2002) Inherent toxicity of aggregates implies a common mechanism for protein misfolding diseases. Nature 416(6880):507–511
17. Walsh DM, Klyubin I, Fadeeva JV, Cullen WK, Anwyl R, Wolfe MS, Rowan MJ, Selkoe DJ (2002) Naturally secreted oligomers of amyloid β protein potently inhibit hippocampal long-term potentiation *in vivo*. Nature 416(6880):535–539
18. Harper JD, Lansbury PT Jr (1997) Models of amyloid seeding in Alzheimer's disease and scrapie: mechanistic truths and physiological consequences of the time-dependent solubility of amyloid proteins. Annu Rev Biochem 66:385–407
19. Jarrett JT, Lansbury PT Jr (1993) Seeding "one-dimensional crystallization" of amyloid: a pathogenic mechanism in Alzheimer's disease and scrapie? Cell 73:1055–1058
20. Ferrone F (1999) Analysis of protein aggregation kinetics. Methods Enzymol 309:256–274
21. Naiki H, Hashimoto N, Suzuki S, Kimura H, Nakakuki K, Gejyo F (1997) Establishment of a kinetic model of dialysis-related amyloid fibril extension in vitro. Amyl Int J Exp Clin Investig 4(4):223–232
22. Serio TR, Cashikar AG, Kowal AS, Sawicki GJ, Moslehi JJ, Serpell L, Arnsdorf MF, Lindquist SL (2000) Nucleated conformational conversion and the replication of conformational information by a prion determinant. Science 289(5483):1317–1321
23. Wright CF, Teichmann SA, Clarke J, Dobson CM (2005) The importance of sequence diversity in the aggregation and evolution of proteins. Nature 438(7069):878–881
24. Wetzel R (2006) Kinetics and thermodynamics of amyloid fibril assembly. Acc Chem Res 39(9):671–679
25. Knowles TPJ, Vendruscolo M, Dobson CM (2014) The amyloid state and its association with protein misfolding diseases. Nat Rev Mol Cell Biol 15(6):384–396
26. Maji SK, Perrin MH, Sawaya MR, Jessberger S, Vadodaria K, Rissman RA, Singru PS, Nilsson KP, Simon R, Schubert D, Eisenberg D, Rivier J, Sawchenko P, Vale W, Riek R (2009) Functional amyloids as natural storage of peptide hormones in pituitary secretory granules. Science 325(5938):328–332
27. Maji SK, Schubert D, Rivier C, Lee S, Rivier JE, Riek R (2008) Amyloid as a depot for the formulation of long-acting drugs. PLoS Biol 6(2):e17
28. MacPhee CE, Dobson CM (2000) Chemical dissection and reassembly of amyloid fibrils formed by a peptide fragment of transthyretin. J Mol Biol 297(5):1203–1215
29. Carulla N, Caddy GL, Hall DR, Zurdo J, Gairi M, Feliz M, Giralt E, Robinson CV, Dobson CM (2005) Molecular recycling within amyloid fibrils. Nature 436(7050):554–558
30. Tipping KW, Karamanos TK, Jakhria T, Iadanza MG, Goodchild SC, Tuma R, Ranson NA, Hewitt EW, Radford SE (2015) pH-induced molecular shedding drives the formation of amyloid fibril-derived oligomers. Proc Natl Acad Sci U S A 112(18):5691–5696
31. Cremades N, Cohen SI, Deas E, Abramov AY, Chen AY, Orte A, Sandal M, Clarke RW, Dunne P, Aprile FA, Bertoncini CW, Wood NW, Knowles TP, Dobson CM, Klenerman D (2012) Direct observation of the interconversion of normal and toxic forms of α-synuclein. Cell 149(5):1048–1059
32. Fowler DM, Koulov AV, Balch WE, Kelly JW (2007) Functional amyloid – from bacteria to humans. Trends Biochem Sci 32(5):217–224
33. Chapman MR, Robinson LS, Pinkner JS, Roth R, Heuser J, Hammar M, Normark S, Hultgren SJ (2002) Role of Escherichia coli curli operons in directing amyloid fiber formation. Science 295(5556):851–855
34. Otzen D, Nielsen PH (2008) We find them here, we find them there: functional bacterial amyloid. Cell Mol Life Sci 65(6):910–927
35. Hall-Stoodley L, Costerton JW, Stoodley P (2004) Bacterial biofilms: from the natural environment to infectious diseases. Nat Rev Microbiol 2(2):95–108

36. Oh J, Kim JG, Jeon E, Yoo CH, Moon JS, Rhee S, Hwang I (2007) Amyloidogenesis of type III-dependent harpins from plant pathogenic bacteria. J Biol Chem 282(18):13601–13609
37. Alteri CJ, Xicohtencatl-Cortes J, Hess S, Caballero-Olin G, Giron JA, Friedman RL (2007) Mycobacterium tuberculosis produces pili during human infection. Proc Natl Acad Sci U S A 104(12):5145–5150
38. Gebbink MF, Claessen D, Bouma B, Dijkhuizen L, Wosten HA (2005) Amyloids – a functional coat for microorganisms. Nat Rev Microbiol 3(4):333–341
39. Talbot NJ (2003) Aerial morphogenesis: enter the chaplins. Curr Biol 13(18):R696–R698
40. Ramsook CB, Tan C, Garcia MC, Fung R, Soybelman G, Henry R, Litewka A, O'Meally S, Otoo HN, Khalaf RA, Dranginis AM, Gaur NK, Klotz SA, Rauceo JM, Jue CK, Lipke PN (2010) Yeast cell adhesion molecules have functional amyloid-forming sequences. Eukaryot Cell 9(3):393–404
41. Hammer ND, Wang X, McGuffie BA, Chapman MR (2008) Amyloids: friend or foe? J Alzheimers Dis 13(4):407–419
42. Osherovich LZ, Weissman JS (2002) The utility of prions. Dev Cell 2(2):143–151
43. True HL, Lindquist SL (2000) A yeast prion provides a mechanism for genetic variation and phenotypic diversity. Nature 407(6803):477–483
44. Uptain SM, Lindquist S (2002) Prions as protein-based genetic elements. Annu Rev Microbiol 56:703–741
45. Chien P, Weissman JS, DePace AH (2004) Emerging principles of conformation-based prion inheritance. Annu Rev Biochem 73:617–656
46. Iconomidou VA, Vriend G, Hamodrakas SJ (2000) Amyloids protect the silkmoth oocyte and embryo. FEBS Lett 479(3):141–145
47. Fowler DM, Koulov AV, Alory-Jost C, Marks MS, Balch WE, Kelly JW (2006) Functional amyloid formation within mammalian tissue. PLoS Biol 4(1):100–107
48. Cherny I, Gazit E (2008) Amyloids: not only pathological agents but also ordered nanomaterials. Angew Chem Int Ed Engl 47(22):4062–4069
49. Knowles TP, Mezzenga R (2016) Amyloid fibrils as building blocks for natural and artificial functional materials. Adv Mater 28(31):6546–6561
50. Palade G (1975) Intracellular aspects of the process of protein synthesis. Science 189(4200):347–358
51. Lacy PE (1975) Endocrine secretory mechanisms. A review. Am J Pathol 79(1):170–188
52. Tooze SA (1998) Biogenesis of secretory granules in the trans-Golgi network of neuroendocrine and endocrine cells. Biochim Biophys Acta 1404(1–2):231–244
53. Arvan P, Castle D (1998) Sorting and storage during secretory granule biogenesis: looking backward and looking forward. Biochem J 332(Pt 3):593–610
54. Kelly RB (1985) Pathways of protein secretion in eukaryotes. Science 230(4721):25–32
55. Kelly RB (1987) Protein transport. from organelle to organelle. Nature 326(6108):14–15
56. Stefan Y, Meda P, Neufeld M, Orci L (1987) Stimulation of insulin secretion reveals heterogeneity of pancreatic B cells in vivo. J Clin Invest 80(1):175–183
57. Loh YP, Kim T, Rodriguez YM, Cawley NX (2004) Secretory granule biogenesis and neuropeptide sorting to the regulated secretory pathway in neuroendocrine cells. J Mol Neurosci 22(1–2):63–71
58. Hide I, Bennett JP, Pizzey A, Boonen G, Bar-Sagi D, Gomperts BD, Tatham PE (1993) Degranulation of individual mast cells in response to Ca2+ and guanine nucleotides: an all-or-none event. J Cell Biol 123(3):585–593
59. Turner MD, Rennison ME, Handel SE, Wilde CJ, Burgoyne RD (1992) Proteins are secreted by both constitutive and regulated secretory pathways in lactating mouse mammary epithelial cells. J Cell Biol 117(2):269–278
60. Kim T, Gondre-Lewis MC, Arnaoutova I, Loh YP (2006) Dense-core secretory granule biogenesis. Physiology (Bethesda) 21:124–133
61. Tooze SA (1991) Biogenesis of secretory granules. Implications arising from the immature secretory granule in the regulated pathway of secretion. FEBS Lett 285(2):220–224

62. Wu MM, Grabe M, Adams S, Tsien RY, Moore HP, Machen TE (2001) Mechanisms of pH regulation in the regulated secretory pathway. J Biol Chem 276(35):33027–33035
63. Steiner DF, Smeekens SP, Ohagi S, Chan SJ (1992) The new enzymology of precursor processing endoproteases. J Biol Chem 267(33):23435–23438
64. Kuliawat R, Prabakaran D, Arvan P (2000) Proinsulin endoproteolysis confers enhanced targeting of processed insulin to the regulated secretory pathway. Mol Biol Cell 11(6):1959–1972
65. Dannies PS (2002) Mechanisms for storage of prolactin and growth hormone in secretory granules. Mol Genet Metab 76(1):6–13
66. Giannattasio G, Zanini A, Meldolesi J (1975) Molecular organization of rat prolactin granules. I. In vitro stability of intact and "membraneless" granules. J Cell Biol 64(1):246–251
67. Morvan J, Tooze SA (2008) Discovery and progress in our understanding of the regulated secretory pathway in neuroendocrine cells. Histochem Cell Biol 129(3):243–252
68. Dannies P (2003) Manipulating the reversible aggregation of protein hormones in secretory granules: potential impact on biopharmaceutical development. BioDrugs 17(5):315–324
69. Dannies PS (1999) Protein hormone storage in secretory granules: mechanisms for concentration and sorting. Endocr Rev 20(1):3–21
70. Dannies PS (2001) Concentrating hormones into secretory granules: layers of control. Mol Cell Endocrinol 177(1–2):87–93
71. Maji SK, Riek R (2010) Formation of secretory granules involves the amyloid structure. In: Rigacci S, Bucciantini M (eds) Functional amyloid aggregation. Kerala, Research Signpost, pp 135–155
72. Hutton JC (1982) The internal pH and membrane potential of the insulin-secretory granule. Biochem J 204(1):171–178
73. Orci L, Ravazzola M, Amherdt M, Madsen O, Perrelet A, Vassalli JD, Anderson RG (1986) Conversion of proinsulin to insulin occurs coordinately with acidification of maturing secretory vesicles. J Cell Biol 103(6 Pt 1):2273–2281
74. Urbe S, Tooze SA, Barr FA (1997) Formation of secretory vesicles in the biosynthetic pathway. Biochim Biophys Acta 1358(1):6–22
75. Huttner WB, Natori S (1995) Regulated secretion. Helper proteins for neuroendocrine secretion. Curr Biol 5(3):242–245
76. Kolset SO, Prydz K, Pejler G (2004) Intracellular proteoglycans. Biochem J 379(Pt 2):217–227
77. Miller F, de Harven E, Palade GE (1966) The structure of eosinophil leukocyte granules in rodents and in man. J Cell Biol 31:349–362
78. Howell SL, Fink CJ, Lacy PE (1969) Isolation and properties of secretory granules from rat islets of Langerhans: I. Isolation of a secretory granule fraction. J Cell Biol 41(1):154–161
79. van Grondelle W, Iglesias CL, Coll E, Artzner F, Paternostre M, Lacombe F, Cardus M, Martinez G, Montes M, Cherif-Cheikh R, Valery C (2007) Spontaneous fibrillation of the native neuropeptide hormone Somatostatin-14. J Struct Biol 160(2):211–223
80. Keeler C, Hodsdon ME, Dannies PS (2004) Is there structural specificity in the reversible protein aggregates that are stored in secretory granules? J Mol Neurosci 22(1–2):43–49
81. Arvan P, Zhang BY, Feng L, Liu M, Kuliawat R (2002) Lumenal protein multimerization in the distal secretory pathway/secretory granules. Curr Opin Cell Biol 14(4):448–453
82. Howell SL, Young DA, Lacy PE (1969) Isolation and properties of secretory granules from rat islets of Langerhans. 3. Studies of the stability of the isolated beta granules. J Cell Biol 41(1):167–176
83. Jacob RS, Das S, Ghosh S, Anoop A, Jha NN, Khan T, Singru P, Kumar A, Maji SK (2016) Amyloid formation of growth hormone in presence of zinc: relevance to its storage in secretory granules. Sci Rep 6:23370
84. Dannies PS (2012) Prolactin and growth hormone aggregates in secretory granules: the need to understand the structure of the aggregate. Endocr Rev 33(2):254–270
85. Gellermann GP, Appel TR, Tannert A, Radestock A, Hortschansky P, Schroeckh V, Leisner C, Lütkepohl T, Shtrasburg S, Röcken C, Pras M, Linke RP, Diekmann S, Fändrich M (2005)

Raft lipids as common components of human extracellular amyloid fibrils. Proc Natl Acad Sci U S A 102(18):6297–6302
86. Perry EK, Oakley AE, Candy JM, Perry RH (1981) Properties and possible significance of substance P and insulin fibrils. Neurosci Lett 25(3):321–325
87. Nespovitaya N, Gath J, Barylyuk K, Seuring C, Meier BH, Riek R (2016) Dynamic assembly and disassembly of functional β-endorphin amyloid fibrils. J Am Chem Soc 138(3):846–856
88. Zhu YL, Conway-Campbell B, Waters MJ, Dannies PS (2002) Prolonged retention after aggregation into secretory granules of human R183H-growth hormone (GH), a mutant that causes autosomal dominant GH deficiency type II. Endocrinology 143(11):4243–4248
89. Deladoey J, Gex G, Vuissoz JM, Strasburger CJ, Wajnrajch MP, Mullis PE (2002) Effect of different growth hormone (GH) mutants on the regulation of GH-receptor gene transcription in a human hepatoma cell line. Eur J Endocrinol 146(4):573–581
90. Deladoey J, Stocker P, Mullis PE (2001) Autosomal dominant GH deficiency due to an Arg183His GH-1 gene mutation: clinical and molecular evidence of impaired regulated GH secretion. J Clin Endocrinol Metab 86(8):3941–3947
91. Anoop A, Ranganathan S, Das Dhaked B, Jha NN, Pratihar S, Ghosh S, Sahay S, Kumar S, Das S, Kombrabail M, Agarwal K, Jacob RS, Singru P, Bhaumik P, Padinhateeri R, Kumar A, Maji SK (2014) Elucidating the role of disulfide bond on amyloid formation and fibril reversibility of somatostatin-14: relevance to its storage and secretion. J Biol Chem 289(24):16884–16903
92. Sambashivan S, Liu Y, Sawaya MR, Gingery M, Eisenberg D (2005) Amyloid-like fibrils of ribonuclease A with three-dimensional domain-swapped and native-like structure. Nature 437(7056):266–269
93. Ivanova MI, Sawaya MR, Gingery M, Attinger A, Eisenberg D (2004) An amyloid-forming segment of β2-microglobulin suggests a molecular model for the fibril. Proc Natl Acad Sci U S A 101(29):10584–10589
94. Baglioni S, Casamenti F, Bucciantini M, Luheshi LM, Taddei N, Chiti F, Dobson CM, Stefani M (2006) Prefibrillar amyloid aggregates could be generic toxins in higher organisms. J Neurosci 26(31):8160–8167
95. Jha NN, Anoop A, Ranganathan S, Mohite GM, Padinhateeri R, Maji SK (2013) Characterization of amyloid formation by glucagon-like peptides: role of basic residues in heparin-mediated aggregation. Biochemistry 52(49):8800–8810
96. Christensen LF, Malmos KG, Christiansen G, Otzen DE (2016) A complex dance: the importance of glycosaminoglycans and zinc in the aggregation of human prolactin. Biochemistry 55(26):3674–3684
97. Langer R (2001) Drug delivery. Drugs on target. Science 293(5527):58–59
98. Orive G, Hernandez RM, Rodriguez Gascon A, Dominguez-Gil A, Pedraz JL (2003) Drug delivery in biotechnology: present and future. Curr Opin Biotechnol 14(6):659–664
99. Brader ML, Sukumar M, Pekar AH, McClellan DS, Chance RE, Flora DB, Cox AL, Irwin L, Myers SR (2002) Hybrid insulin cocrystals for controlled release delivery. Nat Biotechnol 20(8):800–804
100. Jen A, Madorin K, Vosbeck K, Arvinte T, Merkle HP (2002) Transforming growth factor β-3 crystals as reservoirs for slow release of active TGF-β3. J Control Release 78(1–3):25–34
101. Tennent GA, Lovat LB, Pepys MB (1995) Serum amyloid P component prevents proteolysis of the amyloid fibrils of Alzheimer disease and systemic amyloidosis. Proc Natl Acad Sci U S A 92(10):4299–4303
102. Suk JY, Zhang F, Balch WE, Linhardt RJ, Kelly JW (2006) Heparin accelerates gelsolin amyloidogenesis. Biochemistry 45(7):2234–2242
103. Guptabansal R, Frederickson RCA, Brunden KR (1995) Proteoglycan-mediated inhibition of Aβ proteolysis – a potential cause of senile plaque accumulation. J Biol Chem 270(31):18666–18671
104. Cotman SL, Halfter W, Cole GJ (2000) Agrin binds to beta-amyloid (Aβ), accelerates aβ fibril formation, and is localized to Aβ deposits in Alzheimer's disease brain. Mol Cell Neurosci 15(2):183–198

105. Herbst KL (2003) Gonadotropin-releasing hormone antagonists. Curr Opin Pharmacol 3(6):660–666
106. Kirkitadze MD, Bitan G, Teplow DB (2002) Paradigm shifts in Alzheimer's disease and other neurodegenerative disorders: the emerging role of oligomeric assemblies. J Neurosci Res 69(5):567–577
107. Shahnawaz M, Soto C (2012) Microcin amyloid fibrils A are reservoir of toxic oligomeric species. J Biol Chem 287(15):11665–11676
108. Krebs MR, Morozova-Roche LA, Daniel K, Robinson CV, Dobson CM (2004) Observation of sequence specificity in the seeding of protein amyloid fibrils. Protein Sci 13(7):1933–1938
109. Tanaka M, Chien P, Yonekura K, Weissman JS (2005) Mechanism of cross-species prion transmission: an infectious conformation compatible with two highly divergent yeast prion proteins. Cell 121(1):49–62
110. Ghosh D, Singh PK, Sahay S, Jha NN, Jacob RS, Sen S, Kumar A, Riek R, Maji SK (2015) Structure based aggregation studies reveal the presence of helix-rich intermediate during α-Synuclein aggregation. Sci Rep 5:9228
111. Sahay S, Anoop A, Krishnamoorthy G, Maji SK (2014) Site-specific fluorescence dynamics of α-synuclein fibrils using time-resolved fluorescence studies: effect of familial Parkinson's disease-associated mutations. Biochemistry 53(5):807–809
112. Sahay S, Ghosh D, Dwivedi S, Anoop A, Mohite GM, Kombrabail M, Krishnamoorthy G, Maji SK (2015) Familial Parkinson disease-associated mutations alter the site-specific microenvironment and dynamics of α-synuclein. J Biol Chem 290(12):7804–7822
113. Ghosh D, Mondal M, Mohite GM, Singh PK, Ranjan P, Anoop A, Ghosh S, Jha NN, Kumar A, Maji SK (2013) The Parkinson's disease-associated H50Q mutation accelerates α-Synuclein aggregation in vitro. Biochemistry 52(40):6925–6927
114. Ghosh D, Sahay S, Ranjan P, Salot S, Mohite GM, Singh PK, Dwivedi S, Carvalho E, Banerjee R, Kumar A, Maji SK (2014) The newly discovered Parkinson's disease associated Finnish mutation (A53E) attenuates α-synuclein aggregation and membrane binding. Biochemistry 53(41):6419–6421
115. Pertinhez TA, Conti S, Ferrari E, Magliani W, Spisni A, Polonelli L (2009) Reversible self-assembly: a key feature for a new class of autodelivering therapeutic peptides. Mol Pharm 6(3):1036–1039
116. Gupta S, Chattopadhyay T, Pal Singh M, Surolia A (2010) Supramolecular insulin assembly II for a sustained treatment of type 1 diabetes mellitus. Proc Natl Acad Sci U S A 107(30):13246–13251
117. Burn P (2010) Type 1 diabetes. Nature reviews. Drug Discov 9(3):187–188
118. Nelson R, Eisenberg D (2006) Recent atomic models of amyloid fibril structure. Curr Opin Struct Biol 16(2):260–265
119. Calamai M, Kumita JR, Mifsud J, Parrini C, Ramazzotti M, Ramponi G, Taddei N, Chiti F, Dobson CM (2006) Nature and significance of the interactions between amyloid fibrils and biological polyelectrolytes. Biochemistry 45(42):12806–12815
120. Ghosh D, Dutta P, Chakraborty C, Singh PK, Anoop A, Jha NN, Jacob RS, Mondal M, Mankar S, Das S, Malik S, Maji SK (2014) Complexation of amyloid fibrils with charged conjugated polymers. Langmuir 30(13):3775–3786
121. Jha NN, Ghosh D, Das S, Anoop A, Jacob RS, Singh PK, Ayyagari N, Namboothiri IN, Maji SK (2016) Effect of curcumin analogs on α-synuclein aggregation and cytotoxicity. Sci Rep 6:28511
122. Nagai Y, Unsworth LD, Koutsopoulos S, Zhang S (2006) Slow release of molecules in self-assembling peptide nanofiber scaffold. J Control Release 115(1):18–25
123. Koutsopoulos S, Unsworth LD, Nagai Y, Zhang S (2009) Controlled release of functional proteins through designer self-assembling peptide nanofiber hydrogel scaffold. Proc Natl Acad Sci U S A 106(12):4623–4628
124. Mazza M, Notman R, Anwar J, Rodger A, Hicks M, Parkinson G, McCarthy D, Daviter T, Moger J, Garrett N, Mead T, Briggs M, Schatzlein AG, Uchegbu IF (2013) Nanofiber-based delivery of therapeutic peptides to the brain. ACS Nano 7(2):1016–1026

125. Shimanovich U, Efimov I, Mason TO, Flagmeier P, Buell AK, Gedanken A, Linse S, Akerfeldt KS, Dobson CM, Weitz DA, Knowles TP (2015) Protein microgels from amyloid fibril networks. ACS Nano 9(1):43–51
126. Li C, Bolisetty S, Chaitanya K, Adamcik J, Mezzenga R (2013) Tunable carbon nanotube/protein core-shell nanoparticles with NIR- and enzymatic-responsive cytotoxicity. Adv Mater 25(7):1010–1015
127. Shen Y, Posavec L, Bolisetty S, Hilty FM, Nystrom G, Kohlbrecher J, Hilbe M, Rossi A, Baumgartner J, Zimmermann MB, Mezzenga R (2017) Amyloid fibril systems reduce, stabilize and deliver bioavailable nanosized iron. Nat Nanotechnol 12(7):642–647
128. Mains J, Lamprou DA, McIntosh L, Oswald ID, Urquhart AJ (2013) Beta-adrenoceptor antagonists affect amyloid nanostructure; amyloid hydrogels as drug delivery vehicles. Chem Commun (Camb) 49(44):5082–5084
129. Ahn M, Kang S, Koo HJ, Lee JH, Lee YS, Paik SR (2010) Nanoporous protein matrix made of amyloid fibrils of β2-microglobulin. Biotechnol Prog 26(6):1759–1764
130. Bhak G, Lee S, Park JW, Cho S, Paik SR (2010) Amyloid hydrogel derived from curly protein fibrils of α-synuclein. Biomaterials 31(23):5986–5995
131. Raynes JK, Pearce FG, Meade SJ, Gerrard JA (2011) Immobilization of organophosphate hydrolase on an amyloid fibril nanoscaffold: towards bioremediation and chemical detoxification. Biotechnol Prog 27(2):360–367
132. Knowles TP, Oppenheim TW, Buell AK, Chirgadze DY, Welland ME (2010) Nanostructured films from hierarchical self-assembly of amyloidogenic proteins. Nat Nanotechnol 5(3):204–207
133. Gras SL (2009) Surface- and solution-based assembly of amyloid fibrils for biomedical and nanotechnology applications. Adv Chem Eng 35:161–209
134. Mankar S, Anoop A, Sen S, Maji SK (2011) Nanomaterials: amyloids reflect their brighter side. Nano Rev 2:6032
135. Reches M, Gazit E (2003) Casting metal nanowires within discrete self-assembled peptide nanotubes. Science 300(5619):625–627

Chapter 9
Nanozymes: Biomedical Applications of Enzymatic Fe_3O_4 Nanoparticles from *In Vitro* to *In Vivo*

Lizeng Gao and Xiyun Yan

Abstract Fe_3O_4, also called magnetite, is a naturally occurring mineral and has been widely used in biomedical applications. However, in the past, all the applications were based on its excellent magnetic properties and neglected its catalytic properties. In 2007, we found that Fe_3O_4 nanoparticles are able to perform intrinsic enzyme-like activities. A specific term, "nanozyme", is used to describe the new property of intrinsic enzymatic activity of nanomaterials. Since then, Fe_3O_4 nanoparticles have been used as enzyme mimics, which broadens their applications beyond simply their magnetic properties, with applications in biomedical diagnosis and therapy, environmental monitoring and treatment, the food industry and chemical synthesis. In this chapter, we will summarize the basic features of Fe_3O_4 as an enzyme mimetic and its applications in biomedicine.

Keywords Nanozymes · Enzyme-like activity · Enzyme mimetic · Fe_3O_4 · Biomedical application

Abbreviations

ABTS	2, 2′-azino-bis (3-ethylbenzthiazoline-6-sulfonic acid)
AD	Alzheimer's disease
APTES	3-Aminopropyltriethoxysilane
CNT	Carbon nanotube
DAB	3, 3′-Diaminobenzidine

L. Gao
Institute of Translational Medicine, School of Medicine, Yangzhou University, Yangzhou, Jangsu, China

X. Yan (✉)
Key Laboratory of Protein and Peptide Pharmaceutical, Institute of Biophysics, Chinese Academy of Sciences, Beijing, China
e-mail: yanxy@ibp.ac.cn

S. Perrett et al. (eds.), *Biological and Bio-inspired Nanomaterials*, Advances in Experimental Medicine and Biology 1174,
https://doi.org/10.1007/978-981-13-9791-2_9

EBOV	Ebola virus
EPR	Enhanced permeability and retention
Fe_3O_4	Magnetite or iron oxide
GO	Graphene oxide
GOx	Glucose oxidase
H_2O_2	Hydrogen peroxide
HCG	Human chorionic gonadotropin
HFn	Human heavy-chain ferritin
HRP	Horseradish peroxidase
M-HFn	Magnetoferritin nanoparticles
MNPs	Magnetic nanoparticles
MRI	Magnetic resonance imaging
MRSA	Staphylococcus aureus
NPs	Nanoparticles
NTs	Nanotubes
NWs	Nanowires
OPD	o-phenylenediamine
PD	Parkinson's disease
PEG	Polyethylene glycol
RES	Reticuloendothelial system
RGO	Reduced graphene oxide
ROS	Reactive oxygen species
TMB	3, 3′, 5, 5′-Tetramethylbenzidine

9.1 Introduction

Iron (II, III) oxide (Fe_3O_4) is the naturally occurring mineral magnetite, which is a black powder with permanent magnetism. However, this material exhibits special superparamagnetic magnetization when the size is reduced to the nanoscale (e.g. 1–100 nm) [1, 2], so that Fe_3O_4 nanoparticles can be easily aggregated in the presence of an external magnetic field and can be rapidly re-dispersed with the removal of the magnetic field. Based on this unique property, Fe_3O_4 nanoparticles have been applied in various fields, particularly in biomedicine, including bio-separation and purification, biosensors, transfection, MRI, hyperthermia therapy, targeted drug delivery and as a theranostic platform (Fig. 9.1) [3–8].

However, nanoscale Fe_3O_4 has many specific physical and chemical properties besides magnetism, including hyperthermia, photothermal properties, photoacoustic effects, and fluorescence [9]. Its recently reported enzyme-like catalytic properties are of particular interest. In 2007, Fe_3O_4 nanoparticles were discovered to possess intrinsic peroxidase-like activity, similar to that of horseradish peroxidase (HRP) [10]. Since then, many nanomaterials have been found to possess enzymatic activities, which dramatically broadens their applications in biomedicine. The term "nanozyme" [11] was introduced to describe the phenomenon of nanomaterials with intrinsic enzyme-like activities [12–15], with Fe_3O_4 nanoparticles being the first reported example of a nanozyme. In this chapter, we mainly focus on the enzymatic

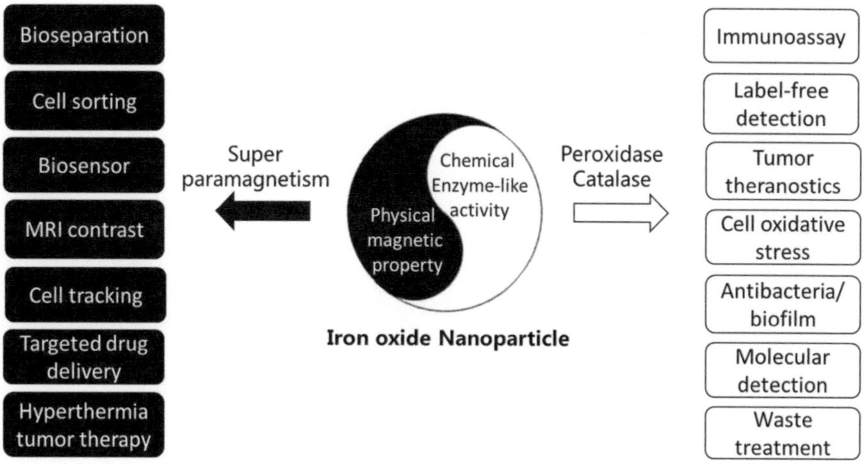

Fig. 9.1 Iron oxide nanoparticles with intrinsic enzyme-like activity allow novel applications in addition to those based on their magnetic properties

properties of Fe_3O_4 nanoparticles and summarize the novel applications of this nanozyme, from *in vitro* bioassays to *in vivo* diagnosis and therapy (Fig. 9.1).

9.2 Basic Features of Fe_3O_4 Nanozymes

As enzyme mimetics, Fe_3O_4 nanozymes show similar catalytic properties to natural enzymes, including substrate specificity, pH and temperature optima, kinetics and mechanism. In general, they are typical nanomaterials which have a diameter on the nanoscale i.e. 1–500 nm and specific nanocrystalline structures, morphologies and surface modifications, providing nanozymes with many advantages over natural enzymes because they are robust inorganic nanomaterials. Therefore, Fe_3O_4 nanozymes have great potential for use as multi-functional nanomaterials for biomedical applications.

9.2.1 Activities of Fe_3O_4 Nanozymes

(1) **Peroxidase-like activity**: Fe_3O_4 nanozymes were first found to have peroxidase-like activity, catalyzing the reaction of H_2O_2 with chromogenic reagents such as TMB, OPD, DAB and ABTS (Fig. 9.2a–c). The optimal conditions for the catalysis are similar to those for HRP i.e. 37–40 C under acidic pH (e.g. sodium acetate buffer pH 3–6.5) which favors the generation of free radicals to oxidize TMB (Fig. 9.2d).
(2) **Catalase-like activity**: Fe_3O_4 nanozymes also show catalase-like activity, decomposing H_2O_2 into oxygen and water under neutral/basic pH conditions

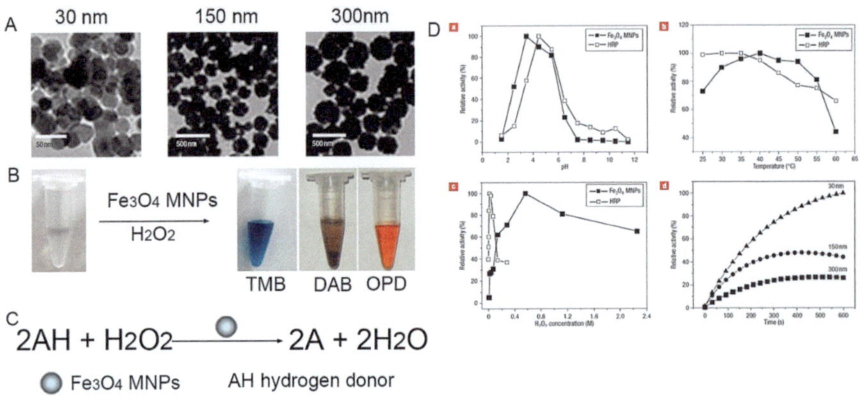

Fig. 9.2 Fe_3O_4 nanoparticles show intrinsic peroxidase-like activity [10]. (**a–c**) Different sizes of Fe_3O_4 nanozymes and colorimetric reactions catalyzed by Fe_3O_4 nanozymes. (**d**) Reaction conditions for Fe_3O_4 nanozymes (300 nm) (*a–c*) and comparison of the activity of Fe_3O_4 nanozymes with different sizes (*d*). (Reprinted from Ref. [10] with permission from the Nature Publishing Group)

Table 9.1 Typical parameters for Michaelis-Menton kinetics [10]

	[E] (M)	Substrate	K_M (mM)	$V_{\max}$ (s^{-1})	k_{cat} (M^{-1} s^{-1})
Fe_3O_4 MNPs	9.402×10^{-13}	TMB	0.098	0.0032	3.37×10^9
Fe_3O_4 MNPs	9.402×10^{-13}	H_2O_2	154	0.0091	9.68×10^9
HRP	2.5×10^{-11}	TMB	0.434	0.0093	3.72×10^8
HRP	2.5×10^{-11}	H_2O_2	3.70	0.0081	3.24×10^8

Reprinted from Ref. [10] with permission from Nature Publishing Group

[16]. Fe_3O_4 nanozymes can decompose H_2O_2 into oxygen and water by this enzymatic activity, which may be used to scavenge H_2O_2 in biosystems.

9.2.2 *Kinetics and Mechanism of Fe_3O_4 Nanozymes*

The catalysis of Fe_3O_4 nanozymes follow Michaelis-Menten kinetics. Curves for both H_2O_2 and TMB in peroxidase-like catalysis fit the equation $v = (V_{max}[S])/(K_M + [S])$, where V_{max} represents the maximum velocity, [S] is the substrate concentration, and K_M is the Michaelis constant (Fig. 9.3). The related parameters are shown Table 9.1. The k_{cat} is equal to V_{max}/[E]. The K_M value of H_2O_2 for Fe_3O_4 nanozymes is higher than that for HRP, indicating lower affinity of H_2O_2 for Fe_3O_4 nanozymes. In contrast, the K_M value of TMB for Fe_3O_4 nanozymes is lower than that for HRP, indicating that Fe_3O_4 nanozymes have higher affinity for TMB. Notably, a single Fe_3O_4 nanozyme with 300 nm diameter showed 40 times higher activity than a single HRP molecule.

Just like HRP, the catalysis of Fe_3O_4 nanozymes follows a ping-pong mechanism. In the peroxidase-like catalysis, a set of parallel lines are produced when a set of *v* against [S] (fixed H_2O_2, varying TMB or fixed TMB, varying H_2O_2) are

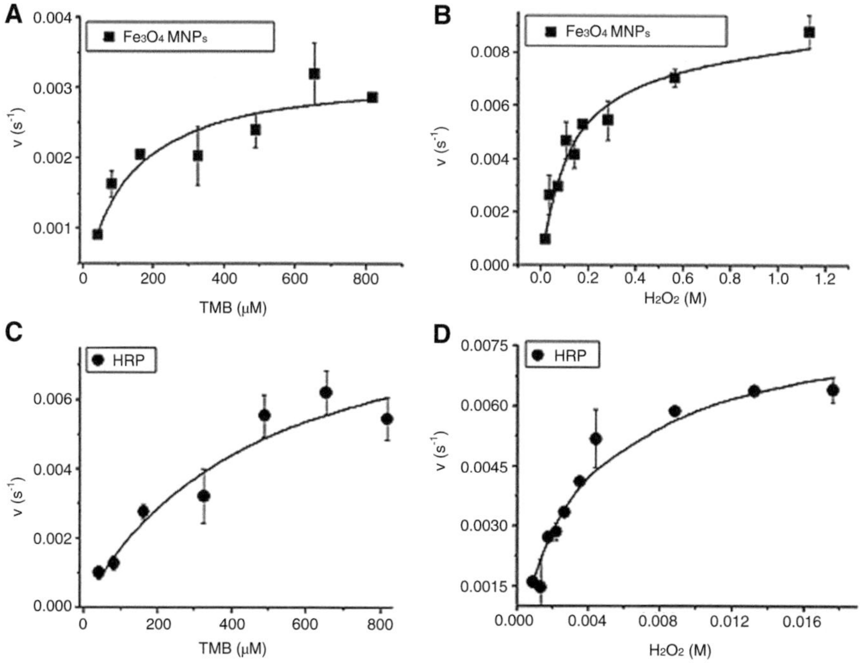

Fig. 9.3 Apparent Michaelis-Menton kinetics for Fe_3O_4 nanozymes in comparison with HRP. (Reprinted from Ref. [10] with permission from the Nature Publishing Group)

plotted in a Lineweaver-Burk plot. That is, H_2O_2 first binds to and react with Fe_3O_4 nanozymes, releasing the first product, then TMB reacts with Fe_3O_4 nanozymes.

It is important to note that the peroxidase activity is not derived from free iron ions via the Fenton reaction. The trace amount of iron released from the surface of Fe_3O_4 nanozymes was around two orders of magnitude lower than the concentration required for the Fenton reaction, only showing negligible catalytic activity [17–20]. These results demonstrate that the observed reaction cannot be attributed to leaching of iron ions into solution, but occurs on the surface of the nanozymes. Importantly, ferrous iron (Fe^{2+}) seems to be more important than ferric iron (Fe^{3+}) in the enzyme-mimicking catalysis, as increasing the ratio of Fe^{2+} in Fe_3O_4 nanoparticles enhances the peroxidase-like activity [10]. Taken together, the intrinsic enzyme-like property of Fe_3O_4 nanozymes arises from the variable valence of iron in the nanoparticles.

9.2.3 *Advantages of Fe_3O_4 Nanozymes*

Compared to natural enzymes, nanozymes show many advantageous features. Fe_3O_4 nanozymes show enhanced stability towards extreme conditions such as temperature (4–90 °C) and pH (2–12) [21]. In contrast, the enzyme HRP did not show any activity after treatment at pH lower than 5 and lost activity rapidly when the temperature was greater than 40 °C. Fe_3O_4 nanoparticles can be stored long term and reused many times [22–26]. In addition, the activity of Fe_3O_4 nanozymes can be tuned by modulating the size, structure or morphology [18, 27–30] (Fig. 9.4), by the addition of dopants [31–34], by surface modifications [35–38] or by hybridizing with other nanomaterials [19, 23, 39–44], which allows the design of nanozymes with optimized activity for the specific purpose required.

Further, Fe_3O_4 nanozymes have both magnetic and catalytic properties and so can simultaneously perform two basic functions: enzyme-like activity and superparamagnetism. Regarding the enzyme-like activity, Fe_3O_4 nanozymes, as mentioned above, can mimic peroxidase and catalase activities [16]. Fe_3O_4 nanozymes can also be used as a carrier to conjugate other functional molecules or groups to construct a cascade reaction [17]. Finally, Fe_3O_4 nanozymes can be synthesized by chemical methods such as solvothermal reaction, sol-gel and co-precipitation. The required chemical reagents are usually much cheaper than biological reagents. Scale up of production is therefore both straightforward and economical. The multifunctionality of Fe_3O_4 nanozymes facilitates a wide range of practical applications, including in biomedicine, such as tumor theranostics, ultrasensitive molecular detection and controlled drug release. The combination of these functions can be used to create versatile state-of-the-art technologies and strategies.

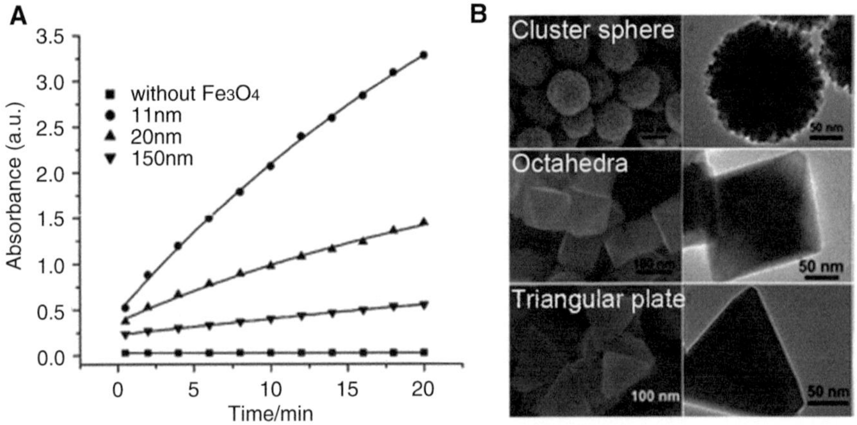

Fig. 9.4 (**a**) Tuning of activity by variation of size. Reprinted from Ref. [28] with permission from Elsevier. (**b**) Tuning of activity by variation of morphology (peroxidase-like activity: cluster spheres > triangular plates > octahedral). (Reprinted from Ref. [18] with permission from John Wiley & Sons Ltd.)

9.3 Biomedical Applications of Fe_3O_4 Nanozymes

The discovery of enzymatic activity of Fe_3O_4 nanozymes has stimulated their use in a range of applications, including immunoassays, tumor diagnosis and therapy, biosensors, antibacterial/antibiofilm reagents, environmental monitoring and pollutant degradation, in the food industry and in chemical synthesis. Here we summarize their state-of-the-art applications particularly in the biomedical field, which span *in vitro* molecular detection to *in vivo* diagnosis and therapy of tumors and diseases.

9.3.1 *In Vitro Bioassays*

The peroxidase-like activity of Fe_3O_4 nanozymes make them an ideal alternative to HRP and thus they can be used to replace HRP in enzyme-linked immunosorbent assay (ELISA) and HRP-related molecular detection [45, 46]. Fe_3O_4 nanozymes [47] can be conjugated with an antibody and applied in ELISA to amplify signals by catalyzing a colorimetric reaction (Fig. 9.5a). Importantly, due to their magnetism, Fe_3O_4 nanozymes with an appropriate ligand or antibody can be employed to capture and enrich very low amounts of sample, which will improve the sensitivity and efficiency of detection [48]. According to this principle, a capture-detection immunoassay has been developed to detect carcinoembryonic antigen (CEA) with a detection limit up to 1 ng mL^{-1} [45]. This kind of nanozyme-based immunoassay can be used to detect multiple antigens including biomarkers and bacteria or cells, such as IgG, hepatocellular carcinoma biomarker GP73 [49], human chorionic gonadotropin (HCG) [50], *Mycoplasma pneumoniae* [51], *Vibrio cholerae*, rotavirus [52], and cancer cells with human epidermal growth factor receptor 2 (HER2) [52, 53].

A nanozyme-strip has been developed by combining the magnetism and peroxidase-like activity of Fe_3O_4 magnetic nanoparticles. This novel strip can be used to detect the glycoprotein of Ebola virus (EBOV) as low as 1 ng/mL, showing a 100-fold increased sensitivity compared to the classic gold-strip method [55] (Fig. 9.6b). Importantly, the nanozyme-strip shows comparable sensitivity and accuracy with ELISA in the detection of EBOV and New Bunyavirus clinical samples. In addition, the nanozyme-strip is much faster (within 30 min) and simpler (observed by naked eye) than ELISA. These results indicate that the nanozyme-strip may be used as a point-of-care test for EBOV detection, providing a simple method for diagnosis of infection in Ebola-affected areas of Africa.

Besides immunoassays via antibody-antigen recognition, other detection methods have been developed based on the specific interaction between DNA molecules or aptamers. Park and coworkers developed a label-free colorimetric detection method for nucleic acids by comparing catalytic activity before and after DNA binding on Fe_3O_4 nanozymes [56]. Thiramanas et al. [57] developed a novel and

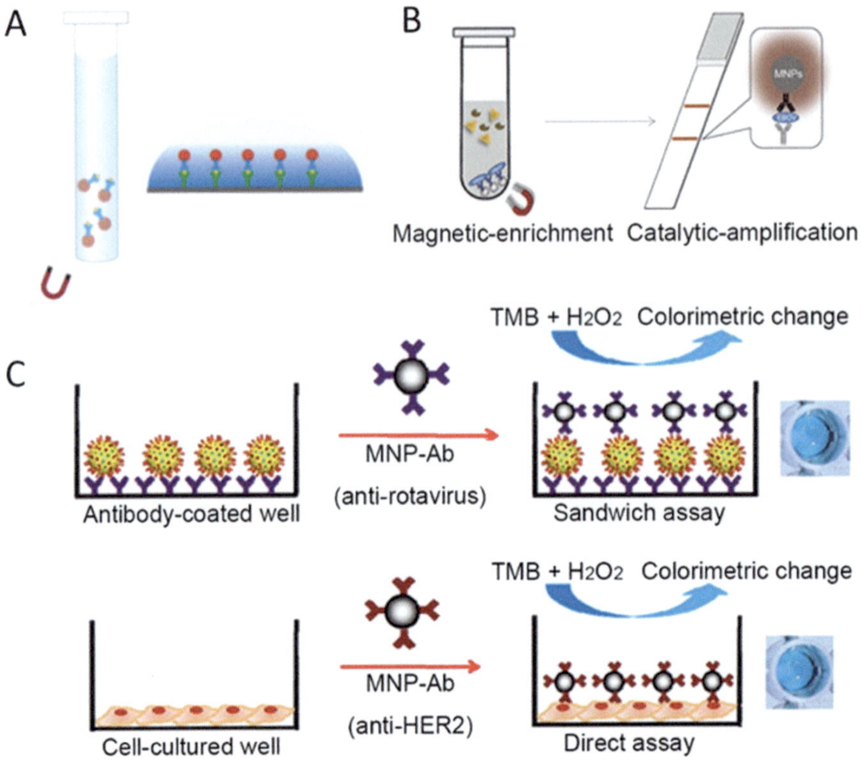

Fig. 9.5 Novel immunoassays based on Fe_3O_4 nanozymes. (**a**) Traditional immunoassay. Reprinted from Ref. [54] with permission from the Nature Publishing Group. (**b**) Nanozyme-strip for Ebola detection. Reprinted from Ref. [55] with permission from Elsevier. (**c**) Virus and cancer cell detection. (Reprinted from Ref. [52] with permission from MDPI AG)

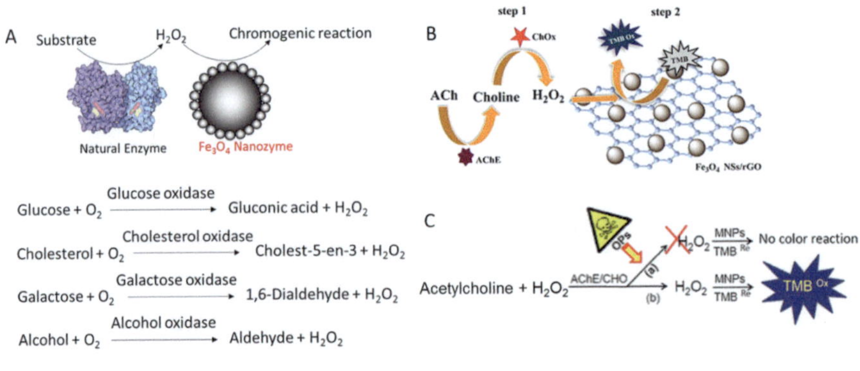

Fig. 9.6 Cascade reactions for molecular detection. (**a**) Dual sequential reactions. (**b** and **c**). Triple sequential reactions for acetylcholine (Ach) [24] and pesticide detection [73]. (Reprinted from Ref. [24, 73] with permission from Elsevier and the American Chemical Society, respectively)

sensitive system for *Vibrio cholerae* detection using magnetic polymeric nanoparticles (MPNPs) composed of a Fe_3O_4 core with peroxidase activity in combination with polymerase chain reaction to integrate magneto–PCR–colorimetry in one system. Aptamers also can be used in similar ways as they can specifically recognize target molecules with high affinity.

Finally, it is feasible to develop cascade catalytic reactions by conjugating natural enzymes onto Fe_3O_4 nanozymes. In a cascade reaction, natural enzyme is responsible to generate H_2O_2 and then Fe_3O_4 nanozymes utilize H_2O_2 to generate colorimetric signals. Wei and Wang [17] first developed a cascade system by conjugating glucose oxidase (GOx) onto Fe_3O_4 nanozymes for glucose detection (Fig. 9.6a). In this hybrid system, GOx catalyzes glucose to generate H_2O_2 and Fe_3O_4 nanozymes then uses H_2O_2 as substrate to produce a colorimetric signal. There is a directly proportional relationship between glucose concentration and the colorimetric signal [17]. In this way, a glucose-response curve can be established with a detection limit for glucose as low as 3×10^{-5} mol L^{-1} and a linear range from 5×10^{-5} to 1×10^{-3} mol L^{-1}. Since then, many groups have used this method by combining GOx with or integrating GOx onto Fe_3O_4 nanozymes for glucose detection [21, 41, 43, 58–72], showing great potential in measuring blood glucose levels for diabetes diagnosis.

Alternatively, other oxidases can be integrated into Fe_3O_4 nanozymes to detect substrates besides glucose, including cholesterol [71], galactose [74], alcohol [75], and acetylcholine (ACh) [24, 73] (Fig. 9.6b, c).

9.3.2 *Ex Vivo* Tracking and Histochemistry Diagnosis

Iron oxide nanoparticles are often used as diagnostic and therapeutic agents for biomedical applications due to their superparamagnetism. Quantitative analysis of their biodistribution, pharmacokinetics and organ clearance in animal models is important to understand their *in vivo* behavior and biosafety. A novel histochemical method for visualizing unlabeled Fe_3O_4 NPs in mouse tissues was developed by our group, which employs the intrinsic peroxidase activity of the NPs to produce a color reaction [76] (Fig. 9.7). It was found that dextran-coated Fe_3O_4 NPs were mainly localized in the liver, spleen and lung rather than the kidney, lymph nodes or thymus. Cellular location was further examined by combining hematoxylin-eosin (H&E) staining. Dextran-coated Fe_3O_4 NPs were taken up mainly by the reticuloendothelial system (RES) in these organs, including Kuppfer macrophage cells in the liver, alveolar macrophages in the lung and macrophage perifollicular areas in the spleen.

Organ clearance could also be evaluated in the same way, showing that the content in liver, spleen and lung increased steadily between 0.25 and 5 h post-injection, and then rapid clearance occurred between 5 and 72 h post-injection. This approach is more sensitive when compared with the traditional Prussian blue staining method because of the highly effective catalytic activity of Fe_3O_4 NPs. Importantly, without

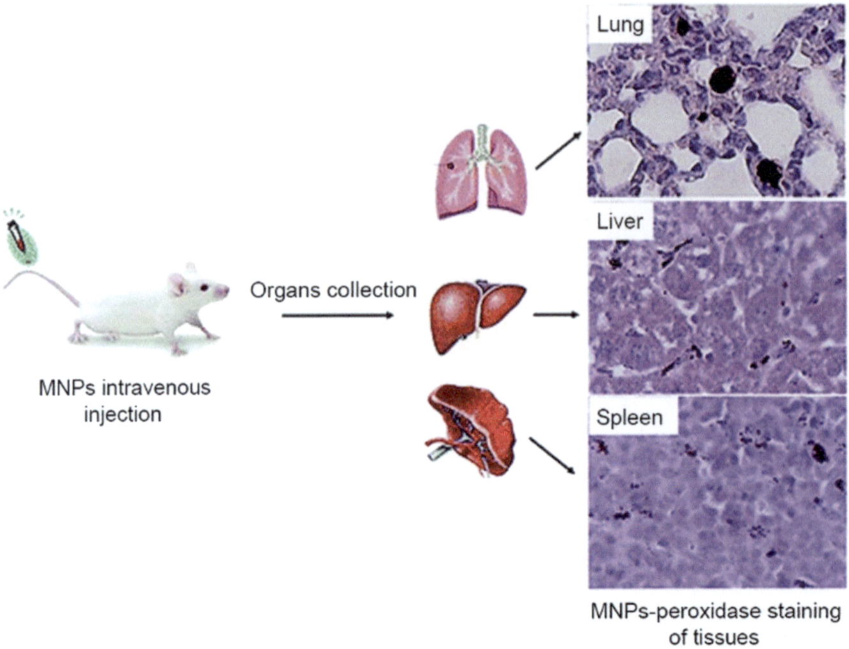

Fig. 9.7 Label-free detection of Fe_3O_4 NP distribution via their peroxidase activity [76]. (Reprinted from Ref. [76] with permission from the American Chemical Society)

labeling with exogenous indicators, it could reduce false signals from background and may provide a better way to understand the real behavior of NPs *in vivo* and thus has significant implications for the clinical translation of Fe_3O_4 NPs. Presumably, other nanoparticles having intrinsic peroxidase activity could also be studied in a similar way to determine their *in vivo* behavior.

Fe_3O_4 nanozymes also can be used for histochemistry diagnosis by integrating with a specific antibody or protein. In particular, Fe_3O_4 nanozymes can be assembled into ferritin to form recombinant magneto ferritin nanoparticles (M-HFn) [77] achieving the capability of tumor targeting and visualization in the same unique system. Fe_3O_4 nanozymes are encapsulated inside and the recombinant human heavy-chain ferritin (HFn) proteins form a protein shell which can recognize tumor cells overexpressing transferrin receptor (TfR1) (Fig. 9.8). Once bound to tumor tissues, the Fe_3O_4 core performs peroxidase-like activity to conduct a colorimetric reaction in the presence of H_2O_2 and then the tumor tissues can be visualized. In this way, 474 clinical specimens from patients with nine types of cancers were examined in a single-step binding and colorimetric reaction. Fe_3O_4 nanozymes successfully distinguished cancerous cells from normal cells with a sensitivity of 98% and specificity of 95%. These results demonstrate that Ferritin- Fe_3O_4 nanozymes can be used as a diagnostic reagent for rapid, low-cost and universal assessment of malignant tumors.

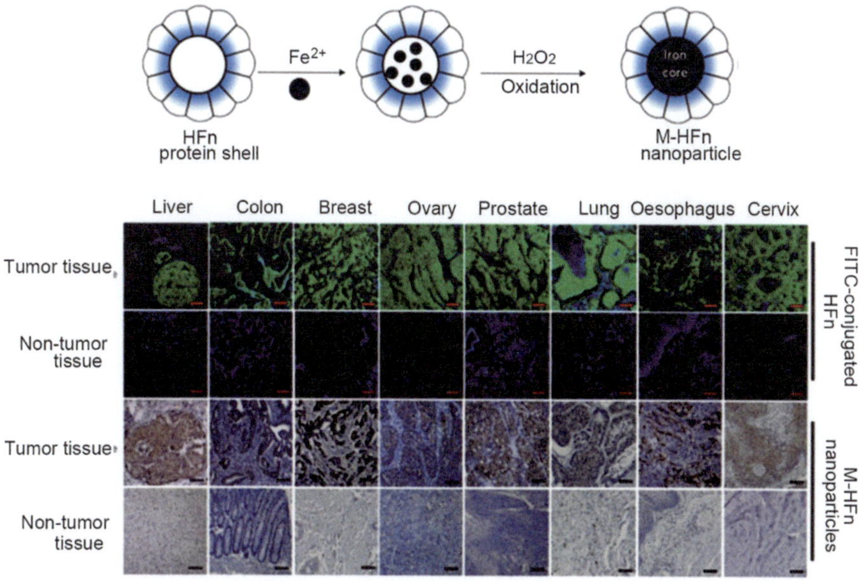

Fig. 9.8 Ferritin with peroxidase activity for tumor diagnosis [77]. (Reprinted from Ref. [77] with permission from the Nature Publishing Group)

9.3.3 *In Vivo* Oxidative Stress Regulation

Since Fe_3O_4 nanoparticles are often used for *in vivo* cancer imaging and therapy, their influence on cell viability needs to be thoroughly considered because there is intracellular H_2O_2 which can be catalyzed by Fe_3O_4 nanozymes. This may lead to changes in the levels of reactive oxygen species (ROS). Gu and coworkers found that Fe_3O_4 nanozymes perform dual enzyme-like activities, peroxidase and catalase, under acidic and neutral pH, respectively [16] (Fig. 9.9). Therefore the location of Fe_3O_4 nanozymes may lead to different outcomes in different intracellular microenvironments. Under conditions mimicking the lysosome, the nanozymes could catalyze H_2O_2 to produce hydroxyl radicals which induce glioma U251 cell damage. However, no hydroxyl radicals are produced under neutral conditions as found in the cytosol because the decomposition of H_2O_2 forms H_2O and O_2 directly under these conditions through catalase-like activity. These results provide a new way to evaluate the cytotoxicity of Fe_3O_4 nanozymes based on the intracellular location of nanoparticles. Besides the potential impact on ROS formation, Fe_3O_4 nanozymes may also cause liposome membrane damage due to lipid oxidation as reported by Wang et al. [20]. Fe_3O_4 nanozymes were found to catalyse pre-existing lipid peroxides (LOOH) or hydrogen peroxide as a substrate to initiate the chain reaction process at acidic pH via peroxidase activity. These results suggest another potential pathway to cellular oxidative damage, but this needs to be further investigated in cell models.

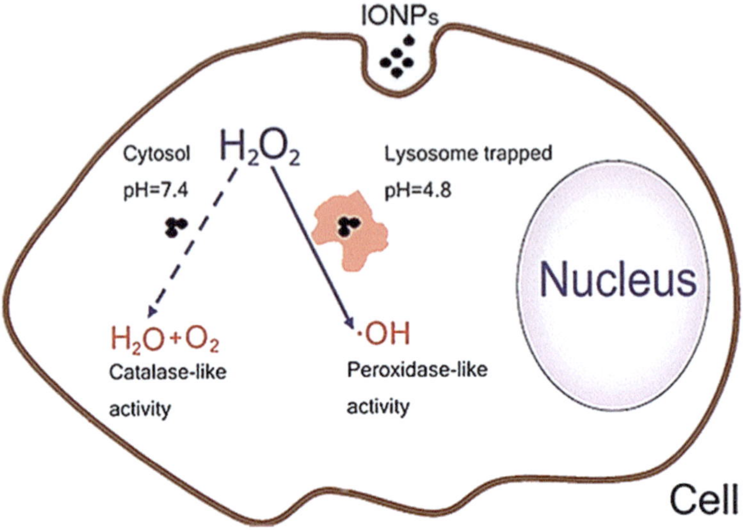

Fig. 9.9 Potential dual activities in the cell [16]. (Reprinted from Ref. [16] with permission from the American Chemical Society)

Despite potential cell damage, Fe_3O_4 nanozymes may provide cell protection by regulating ROS related oxidative stress. Huang et al. found that iron oxide nanoparticles could promote human mesenchymal stem cell (hMSC) proliferation [78]. They found that the commercial Ferucarbotran, an ionic superparamagnetic iron oxide (SPIO), could promote cell proliferation by diminishing intracellular H_2O_2. These nanozymes could also accelerate cell cycle progression. Similarly, Wang et al. found that poly(L-lysine)-modified Fe_3O_4 nanozymes could promote the proliferation of cancer stem cells from U251 glioblastoma multiform by reducing intracellular H_2O_2 [79]. Interestingly, Zhang et al. found that dietary Fe_3O_4 nanozymes could delay aging and ameliorate neurodegeneration in *Drosophila* through their intrinsic catalase-like activity [80] (Fig. 9.10). Fe_3O_4 nanozymes demonstrated the ability to protect cells from H_2O_2 induced oxidative stress and apoptosis. Furthermore, intracellular Fe_3O_4 nanozymes showed a neuroprotective effect in a Parkinson's disease (PD) cell model (PC12 cells originated from rat), which effectively inhibited α-synuclein accumulation and blocked caspase-3 activation. Fe_3O_4 nanozymes as a dietary supplement could enhance the climbing ability and prolong life span of aged *Drosophila* by reducing *in vivo* ROS. These nanozymes also alleviated neurodegeneration and increased longevity in an Alzheimer's disease (AD) model of *Drosophila*. All these investigations demonstrate that Fe_3O_4 nanozymes may perform beneficial functions including diminishing intracellular oxidative stress, delaying animal aging and protecting against neurodegeneration.

The cellular roles of Fe_3O_4 nanozymes relate to their biosafety when they are used for *in vivo* imaging or drug delivery. Although most studies indicate that Fe_3O_4 NPs have very low or negligible cytotoxicity, there is still concern about

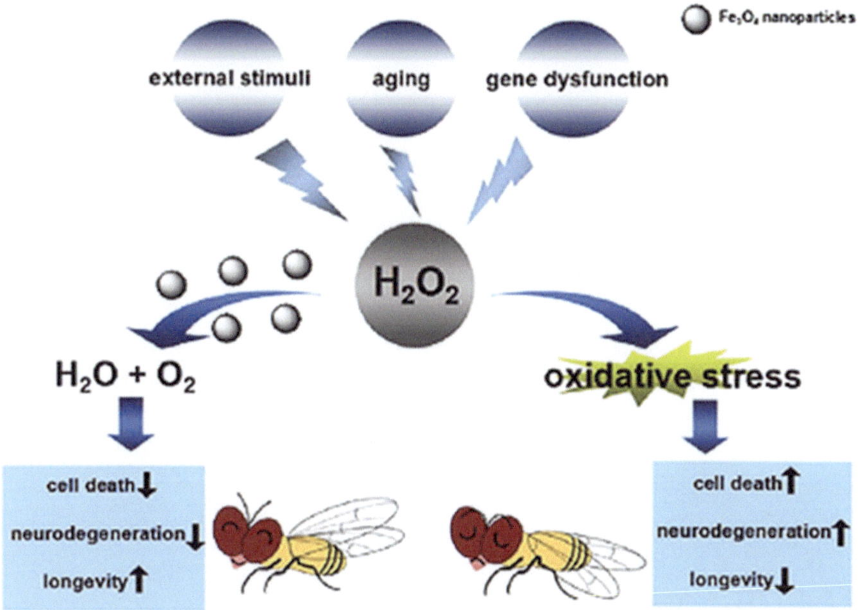

Fig. 9.10 Fe_3O_4 nanozymes delay aging and ameliorate neurodegeneration in *Drosophila* [80]. (Reprinted from Ref. [80] with permission from and John Wiley & Sons Ltd.)

their biosafety, especially for long-term use. The pH-dependent enzyme activities of Fe_3O_4 NPs provide new understanding of their potential functions when taken up by the cell and it may be possible to control cell viability and fate via Fe_3O_4 nanozymes.

Besides acting as anti-ROS agents for neuronal protection, Fe_3O_4 nanozymes also show the potential for direct tumor destruction. Generally Fe_3O_4 nanoparticles are used for cancer imaging or targeted drug delivery. The enzyme-like activity of Fe_3O_4 nanoparticles is usually neglected in tumor therapy, but theoretically, this activity could affect tumor viability by catalyzing H_2O_2 to generate toxic radicals (Fig. 9.11). Zhang et al. have shown that magnetite Fe_3O_4 nanozymes can catalyze the decomposition of hydrogen peroxide to generate reactive oxygen species (ROS) to inhibit tumors *in vivo*, indicating their potential as a theranostic reagent for tumor therapy when combined with an enhanced T2-weighted signal in magnetic resonance imaging to target the tumor [81]. Here Fe_3O_4 nanozymes (13 nm in diameter) can be retained in the tumor microenvironment by an enhanced permeability and retention effect (EPR) and internalized by tumor cells via non-specific endocytosis. The combination of Fe_3O_4 nanozymes and H_2O_2 showed a significant inhibition effect on cell viability and more than 80% of HeLa cells died after treatment at different pH values. Furthermore, treatment with the combination of Fe_3O_4 NPs and H_2O_2 showed significant inhibition of tumor growth when applied to mice bearing subcutaneous HeLa tumors which often possess an acidic

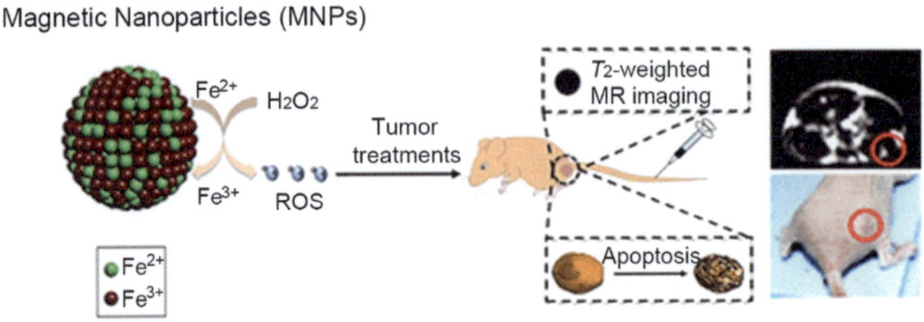

Fig. 9.11 Fe_3O_4 nanozymes for *in vivo* tumor diagnosis and therapy [81]. (Reprinted from Ref. [81] with permission from The Royal Society of Chemistry)

microenvironment which favors peroxidase-like catalysis, indicating that Fe_3O_4 nanozymes could be used for cancer theranostics, especially in epidermal diseases.

9.3.4 Hygiene and Dental Therapy

Hydrogen peroxide is a biocidal chemical that has various cleaning and disinfectant uses, including use as an anti-bacterial agent for hygiene and medical treatments. The mechanism is that H_2O_2 can release radicals slowly which damages cell membranes, proteins and nucleic acids. The addition of Fe_3O_4 nanozymes in the presence of its substrate H_2O_2 can boost the generation of hydroxyl radicals and thus enhance the antibacterial efficiency under acidic conditions.

Zhang et al. reported that Fe_3O_4 nanozymes combined with H_2O_2 have anti-bacterial activity towards *E. coli* [81]. A complete inhibition of *E. coli* proliferation (1×10^6 CFU mL^{-1}) was achieved in the presence of 20 μg mL^{-1} of iron oxide with diameter at 6 nm and 13.5 μg mL^{-1} of H_2O_2. In addition, Pan et al. designed a synergistic system by hybridizing reduced graphene with iron oxide nanoparticles (rGO-IONP). The rGO-IONP can effectively kill methicillin-resistant *Staphylococcus aureus* (MRSA) upon exposure to a near-infrared laser generating heat and hydroxyl radicals [82]. Animal experiments showed that the rGO-IONP promoted wound healing in the model infected with MRSA, indicating the system can be used as a general antibacterial strategy against drug- resistant bacteria.

The catalysis of H_2O_2 reduction to generate radicals also provides the opportunity to eliminate biofilms, which are generated by bacterial communities leading to drug resistance by limiting the penetration of antibiotics or other biocides into the protective, organic matrix of the biofilm (Fig. 9.12). Gao et al. found that Fe_3O_4 NPs with peroxidase-like activity could potentiate the efficacy of H_2O_2 in biofilm degradation and prevention via enhanced oxidative cleavage of biofilm components (model nucleic acids, proteins, and oligosaccharides) in the presence of H_2O_2 [84]. When challenged with live biofilm-producing bacteria, the Fe_3O_4 NP–H_2O_2

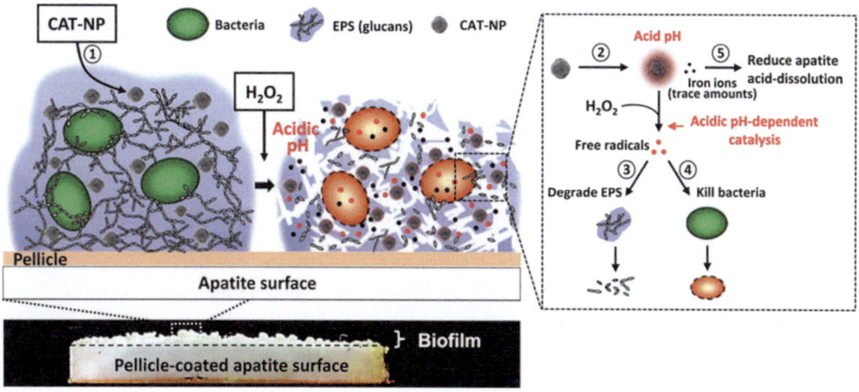

Fig. 9.12 Fe_3O_4 nanozymes for oral biofilm elimination and dental caries prevention [83]. (Reprinted from Ref. [83] with permission from Elsevier)

system efficiently broke down the existing biofilm and prevented new biofilms from forming, killing both planktonic bacteria and those within the biofilm, providing a novel strategy for biofilm elimination, and other applications utilizing oxidative breakdown. This strategy was successfully applied to dental biofilm elimination and caries prevention [83].

9.3.5 Eco Environment Applications

Besides biomedical applications, Fe_3O_4 nanozymes show great potential in environmental engineering, especially in hazard detection and removal. According to the catalytic reaction, the colorimetric signal is generated in proportion to the amount of H_2O_2 in the presence of chromogenic substrates. Therefore numerous applications have been focused on H_2O_2 detection since Wei and Wang first reported it [17, 59, 65, 72, 85–89]. H_2O_2 can be detected from many sources, including acidic rain [90], food, and living cells [91]. On the other hand, the signal is proportional to the amount of chromogenic substrates for a given amount of H_2O_2, which allows the detection of dyes in samples, such as certain pesticides [92], arsenic and antimony [93], 2,4-dinitrotoluene [94], beta-estradiol (beta-E-2) [95] and glutathione (GSH) [62]. Besides detection, catalysis of H_2O_2 decomposition by Fe_3O_4 nanozymes can also be used to degrade organic substrates for removal of pollutants, including phenol [23, 96–98], bisphenol [99, 100], aniline [22], methylene blue [101, 102], norfloxacin [103], xylenol orange [104], sulfathiazole [105] and Rhodamine B (RhB) [27, 106].

9.4 Summary and Future Perspectives

Fe_3O_4 nanoparticles show intrinsic enzyme-like activities including peroxidase- and catalase-like activities under acidic and neutral/basic pH, respectively. It should be noted that other types of iron oxide, such as Fe_2O_3, also possesses enzyme-like activities, but at lower levels than nanoscale Fe_3O_4 [9]. Therefore, Fe_3O_4 nanozymes are the main example of iron oxide nanozymes. Representing a new generation of mimetics, the catalytic properties and reaction kinetics of Fe_3O_4 nanozymes resemble natural enzymes. However, Fe_3O_4 nanozymes are much more robust and stable. More importantly, the activities of Fe_3O_4 nanozymes are tunable by controlling the size, morphology, nanostructure, dopants and surface modifications or by integration with other nanomaterials, which enables the rational design of nanozymes appropriate to the application of interest. Compared to traditional enzyme mimetics or natural enzymes, Fe_3O_4 nanozymes are multifunctional with intrinsic superparamagnetism and are readily functionalized with other molecules or labels. These features facilitate their use in a broad range of applications.

However, despite these many advantages, the activity of Fe_3O_4 nanozymes is still lower than natural enzymes and few studies on selectivity have so far been carried out. Therefore, the catalytic mechanisms need to be further investigated, especially in terms of nanoscale effects, in order to improve activity and selectivity by mimicking further features of the active site of natural enzymes. In addition, rational surface modifications are required to balance the activity and biocompatibility in biomedical applications. Finally, considering that more and more *in vivo* applications of Fe_3O_4 nanoparticles are being cdeveloped, their potential influence at the biochemical and cellular levels will be a new focus for nanozyme research (Fig. 9.13). We believe Fe_3O_4 nanozymes will have broad applications in biomedicine, industry and the environment, utilizing their intrinsic features and nanoscale effects synergistically.

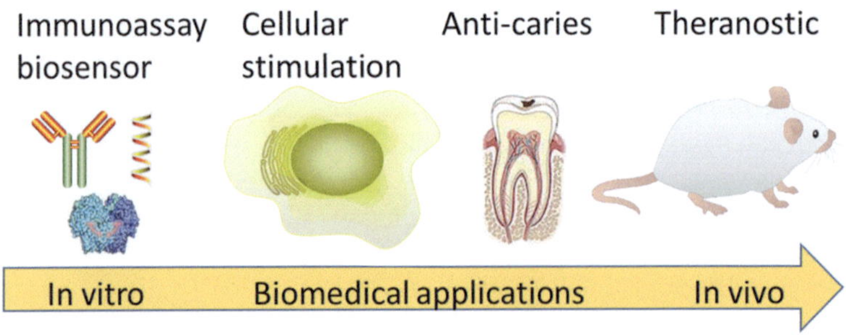

Fig. 9.13 The trend of Fe_3O_4 nanozyme applications from *in vitro* to *in vivo*

Acknowledgement This work was supported in part by the Foundation of the Thousand Talents Plan for Young Professionals and Jiangsu Specially-Appointed Professor, the Interdisciplinary Funding at Yangzhou University, Strategic Priority Research Program of the Chinese Academy of Sciences (Grant No. XDA09030306), National Natural Science Foundation of China (Grant No. 31530026 and 81671810), Natural Science Foundation of Jiangsu (Grant No. BK20161333).

References

1. Lu AH, Salabas EL, Schuth F (2007) Magnetic nanoparticles: synthesis, protection, functionalization, and application. Angew Chem 46(8):1222–1244
2. Polshettiwar V, Luque R, Fihri A, Zhu HB, Bouhrara M, Bassett JM (2011) Magnetically recoverable nanocatalysts. Chem Rev 111(5):3036–3075
3. Unsoy G, Gunduz U, Oprea O, Ficai D, Sonmez M, Radulescu M, Alexie M, Ficai A (2015) Magnetite: from synthesis to applications. Curr Top Med Chem 15(16):1622–1640
4. Xie J, Huang J, Li X, Sun S, Chen X (2009) Iron oxide nanoparticle platform for biomedical applications. Curr Med Chem 16(10):1278–1294
5. Pan Y, Du XW, Zhao F, Xu B (2012) Magnetic nanoparticles for the manipulation of proteins and cells. Chem Soc Rev 41(7):2912–2942
6. Frimpong RA, Hilt JZ (2010) Magnetic nanoparticles in biomedicine: synthesis, functionalization and applications. Nanomedicine-UK 5(9):1401–1414
7. Colombo M, Carregal-Romero S, Casula MF, Gutierrez L, Morales MP, Bohm IB, Heverhagen JT, Prosperi D, Parak WJ (2012) Biological applications of magnetic nanoparticles. Chem Soc Rev 41(11):4306–4334
8. Ho D, Sun XL, Sun SH (2011) Monodisperse magnetic nanoparticles for theranostic applications. Accounts Chem Res 44(10):875–882
9. Gao LZ, Fan KL, Yan XY (2017) Iron oxide nanozyme: a multifunctional enzyme mimetic for biomedical applications. Theranostics 7(13):3207–3227
10. Gao L, Zhuang J, Nie L, Zhang J, Zhang Y, Gu N, Wang T, Feng J, Yang D, Perrett S, Yan X (2007) Intrinsic peroxidase-like activity of ferromagnetic nanoparticles. Nat Nanotechnol 2(9):577–583
11. Wei H, Wang E (2013) Nanomaterials with enzyme-like characteristics (nanozymes): next-generation artificial enzymes. Chem Soc Rev 42(14):6060–6093
12. Gao L, Yan X (2016) Nanozymes: an emerging field bridging nanotechnology and biology. Sci China Life Sci 59(4):400–402
13. Gao LZ, Yan XY (2013) Discovery and current application of nanozyme. Prog Biochem Biophys 40(10):892–902
14. Shin HY, Park TJ, Kim MI (2015) Recent research trends and future prospects in nanozymes. J Nanomater
15. Lin Y, Ren J, Qu X (2014) Catalytically active nanomaterials: a promising candidate for artificial enzymes. Acc Chem Res 47(4):1097–1105
16. Chen Z, Yin JJ, Zhou YT, Zhang Y, Song L, Song M, Hu S, Gu N (2012) Dual enzyme-like activities of iron oxide nanoparticles and their implication for diminishing cytotoxicity. ACS Nano 6(5):4001–4012
17. Wei H, Wang E (2008) Fe3O4 magnetic nanoparticles as peroxidase mimetics and their applications in H2O2 and glucose detection. Anal Chem 80(6):2250–2254
18. Liu S, Lu F, Xing R, Zhu JJ (2011) Structural effects of Fe3O4 nanocrystals on peroxidase-like activity. Chemistry 17(2):620–625
19. Wang H, Jiang H, Wang S, Shi WB, He JC, Liu H, Huang YM (2014) Fe3O4-MWCNT magnetic nanocomposites as efficient peroxidase mimic catalysts in a Fenton-like reaction for water purification without pH limitation. RSC Adv 4(86):45809–45815

20. Wang LJ, Min Y, Xu DD, Yu FJ, Zhou WZ, Cuschieri A (2014) Membrane lipid peroxidation by the peroxidase-like activity of magnetite nanoparticles. Chem Commun 50(76):11147–11150
21. Liu Y, Yuan M, Qiao LJ, Guo R (2014) An efficient colorimetric biosensor for glucose based on peroxidase-like protein-Fe3O4 and glucose oxidase nanocomposites. Biosens Bioelectron 52:391–396
22. Zhang SX, Zhao XL, Niu HY, Shi YL, Cai YQ, Jiang GB (2009) Superparamagnetic Fe3O4 nanoparticles as catalysts for the catalytic oxidation of phenolic and aniline compounds. J Hazard Mater 167(1–3):560–566
23. Wu XC, Zhang Y, Han T, Wu HX, Guo SW, Zhang JY (2014) Composite of graphene quantum dots and Fe3O4 nanoparticles: peroxidase activity and application in phenolic compound removal. RSC Adv 4(7):3299–3305
24. Qian J, Yang XW, Jiang L, Zhu CD, Mao HP, Wang K (2014) Facile preparation of Fe3O4 nanospheres/reduced graphene oxide nanocomposites with high peroxidase-like activity for sensitive and selective colorimetric detection of acetylcholine. Sens Actuat B-Chem 201:160–166
25. Qi CC, Zheng JB (2015) Novel nonenzymatic hydrogen peroxide sensor based on Fe3O4/PPy/Ag nanocomposites. J Electroanal Chem 747:53–58
26. Yang X, Wang LN, Zhou GZ, Sui N, Gu YX, Wan J (2015) Electrochemical detection of H2O2 based on Fe3O4 nanoparticles with graphene oxide and polyamidoamine dendrimer. J Clust Sci 26(3):789–798
27. Wang N, Zhu LH, Wang DL, Wang MQ, Lin ZF, Tang HQ (2010) Sono-assisted preparation of highly-efficient peroxidase-like Fe3O4 magnetic nanoparticles for catalytic removal of organic pollutants with H2O2. Ultrason Sonochem 17(3):526–533
28. Peng FF, Zhang Y, Gu N (2008) Size-dependent peroxidase-like catalytic activity of Fe3O4 nanoparticles. Chin Chem Lett 19(6):730–733
29. Cheng XL, Jiang JS, Jiang DM, Zhao ZJ (2014) Synthesis of rhombic dodecahedral Fe3O4 nanocrystals with exposed high-energy {110} facets and their peroxidase-like activity and Lithium storage properties. J Phys Chem C 118(24):12588–12598
30. Zhang K, Zuo W, Wang ZY, Liu J, Li TR, Wang BD, Yang ZY (2015) A simple route to CoFe2O4 nanoparticles with shape and size control and their tunable peroxidase-like activity. RSC Adv 5(14):10632–10640
31. Ma M, Xie J, Zhang Y, Chen ZP, Gu N (2013) Fe3O4@Pt nanoparticles with enhanced peroxidase-like catalytic activity. Mater Lett 105:36–39
32. Wang CQ, Qian J, Wang K, Yang XW, Liu Q, Hao N, Wang CK, Dong XY, Huang XY (2016) Colorimetric aptasensing of ochratoxin A using Au@Fe3O4 nanoparticles as signal indicator and magnetic separator. Biosens Bioelectron 77:1183–1191
33. Lee Y, Garcia MA, Frey Huls NA, Sun S (2010) Synthetic tuning of the catalytic properties of Au-Fe3O4 nanoparticles. Angew Chem 49(7):1271–1274
34. Sun HY, Jiao XL, Han YY, Jiang Z, Chen DR (2013) Synthesis of Fe3O4-Au Nanocomposites with enhanced peroxidase-like activity. Eur J Inorg Chem 1:109–114
35. Fan KL, Wang H, Xi JQ, Liu Q, Meng XQ, Duan DM, Gao LZ, Yan XY (2017) Optimization of Fe3O4 nanozyme activity via single amino acid modification mimicking an enzyme active site. Chem Commun 53(2):424–427
36. Zhang XQ, Gong SW, Zhang Y, Yang T, Wang CY, Gu N (2010) Prussian blue modified iron oxide magnetic nanoparticles and their high peroxidase-like activity. J Mater Chem 20(24):5110–5116
37. Hu SL, Zhang XQ, Zang FC, Zhang Y, Zhang W, Wu YH, Song MJ, Wang YH, Gu N (2016) Surface modified Iron oxide nanoparticles as Fe source precursor to induce the formation of Prussian blue nanocubes. J Nanosci Nanotechnol 16(2):1967–1974
38. Zhang Z, Zhang X, Liu B, Liu J (2017) Molecular imprinting on inorganic nanozymes for hundred-fold enzyme specificity. J Am Chem Soc 139(15):5412–5419
39. Lee JW, Jeon HJ, Shin HJ, Kang JK (2012) Superparamagnetic Fe3O4 nanoparticles-carbon nitride nanotube hybrids for highly efficient peroxidase mimetic catalysts. Chem Commun 48(3):422–424

40. An Q, Sun C, Li D, Xu K, Guo J, Wang C (2013) Peroxidase-like activity of Fe3O4@carbon nanoparticles enhances ascorbic acid-induced oxidative stress and selective damage to PC-3 prostate cancer cells. ACS Appl Mater Interfaces 5(24):13248–13257
41. Li Q, Tang GG, Xiong XW, Cao YL, Chen LL, Xu FG, Tan HL (2015) Carbon coated magnetite nanoparticles with improved water-dispersion and peroxidase-like activity for colorimetric sensing of glucose. Sens Actuat B-Chem 215:86–92
42. Zubir NA, Yacou C, Motuzas J, Zhang X, Diniz da Costa JC (2014) Structural and functional investigation of graphene oxide-Fe3O4 nanocomposites for the heterogeneous Fenton-like reaction. Sci Rep 4:4594
43. Dong YL, Zhang HG, Rahman ZU, Su L, Chen XJ, Hu J, Chen XG (2012) Graphene oxide-Fe3O4 magnetic nanocomposites with peroxidase-like activity for colorimetric detection of glucose. Nanoscale 4(13):3969–3976
44. Song Y, Qu K, Zhao C, Ren J, Qu X (2010) Graphene oxide: intrinsic peroxidase catalytic activity and its application to glucose detection. Adv Mater 22(19):2206–2210
45. Gao LZ, Wu JM, Lyle S, Zehr K, Cao LL, Gao D (2008) Magnetite nanoparticle-linked immunosorbent assay. J Phys Chem C 112(44):17357–17361
46. Wang X, Niessner R, Tang DP, Knopp D (2016) Nanoparticle-based immunosensors and immunoassays for aflatoxins. Anal Chim Acta 912:10–23
47. Tang ZW, Wu H, Zhang YY, Li ZH, Lin YH (2011) Enzyme-mimic activity of ferric nano-core residing in ferritin and its biosensing applications. Anal Chem 83(22):8611–8616
48. Bhattacharya D, Baksi A, Banerjee I, Ananthakrishnan R, Maiti TK, Pramanik P (2011) Development of phosphonate modified Fe(1-x)MnxFe2O4 mixed ferrite nanoparticles: novel peroxidase mimetics in enzyme linked immunosorbent assay. Talanta 86:337–348
49. Yang MZ, Guan YP, Yang Y, Xie L, Xia TT, Xiong WB, Guo C (2014) Immunological detection of hepatocellular carcinoma biomarker GP73 based on dissolved magnetic nanoparticles. Colloid Surf A 443:280–285
50. Liu Y, Du JJ, Yan M, Lau MY, Hu J, Han H, Yang OO, Liang S, Wei W, Wang H, Li JM, Zhu XY, Shi LQ, Chen W, Ji C, Lu YF (2013) Biomimetic enzyme nanocomplexes and their use as antidotes and preventive measures for alcohol intoxication. Nat Nanotechnol 8(3):187–192
51. Yang MZ, Guan YP, Yang Y, Xia TT, Xiong WB, Guo C (2014) A sensitive and rapid immunoassay for mycoplasma pneumonia based on Fe3O4 nanoparticles. Mater Lett 137:113–116
52. Woo MA, Kim MI, Jung JH, Park KS, Seo TS, Park HG (2013) A novel colorimetric immunoassay utilizing the peroxidase mimicking activity of magnetic nanoparticles. Int J Mol Sci 14(5):9999–10014
53. Il Kim M, Kim MS, Woo MA, Ye Y, Kang KS, Lee J, Park HG (2014) Highly efficient colorimetric detection of target cancer cells utilizing superior catalytic activity of graphene oxide-magnetic-platinum nanohybrids. Nanoscale 6(3):1529–1536
54. Perez JM (2007) Iron oxide nanoparticles – hidden talent. Nat Nanotechnol 2(9):535–536
55. Duan DM, Fan KL, Zhang DX, Tan SG, Liang MF, Liu Y, Zhang JL, Zhang PH, Liu W, Qiu XG, Kobinger GP, Gao GF, Yan XY (2015) Nanozyme-strip for rapid local diagnosis of Ebola. Biosens Bioelectron 74:134–141
56. Park KS, Kim MI, Cho DY, Park HG (2011) Label-free colorimetric detection of nucleic acids based on target-induced shielding against the peroxidase-mimicking activity of magnetic nanoparticles. Small 7(11):1521–1525
57. Thiramanas R, Jangpatarapongsa K, Tangboriboonrat P, Polpanich D (2013) Detection of Vibrio cholerae using the intrinsic catalytic activity of a magnetic polymeric nanoparticle. Anal Chem 85(12):5996–6002
58. Liu QY, Zhang LY, Li H, Jia QY, Jiang YL, Yang YT, Zhu RR (2015) One-pot synthesis of porphyrin functionalized gamma-Fe2O3 nanocomposites as peroxidase mimics for H2O2 and glucose detection. Mater Sci Eng C-Mater 55:193–200
59. Gao Y, Wang GN, Huang H, Hu JJ, Shah SM, Su XG (2011) Fluorometric method for the determination of hydrogen peroxide and glucose with Fe3O4 as catalyst. Talanta 85(2):1075–1080

60. Kim MI, Shim J, Li T, Lee J, Park HG (2011) Fabrication of nanoporous nanocomposites entrapping Fe3O4 magnetic nanoparticles and oxidases for colorimetric biosensing. Chem-Eur J 17(38):10700–10707
61. Liu CH, Tseng WL (2011) Oxidase-functionalized Fe3O4 nanoparticles for fluorescence sensing of specific substrate. Anal Chim Acta 703(1):87–93
62. Ma YH, Zhang ZY, Ren CL, Liu GY, Chen XG (2012) A novel colorimetric determination of reduced glutathione in A549 cells based on Fe3O4 magnetic nanoparticles as peroxidase mimetics. Analyst 137(2):485–489
63. Wang H, Li S, Si YM, Sun ZZ, Li SY, Lin YH (2014) Recyclable enzyme mimic of cubic Fe3O4 nanoparticles loaded on graphene oxide-dispersed carbon nanotubes with enhanced peroxidase-like catalysis and electrocatalysis. J Mater Chem B 2(28):4442–4448
64. Yang ZH, Chai YQ, Yuan R, Zhuo Y, Li Y, Han J, Liao N (2014) Hollow platinum decorated Fe3O4 nanoparticles as peroxidase mimetic couple with glucose oxidase for pseudobienzyme electrochemical immunosensor. Sens Actuat B-Chem 193:461–466
65. Chang Q, Tang HQ (2014) Optical determination of glucose and hydrogen peroxide using a nanocomposite prepared from glucose oxidase and magnetite nanoparticles immobilized on graphene oxide. Microchim Acta 181(5–6):527–534
66. Liu QY, Li H, Zhao QR, Zhu RR, Yang YT, Jia QY, Bian B, Zhuo LH (2014) Glucose-sensitive colorimetric sensor based on peroxidase mimics activity of porphyrin-Fe(3)o(4) nanocomposites. Mater Sci Eng C-Mater 41:142–151
67. Shi Y, Su P, Wang YY, Yang Y (2014) Fe3O4 peroxidase mimetics as a general strategy for the fluorescent detection of H2O2-involved systems. Talanta 130:259–264
68. Pan Y, Li N, Mu JS, Zhou RH, Xu Y, Cui DZ, Wang Y, Zhao M (2015) Biogenic magnetic nanoparticles from Burkholderia sp. YN01 exhibiting intrinsic peroxidase-like activity and their applications. Appl Microbiol Biotechnol 99(2):703–715
69. Wang YH, Zhou B, Wu S, Wang KM, He XX (2015) Colorimetric detection of hydrogen peroxide and glucose using the magnetic mesoporous silica nanoparticles. Talanta 134:712–717
70. Shi Y, Huang J, Wang JN, Su P, Yang Y (2015) A magnetic nanoscale Fe3O4/P beta-CD composite as an efficient peroxidase mimetic for glucose detection. Talanta 143:457–463
71. Kim MI, Cho D, Park HG (2015) Colorimetric quantification of glucose and cholesterol in human blood using a nanocomposite entrapping magnetic nanoparticles and oxidases. J Nanosci Nanotechnol 15(10):7955–7961
72. Zhang J, Yang C, Chen CX, Yang XR (2013) Determination of nitrite and glucose in water and human urine with light-up chromogenic response based on the expeditious oxidation of 3,3′, 5,5′-tetramethylbenzidine by peroxynitrous acid. Analyst 138(8):2398–2404
73. Liang MM, Fan KL, Pan Y, Jiang H, Wang F, Yang DL, Lu D, Feng J, Zhao JJ, Yang L, Yan XY (2013) Fe3O4 magnetic nanoparticle peroxidase mimetic-based colorimetric assay for the rapid detection of organophosphorus pesticide and nerve agent. Anal Chem 85(1):308–312
74. Kim MI, Shim J, Li T, Woo MA, Cho D, Lee J, Park HG (2012) Colorimetric quantification of galactose using a nanostructured multi-catalyst system entrapping galactose oxidase and magnetic nanoparticles as peroxidase mimetics. Analyst 137(5):1137–1143
75. Kim MI, Shim J, Parab HJ, Shin SC, Lee J, Park HG (2012) A convenient alcohol sensor using one-pot nanocomposite entrapping alcohol oxidase and magnetic nanoparticles as peroxidase mimetics. J Nanosci Nanotechnol 12(7):5914–5919
76. Zhuang J, Fan KL, Gao LZ, Lu D, Feng J, Yang DL, Gu N, Zhang Y, Liang MM, Yan XY (2012) Ex vivo detection of Iron oxide magnetic nanoparticles in mice using their intrinsic peroxidase-mimicking activity. Mol Pharm 9(7):1983–1989
77. Fan KL, Cao CQ, Pan YX, Lu D, Yang DL, Feng J, Song LN, Liang MM, Yan XY (2012) Magnetoferritin nanoparticles for targeting and visualizing tumour tissues. Nat Nanotechnol 7(7):459–464
78. Huang DM, Hsiao JK, Chen YC, Chien LY, Yao M, Chen YK, Ko BS, Hsu SC, Tai LA, Cheng HY, Wang SW, Yang CS, Chen YC (2009) The promotion of human mesenchymal stem cell proliferation by superparamagnetic iron oxide nanoparticles. Biomaterials 30(22):3645–3651

79. Wang XQ, Tu Q, Zhao B, An YF, Wang JC, Liu WM, Yuan MS, Ahmed SM, Xu J, Liu R, Zhang YR, Wang JY (2013) Effects of poly(L-lysine)-modified Fe3O4 nanoparticles on endogenous reactive oxygen species in cancer stem cells. Biomaterials 34(4):1155–1169
80. Zhang Y, Wang ZY, Li XJ, Wang L, Yin M, Wang LH, Chen N, Fan CH, Song HY (2016) Dietary Iron oxide nanoparticles delay aging and ameliorate neurodegeneration in Drosophila. Adv Mater 28(7):1387–1393
81. Zhang D, Zhao YX, Gao YJ, Gao FP, Fan YS, Li XJ, Duan ZY, Wang H (2013) Anti-bacterial and in vivo tumor treatment by reactive oxygen species generated by magnetic nanoparticles. J Mater Chem B 1(38):5100–5107
82. Pan WY, Huang CC, Lin TT, Hu HY, Lin WC, Li MJ, Sung HW (2016) Synergistic antibacterial effects of localized heat and oxidative stress caused by hydroxyl radicals mediated by graphene/iron oxide-based nanocomposites. Nanomed-Nanotechnol 12(2):431–438
83. Gao LZ, Liu Y, Kim D, Li Y, Hwang G, Naha PC, Cormode DP, Koo H (2016) Nanocatalysts promote Streptococcus mutans biofilm matrix degradation and enhance bacterial killing to suppress dental caries in vivo. Biomaterials 101:272–284
84. Gao L, Giglio KM, Nelson JL, Sondermann H, Travis AJ (2014) Ferromagnetic nanoparticles with peroxidase-like activity enhance the cleavage of biological macromolecules for biofilm elimination. Nanoscale 6(5):2588–2593
85. You X, Kim J, Pak YK, Pak JJ (2013) Preparation and application of graphene-poly (diallyldimethylammoniumchloride)-Iron oxide nanoparticles buckypaper for hydrogen peroxide detection. J Nanosci Nanotechno 13(11):7349–7357
86. Gao Y, Wei Z, Li F, Yang ZM, Chen YM, Zrinyi M, Osada Y (2014) Synthesis of a morphology controllable Fe3O4 nanoparticle/hydrogel magnetic nanocomposite inspired by magnetotactic bacteria and its application in H2O2 detection. Green Chem 16(3):1255–1261
87. Ye YP, Kong T, Yu XF, Wu YK, Zhang K, Wang XP (2012) Enhanced nonenzymatic hydrogen peroxide sensing with reduced graphene oxide/ferroferric oxide nanocomposites. Talanta 89:417–421
88. Jiang ZL, Kun L, Ouyang HX, Liang AH, Jiang HS (2011) A simple and sensitive fluorescence quenching method for the determination of H2O2 using Rhodamine B and Fe3O4 nanocatalyst. J Fluoresc 21(5):2015–2020
89. Chang Q, Deng KJ, Zhu LH, Jiang GD, Yu C, Tang HQ (2009) Determination of hydrogen peroxide with the aid of peroxidase-like Fe3O4 magnetic nanoparticles as the catalyst. Microchim Acta 165(3–4):299–305
90. Zhuang J, Zhang JB, Gao LZ, Zhang Y, Gu N, Feng J, Yang DL, Yan XY (2008) A novel application of iron oxide nanoparticles for detection of hydrogen peroxide in acid rain. Mater Lett 62(24):3972–3974
91. Fang HT, Pan YL, Shan WQ, Guo ML, Nie Z, Huang Y, Yao SZ (2014) Enhanced nonenzymatic sensing of hydrogen peroxide released from living cells based on Fe3O4/self-reduced graphene nanocomposites. Anal Method-UK 6(15):6073–6081
92. Guan GJ, Yang L, Mei QS, Zhang K, Zhang ZP, Han MY (2012) Chemiluminescence switching on peroxidase-like Fe3O4 nanoparticles for selective detection and simultaneous determination of various pesticides. Anal Chem 84(21):9492–9497
93. Jia Y, Yu HM, Wu L, Hou XD, Yang L, Zheng CB (2015) Three birds with one Fe3O4 nanoparticle: integration of microwave digestion, solid phase extraction, and magnetic separation for sensitive determination of arsenic and antimony in fish. Anal Chem 87(12):5866–5871
94. Nie DX, Shi GY, Yu YY (2016) Fe3O4 magnetic nanoparticles as peroxidase mimetics used in colorimetric determination of 2,4-Dinitrotoluene. Chin J Anal Chem 44(2):179–184
95. Wei SL, Li JW, Liu Y (2015) Colourimetric assay for beta-estradiol based on the peroxidase-like activity of Fe3O4@mSiO(2)@HP-beta-CD nanoparticles. RSC Adv 5(130):107670–107679
96. Wang W, Liu Y, Li TL, Zhou MH (2014) Heterogeneous Fenton catalytic degradation of phenol based on controlled release of magnetic nanoparticles. Chem Eng J 242:1–9

97. Zhang JB, Zhuang J, Gao LZ, Zhang Y, Gu N, Feng J, Yang DL, Zhu JD, Yan XY (2008) Decomposing phenol by the hidden talent of ferromagnetic nanoparticles. Chemosphere 73(9):1524–1528
98. Wang W, Mao Q, He HH, Zhou MH (2013) Fe3O4 nanoparticles as an efficient heterogeneous Fenton catalyst for phenol removal at relatively wide pH values. Water Sci Technol 68(11):2367–2373
99. Huang RX, Fang ZQ, Fang XB, Tsang EP (2014) Ultrasonic Fenton-like catalytic degradation of bisphenol A by ferroferric oxide (Fe3O4) nanoparticles prepared from steel pickling waste liquor. J Colloid Interf Sci 436:258–266
100. Huang RX, Fang ZQ, Yan XM, Cheng W (2012) Heterogeneous sono-Fenton catalytic degradation of bisphenol A by Fe3O4 magnetic nanoparticles under neutral condition. Chem Eng J 197:242–249
101. Wang XS, Huang H, Li GQ, Liu Y, Huang JL, Yang DP (2014) Hydrothermal synthesis of 3D hollow porous Fe3O4 microspheres towards catalytic removal of organic pollutants. Nanoscale Res Lett 9:648
102. Zhang XL, He ML, Liu JH, Liao R, Zhao LQ, Xie JR, Wang RJ, Yang ST, Wang HF, Liu YF (2014) Fe3O4@C nanoparticles as high-performance Fenton-like catalyst for dye decoloration. Chin Sci Bull 59(27):3406–3412
103. Niu HY, Dizhang NH, Meng ZF, Cai YQ (2012) Fast defluorination and removal of norfloxacin by alginate/Fe@Fe3O4 core/shell structured nanoparticles. J Hazard Mater 227:195–203
104. Zhu MY, Diao GW (2011) Synthesis of porous Fe3O4 nanospheres and its application for the catalytic degradation of Xylenol Orange. J Phys Chem C 115(39):18923–18934
105. Niu HY, Zhang D, Zhang SX, Zhang XL, Meng ZF, Cai YQ (2011) Humic acid coated Fe3O4 magnetic nanoparticles as highly efficient Fenton-like catalyst for complete mineralization of sulfathiazole. J Hazard Mater 190(1–3):559–565
106. Wang N, Zhu LH, Wang MQ, Wang DL, Tang HQ (2010) Sono-enhanced degradation of dye pollutants with the use of H2O2 activated by Fe3O4 magnetic nanoparticles as peroxidase mimetic. Ultrason Sonochem 17(1):78–83

Chapter 10
Self-Assembly of Ferritin: Structure, Biological Function and Potential Applications in Nanotechnology

Soumyananda Chakraborti and Pinak Chakrabarti

Abstract Protein cages are normally formed by the self-assembly of multiple protein subunits and ferritin is a typical example of a protein cage structure. Ferritin is a ubiquitous multi-subunit iron storage protein formed by 24 polypeptide chains that self-assemble into a hollow, roughly spherical protein cage. Ferritin has external and internal diameters of approximately 12 nm and 8 nm, respectively. Functionally, ferritin performs iron sequestration and is highly conserved in evolution. The interior cavity of ferritin provides a unique reaction vessel to carry out reactions separated from the exterior environment. In nature, the cavity is utilized for sequestration of iron and bio-mineralization as a mechanism to render iron inert and safe from the external environment. Material scientists have been inspired by this system and exploited a range of ferritin superfamily proteins as supramolecular templates to encapsulate different carrier molecules ranging from cancer drugs to therapeutic proteins, in addition to using ferritin proteins as well-defined building blocks for fabrication. Besides the interior cavity, the exterior surface and sub-unit interface of ferritin can be modified without affecting ferritin assembly.

Keywords Iron storage protein · Protein self assembly · Ferritin structure and function · Nanotechnology application of ferritin

---

Brief Description This chapter will describe the self-assembly properties of the iron-carrying protein ferritin into nanoscale structures and their biological properties as well as their applications.

S. Chakraborti (✉)
Department of Biochemistry, Bose Institute, Kolkata, India

Malopolska Centre of Biotechnology, Jagiellonian University, Krakow, Poland
e-mail: soumyananda.chakraborti@uj.edu.pl

P. Chakrabarti
Department of Biochemistry, Bose Institute, Kolkata, India

S. Perrett et al. (eds.), *Biological and Bio-inspired Nanomaterials*, Advances in Experimental Medicine and Biology 1174,
https://doi.org/10.1007/978-981-13-9791-2_10

10.1 Introduction

Nature uses the self-assembly properties of proteins to produce a wide variety of large, complex, and highly symmetric protein architectures [1, 2]. Understanding the details of self-assembly is not only important for basic science, but this knowledge is also essential to build ambitious and challenging technologies, such as programmable nano-machines [3].

Naturally occurring proteins consist of amino acids forming a polypeptide chain which is then folded in an energy efficient manner to produce the stable three-dimensional structure of the protein [4]. The thermodynamics of protein folding is such that some of its residues orientate and project themselves on the surface such that they can interact with other proteins [4]. Thus, the information flow defined by the Central Dogma in molecular biology that begins with the genetic code can be further expanded through protein folding into the realm of protein-protein interactions [5]. These protein-protein interactions are the basis of self-assembly events of large, complex macromolecular protein structures [6].

Self-assembly of proteins can generate a variety of supramolecular structures with a broad range of biological functions. These complexes include filaments, protein lattices and symmetric cages [7]. Among different supramolecular protein complexes, protein cages are probably the most sophisticated protein-based architectures. Their self-assembly from a small number of subunits into symmetrical, monodisperse architectures has inspired scientists from many disciplines. Cage architecture is abundant in nature ranging from virus capsids [8] to bacterial micro-compartments such as the carboxysome [9]. Cage architectures have also been observed in chaperones [10], DNA binding proteins [11]and ferritin proteins [12] (Fig. 10.1). It has been found that the majority of protein cages are hollow and spherical though there are some notable exceptions, and they often possess internal symmetry, either icosahedral, octahedral or tetrahedral, which plays an important role in controlling their inter-subunit interactions. In the last two decades, protein cages have been developed as platforms for nanomaterial synthesis, mostly because of their remarkable diversity in size, shape and structures. Furthermore, their ease of production in large quantities using biological systems, as well as their biocompatibility, makes them an attractive choice for drug delivery, cell specific targeting and enzyme catalysis [13]. Among protein cages, ferritin is one of the most commonly used because of its unique bio-mineralization ability and extraordinary stability [12].

10.2 Historical Perspective

Ferritin was first discovered in 1937 and was isolated from horse spleen. However, the presence of ferritin in human blood serum was determined even much later [14]. To date, the ferritin family of proteins is one of the most studied (probably the

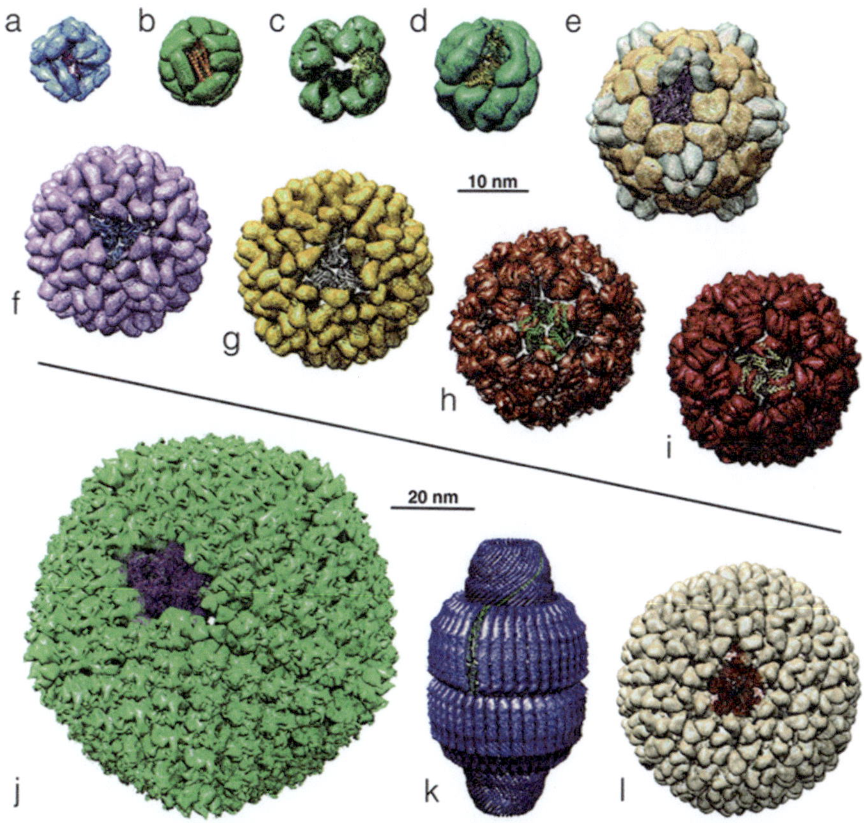

Fig. 10.1 Structures of different protein cages. (**a**) Small heat-shock protein (HSP), (**b**) Apoferritin, (**c**) Pyruvate dehydrogenase multi-enzyme complex, (**d**) Thermosome, (**e**) Cowpea mosaic virus (CPMV), (**f**) Brome mosaic virus, (**g**) Cowpea chlorotic mottle virus (CCMV), (**h**) Bacteriophage lambda, (**i**) Bacteriophage MS2, (**j**) Human adenovirus (AdV), (**k**) Vault particle, and (**l**) Bacteriophage P22. Adapted with permission from ref. [8] copyright Royal Society of Chemistry, 2016

second most studied protein after hemoglobin), and serum ferritin is still considered as an important clinical marker for inflammation, infection and malignancy. Although ferritin has been extensively studied and its clinical relevance has been established for many years, there are still many fundamental questions regarding ferritin biology that remain to be answered, such as its tissue of origin, secretory pathway, interaction partners, cell surface receptors and its degradation pathways. The first crystal structure of ferritin was determined in 1991 [15], and since then ferritins from different organisms including animals, plants and bacteria have been isolated, purified and crystallized.

Ferritin is generally found in the cytosol; however, a mitochondrial form of ferritin has been isolated recently [14]. Studies have further shown that extracellular ferritin can function as an iron carrier to provide iron to cells. Compared to transferrin (another iron storage protein), which carries a maximum of two iron atoms, a single ferritin molecule can encapsulate up to 4500 iron atoms, thus making it potentially a very effective iron storage and delivery system [16].

10.3 Ferritin: Basic Biology

The ferritin proteins are ubiquitous in all forms of life (eukaryotes, archaea and bacteria); the only notable exception is yeast. Functionally, ferritin is an iron storage protein whose *in vivo* role is iron storage to prevent metal toxicity in the cell. Studies have further revealed that ferritin not only stores iron but also mineralizes excess iron in the form of hydrous ferric oxide in its cavity. The ability to sequester iron allows ferritin to perform dual functions in both iron detoxification and maintaining the cellular iron reserve [12, 14]. It is noteworthy that most mammalian ferritin can be constituted either from heavy chain (H) or light chain (L) or from both (Fig. 10.2). The heavier H-isoform and the lighter L-isoform are most abundant in heart and liver, respectively [12, 14]. The molecular weight of the H and L subunits are 21 kDa and 19 kDa respectively. The genes responsible for H and L ferritin are found in chromosomes 11q and 19q of the human genome [14]. [17], Amino acid sequence similarity between the H and L subunits of ferritin is around 50% in mammals. The H-subunit normally contains the active and highly conserved "ferroxidase" oxidation site, which binds and oxidizes ferrous ions. The L-chain ferritins oxidize iron very slowly as they do not possess a catalytic ferroxidase center. Individual subunits are arranged with tetrahedral symmetry to form a protein cage with twofold, threefold and fourfold rotational symmetry axes [18]. The protein shell of mammalian ferritin is usually heterogeneous. The ratio of H and L subunits in the complex depends on the relative expression of the two genes, and the ratio varies depending on tissue type. In contrast, plant and bacterial ferritins tend to be homo-polymers. Interestingly the most commonly used equine (horse) spleen ferritin contains approximately 90% L-subunit ferritin. Amphibians have an additional ("M") type ferritin [19]; the ferritin from plants and bacteria closely resembles the vertebrate H-type.

Generally, ferritin cages contain 24 protein subunits, with subunit self-assembly occurring through a dimeric intermediate, subsequently forming a dodecameric cage with an outer diameter of approximately 12 nm and an inner cavity of 8 nm, which is usually filled with a ferric oxo-hydroxy core. Ferritin without the inorganic core is called apoferritin [12]. The ferritin protein cage is stable upto 85 °C and tolerates reasonably high levels of urea, guanidinium chloride, and many other denaturants at neutral pH. Ferritin isolation from natural sources most of the time results in heterogeneity, generating aggregated dimers and trimers [20]. In nature, the ferritin cavity is used to mineralize iron, although it has been demonstrated recently that this cavity can be effectively utilized as a reaction vessel for generating different metal nanoparticles with definite size distributions [12].

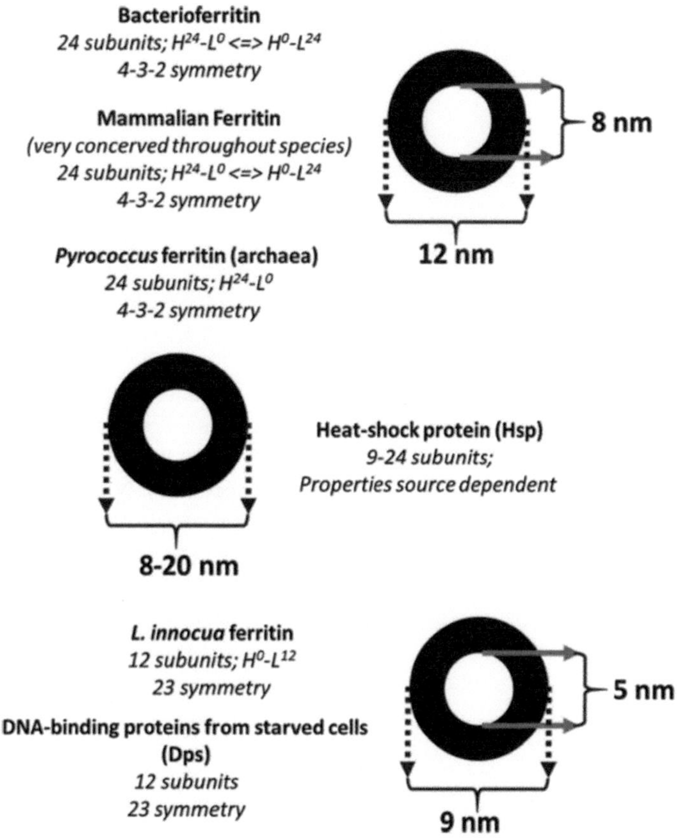

Fig. 10.2 Ferritin and ferritin-like protein cages explored from different sources. Here, the ferritins isolated from different organisms are illustrated; descriptions are related to number of subunits, symmetry and composition. Adapted with permission from ref. [12] copyright American Chemical Society, 2015

10.4 Ferritin Protein Family

The ferritin protein superfamily can be divided into three major sub-classes: the classical ferritins (Ftn), the bacterioferritins (Bfr) and the DNA-binding proteins from starved cells, also known as Dps (Fig. 10.2) [21]. The classical Ftn and Bfr proteins belong to the maxi-ferritin family, as structurally they are slightly larger. In contrast, Dps proteins belong to the mini-ferritin protein family. Interestingly all three protein sub-families share a four-helix bundle fold. The Bfr proteins seem to possess almost identical quaternary structure to the classical Ftn proteins and both are assembled from 24 subunits; however, bacterioferritins are only found in archaea and bacteria. The major difference between Bfr and Ftn lies in possession of heme moieties: Bfr contains 12 heme groups, compared to classical ferritins, which lack heme molecules in their structure. The other major member of the ferritin superfamily are the Dps proteins, which form a smaller cage with a lower iron storage capacity compared to Ftn and also contain a unique ferroxidase site. Dps proteins are formed from 12 subunits and their major function is the prevention of harmful Fenton reaction by peroxide and iron, thus protecting DNA against iron-induced oxidative damage [22]. Each monomer of Dps also possesses a four-helix bundle fold, typically characteristic of the ferritin super family (Fig. 10.3a). Dps proteins not only protect bacteria from oxidative damages but also form high affinity complexes with DNA without apparent sequence specificity. The crystal structure of *E. coli* Dps (Protein Data Bank accession number: 1DPS) shows a hollow protein with tetrahedral point group symmetry, confirming that Dps is a structural analogue of the ferritin family [23].

In general, the Dps dodecamer cages measure around 9 nm in diameter, with a central cavity of around 4.5 nm which can accomodate an iron core of up to 500 Fe^{3+} iron ions. The principal function of Dps is not iron storage, but rather to protect the cell from iron-mediated oxidative damaged [22]. N-terminal negatively charged residues line the threefold channel of the Dps pores and generate an electrostatic potential, which facilitates entry of iron into the protein inner cavity; this feature is observed in both maxi- and mini-ferritins [22, 23]. The pore formed at the C-terminal of Dps contains fewer negative charges and is also smaller due to hydrophobic constriction (Fig. 10.3b) [23], therefore probably does not play a role in iron transport. The maxi- and mini-ferritin proteins fold into very similar monomer structures, containing 4-helix bundles, but their assembled architectures

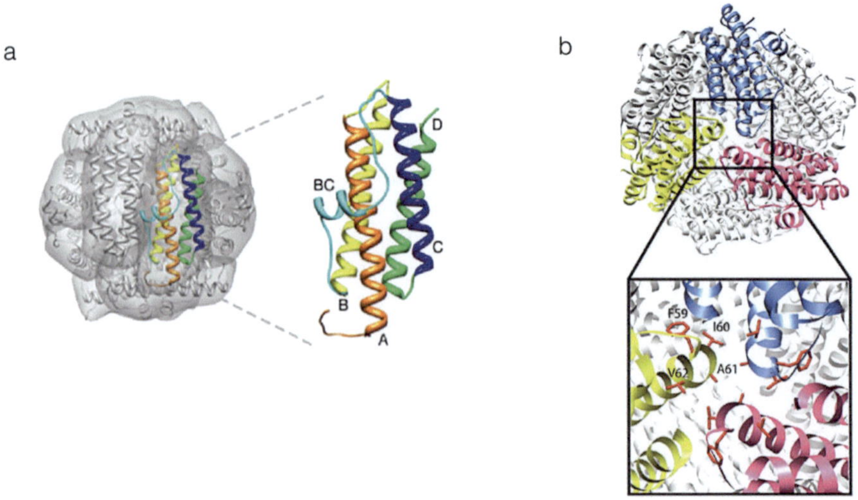

Fig. 10.3 (**a**) Typical tetrahedral structure of mini-ferritin (Dps). The four-helix bundle monomers are shown as ribbons. (**b**) Dps mini-ferritin protein cages viewed across the threefold axis; the image shows the positioning of hydrophobic amino acids along the pore. Adapted with permission from ref. [22], copyright mdpi.com, 2011

are different. Although they both form cages, but they contain different numbers of subunits and have different symmetries [22, 23].

10.5 Structure of Ferritin

At the nucleic and amino acid level, the sequence identity of the ferritin proteins is low, whereas their overall fold (tertiary structure) is highly conserved, possibly indicating identical/similar monomer folding [22]. However, their self-assembled cage structures can differ considerably, as mentioned above. According to its crystal structure, H-ferritin is a hollow globular protein machinery of ~500 kDa consisting of 24 subunits with octahedral symmetry (PDB accession number: 2FHA), and is negatively charged at physiological pH (Fig. 10.4a, b) [12]. The crystal structure further reveals that ferritin is an α-helical protein with an α-helix content of ~70% [24].

Each H-ferritin subunit normally consists of 174 residues which can be further subdivided into five α-helices known as A–E; and residues corresponding to each helix are numbered 14–40, 49–76, 96–123, 127–161 and 163–173. Helices A-D constitute a long central bundle consisting of four parallel and anti-parallel helices which is a signature of the ferritin superfamily, with the fifth short helix, E, butting on to one end of the α-helical bundle (Fig. 10.5a) [18]. There are additional short non-helical regions at both N- and C-termini and at the right-handed turns between the AB, CD and DE helices (the latter is particularly tight). The long BC loop (residues 78–94) connects the B and C helices and provides flexibility to the structure. The crystal structure further reveals that most parts of helices A and C, loop BC and the N-terminus, are exposed exterior of the molecular machine, whereas helices B and D faces inwards. A break in hydrogen bonding at His 136 is responsible for kink generation in the long D helix [18]. The D helix kink occurs

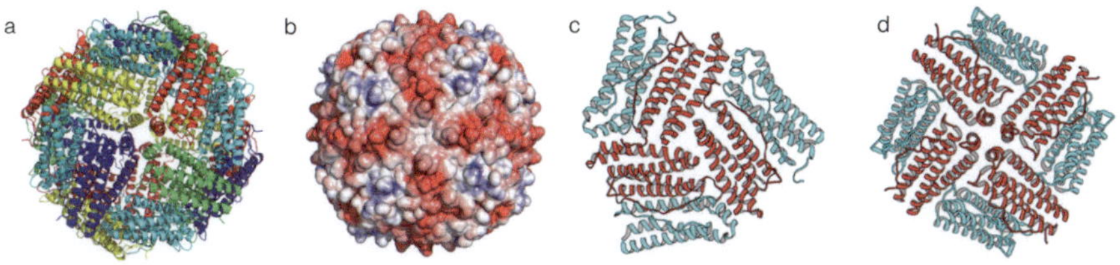

Fig. 10.4 (**a**) Crystal structure of ferritin (PDB 2ffx). Each of the 24 chains are individually colored. The structure is shown looking down the fourfold axis. (**b**) Structure of ferritin with electrostatic potential mapped onto the solvent-accessible surface. Red: $-5\ e^{-1}$ kT, blue: $+5\ e^{-1}$ kTe^{-1}. Adapted with permission from ref. [56] copyright Wiley Press, 2014. (**c**) Cartoon representation of the three-fold (polar) channel in the ferritin protein. Fe(II) normally enter the ferritin shell through this channel. (**d**) Cartoon representation of the fourfold (nonpolar) channel in the ferritin protein; mostly electrons are transported using this channel. Adapted with permission from ref. [57] copyright The American Society for Biochemistry and Molecular Biology, 2016

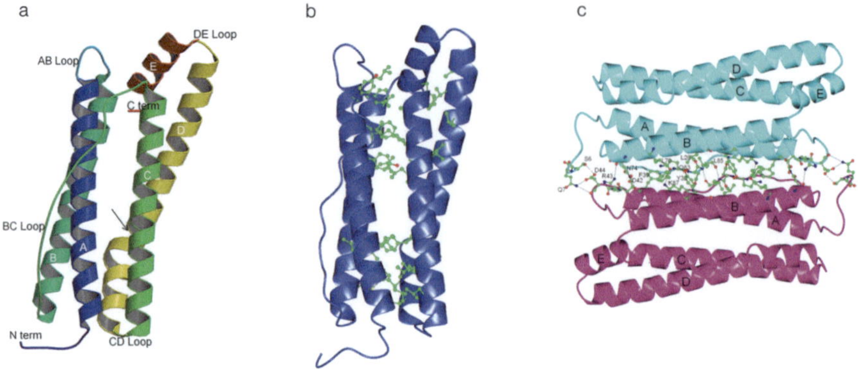

Fig. 10.5 (**a**) Ribbon representation of secondary structure elements of recombinant horse L apoferritin, (PDB 2v2i). The outer surface (front) contains helices A and C and the loop BC, while the inner surface (back) consists of helices B and D. Helix E is positioned at ~60° to the four-helix bundle. The kink formed by the D helix is shown by the arrow. (**b**) Representative hydrophobic interactions which stabilize the hydrophobic cores at the two ends of the four-helix bundle. (**c**) Representative hydrogen bond interactions occurring at the interface between subunits I and II. Adapted with permission from ref. [58] copyright Elsevier, 2010

at an important position where all three subunits intersect near the threefold axis, allowing a channel to form without disrupting the packing of the helices in the remainder of the structure (Fig. 10.4c) [25].

Each subunit of ferritin is roughly cylindrical, 50 Å in length and 25 Å in diameter and within each subunit, there are extensive side chain interactions [12, 26]. In both classical and bacterial ferritin, the four-helix bundle does not have a uniform hydrophobic core. At the two ends of the helix bundle, many side chains cross over and form a tightly packed hydrophobic interior (Fig. 10.5b). The central hydrophilic region of each ferritin subunit is situated between these two hydrophobic cores [18]. In this region, there are a number of buried polar and 181 hydrophilic residues, which form a hydrogen bond network, and interestingly most differences in structure between H and L subunits are in this part [18].

A second characteristic structural feature of the ferritin superfamily is the propensity to form dimers. Although the overall pathway of oligomerization remains unclear, there is general consensus that the dimer is almost certainly the first intermediate. Dimers further assemble to form 6-mers as the next intermediate, with equal probability of direct formation of 6-mers versus the formation of 6-mers via 4-mers. The next prominent intermediate in the assembly pathway is the 12-mer, formed via docking of two 6-mers [27]. The dimer interface involves the helices A and B; the BC loops, N-terminus and AB turn also participate in the interaction, mainly through extensive hydrogen bonding interactions between the two subunits (Fig. 10.5c).

As described earlier, the 24 ferritin subunits each contain a four-helix bundle with a left-handed twist arranged in 12 antiparallel pairs, forming a rhombic dodecahedron structure (space group F432). The resulting eight funnel-like hydrophilic channels are formed at the threefold axis. The remaining six channels at the fourfold axis are surrounded by the short E-helices at the C-terminus, which lie at close to a 60° angle with respect to the other four long helices and provide a hydrophobic patch of leucine residues. The channel present at the fourfold axis is not involved in iron exchange although it can initiate the transit of protons. Channels found at both the threefold and fourfold axes are approximately 0.4 nm wide (Fig. 10.4d) [27]. Iron entry occurs almost exclusively through the threefold interior. Compared to H-chains, L-chain ferritins are more highly structured, which is probably a reason for the preferred location of nucleation sites. Ferritin can be disassembled and reassembled by changing the pH of the buffer to a value as low as pH 2 and then increasing it above pH 7 [28, 29]. During the reassembly process, the solution and dissolved molecules as well as nanoparticles can be easily entrapped in the interior. This is an efficient way to encapsulate materials inside ferritin, which would normally be unable to penetrate into the interior of the protein shell through one of the pores.

In human ferritin, three monomers at the N-termini form threefold local symmetry on the surface of the protein cage with a distance of 5 nm between the N-terminal amino acids of the three monomers [30]. At the C-terminus, four monomers further form fourfold local symmetry on the cage surface with a distance of about 1 nm between the identical amino acids of the four monomeric loops. Hence, receptors or ligands that are fused to the N- or C-termini of ferritin monomers are always held in close proximity to each other when the monomers self-assemble into the protein cage, leading to enhanced avidity in a synergistic manner [30]. Recently a very interesting ferritin variant was obtained from archaea (*Archaeoglobus fulgidus*). On characterization it was found that in solution, it remains as dimeric species and it only assembles to a non-canonical 24-mer cage in the presence of high salt concentration (mono or divalent cations). Arrangement of 24 subunits in this ferritin is unique, as they assemble into a tetrahedrally symmetric structure. It also possesses four large triangular openings, ~45 Å in diameter [31], which is also unique in the ferritin protein family.

10.6 Application of Ferritin in Nanotechnology

Ferritin proteins have been subjected to intense investigation for various applications in bio-nanotechnology, as they offer numerous advantages (Fig. 10.6a) [12]. First, ferritins have a highly symmetrical structure with remarkable chemical and thermal stability. Second, reconstitution of the ferritin cage is possible through controlled reassembly. Third, a broad range of metals can be loaded and mineralized inside the cavity of ferritin [12]. Fourth, it is easy to modify the interior and outer surface of the ferritin cage, through the addition of peptides or protein tags using recombinant genetic methods. Finally, ferritins are highly biocompatible and less immunogenic compared to any other protein cage [32]. These characteristics have made ferritins

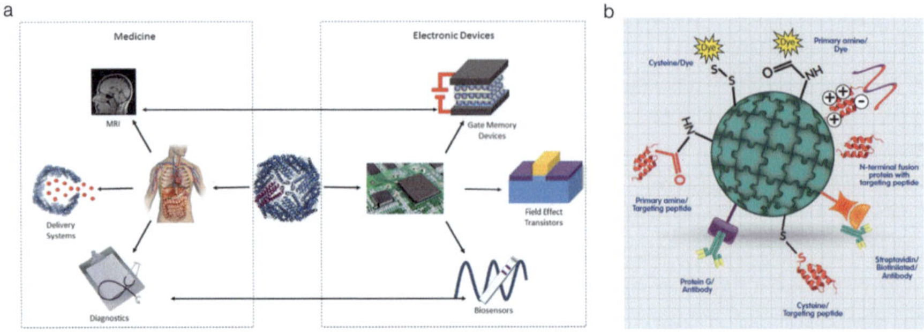

Fig. 10.6 (**a**) Various applications of ferritin in (bio) medical science, chemistry, materials science, (bio)chip, and electronic device manufacturing. Adapted by permission from ref. [12] copyright the American Chemical Society, 2015. (**b**) Schematic representation of strategies used to modify the ferritin surface with peptides, antibodies, siRNA, small molecules and fluorescent dyes. Adapted with permission from ref. [59] copyright Elsevier, 2016

attractive vehicles for drug delivery and as scaffolds for vaccine development. In addition, the plasticity of *in vitro* mineralization makes ferritin an ideal candidate for cellular and medical imaging [33].

Nanoparticles obtained from mineralization of semiconductors (quantum dots) are promising, since their fluorescence properties are related to nanoparticle size and shape. In this setting, ferritin is highly relevant, as it has been used for the synthesis of semi-conductor nanoparticles for a long time [12]. However, the main disadvantage of this procedure is that ion aggregation is induced by high concentrations of transition metal ions during the chemical reaction. Recently, Yamashita et al. improved this methodology and performed successful mineralization of CdSe and ZnSe inside the ferritin cavity by utilizing a slow chemical reaction method [34, 35]. This method bypasses excessive ion aggregation since Cd^{2+} and Zn^{2+} are attracted by the negative inner residues of apoferritin thus preventing the aggregation process; therefore, mineralization inside an apoferritin cavity slows down the nucleation of CdSe and ZnSe [34, 35].

10.7 Drug Delivery and Ferritin

It has been found that drugs, which have a natural tendency to bind metals, such as cisplatin, carboplatin, desferrioxamine B and daunomycin, can be easily entrapped inside a ferritin shell [14, 31]. Cisplatin encapsulation was first reported in 2007 by Yang et al. [36], and the same group also studied the details of cellular internalization of these nanocages, as well as different applications of drug-loaded nanoparticles in tumor treatment [37]. Other studies have also used the same strategy and showed that cisplatin-loaded apoferritin was capable of inducing apoptosis in gastric cancer cells [38]. More recently, a drug delivery device targeted to melanomas has been developed, using cisplatin-loaded ferritin, chemically functionalized with an antibody specifically targeting against the melanoma antigen CSPG4 [39]. In a separate study, Xie et al. have shown that doxorubicin can also be loaded onto RGD modified apo-ferritin nanocages with the assistance of Cu ions [40]. According to their results, this doxorubicin-loaded ferritin showed a longer circulation half-life, higher tumor uptake, better tumor (glioblastoma) inhibition, and less cardiotoxicity than free doxorubicin. Other targeting moieties, such as small molecules, peptides, antibodies, and even fusion proteins have been successfully expressed mainly as N-terminal fusions of ferritin, allowing 24 targeting domains to be exposed on the ferritin surface, with uniform and precise orientation (Fig. 10.6b) [14, 41, 42]. These studies have provided a strong foundation for superior therapeutic options using ferritin, particularly in terms of targeting cancer cells such as melanoma, a type of cancer totally resistant to chemotherapy in later stages.

The 8 nm diameter interior cavity of apoferritin can be effectively used as a nano-vessel and can be loaded with different cargo molecules. Cutrin et al. have shown efficient loading of curcumin as well as Gd-HPDO3A inside ferritin and its controlled release to its target cells [43]. The loading procedure is simple and consists of lowering the pH of the apoferritin solution followed by addition of the solution containing curcumin and Gd-HPDO3A. However, disassembling the ferritin nanocage at extreme pH (pH 2) often causes irreversible damage to the protein cage architecture and forms a hole in the spherical protein [44]. The irreversible damage will severely impair its *in vivo* stability and drug delivery efficiency. Recently, an alternative and improved method was developed, which causes minimal structural damage to ferritin. In this method, disassembly of ferritin was performed by treating the protein with 8 M urea [45]. Upon urea treatment, the supramolecular structure of apoferritin collapses because of weakening of electrostatic interactions between subunits, and removal of the urea by stepwise dialysis then allows reassembly of the ferritin cage.

10.8 Surface Modification and Cellular Interactions of Ferritin Nanoparticles

Cell internalization of human H-ferritin occurs via the transferrin 1 (TfR1) receptor. After binding to the cell surface, the H-ferritin-TfR1 receptor complex is internalized and can be easily detected in early and recycling endosomes [46]. H-ferritin is also involved in interaction with other receptors such as TIM-2 (T-cell Immunoglobulin-domain and mucin-domain protein) of mouse, which is normally overexpressed in B-cells [47]. Recently, target receptors for L-ferritin (hepatic SCARA 5 receptors, L-chain specific) were also identified, and their cell internalization pathway was studied in detail [48]. Despite the intrinsic ability of ferritin to target cancer cells, a number of research groups have modified the surface of ferritin either chemically or using recombinant genetic techniques.

Ferritins have been modified with various designed motifs including antibodies, peptides and antibody fragments, in order to selectively recognize and target specific cells. Lysine or cysteine residues on the ferritin surface can be subjected to chemical conjugation mainly by using different heterobifunctional cross-linkers such as N-hydroxysuccinimide (NHS) ester and maleimide groups [12]. Lys and Cys residues have also been chemically modified to attach dyes, quencher molecules or polyethylene glycol (PEG) [12, 14]. Dependent on the species, the N- and C-termini of some ferritins are surface exposed (for example, in human ferritin the N-terminus is solvent exposed) and numerous studies have shown that genetic modification of the N- and C-termini of the protein have no impact on protein cage architecture [16, 48]. In fact Kim et al. have shown that ferritin can be modified at both N- and C-termini to produce so called "double chambered ferritin" [49]. In double chambered ferritin, a tumor targeting pro-apoptotic peptide was fused to the N-terminus of the ferritin protein and GFP was attached at the C-terminus. Protein fusion at the N-terminus of ferritin can allow display of large proteins such as viral hemagglutinin on the ferritin surface, which is then effective at neutralizing H1N1 virus [50]. C-terminal fusion of peptides to ferritin nanoparticles has also been exploited for various applications including the development of dendritic cell-based vaccines [51]. In addition to these fusion approaches, other strategies for the conjugation of targeting groups on the ferritin surface have also been investigated, including biotinylation of the ferritin nanocage to allow insertion of targeting functions [52], conjugation of a small aptamer or antibody to ferritin and then use of a selective antigen for capture, as well as genetic engineering of ferritin to generate a fusion protein, which exploits N-terminal attachment of protein G for antibody immobilization [53].

10.9 Other Potential Applications of Ferritin

Other than drug delivery, ferritin proteins have been widely explored for Magnetic resonance imaging (MRI) applications. MRI is based on the relaxation process of protons after they have been perturbed by a radio frequency pulse from their aligned state in an external field [12]. To perform MRI, a contrast agent is needed which can be further classified into two types, paramagnetic contrast agents such as Gd(III), and superparamagnetic contrast agents such as iron oxide nanoparticles (Fe_3O_4). Paramagnetic agents appear bright in T1-type images because they induce an increase in signal intensity, while superparamagnetic agents appear dark in T2-type images because they result in a decrease in signal intensity [12]. Normally, the contrast of biological specimens depends on the varying water concentration of the local environment and the concentration of contrast agent. Ferritin with an iron oxide core has been used for a long time as a high-relaxivity (T2) probe for MRI imaging [12, 14, 31]. The laboratory synthesis of magneto ferritin is straightforward and is normally carried out under anaerobic conditions in the presence of either Ar or N_2. The reaction product is a homogeneous black-brown solution, in contrast to the blood-red color of native ferritin, which can be precipitated by high magnetic fields. The formation of magneto ferritin can be best detected by transmission electron microscopy. Recently, using magneto-ferritin, direct *in vivo* vascular imaging of atherosclerotic plaques was successfully performed [54]. Furthermore, it was shown that ferritin can also be used in visualizing tumor tissue without use of any contrast agent [46]. In this study highly crystalline iron oxide nanoparticles were synthesized within the interior cavity of ferritin and high quality MRI contrast images were generated of various tumor tissues, including breast cancer [46].

10.10 Conclusions and Future Perspectives of Ferritin in Nano-biology

Protein nanotechnology has emerged as a promising frontier in medical, diagnostic and biotechnological applications. Protein cages are probably the most sophisticated design of protein nanotechnology and they are very effective in improving bioavailability and reducing toxicity of some known cytotoxic drugs. In addition, protein cage particles, particularly generated from virus origin, have been found to be extremely useful in treating disease like influenza, cancer and AIDS [55]. The advantages of protein cages over other carriers include their high programmability, low immunogenicity, high stability and biocompatibility. This means that specific targeting can be achieved with protein cages, with minimum toxicity. Among different available protein cages, ferritin appears to be extremely powerful in both imaging and drug delivery. In recent years, different strategies for loading ferritin with drugs, molecules and metals have been successfully explored. Extensive research has also been carried out to increase the types of molecules encapsulated, particularly in the case of non-metal-containing drugs. However, some aspects need further investigation, such as the kinetic mechanism of drug release from ferritin *in vivo*. Recently it has also been established that H-ferritin recognizes and binds specifically to the transferrin (TfR1) receptor [14, 45]. Interestingly, TfR1 is over-expressed in many different cancer cells. Specific targeting of cancer cell receptors enhances anticancer activity by selective inhibition of the affected cells, and mediates a selective release of cytotoxic drugs. There are also reports suggesting that cancer receptor targeting peptides or antibodies can be genetically fused with the ferritin, and genetically modified ferritin remains active [12, 40]. However, we still do not know how these modifications of ferritin alter the immunogenicity of encapsulated nanoparticles. Another aspect which needs to be explored rigorously is the possibility of combination therapy. Encapsulation of various drugs into the same cavity may provide synergistic action against different diseases. Ferritin also has important applications in the field of *in vivo* imaging. Many different ferritin-based contrast agents have been developed in the last 10 years and used for MRI, providing more sensitive imaging platforms with improved tumor detection ability, in comparison to the contrast agents presently available in the clinic [31].

However, toxicity, bioavailability and clearance of these nanocages loaded with contrast agents needs to be understood more thoroughly before their translation into the clinic. Overall, ferritin emerges as a very promising platform for imaging and drug delivery. Physiological features (especially stability) and high programmability makes the translation of ferritin nanocages from the bench to the clinic a realistic possibility. Finally, the combination of imaging and therapeutic functionality into ferritin nanocages seems to be the most prominent and fascinating frontier in therapeutics.

References

1. Norn CH, André I (2016) Computational design of protein self-assembly. Curr Opin Struct Biol 39:39–45
2. Marsh JA, Teichmann SA (2015) Structure, dynamics, assembly, and evolution of protein complexes. Annu Rev Biochem 84:551–575
3. Gradišar H, Jerala R (2014) Self-assembled bionanostructures: proteins following the lead of DNA nanostructures. J Nanobiotechnol 12:4
4. Englander SW, Mayne L, Krishna MM (2007) Protein folding and misfolding: mechanism and principles. Q Rev Biophys 40(4):287–326
5. Davis L, Chin JW (2012) Designer proteins: applications of genetic code expansion in cell biology. Nat Rev Mol Cell Biol 13(3):168–182
6. Daube SS, Bar-Ziv RH (2013) Protein nanomachines assembly modes: cell-free expression and biochip perspectives. Wiley Interdiscip Rev Nanomed Nanobiotechnol 5(6):613–628
7. Luo Q, Hou C, Bai Y, Wang R, Liu J (2016) Protein assembly: versatile approaches to construct highly ordered nanostructures. Chem Rev. https://doi.org/10.1021/acs.chemrev.6b00228
8. Rother M, Nussbaumer MG, Renggli K, Bruns N (2016) Protein cages and synthetic polymers: a fruitful symbiosis for drug delivery applications, bionanotechnology and materials science. Chem Soc Rev 45(22):6213–6249
9. Corchero JL, Cedano J (2011) Self-assembling, protein-based intracellular bacterial organelles: emerging vehicles for encapsulating, targeting and delivering therapeutical cargoes. Microb Cell Factories 3(10):92
10. Kim YE, Hipp MS, Bracher A, Hayer-Hartl M, Hartl FU (2013) Molecular chaperone functions in protein folding and proteostasis. Annu Rev Biochem 82:323–355
11. Pieters BJ, van Eldijk MB, Nolte RJ, Mecinović J (2016) Natural supramolecular protein assemblies. Chem Soc Rev 45(1):24–39
12. Jutz G, van Rijn P, Santos Miranda B, Böker A (2015) Ferritin: a versatile building block for bionanotechnology. Chem Rev 115(4):1653–1701
13. Zhang Y, Ardejani MS, Orner BP (2016) Design and applications of protein-cage-based nanomaterials. Chem Asian J 11(20):2814–2828
14. Truffi M, Fiandra L, Sorrentino L, Monieri M, Corsi F, Mazzucchelli S (2016) Ferritin nanocages: a biological platform for drug delivery, imaging and theranostics in cancer. Pharmacol Res 107:57–65
15. Lawson DM, Artymiuk PJ, Yewdall SJ, Smith JM, Livingstone JC, Treffry A, Luzzago A, Levi S, Arosio P, Cesareni G et al (1991) Solving the structure of human H ferritin by genetically engineering intermolecular crystal contacts. Nature 349(6309):541–544
16. He D, Marles-Wright J (2015) Ferritin family proteins and their use in bio nanotechnology. New Biotechnol 32(6):651–657

17. Worwood M, Brook JD, Cragg SJ, Hellkuhl B, Jones BM, Perera P, Roberts SH, Shaw DJ (1985) Assignment of human ferritin genes to chromosomes 11 and 19q13.3-19qter. Hum Genet 69(4):371–374
18. Crichton RR, Declercq JP (2010) X-ray structures of ferritins and related proteins. Biochim Biophys Acta 1800(8):706–718
19. Ha Y, Shi D, Small GW, Theil EC, Allewell NM (1999) Crystal structure of bullfrog M ferritin at 2.8A resolution: analysis of subunit interactions and the binuclear metal center. J Biol Inorg Chem 4(3):243–256
20. Jutz G, Böker A (2011) Bionanoparticles as functional macromolecular building blocks – a new class of nanomaterials. Polymer 52(2):211–232
21. Andrews SC (2010) The ferritin-like superfamily: evolution of the biological iron storeman from a rubrerythrin-like ancestor. Biochim Biophys Acta 1800(8):691–705
22. Zhang Y, Orner BP (2011) Self-assembly in the ferritin nano-cage protein superfamily. Int J Mol Sci 12:5406–5421
23. Grant RA, Filman DJ, Finkel SE, Kolter R, Hogle JM (1998) The crystal structure of Dps, a ferritin homolog that binds and protects DNA. Nat Struct Biol 5(4):294–303
24. Hasan MR, Tosha T, Theil EC (2008) Ferritin contains less iron (59Fe) in cells when the protein pores are unfolded by mutation. J Biol Chem 283(46):31394–31400
25. Zang J, Chen H, Zhao G, Wang F, Ren F (2016) Ferritin cage for encapsulation and delivery of bioactive nutrients: from structure, property to applications. Crit Rev Food Sci Nutr. https://doi.org/10.1080/10408398.2016.1149690
26. Plath LD, Ozdemir A, Aksenov AA, Bier ME (2015) Determination of iron content and dispersity of intact ferritin by superconducting tunnel junction cryo-detection mass spectrometry. Anal Chem 87(17):8985–8993
27. Sato D, Ohtomo H, Yamada Y, Hikima T, Kurobe A, Fujiwara K, Ikeguchi M (2016) Ferritin assembly revisited: a time-resolved Small-angle X-ray scattering study. Biochemistry 55(2):287–293
28. Kim M, Rho Y, Jin KS, Ahn B, Jung S, Kim H, Ree M (2011) pH-dependent structures of ferritin and apoferritin in solution: disassembly and reassembly. Biomacromolecules 12(5):1629–1640
29. Ghisaidoobe AB, Chung SJ (2015) Functionalized protein nanocages as a platform of targeted therapy and immune detection. Nanomedicine 10(24):3579–3595
30. Uchida M, Kang S, Reichhardt C, Harlen K, Douglas T (2010) The ferritin superfamily: supramolecular templates for materials synthesis. Biochim Biophys Acta 1800(8):834–845
31. Tetter S, Hilvert D (2017) Enzyme encapsulation by a ferritin cage. Angew Chem Int Ed Engl 56(47):14933–14936
32. Kim S, Jeon JO, Jun E, Jee J, Jung HK, Lee BH, Kim IS, Kim S (2016) Designing peptide bunches on nanocage for bispecific or super-affinity targeting. Biomacromolecules 17(3):1150–1159
33. Chasteen ND, Harrison PM (1999) Mineralization in ferritin: an efficient means of iron storage. J Struct Biol 126(3):182–194
34. Yamashita I, Hayashi J, Hara M (2004) Bio-template synthesis of uniform CdSe nanoparticles using cage-shaped protein apoferritin. Chem Lett 33:1158–1159
35. Iwahori K, Yoshizawa K, Muraoka M, Yamashita I (2005) Fabrication of ZnSe nanoparticles in the apoferritin cavity by designing a slow chemical reaction system. Inorg Chem 44:6393–6400
36. Yang Z, Wang X, Diao H, Zhang J, Li H, Sun H, Guo Z (2007) Encapsulation of platinum anticancer drugs by apoferritin. Chem Commun 33:3453–3455
37. Xing R, Wang X, Zhang C, Zhang Y, Wang Q, Yang Z, Guo Z (2009) Characterization and cellular uptake of platinum anticancer drugs encapsulated in apoferritin. J Inorg Biochem 103(7):1039–1044

38. Ji XT, Huang L, Huang HQ (2012) Construction of nanometer cisplatin core-ferritin (NCC-F) and proteomic analysis of gastric cancer cell apoptosis induced with cisplatin released from the NCC-F. J Proteome 75(11):3145–3157
39. Falvo E, Tremante E, Fraioli R, Leonetti C, Zamparelli C, Boffi A, Morea V, Ceci P, Giacomini P (2013) Antibody-drug conjugates: targeting melanoma with cisplatin encapsulated in protein-cage nanoparticles based on human ferritin. Nanoscale 5:12278–12285
40. Zhen Z, Tang W, Chen H, Lin X, Todd T, Wang G, Cowger T, Chen X, Xie J (2013) RGD modified apoferritin nanoparticles for efficient drug delivery to tumors. ACS Nano 7(6):4830–4837
41. Vannucci L, Falvo E, Fornara M, Micco P, Benada O, Krizan J, Svoboda J, Hulikova-Capkova K, Morea V, Boffi A, Ceci P (2012) Selective targeting of melanoma by PEG-masked protein-based multifunctional nanoparticles. Int J Nanomed 7:1489–1509
42. Lee JH, Seo HS, Song JA, Kwon KC, Lee EJ, Kim HJ, Lee EB, Cha YJ, Lee J (2013) Proteinticle engineering for accurate 3D diagnosis. ACS Nano 7:10879–10886
43. Cutrin JC, Crich SG, Burghelea D, Dastrù W, Aime S (2013) Curcumin/Gd loaded apoferritin: a novel "theranostic" agent to prevent hepatocellular damage in toxic induced acute hepatitis. Mol Pharm 10(5):2079–2085
44. Kim M, Rho Y, Jin KS, Ahn B, Jung S, Kim H, Ree M (2011) pH-dependent structures of ferritin and apoferritin in solution: disassembly and reassembly. Biomacromolecules 12(5):1629–1640
45. Liang M, Fan K, Zhou M, Duan D, Zheng J, Yang D, Feng J, Yan X (2014) H-ferritin-nanocaged doxorubicin nanoparticles specifically target and kill tumors with a single-dose injection. Proc Natl Acad Sci USA 111(41):14900–14905
46. Zhao Y, Liang M, Li X, Fan K, Xiao J, Li Y, Shi H, Wang F, Choi HS, Cheng D, Yan X (2016) Bioengineered magnetoferritin nanoprobes for single-dose nuclear-magnetic resonance tumor imaging. ACS Nano 10(4):4184–4191
47. Chen TT, Li L, Chung DH, Allen CD, Torti SV, Torti FM, Cyster JG, Chen CY, Brodsky FM, Niemi EC, Nakamura MC, Seaman WE, Daws MR (2005) TIM-2 is expressed on B cells and in liver and kidney and is a receptor for H-ferritin endocytosis. J Exp Med 202(7):955–965
48. Mendes-Jorge L, Ramos D, Valença A, López-Luppo M, Pires VMR, Catita J, Nacher V, Navarro M, Carretero A, Rodriguez-Baeza A, Ruberte J (2014) L-ferritin binding to Scara5: a new Iron traffic pathway potentially implicated in retinopathy. PLoS One 9(9):e106974. https://doi.org/10.1371/journal.pone.0106974
49. Kim S, Kim GS, Seo J, Gowri Rangaswamy G, So IS, Park RW, Lee BH, Kim IS (2016) Double-chambered ferritin platform: dual-function payloads of cytotoxic peptides and fluorescent protein. Biomacromolecules 17(1):12–19
50. Kanekiyo M, Wei CJ, Yassine HM, McTamney PM, Boyington JC, Whittle JR, Rao SS, Kong WP, Wang L, Nabel GJ (2013) Self-assembling influenza nanoparticle vaccines elicit broadly neutralizing H1N1 antibodies. Nature 499(7456):102–106
51. Han JA, Kang YJ, Shin C, Ra JS, Shin HH, Hong SY, Do Y, Kang S (2014) Ferritin protein cage nanoparticles as versatile antigen delivery nanoplatforms for dendritic cell (DC)-based vaccine development. Nanomedicine 10(3):561–569
52. Crich SG, Bussolati B, Tei L, Grange C, Esposito G, Lanzardo S, Camussi G, Aime S (2006) Magnetic resonance visualization of tumor angiogenesis by targeting neural cell adhesion molecules with the highly sensitivegadolinium-loaded apoferritin probe. Cancer Res 66:9196–9201
53. Hwang M, Lee JW, Lee KE, Lee KH (2013) Think modular: a simple apoferritin-based platform for the multifaceted detection of pancreatic cancer. ACS Nano 9:8167–8174
54. Sosnovik DE, Caravan P (2009) Molecular MRI of atherosclerotic plaque with targeted contrast agents. Curr Cardiovasc Imaging Rep 2(2):87–94
55. López-Sagaseta J, Malito E, Rappuoli R, Bottomley MJ (2015) Self-assembling protein nanoparticles in the design of vaccines. Comput Struct Biotechnol J 14:58–68
56. Beck T, Tetter S, Künzle M, Hilvert D (2015) Construction of Matryoshka-type structures from supercharged protein nanocages. Angew Chem Int Ed Engl 54(3):937–940

57. Chandramouli B, Bernacchioni C, Di Maio D, Turano P, Brancato G (2016) Electrostatic and structural bases of Fe2+ translocation through ferritin channels. J Biol Chem 291(49):25617–25628
58. Crichton RR, Declercq JP (2010) Xray structures of ferritins and related proteins. Biochim Biophys Acta 1800(8):706–718
59. Truffi M, Fiandra L, Sorrentino L, Monieri M, Corsi F, Mazzucchelli S (2016) Ferritin nanocages: a biological platform for drug delivery, imaging and theranostics in cancer. Pharmacol Res 107:57–65

Chapter 11
DNA Nanotechnology for Building Sensors, Nanopores and Ion-Channels

Kerstin Göpfrich and Ulrich F. Keyser

Abstract DNA nanotechnology has revolutionised the capabilities to shape and control three-dimensional structures at the nanometre scale. Designer sensors, nanopores and ion-channels built from DNA have great potential for both cross-disciplinary research and applications. Here, we introduce the concept of structural DNA nanotechnology, including DNA origami, and give an overview of the work flow from design to assembly, characterisation and application of DNA-based functional systems. Chemical functionalisation of DNA has opened up pathways to transform static DNA structures into dynamic nanomechanical sensors. We further introduce nanopore sensing as a powerful label-free single-molecule technique and discuss how it can benefit from DNA nanotechnology. Especially exciting is the possibility to create membrane-inserted DNA nanochannels that mimic their protein-based natural counterparts in form and function. In this chapter we review the status quo of DNA sensors, nanopores and ion channels, highlighting opportunities and challenges for their future development.

Keywords DNA nanotechnology · DNA origami · Nanopores · Single-molecule sensing · Synthetic ion-channels

---

K. Göpfrich
Cavendish Laboratory, University of Cambridge, Cambridge, UK

Department of Cellular Biophysics, Max Planck Institute for Medical Research, Heidelberg, Germany
e-mail: kerstin.goepfrich@mr.mpg.de

U. F. Keyser (✉)
Cavendish Laboratory, University of Cambridge, Cambridge, UK
e-mail: ufk20@cam.ac.uk

S. Perrett et al. (eds.), *Biological and Bio-inspired Nanomaterials*, Advances in Experimental Medicine and Biology 1174,
https://doi.org/10.1007/978-981-13-9791-2_11

11.1 Self-Assembly with DNA

Desoxyribonucleic acid (DNA) is predominantly known as the molecule that carries genetic information. Two complementary DNA strands are held together by the Watson-Crick basepairing of adenine (A) with thymine (T) and cytosine (C) with guanine (G) forming the iconic double helix [1]. While nature exploits the base pairing in the replication and expression of genetic information, the same property makes DNA an excellent material for molecular self-assembly. With a full helical turn of 10.5 nm, a pitch of 3.3 nm and an anhydrated diameter of 2 nm, the dimensions of the DNA double helix are inherently suited for precise architectural control on the nanometre length scale. A linear DNA double helix, however, is of not much use for construction. Transient branched DNA structures have been found in nature, for instance as intermediates occurring during the process of genetic recombination (Holliday junctions) [2, 3]. But only the realisation that it is possible to engineer stable DNA branches through intelligent sequence matching opened up the possibility to use DNA beyond its natural purpose as a nanomaterial for design and construction [4, 5]. Over the past 30 years, structural DNA nanotechnology, the science of building with DNA, has been developed tremendously and applied to research questions across disciplines thanks to multiple developments:

1. Automated and high throughput synthesis of short DNA strands has become available.
2. Off-the-shelf and custom-made chemical modifications of DNA have been developed broadening the scope of DNA nanostructures.
3. The self-assembly of DNA nanostructures does not require specialised equipment.
4. Computer-aided design tools specifically developed for DNA nanotechnology make prototyping quick and accessible.

The following sections will give an overview of some of the most relevant achievements in the field of DNA nanotechnology, including DNA origami, and provide guidance for the design and characterisation of DNA nanostructures. We will further discuss possibilities for the functionalisation of DNA.

11.1.1 DNA Lattices and Tiles

In 1982, Nadrian C. Seeman laid the intellectual foundation for the creation of two- and three-dimensional lattices from DNA. He realised the possibility to create immobile nucleic acid junctions by intelligent sequence design as shown in Fig. 11.1a.

If complementary single-stranded DNA overhangs, so-called sticky ends, are extended from this branched structure, it should be possible to link them up as illustrated in Fig. 11.1b, and eventually to construct crystalline “macromolecular valence

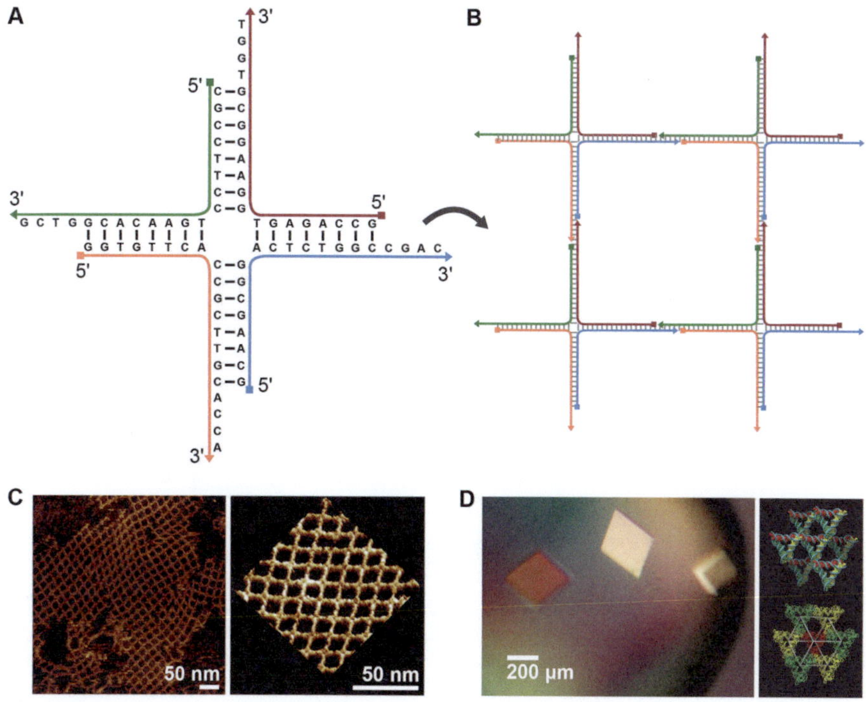

Fig. 11.1 Milestone achievements implementing DNA lattices and crystals. (**a**) Schematic drawing of an immobile nucleic acid junction of rank four as proposed by Seeman [4]. Each of the four single-strands (green, red, blue and orange) is partially complementary to two others, complementary single-stranded "sticky ends" induce multimerisation (**b**). Squares indicate 5' ends of the DNA, triangles indicate 3' ends. (**b**) DNA lattice formed from the interconnected nucleic acid junctions from A. By changing the rank of the junction, it should in principle be possible to create versatile two-dimensional lattices and three-dimensional crystals from DNA [4]. (**c**) AFM images of a DNA lattice by Yan et al. Each section consists of two parallel double-strands of DNA, the nodes were functionalised with proteins. (Adapted from [6]. Reprinted with permission from AAAS). (**d**) DNA crystal at 4 Å resolution based on a DNA tensegrity triangle. (Adapted by permission from Springer Customer Service Centre GmbH: Springer Nature, *Nature* [7], copyright 2009)

clusters" from DNA [4]. This idea has in the meantime been employed to construct two-dimensional DNA lattices characterised with atomic force microscopy (AFM) [6, 8–10], see Fig. 11.1c. It has fully been realised with a DNA-based crystal structure at 4 Å resolution [7] of an earlier DNA tensegrity triangle [11], Fig. 11.1d. Since the location of an individual DNA sequence within the lattice is known, positions are spatially addressable and DNA strands with functional groups like proteins [6, 10] can be incorporated as required, see Sect. 11.1.4.7. Additionally, self-assembly of branched DNA structures with a controllable number of binding sites is a powerful tool to probe theories on algorithmic cluster growth and emergence of fractal patterns [9, 12].

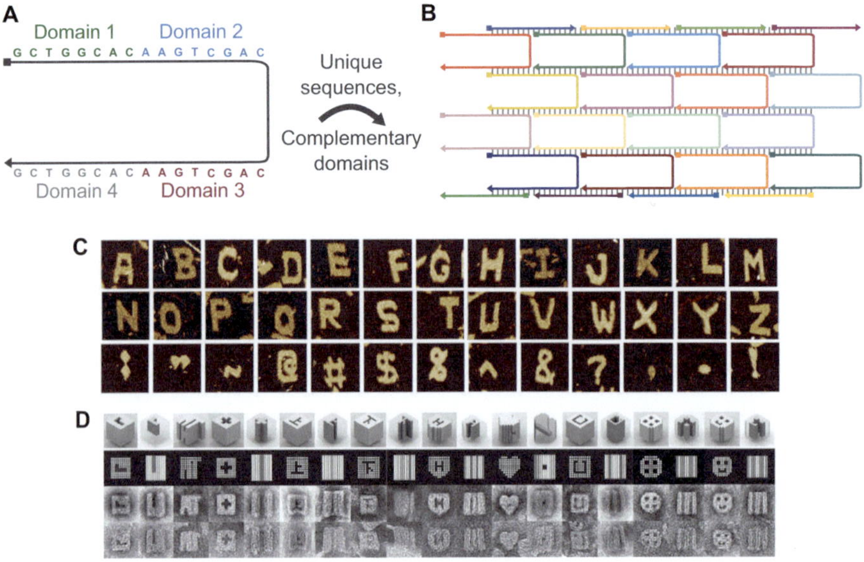

Fig. 11.2 Self-assembly from single-stranded DNA tiles. (**a**) Single-stranded DNA tile motive with four modular domains as proposed by Yin et al. [13]. A square indicates the 5' end, a triangle the 3' end. (**b**) Unique tiles assemble into the designed secondary structure, whereby each tile binds a complementary domain of four neighbouring tiles. Base pairs are indicated as grey bars. (**c**) AFM images of two-dimensional DNA tile assemblies designed by Wei et al. (Adapted by permission from Springer Customer Service Centre GmbH: Springer Nature, *Nature* [14], copyright 2012). (**d**) Ke et al. used single-stranded tiles to construct modular three-dimensional DNA bricks. (Adapted from [15]. Reprinted with permission from AAAS)

Apart from branched junctions, single-stranded DNA bricks or tiles are popular motifs in DNA nanotechnology. 32–42 base long single-strands, divided into four domains as shown in Fig. 11.2a, bind to their local neighbours, Fig. 11.2b. Such DNA tiles have been used to program DNA tube circumferences [13] or to construct complex finite-size two- [14] or three-dimensional structures [15], see Fig. 11.2c, d.

11.1.2 DNA Origami

The basic element of most common forms of structural DNA nanotechnology is the four-way DNA junction that allows for the construction of complex objects. One of the most promising techniques for creating large finite-size structures was invented by realising that a long DNA single-strand can be forced into almost any configuration by using short complementary DNA strands. Inspired by previous work [16, 17], Paul Rothemund folded a seven kilobase long single-strand of DNA into a predefined shape by choosing around 200 short complementary

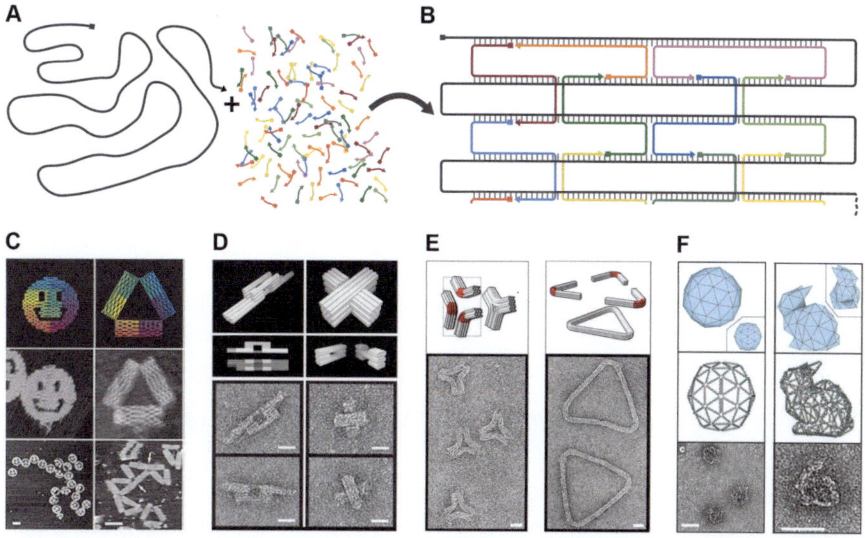

Fig. 11.3 Self-assembly of DNA origami. (**a**) A Viral single-strand of DNA ("scaffold", grey) is mixed with an excess of short synthetic oligonucleotides ("staples", multicoloured). (**b**) Via complementary base-pairing, the staples fold the scaffold into the pre-designed shape. Base pairs are indicated as grey bars. Only a small section of the DNA origami is shown. To fold the entire scaffold, around 200 staples are required. (**c**) Paul Rothemund's first demonstration of two-dimensional DNA origami. (Adapted by permission from Springer Customer Service Centre GmbH: Springer Nature, *Nature* [18], copyright 2006). (**d**) First three-dimensional DNA origami nanostructures by Douglas et al. (Adapted by permission from Springer Customer Service Centre GmbH: Springer Nature, *Nature* [19], copyright 2009). (**e**) Extension to curved DNA origami by Dietz et al. (Adapted from [20]. Reprinted with permission from AAAS). (**f**) Rendering of arbitrary polyhedral meshes from DNA origami by Benson et al. (Adapted by permission from Springer Customer Service Centre GmbH: Springer Nature, *Nature* [21], copyright 2015). All structures are tens of nanometres across

oligonucleotide staples. The DNA origami concept is illustrated in Fig. 11.3a, b. The long scaffold is normally obtained by extracting the genomic DNA of a virus, often the M13mp18 phage with its single-stranded genome, while the short staples can be synthesised as required.

Both scaffold and staples are commercially available, making DNA origami accessible to a broad community of scientists. The development of computer-aided design tools for DNA origami [22], described in detail in Sect. 11.1.3, streamlined the process of prototyping making it achievable in four simple steps: In the first step, the desired shape is approximated by a lattice-based scaffold path. Secondly, staple pathways complementary to the scaffold are assigned in such a way that the target structure is the single most stable configuration, see Fig. 11.3b. Since the base sequence of the scaffold is known, this step defines the staple sequences. Thirdly, these 18–50 bases long sequences are chemically synthesised, often using commercial services. As a last step, the scaffold is mixed with an excess of staples and exposed to a distinct temperature gradient to enable correct hybridisation.

This process reliably allows for the creation of microgram quantities of identical nanostructures in a simple and robust one-pot reaction. With all these tools at hand, it is not surprising that the DNA origami community grew quickly, contributing to new developments and notable achievements across different disciplines.

Paul Rothemund invented DNA origami in 2006, famously exemplified with the nanoscale smiley face shown in Fig. 11.3c [18]. Douglas et al. were the first to create three-dimensional structures from DNA origami, Fig. 11.3d [22]. Possible DNA geometries were expanded further by Dietz et al. who created the curved DNA structures in Fig. 11.3e [20], and Benson et al. who demonstrated a general method for rendering of polyhedral meshes with DNA origami, Fig. 11.3f [21].

11.1.3 Design and Assembly of DNA Nanostructures

In this section, we discuss the design rules for structural DNA nanotechnology, introduce computational frameworks that facilitate the prototyping process and discuss conditions for assembly and storage of DNA nanostructures.

11.1.3.1 Conceiving the Target Shape

The design of a new DNA nanostructure will depend on the functional requirements for its envisioned application. The first decision to make is whether to use scaffolded DNA origami, see Sect. 11.1.2, or a DNA tile approach where the structure is assembled from short single-strands, Sect. 11.1.1. When using a kilobase long scaffold, the size of the DNA nanostructure immediately falls into the megadalton regime. The achievable concentrations are limited by the scaffold concentration and placing modifications on the scaffold strand is challenging. Therefore, assembly from short single-strands is useful when the target structure is smaller or has repeating structural units like lattices [10] or tubes [13]. Assembly from single-strands, however, becomes less efficient (12–17% yield [14]) for large finite-size assemblies, which is why scaffolded DNA origami is then often the method of choice. Additionally, many of the software tools described in Sect. 11.1.3.3 have been developed specifically for DNA origami making its design relatively straightforward. The final yield of a DNA origami structure will depend on the design, the assembly conditions and the chosen scaffold. A good scaffold should have little internal sequence complementarity and secondary structure, like the M13mp18 viral DNA [18]. An additional benefit of selecting one of a few commonly used scaffolds is that they can be purchased from vendors like New England Biolabs or Tilibit. To obtain smaller structures, one can either cut the scaffold to the desired length using restriction enzymes [23, 24] or rely on the synthesis of a long custom DNA sequence (e.g. Ultramers from Integrated DNA Technologies), which can then be employed as a scaffold strand.

11.1.3.2 Crossover Rules for DNA Nanostructures

The natural geometry of B-form DNA sets several rules for the design of DNA nanostructures, specifically for the placement of crossovers interconnecting adjacent DNA helices. 10.5 base pairs complete a full helical turn meaning that the angle between two base pairs is approximately 34.3°. If arranged on a hexagonal lattice, there is a 240° angle between neighbouring helices. In order to avoid twist and axial strain [20], a crossover should be placed where the base is pointing in the direction of the neighbour. This happens in constant intervals of 240° or equivalently seven base pairs as illustrated in Fig. 11.4a. On a square lattice, where each helix has four immediate neighbours, crossover positions occur every 270° or 7.875 base pairs, see Fig. 11.4b. For convenience, a constant eight base pair crossover spacing is often chosen. Global twisting torques can be avoided by either deviating from constant crossover spacing or by introducing base-skips leaving one base of the scaffold unpaired [25]. For other helical geometries similar considerations have to be made. By violating crossover spacing rules on purpose, one can create twisted and curved structures from DNA [20].

Both scaffold and staple strands are available for crossovers. A common approach is to interconnect helices with a high density of staple crossovers and reduce the number of scaffold crossovers. Generally, scaffold crossovers should be shifted by 180° or five to six bases relative to the staple crossovers [26]. It is possible that staples interconnect more than two DNA helices while undergoing multiple crossovers, as long as their length is roughly between 18–50 bases. While the lower limit ensures stable binding at room temperature (RT), convenient synthesis at low cost and high purity set the upper limit. Equally, scaffold or staple regions within

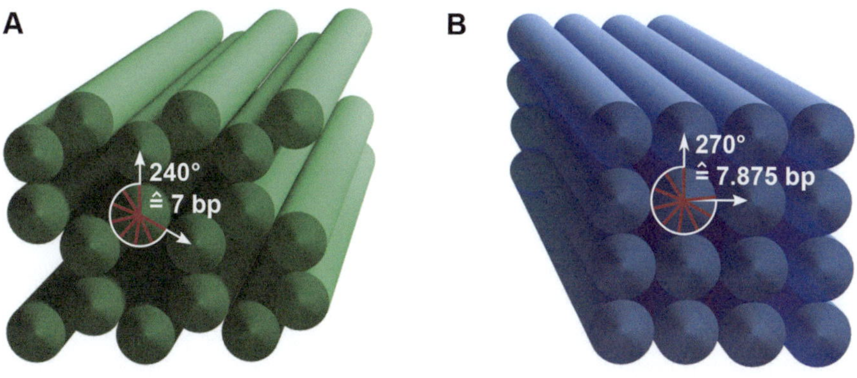

Fig. 11.4 Illustration of crossover rules for structural DNA nanotechnology. (a) Cross-sectional view of a three-dimensional DNA origami object with hexagonal packing. Suitable crossover positions exist every 7 base pairs along the helical axis, interconnecting a duplex with its three nearest neighbours. (**b**) Cross-sectional view of a three-dimensional DNA origami object on a square lattice. Crossover positions to the four nearest neighbours can be found every 7.875 base pairs. Equidistant crossover spacing will hence induce a global twist

Table 11.1 Overview of geometrical and molecular features of B-DNA for structural DNA nanotechnology

Geometry attribute	Approximate value for B-DNA
Helix sense	Right-handed
Rotation per base pair	34.3°
Rise per base pair	3.32 Å
Basepairs per helical turn	10.5
Width of anhydrated DNA duplex	2 nm
Average weight of nucleotide (phosphate group + sugar + base)	650 Da
Persistence length	35 nm
Absorbance maximum	260 nm
Ratio of absorbance at 260 and 280 nm ($A_{260/280}$)	1.8
Average weight of nucleotide (phosphate group + sugar + base)	650 Da
Charge of nucleotide	$-e$ (-1.6×10^{-19} C)

the structure can be left single-stranded, for instance for the later attachment of functionalised oligomers, as flexible hinges [27, 28], to support tensegrity structures [29] or, most commonly, to prevent unwanted base-stacking interactions between structures [18, 19]. An overview of useful values for structural DNA nanotechnology is presented in Table 11.1.

11.1.3.3 Computational Tools for DNA Nanotechnology

Multiple software tools have been developed to aid the design of DNA nanostructures. Without any claim to completeness, the following paragraphs describe some of them.

caDNAno

caDNAno is an open-source DNA origami design software developed by Douglas et al., available at http://cadnano.org/ [22]. Multiple online tutorials offer guidance for inexperienced users and template designs are available for download. caDNAno exists as a stand-alone programme as well as a plug-in for Autodesk Maya providing an interface for three-dimensional visualisation. Possible DNA geometries are limited to hexagonal and square lattices. While caDNAno has been developed for scaffolded DNA origami, the scaffold path can later be broken up into short strands if the target structure is scaffold-free. Once a known DNA sequence has been assigned to the scaffold, a set of complementary staples is generated upon a single mouse-click. Having placed crossovers by following the embedded rules, the staple sequences can be exported as a spreadsheet and the design-file can be submitted to CanDo for structural analysis.

SARSE

SARSE is an earlier software package available at http://www.cdna.dk/index.php/software.html for the automatic generation of two-dimensional DNA origami structures [30]. It includes a bitmap reader to import any shape as a design object and generates an atomic model. Conveniently, it tracks the design history and has a fully extendible toolbox.

Tiamat

Tiamat is a graphical user interface for efficient modelling of large DNA nanostructures beyond DNA origami [31]. DNA duplexes can be placed freely in the three-dimensional space offering more flexibility than caDNAno. It includes a convenient visualisation tool.

CanDo

CanDo, http://cando-dna-origami.org/, is a free online tool to predict the three-dimensional shape and flexibility of scaffolded and scaffold-free DNA nanostructures in solution using finite-element analysis [26]. It can thus help to make design choices before the more cost- and time-intensive DNA synthesis. The structural predictions are regularly improved and updated [32]. Combining caDNAno and CanDo, one can be relatively certain that a new DNA origami design will assemble as designed – merely the yield remains uncertain.

NUPACK

NUPACK, http://www.nupack.org/, is a free online tool for the analysis and the design of secondary structures of one or more interacting DNA sequences [33]. It is especially useful to inform the choice of DNA sequences for small scaffold-free DNA nanostructures. For such structures, NUPACK's thermodynamic analysis can help select sequences without stable secondary structures and unwanted complementaries with other strands. A favourable set of DNA sequences will reduce the assembly times and produce a higher yield of the target structure.

vHelix

vHelix, http://www.vhelix.net/, is a free plug-in for Autodesk Maya, similar to caDNAno, which enables straightforward rendering of polyhedral meshes from DNA [21].

11.1.3.4 Assembly and Stability of DNA Nanostructures

Design and assembly conditions are equally important to obtain uniform DNA nanostructures at high yield. Once a DNA nanostructure has been designed, the DNA needs to be synthesised, often relying on commercial services which include purification – standard-desalting is sufficient for unmodified staples, e.g. from Integrated DNA Technologies or Biomers. The first step in the assembly process is the preparation of a staple mix with equimolar concentrations of all constituent DNA staples. For scaffolded DNA origami, staples are added to the scaffold strand in five to tenfold excess. Excess staples can displace unwanted secondary structure by strand invasion helping to prevent misfolding [18] or cross-linking of two scaffolds. For the folding reaction, the DNA mix is supplemented with magnesium chloride and pH-stabilising buffer, often 40 mM Tris-HCl, 45 mM boric acid, 1 mM EDTA, pH 8.2 (0.5× TBE). The magnesium chloride acts as a charge-screening agent to reduce electrostatic repulsion between the densely packed DNA duplexes. The required magnesium chloride concentration will depend on the architecture of the DNA nanostructure. It is advisable to assemble a new structure at a range of magnesium chloride concentrations and determine the optimum concentration via gel electrophoresis as described in Sect. 11.1.4.3. For small simple DNA tiles, 2 mM $MgCl_2$ can be sufficient, whereas multilayer DNA origami often requires 10–20 mM. Magnesium chloride can be replaced with increased concentrations of monovalent salts [34], allowing for DNA nanostructures to be assembled in a variety of buffers including phosphate-buffered saline (PBS) [35]. Assembly of DNA nanostructures is normally achieved via thermal annealing. The folding mix is heated to 80 °C to melt any secondary structures and subsequently cooled to RT over a few hours [18] or days [19]. UV melting profiles can help to optimise the annealing protocol, see Sect. 11.1.4.2. Simple scaffold-free designs with optimised sequences can be assembled at RT within minutes [36]. Isothermal assembly is also achievable for more complex structures under the right conditions [37, 38]. After assembly, buffer conditions are less crucial than during the folding process, increasing the possibilities for applications. DNA nanostructures are often stable in a range of buffer conditions for a certain period of time, including overnight-incubation in cell culture medium or acidic buffers of pH 2.0 [26]. Unmodified DNA nanostructures can normally be stored at 4 °C for weeks. For longer time periods, it is beneficial to freeze a sample whilst avoiding too many freeze-thaw cycles.

For most applications, scaffolded DNA origami requires purification from excess staples and from potentially misfolded structures. This can be achieved in many different ways, for instance via ultrafiltration, gel extraction, PEG precipitation or magnetic bead capture. Shaw et al. provide guidance to obtain the best recovery yield depending on the target structure and its functionalisations [39].

11.1.4 Experimental Characterisation of DNA Nanostructures

After assembly, it is essential to assess and to quantify the quality of the DNA product. While the level of detail of the characterisation will depend on the target application, there is a number of routine experiments to demonstrate successful folding.

11.1.4.1 UV-Vis Spectroscopy

After purification of the DNA origami, the ultraviolet-visible (UV-vis) spectrum gives a first insight into the obtained yield and the purity of the sample. DNA has an absorption maximum at 260 nm [40], a typical UV spectrum is shown in Fig. 11.5a. Using the Beer-Lambert law [41], one can directly relate the amount of light absorbed to the concentration of the absorbing molecule. At a wavelength of 260 nm, double-stranded DNA has an average extinction coefficient of $0.020\,(\mu g/ml)^{-1}\,cm^{-1}$. Thus, an Absorbance of $A = 1$ corresponds to a DNA concentration of $50\,\mu g/ml$. This method for the determination of concentration is valid up to at least $A \leq 2$ [42]. Low-volume UV-vis spectrophotometers can reliably detect DNA at low concentrations down to a few ng/μL. Since the light is absorbed by the aromatic bases, the extinction coefficient of single-stranded DNA with unpaired bases is higher, approximately $0.027\,(\mu g/ml)^{-1}\,cm^{-1}$. This number is, however, affected by the length and the base composition for short oligonucleotides [42]. When preparing a DNA sample, it is possible to introduce impurities, like proteins or other organic molecules. While good laboratory practice will help minimise the risk of contamination, the UV spectrum offers a useful control measure. The 260/280 absorbance ratio, $A_{260/280}$, of pure DNA lies around 1.8 [40]. Protein contamination, for instance, can be detected in the UV spectrum because it will result in $A_{260/280} < 1.8$. UV spectroscopy should be repeated every time a DNA nanostructure is assembled, as it is a convenient way of obtaining some information about a DNA sample. It does, however, not give any insights about the shape of the DNA nanostructures or the homogeneity of the sample.

11.1.4.2 UV Melting Profile

On a UV spectrophotometer with temperature control, a measurement of the melting profile can directly follow the determination of the UV spectrum. While increasing the temperature, typically at a rate of around 0.25 °C/min to allow for equilibration inside the measurement cuvette, the absorption at 260 nm is determined. At a certain temperature, when the double-stranded DNA melts into its constituent single-strands, a pronounced increase in absorption can be observed. A typical melting profile is shown in Fig. 11.5b. The melting temperature is defined as the inflection point of the melting curve, which is why it is common to plot the first derivative dA/dT. The melting temperature can then directly be determined via peak fitting. While the melting temperature is concentration and buffer dependent, it typically lies between 50 and 65 °C for DNA origami and other DNA nanostructures [26], which is below the melting temperature of the individual strands. This is caused by electrostatic repulsion and mechanical strain resulting from the tight packing.

Measuring the melting profile is useful for two reasons: First, a well-defined melting transition is indicative of cooperative melting behaviour, which is expected for homogeneous and well-folded DNA nanostructures [44]. It can, however, not be understood as a definite proof for correct folding. Second, the melting point can inform the process of optimising the annealing protocol of DNA nanostructures to achieve better yield and shorter annealing times. The percentage of correctly folded structures can be increased by reducing the temperature gradient just before reaching the melting temperature, while the rest of the annealing process can be sped up [45].

11.1.4.3 Gel Electrophoresis

As one of the most common experimental techniques in molecular biology, gel electrophoresis is a powerful and essential tool in the characterisation of new DNA nanostructure designs. The negatively charged DNA migrates across the porous gel against the direction of the applied electric field. The migration speed will depend on the applied voltage [46], the pore size of the gel [47, 48], the ion concentration and composition of the buffer as well as the size and the shape of the DNA nanostructure [49]. In order to gain the most useful information from a gel, the pore size has to be chosen according to the size of the nanostructure. Generally, two types of gels

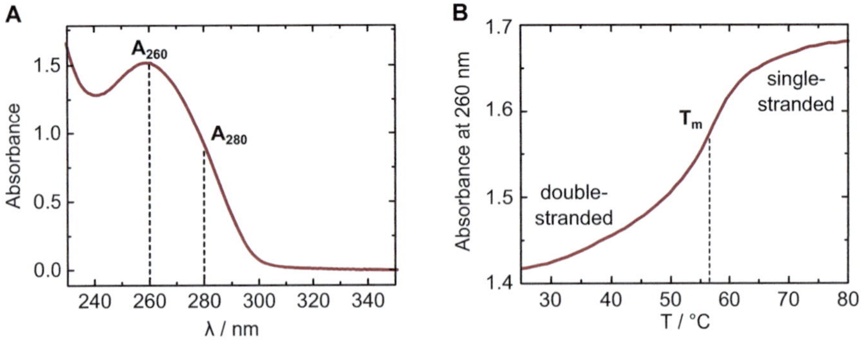

Fig. 11.5 (**a**) Typical UV-vis spectrum of a DNA sample with its absorbance maximum at $A = 260$ nm. A_{260} is used to quantify the DNA concentration, for a sample without impurities $A_{260}/A_{280} = 1.8$. (**b**) UV melting profile, obtained by applying a linear heating ramp and monitoring the absorbance at 260 nm. The melting temperature T_m is defined as the inflection point of the melting profile. The displayed data was obtained from a four-helix DNA tile structure [43], the background absorbance was subtracted

Table 11.2 Recommended gel percentages for optimal resolution of linear DNA. (Adapted from: Agilent technologies)

Polyacrylamide gels		Agarose gels	
DNA size range	Gel percentage	DNA size range	Gel percentage
100–1,000 bp	3.5%	1,000–30,000 bp	0.5%
75–500 bp	5.0%	800–12,000 bp	0.7%
50–400 bp	8.0%	500–10,000 bp	1.0%
35–250 bp	12%	400–7000 bp	1.2%
20–150 bp	15%	200–3,000 bp	1.5%
5–100 bp	20%	50–2,000 bp	2.0%

are used: Polyacrylamide gel electrophoresis (PAGE) is optimal to separate small DNA nanostructures of 5–1,000 base pairs, whereas scaffolded DNA origami and nanostructures between 50 and 30,000 base pairs, are studied in agarose gels. The polyacrylamide or agarose concentration will determine the pore size of the gel [47, 48]. Table 11.2 offers guidance to select a suitable gel concentration for the designed nanostructure.

In a uniform sample, all DNA nanostructures are expected to migrate at the same speed, whereby small structures migrate faster than larger ones. After staining the gel in a bath solution supplemented with an UV-sensitive intercalating dye, the position of the DNA within the gel can be visualised using UV transillumination. Structures of similar size and shape are visible as a clear distinct band. Gels of DNA nanostructures sometimes show a second band at a slower migration speed, which can often be attributed to the formation of dimers. The relative intensity of the bands allows for conclusions on the abundance of different structures within a DNA sample. By cutting individual bands out of a gel and extracting the DNA from them, it is possible to purify a DNA sample from unwanted structures. It is essential to run gels in the presence of magnesium ions to ensure structural integrity of the sample. A DNA ladder with linear DNA fragments of different length can be used for reference. The gel should typically be loaded with 5–100 ng of DNA, whereby the optimum amount will depend on gel composition, running conditions and staining. For detailed advice and troubleshooting on gel electrophoresis of DNA samples see [42].

11.1.4.4 Dynamic Light Scattering

Like gel electrophoresis, dynamic light scattering (DLS) allows for conclusions on the size distribution of DNA nanostructures in solution. It is mainly useful for small DNA structures, where direct imaging via AFM and TEM is more difficult and less conclusive. The hydrodynamic radius of particles is determined by shining a laser onto a sample, collecting the scattered light and analysing temporal fluctuations of the scattered intensity trace. The autocorrelation function in the time domain decays faster for smaller particles, because Brownian motion causes them to move faster than large particles. DLS does, however, not allow for conclusions regarding the shape of a DNA nanostructure, as all particles are approximated as spheres with a certain hydrodynamic radius. For details regarding DLS measurement and analysis see [50].

11.1.4.5 AFM and TEM

Due to the dimensions of DNA origami and other DNA nanostructures, only atomic force microscopy (AFM) and transmission electron microscopy (TEM) offer sufficient resolution to obtain precise structural information. At the same time, these imaging techniques are the most convincing and reliable evidence for the correct assembly. In AFM, a sample is raster-scanned with a cantilever that has a very sharp tip attached to it. The interactions between the tip and features on the sample can be measured through changes in cantilever bending or vibrational frequency, which are detected by collecting the reflection of a laser beam directed onto the cantilever. This information is used to reconstruct the surface topology. AFM imaging of DNA nanostructures is carried out in tapping mode, where the cantilever is driven to oscillate up and down near its resonance frequency. The height of the cantilever is controlled to maintain constant oscillation amplitude. The image is thus produced by determining the force of the intermittent contacts of the tip with the sample. It is possible to use a dry sample, where the DNA is deposited onto a mica surface. While this is often sufficient to confirm the structural integrity, it should be noted that hollow DNA structures can collapse during the drying process [51]. The apparent height of a DNA duplex imaged in air is only around 0.7–1 nm [52] likely because the DNA collapses and is immersed in a salt layer [53]. When imaging in liquid, DNA nanostructures are visualised in their native environment. Tip compression and tip deconvolution can still alter the observed dimensions. With high-speed AFM, dynamic processes can be monitored. Nanomechanical switching [54, 55], attachment of ligands to a DNA platform [6, 10] or cluster growth [56] have been studied using AFM imaging. A detailed protocol can be found in [57].

While the working principle of a TEM is completely different from an AFM, both imaging techniques obtain comparable resolution for DNA nanostructures [58]. In TEM, a focussed electron beam is transmitted through the sample and an image is generated by detecting the interaction of the transmitted electrons with the sample [59]. Due to the small de Broglie wavelength of electrons, the resolution is greatly enhanced compared to light microscopes. With high-resolution cryo-TEM images, it is possible to identify individual crossovers within a DNA origami structure [60] and to visualise DNA origami together with organic structures like lipid vesicles [61]. TEM is well-suited for hollow three-dimensional structures, as it circumvents problems with tip-compression in AFM. For two-dimensional structures, however, the contrast of TEM images is often not as good, which is why AFM is then the method of choice. Moreover, TEM cannot be used to image living cells, as it requires a vacuum. Details on TEM imaging can be found in [59].

11.1.4.6 DNA-PAINT

DNA-PAINT, or "DNA-based point accumulation for imaging in nanoscale topography", is a super-resolution imaging method. Super-resolution is obtained by temporally switching each target between a fluorescence on-state and an off-state, and determining their respective positions with sub-diffraction limit precision. DNA-PAINT uses transient binding of short fluorescently labelled single-strands of DNA to achieve a resolution down to 5–10 nm [62]. Unlike in AFM and TEM, DNA-PAINT offers chemical specificity due to the required hybridisation and even single-stranded DNA can be visualised [63].

11.1.4.7 Functionalisation of DNA

With all the tools at hand, the creation of new shapes from DNA is no longer the main challenge in the field of structural DNA nanotechnology. We are experiencing a shift from mere structures to functional nano-objects with diverse applications. This is only achievable thanks to the possibility of site-directed chemical modification of DNA with functional groups. Commercially available amino- or thiol-modified oligonucleotides often serve as a starting point for the covalent attachment of the selected crosslinker. Proteins and aptamers have been attached to DNA nanostructures for biomimetic enzyme cascades [64, 65], single molecule detection [66], cellular recognition [67, 68] or controlled substrate release [67, 69]. Gold [70–72] and silver nanoparticles [73, 74] and quantum dots [75] have been incorporated into DNA nanostructures for plasmonic applications [76]. DNA-based polyhedral meshes [21] have recently been used to assemble crystalline nanoparticle-DNA frameworks [77]. Attachment of carbon nanotubes [78] and metallisation of DNA structures [79] may lead to the fabrication of new nanomaterials for electronics. Responsive polymers offer scope for the creation of DNA nanostructures that reversibly change conformation as a response to external stimuli such as light [72, 80] or pH [81]. Fluorescent dyes and quenchers are commonly incorporated for visualisation purposes or to monitor conformational changes of DNA nanostructures, see Sect. 11.2.1. They have also been used to study light harvesting on DNA platforms [82]. Fluorescently tagged DNA nanostructures have found commercial use as standards for superresolution microscopy [83]. Lipophilic modifications, such as cholesterol or porphyrin tags, will be discussed in more detail in the context of the synthetic DNA-based membrane pores in Sect. 11.3.3. They have further been employed to promote cellular uptake, for two-dimensional assembly of DNA nanostructure arrays [56], or to scaffold [84] and bend vesicles [85]. Some functionalised custom oligomers, including fluorescent or lipophilic tags, can be purchased from vendors like Integrated DNA Technologies or Biomers. Combination of synthetic nucleic acids, like peptide nucleic acid (PNA) [86] XNA [87], with DNA nanostructures may protect them from enzymatic digestion in living systems and offers control over the backbone charge [88]. Achieving a sufficient yield of the final purified product is still the main challenge for the integration of many heteroelements into DNA nanostructures. An additional challenge for some applications, like DNA-based synthetic light harvesting systems [82], is the lack of control over the orientation of the functional groups. Without chemical modification, dynamic motion of DNA nanostructures can be achieved by strand displacement, where the target strand is equipped with a single-stranded DNA toehold. The addition of a DNA strand which binds to the toehold and has higher sequence complementarity to the target strand will then replace a previously bound strand [89]. This mechanism has been employed for DNA-based logic gates [90], catalysed reaction circuits [91] or synthetic molecular motors [92].

The long-term prospects of structural DNA nanotechnology will depend on its success in constructing sophisticated nano-machines driving active and functional behaviour at a capacity that is unmatched by competing forms of nanotechnology. Until stability, scalability and cost issues are addressed, DNA nanotechnology will mainly be employed to address specific research questions. Yet thanks to the scope of chemical functionalisation, DNA sensors and nanopores have made remarkable progress towards viable applications beyond the research community. These will be discussed in the following section.

11.2 DNA Sensors and Nanopores

Living cells have developed an intriguingly complex sensory machinery to detect changes in their environment. Molecular control circuits help convert the sensor information into a reliable motor response adapting to environmental factors. Remarkably, in vitro DNA nanotechnology has made progress towards constructing molecular sensory devices. Applications of DNA sensors range from disease diagnostics to environmental screening or fundamental research. Single-molecule resolution can be achieved and is highly desirable for precise quantification and when a sample is scarce. As of today, DNA-based sensors are one of the most successful applications of structural DNA nanotechnology. They are either based on nanomechanical actuation or nanopores for biosensing.

In this section, we will describe the working principle of nanomechanical DNA sensors, review excising devices and discuss the potential for future ones. We will then introduce nanopore sensors as a label-free single-molecule sensing technique and discuss how DNA nanotechnology can enhance this type of sensors. DNA origami nanopores are finally presented as an addition to existing types of nanopores. Liu et al. [93] and Chandrasekaran et al. [94] have written detailed and comprehensive overviews on DNA-based sensors, we just aim to give a brief overview.

11.2.1 Nanomechanical DNA-Based Sensors

In a common approach for the fabrication of DNA-based sensors, the DNA nanostructure is designed to undergo a detectable conformation change in response to a signal. This requires a shift from static structures towards dynamic DNA nanotechnology achieved by chemical functionalisation or strand displacement, see Sect. 11.1.4.7 [92]. Sections of single-stranded DNA can serve as flexible hinges interconnecting stiff regions of double-stranded DNA. Since the DNA nanostructure can only amplify the signal, a detection mechanism is still required. The conformational change of a larger DNA structure can easily be observed in AFM and TEM as described in Sect. 11.1.4.5. Alternatively, a FRET (Förster resonance energy transfer) pair of fluorescent dyes can be placed on the DNA nanostructure. The donor fluorophore is excited and emits at the right wavelength to excite the acceptor. The Stokes-shifted longer-wavelength emission of the acceptor is then detected. Since the energy transfer efficiency between the two fluorophores depends on their distance with an inverse 6th-power law, the fluorophore emission is a sensitive measure for the conformation of a DNA nanostructure if the dyes are placed appropriately. Using this approach, a straightforward bulk measurement of the emission spectrum is sufficient to detect conformational changes, and surface immobilisation is not required. Single-molecule FRET can be carried out in addition. Other detection methods for nanomechanical DNA switches involve fluorescent quenchers [95], nanoparticles, colorimetric and electrochemical sensing techniques, gel electrophoresis [96], microfluidics, optical tweezers [97] or magnetic resonance imaging (MRI) contrast agents. DNA sensors with picomolar detection limits and over million-fold selectivity have been reported [95].

Nanomechanical DNA sensors can be based on large DNA origami structures or small functional nucleic acids. Regarding their functionality, they can broadly be divided into two classes: sensors that detect molecules or ions and sensors that detect environmental factors, like temperature or pH.

11.2.1.1 Molecular Sensors

Proteins, DNA and ribonucleic acid (RNA) have been detected employing variations of DNA origami or smaller nucleic acid-based sensors [98, 99]. The mechanical response of a DNA nanostructure has also been used to assess DNA repair activity [100] or to detect single-base polymorphisms in double-stranded genomic DNA [101]. Not just molecules, also ions have been detected, employing salt-dependent sequence motifs, like the G-quadruplex or metal-ion-bridged duplex nucleic acids structures [98, 99, 102–104]. A mercury sensor based on functional nucleic acids was integrated into dip-stick tests, making it practical for point-of-care diagnostics [95]. Some DNA-based sensors have been employed in vivo for the detection of messenger RNA as disease markers [105] or cell surface markers [99, 106]. These approaches may lead to the development of "smart drugs" capable of making autonomous decisions and releasing a molecular payload if appropriate. The cancer-targeting DNA origami drug delivery vehicle by Douglas et al. [69] and the DNA logic gates for drug release tested in living animals [107] can be seen as proof of principle for such devices.

11.2.1.2 Environmental Sensors

Environmental factors, such as temperature, pH, light or electric field can directly affect the conformation of DNA nanostructures or indirectly prompt a mechanical response mediated by functional groups. DNA-based sensors with pH-sensitive i-motifs have successfully been used to map pH changes inside living cells [108, 109].

All sensors described until now can be classified as nanomechanical sensors, where the sensing molecule triggers a mechanical switch. The sensing mechanism of nanopores, described in the next section, does not rely on such a conformation change.

11.2.2 Nanopores for Single-Molecule Detection

In the first instance, a nanopore is nothing but a nanometre-sized hole in an insulating material. But by applying a voltage across a nanopore immersed in an electrolyte solution, one employs it as a powerful, albeit simple single-molecule sensor. The applied voltage induces an ionic current. Charged analytes are driven through the pore via electrophoresis, stochastically blocking part of the passage for ions. The current blockade signal can be used to detect the analyte, while its duration and magnitude provide information about charge, length and conformation of the molecule [110]. In this way, nanopores can be employed for the label-free detection of nanometre-sized particles, but the sensing mechanism is applicable across scales – as long as the size of the pore matches the size of the analyte. A Coulter counter, invented by Wallace H. Coulter in 1953 [111], uses the described principle to detect, size and count cells and other particles between 400 nm and 1 mm [112, 113]. The brilliantly simple and broadly applicable sensing mechanism based on this resistive pulse principle is illustrated in Fig. 11.6a, b. A big advantage of nanopore sensors compared to the nanomechanical sensors described in Sect. 11.2.1 is that they can be employed as multi-purpose sensors. Whereas previous sensors had to be designed for a specific molecule, nanopores can in principle detect a broad range of different molecules in solution.

Bezrukov et al. were the first to demonstrate that a molecular-scale Coulter counter can indeed detect single polymers (polyethylene glycol) in solution [114]. The sensing nanopore itself was taken from nature, where nanopores are ubiquitous as molecular gatekeepers of cellular function. Biological protein nanopores are embedded in the membranes of all cells, where they control the selective transport of ions and biomolecules, see Sect. 11.3.1. The challenge for the field of nanopore

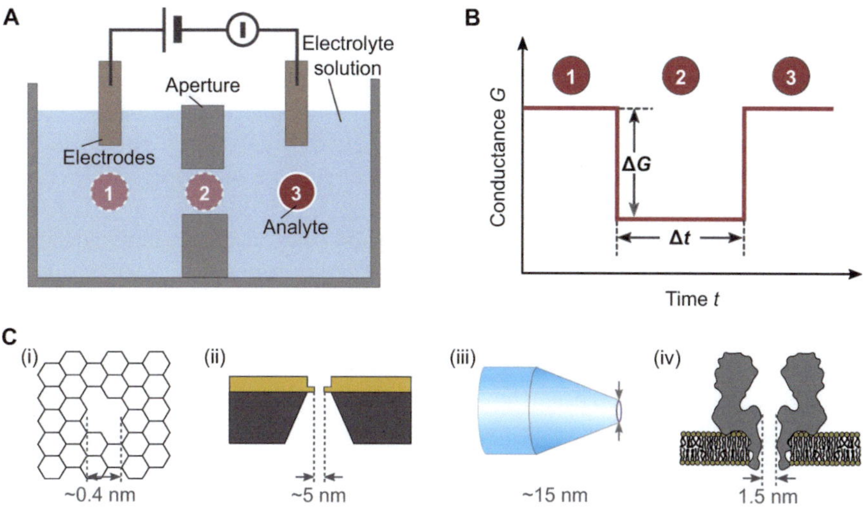

Fig. 11.6 Resistive pulse sensing. (**a**) Schematic illustration of a general Coulter counter and its key components. Two reservoirs containing an analyte species, immersed in an electrolyte solution, are connected via an aperture. A voltage is applied across the aperture while the ionic conductance is recorded. The passage of an analyte across the aperture via the positions 1,2 and 3 causes a characteristic resistive pulse signal as illustrated in (**b**). Conductance drop ΔG and passage time Δt contain information about the dimensions and the charge of the analyte. (**c**) Engineered solid-state nanopores, like graphene (i), silicon nitride (ii) or glass (iii) nanopores, and natural protein pores, like α-hemolysin (iv), have been employed for resistive pulse sensing on the nanoscale to detect single molecules

sensing was to find a protein nanopore which is large enough for the passage of macromolecules and remains open for a sufficiently long period of time. Most protein pores show stochastic opening and closing which could potentially be misinterpreted as molecular translocation events. The bacterial toxin α-hemolysin, however, proved to be ideally suited for the purpose of nanopore sensing and became the most widely used sensing pore [110]. Its crystal structure showed a 10 nm long solvent-filled channel with a constriction of 1.4 nm diameter [115]. Kasianowicz et al. demonstrated that α-hemolysin can be used to detect single-stranded DNA and RNA [116]. This result quickly inspired researchers to refine the technology for the purpose of rapid DNA sequencing [117–119], even though the idea of nanopore sequencing had already been sketched out much earlier by David Deamer [120]. The assumption here is that the interaction between the nanopore and the DNA is base-specific [117]. While different oligonucleotides could indeed be distinguished [121], single-base resolution required mechanisms for slowing down and controlling the translocation process. Genetic engineering of the protein pore proved to be a promising approach [122]. The attachment of DNA polymerases [123, 124] and exonucleases [125] to the pore were major steps towards nanopore sequencing. Today, nanopore arrays for real-time sequencing are commercially available [126–

128] and have been employed for several sequencing tasks [129–131]. While competing technologies remain the state of the art, the advantages of portable real-time sequencing devices have been demonstrated for Ebola surveillance [131] and aboard the International Space Station [132]. While nanopore-based DNA sequencing has attracted much attention, the native or genetically engineered α-hemolysin has also been used to detect many other analytes including small organic molecules [133] or even ions [134].

Apart from protein nanopores, pores in solid-state materials have advanced the field of nanopore sensing. Using nanofabrication techniques, such as ion or electron-beam lithography, single pores can be obtained in materials like silicon nitride [135, 136]. Nanopores in graphene, reducing the material thickness down to a single atomic layer, promise high sensitivity [137, 138]. Even simple low-cost glass capillaries have been employed to detect DNA [139] and proteins [140]. By combining ionic current measurements with force measurements using optical tweezers, it is possible to obtain additional information about the translocating molecule [141]. The advantage of solid-state pores are their high stability under a wide range of conditions and their tunable diameters. Even though surface modification with polymers [142], DNA [143] or proteins [144] offers routes to control translocation, the atomic precision and reproducibility of protein pores with the possibility of site-specific mutation remains unmatched. Protein sequencing with nanopores remains an interesting challenge in the field [145].

11.2.3 DNA Nanotechnology for Enhanced Nanopore Sensing

In principle, nanopores are ideally suited as general-purpose sensors capable of identifying a range of different molecules, beyond DNA, in minimal sample volumes [146]. Yet despite all achievements in the field of nanopore sensing, detecting a molecule in a mixture of others remains challenging. Given that even an isolated molecule can exhibit different current signatures depending on its folding state in the moment of translocation [139], a multitude of similar complexes becomes virtually indistinguishable. This challenge is sketched out in Fig. 11.7a, b. While receptor-modified sensing pores offer molecular specificity [133], the pore can then no longer be used as a general sensor for simultaneous detection of a variety of analytes.

Structural DNA nanotechnology offers a route to enhance the capabilities of nanopores, while maintaining the simplicity of the sensing technique [146]. As discussed in Sect. 11.1.4.7, DNA nanostructures can be functionalised to bind a range of proteins or other molecules of interest. Before translocating through the pore, a DNA carrier can then fish out analytes in solution. Empty carriers show different current signatures than loaded ones, indicating the presence or absence of the molecule of interest. This concept has been demonstrated with a linear DNA carrier containing aptamer binding sites for proteins at predefined positions along the DNA [147, 148]. Different positions could encode for different proteins. The

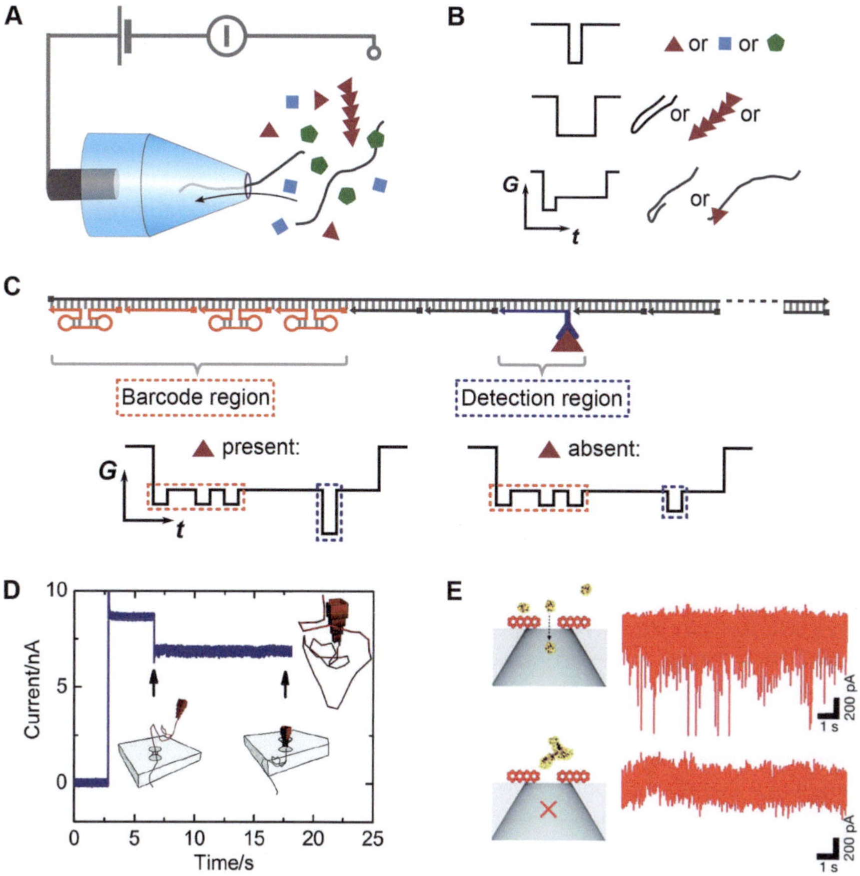

Fig. 11.7 (**a**) Schematic illustration of a glass capillary-based nanopore sensor. Various molecules can be driven through the pore via electrophoresis. (**b**) Resistive pulse signals for the passage of different molecules and their ambiguous interpretation. Different proteins (square, triangle, pentagon) are often indistinguishable, signals of protein complexes or DNA-bound proteins may be mistaken for DNA translocating in a non-linear conformation. (**c**)–(**e**) Approaches to enhance the specificity of nanopore sensors with DNA nanotechnology. (**c**) DNA carrier idea demonstrated by Bell & Keyser. The molecule of interest binds to a specific binding site on the DNA carrier, different carriers for different molecules can be identified via their barcode made from protrusions of DNA. The resistive pulse signal now gives a read-out for the barcode and the presence or absence of the target molecule [149]. (**d**)–(**e**) DNA origami hybrid nanopores. (**d**) The funnel-shaped DNA origami nanopore by Bell et al. was inserted into a solid-state support. The long leash helped guide the voltage-driven formation of the hybrid nanopore. A recorded current trace is shown in blue. The formation of the hybrid pore can be detected as a drop in current (at around $t = 6\,\mathrm{s}$). (Adapted with permission from [150]. Copyright 2012 American Chemical Society). (**e**) Wei et al. used a flat DNA origami nanopore to demonstrate size-selective transport of proteins. The smaller streptavidin (top) can translocate through the DNA pore, visible as downward spikes in the current trace (right). The larger immunoglobin G (bottom) does not fit through the hybrid nanopore. The bare solid-state pore, however, would have transported both species [151]. (Copyright Wiley-VCH Verlag GmbH & Co. KGaA. Reproduced with permission)

fabrication of the linear carrier relies on a simplified version of DNA origami: The kilobase long DNA scaffold is hybridised with hundreds of short addressable staples without crossovers. Bell & Keyser designed a library of DNA nanostructures, where each member has a binding site for a specific antibody and produces a unique current signal due to a barcode of DNA hairpins [149]. In this manner, nanopore sensing gained unprecedented specificity: An occupied or unoccupied binding site on the DNA carrier indicates the presence or absence of the protein, while the barcode provides information on the type of protein as illustrated in Fig. 11.7c. Relying on DNA nanotechnology, more complex DNA origami carriers for nanopore sensing will contribute to the development of highly portable general-purpose lab-on-the-chip devices.

11.2.4 DNA Origami Hybrid Nanopores

Another route to enhance nanopore sensing with DNA nanotechnology is to use the DNA nanostructure itself as the sensing pore. Relying on the ability to create arbitrary three-dimensional structures from DNA [18], it should be relativelz straightforward to model the shape of a nanopore. DNA origami nanopores have indeed been created and used for sensing purposes by integrating them into solid-state nanopore systems [150, 151] and into lipid bilayers, see Sect. 11.3.3. To achieve the former, an appropriate electric field is applied across the solid-state pore in the presence of the DNA origami pores. The negatively charged DNA origami is thus driven towards the pore via electrophoresis, where it is trapped if its diameter is too large for translocation. A double-stranded leash on the DNA origami helps guide the entrance into the solid-state support and to achieve correct positioning. The formation of a hybrid nanopore can be monitored with the standard resistive-pulse measurement. If a DNA origami nanopore is trapped, it partially blocks the flow of ions resulting in a long-lasting stepwise decrease in the ionic current accompanied by a variable increase in noise, see Fig. 11.7d. A similar method was previously used to trap α-hemolysin in a solid-state support [152], highlighting the applicability of the approach for a range of different pore geometries.

By reversing the voltage, the DNA origami pore can subsequently be ejected and the solid-state support is available for further trapping cycles. The reversibility of the system provides statistical information about the inherent variability of the process, caused by different trapping orientations, mechanical conformations or structural damage [153]. Conical glass nanopores [154] as well as silicon nitride nanopores [150, 151] have been used as solid-state supports capable of trapping DNA origami funnels, Fig. 11.7d [150], or flat DNA origami plates, Fig. 11.7e [151, 154]. The compatibility with glass capillaries is advantageous, as they can be combined with optical techniques to monitor the trapping [154], they allow for multiplexing [155] and their fabrication is straight-forward [156]. In addition to double-stranded DNA [150, 157], single-stranded DNA [151, 154] and proteins [151] were detected in DNA origami nanopore systems, whereby the voltage required to keep the DNA origami trapped is also used to drive the translocation of the analyte, see Fig. 11.7e. Since the detection of all these molecules has been achieved before, without DNA origami, one is prompted to ask what the DNA origami hybrid system adds to the existing technologies. Its advantages are listed below.

11.2.4.1 Nanopore Architecture By Design

First of all, the diversity of possible pore architectures achievable with DNA origami is unmatched by the conventional types of nanopores. Nanofabrication of solid-state pores offers still limited control, especially regarding the pore geometry on atomic length scales. Genetic engineering enables the shape of a protein pore to be adapted to some extend, but not in the same way as de novo design with DNA nanotechnology. Due to the specific base pairing interactions, DNA pores can be made reproducibly with high precision.

11.2.4.2 Tunable Pore Diameter

Precise control over the pore geometry enables fine-tuning of the pore size. While the diameter of α-hemolysin is well suited for the translocation of single-stranded DNA, detection of double-stranded DNA or folded proteins would require a larger pore. Large protein pores which are at the same time stable under high salt conditions and remain open for extended periods of time are rare in nature. While double-stranded DNA has been translocated through the phi29 pore [158], protein pores large enough for the passage of kilodalton-sized protein complexes are under active development with promising recent results [159]. The diameters of DNA-based nanopores, on the other hand, can easily be designed to match the size of the analyte, from ions to macromolecules. Published DNA nanopores span over one order of magnitude in diameter from sub-nanometre pores embedded in lipid membranes, see Sect. 11.3.3 [43], to 15 nm pores in solid-state supports [154]. They are thus ideally suited for size-selective sensing. DNA origami nanopores with different diameters have for instance been used to sort two proteins of different sizes, Fig. 11.7e [151] and to control the conformation of a linear double-strand of DNA [154].

11.2.4.3 High Specificity

Probably the most notable advantage of DNA nanopores are the diverse possibilities to add chemical functionality to the pore. The DNA structure is thereby used as an addressable skeleton, which can be functionalised as described in Sect. 11.1.4.7. Specific binding sites inside a nanopore can help to obtain information about the chemical identity of a molecule. While proteins or DNA hairpins have been used to functionalise solid-state nanopores, controlling the number and the exact location of these modifications within the sensing volume remains challenging. Genetic engineering solves this problem in the case of protein pores – yet only DNA pores have tunable diameters and come with the convenience of commercial synthesis for a wide range of chemical tags. Binding sites obtained by extending single-stranded DNA overhangs into the mouth of the pore have been used for the sequence-specific detection of complementary single-stranded DNA [151, 154]. Attachment of gold or silver nanoparticles to a DNA origami pore offer the possibility of plasmonic sensing in addition to the resistive pulse signal [71].

11.2.4.4 Stimuli Response

Chemical functionalisation further allows for the creation of stimuli responsive DNA origami nanopores. This offers a route to generalise DNA-based nanopore sensors beyond the detection of molecules. Responsive polymers attached to the DNA nanopore could trigger a resistive pulse signal in response to a change in physical environment. Light, temperature, pressure or pH could induce a conformational change of the DNA nanopore, similar to gating of natural ion channels. While these opportunities have not yet been fully exploited, voltage-response of DNA nanopores has been reported even without added chemical functionality in solid-state supports [157] and in lipid bilayers [61, 160], see Sect. 11.3.3.

11.2.4.5 Ease of Fabrication

With the recent advances in DNA nanotechnology, we can access an elaborate toolbox for the design, assembly and modification of DNA nanopores, see Sect. 11.1.3. While solid-state pores usually have to be fabricated one by one, billions of DNA nanostructures are obtained from a simple one-pot reaction. All components can be purchased off the shelf and the structural characterisation of the pores is straightforward, see Sect. 11.1.4.

DNA origami hybrid nanopores thus possess advantageous properties of both, protein and solid-state nanopores. But to make them really competitive, several challenges have to be addressed – a major one being the leakage currents reducing the signal-to-noise ratio for the detection of translocating molecules. Comparing the experimental conductance values of the hybrid pore with simple geometrical models, it is obvious that the DNA structures are permeable to ions [161]. To address this issue, Li et al. carried out molecular dynamics (MD) simulations and showed that the ionic conductivity can be reduced by increasing the concentration of magnesium ions in solution. They verified their results experimentally [162]. A DNA origami plate designed on a hexagonal lattice showed lower ion permeability than a square lattice plate [163]. Strategies to chemically attach the DNA nanopore to its solid-state support and thus tighten the seal will reduce leakage on the one hand, but make repeated trapping and ejection more difficult. Major grove binders [161] or hydrophobic coating [153] have been suggested as additional strategies. If

DNA nanopores are inserted into lipid membranes, the leakage should be reduced further and it should in principle be possible to study passive transport, because the trapping voltage is no longer required.

A second concern is the mechanical stability of the DNA origami. MD simulations have shown bending and reversible layer-separation for DNA origami nanoplates [162]. Also experiments showed buckling and voltage-dependent conformational changes [157, 163]. This may not be surprising given the high electric field strength of around 10^7 V/m in typical nanopore experiments [161].

Therefore, while the established types of nanopores are unlikely to be replaced entirely, DNA origami nanopores are useful for more challenging sensing applications where precise control over surface chemistry and geometry is an absolute necessity.

11.3 Synthetic Membrane Nanopores

In the previous section, we presented synthetic nanopores in solid-state materials as well as biological nanopores in organic lipid membranes. Now, we want to go one step further and ask if it is possible to create an entirely organic nanopore system from scratch, where a man-made pore self-assembles into a lipid bilayer membrane. The scope of synthetic membrane pores reaches from nanopore sensing to biomimicry and beyond, offering broader possibilities for biomedical applications. From the perspective of basic science, a synthetic membrane-embedded pore is a biomimetic system to study transport on the molecular scale. Realising that 50% of the currently authorised drugs target membrane pores [164] adds another dimension to the research, one that is based on the hope that a better understanding of membrane pores will lead to the design of better drugs. It fuels the dream that designer pores could replace defective channels in cells and thereby potentially offer a cure to the many diseases related to ion-channel malfunction – from cystic fibrosis [165] to autoimmune diseases [166]. Since many toxins are protein nanopores or destabilise the membrane, pore-forming synthetic compounds may be used as antimicrobial agents [167]. If the synthetic channel does not act as a therapeutic itself, it could be coupled to a pharmaceutically active compound which is directly released into the target cell upon membrane insertion. Drug permeation into the cell, and specifically selective permeation into the diseased cell, is often a challenge. Synthetic membrane pores as drug carriers would be a step away from the current small-molecule drugs towards macro-molecular delivery machines with the ability to recognise the target cell, perforate its membrane and release the molecular payload. A quest for synthetic pores could also come from the rising field of synthetic bottom-up biology: A synthetic cell built from molecular components will eventually require structures for membrane transport [168]. It is thus not surprising that synthetic membrane pores have been an active field of research for the past 30 years.

In this section, we start by presenting characteristics of natural membrane pores as gold-standards for creating artificial ones. Having reviewed milestone achievements in the history of synthetic pores, we present the state of the art of transmembrane-spanning nanopores constructed from DNA.

11.3.1 Membrane Pores in Nature

Before creating a synthetic membrane pore, a closer look at their natural counterparts, protein membrane pores, is illuminating. Their functions in living organisms are as diverse as their architectures: Small kilodalton ion channels control the selective transport of ions across the lipid membrane and are thus key players in electrically excitable tissue, like neurons or muscles [169]. While ion channels rely on passive transport of ions following an electrochemical or a concentration gradient, ion pumps are active, fueled by the hydrolysis of adenosine triphosphate (ATP) [169]. Larger porins, found in many prokaryotes and the mitochondria of eukaryotic cells, open selective transmembrane pathways for polar solutes, like sugars or water molecules [170]. The megadalton-sized nuclear pore complex, embedded in the nuclear envelope, can be considered the most sophisticated among all membrane pores. It enables the spacial separation of transcription and translation in eukaryotic cells. The transport of larger substrates is highly selective. For instance, only correctly processed messenger RNA can exit the nucleus, while proteins and lipids are imported [171]. Protein membrane pores span one order of magnitude in diameter and three orders of magnitude in ionic conductance and molecular weight. The common feature of all these protein pores is that they span the lipid bilayer membrane, which defines the cell relative to its environment and divides it into functional compartments to sequester biochemical processes. This membrane is made from amphiphilic phospholipids forming an approximately 4 nm thick bilayer sheet. Hydrophilic lipid head groups face the cytoplasmic environment and shield the hydrophobic tails from polar water molecules creating a semipermeable barrier. While small apolar molecules can pass via free diffusion, the membrane is highly impermeable to ions and polar biomolecules, like sugars, DNA or proteins. The energy cost for ions to cross the lipid membrane via free diffusion lies in the range of hundreds of kJ/mol. This process is thus by far not efficient enough to maintain metabolic function requiring efficient molecular exchange mechanisms between membranous compartments [172]. Membrane proteins serve as catalysts of translocation, often switching conformation as a response to external stimuli like membrane potential [173] or specific ligands [174]. For some ion channels, the existence of an open and a closed state has been demonstrated with the corresponding crystal structures [175]. Generally, X-ray crystallography of membrane proteins is challenging because of their hydrophobicity and structural flexibility [176]. As of July 2016, only about 600 membrane protein structures were known [177], although 30% of human genes encode for them [164]. Therefore, functional studies via ionic current recordings of lipid membranes can provide crucial insights.

To summarise, protein membrane pores show four main characteristics [178]:

1. They can have a variety of structures depending on their function.
2. They span the lipid bilayer membrane.
3. They mediate the rapid flux of ions without moving themselves.
4. They are often responsive to physical or chemical stimulation.

These are the benchmark criteria a synthetic pore should fulfil.

11.3.2 Milestones of Synthetic Membrane Pores

Early attempts to mimic the behaviour of protein pores relied on synthetic ionophores, carriers which encapsulated ions before diffusing across the membrane. Prominent ionophores are based on cryptands, calixarenes and crown ethers [179]. Even though this approach mimics some of the biological functions, carrier-mediated transport is slow compared to the passage through an ion channel and does not fulfil our criteria for the classification as a synthetic pore presented in Sect. 11.3.1. Nevertheless, natural ionophores, like indole, play an important role in biological systems [180].

Ion channel activity has first been reported for a synthetic peptide in 1977 [181]. Synthetic pore-forming peptides remain an active field of research [182], but when using nature's own building blocks, a channel can only be called semi-synthetic. The first purely synthetic de novo designed channel is commonly attributed to Tabushi et al. [183]. The dimer of amphiphilic cyclic oligosaccharides (β-cyclodextrins) still inspires current research [184]. In 1998, Fyles et al. created the first stimuli-responsive artificial nanopore: a voltage-sensitive channel [185]. The voltage sensitivity emerged from the asymmetrically positioned charges. The pore design was based on an earlier crown-ether based pore, which can be considered as a milestone achievement itself because of its excellent single-channel activity [186]. Tanaka et al. were the first to report ion selectivity for their artificial membrane pore [187]. Moving away from the traditional chemical synthesis approaches for synthetic pores, ion channel behaviour has also been reported for membrane-piercing carbon nanotubes [188]. While all these accounts of synthetic ion channels still fall behind their natural counterparts in terms of structural complexity and specificity of function, it is striking that they achieve the same high ion flux. Catalysis of translocation seems easier to achieve compared to synthetic enzymes, which still fall orders of magnitude behind their natural benchmarks [172].

There are several key challenges to address in taking the field of synthetic membrane pores forward: All synthetic pores presented above are minuscule in size compared to natural protein pores. Their molecular weight falls one to two orders of magnitude short of protein pores which typically have a molecular weight in the megadalton range. Chemical synthesis of large constructs is hard to achieve and

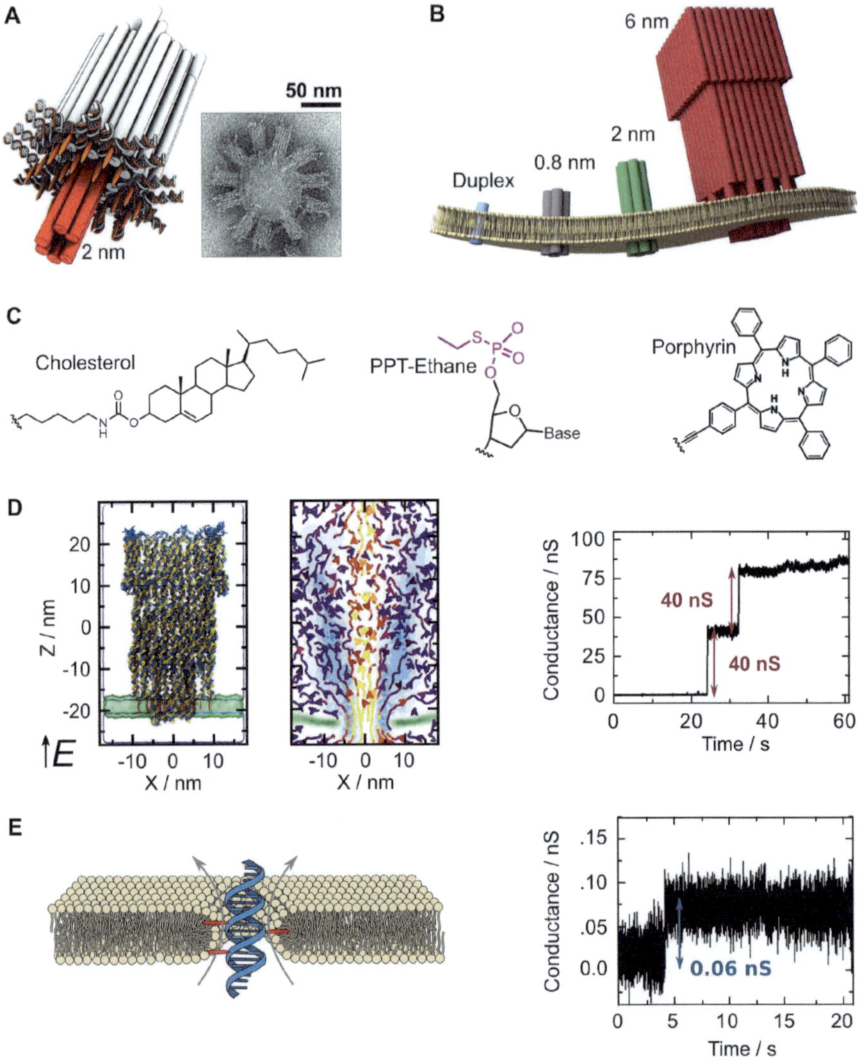

Fig. 11.8 Variety of DNA-based membrane pores. (**a**) Sketch of first membrane-spanning DNA origami pore by Langecker et al. with a nominal diameter of 2 nm. TEM images gave convincing evidence of membrane insertions (right). The publication contains excellent single-channel data. (Adapted from [61]. Reprinted with permission from AAAS). (**b**) Sketches of subsequent DNA membrane pore architectures and their diameters. From left to right: A single transmembrane-spanning DNA duplex (blue) [189], a four-helix structure with an 0.8 nm channel [43], the archetypal 2 nm DNA membrane pore (green, a scaffold-free assembly first presented by Burns et al. is shown here [191]) and a large funnel-shaped DNA origami porin with a 6 nm constriction (red) [190]. (**c**) Chemical structure of three hydrophobic tags that have been used to achieve membrane anchoring of DNA pores: cholesterol [43, 61], ethane-phosphorothioate [191] and tetra-phenyl porphyrin [44]. (**d**) MD simulations (left) and ionic current recordings (right) of the largest

inefficient. While all natural pores are based on polypeptides, independent of their size and function, synthetic pores do not yet have a common molecular basis. This makes it more difficult to compare different synthetic pores and to make functional adaptations. The preparation of many of the described constructs is labour-intensive, often not scalable and requires expert knowledge.

11.3.3 DNA-Based Membrane Pores

With the recent advances in structural DNA nanotechnology described in Sect. 11.1, the creation of megadalton sized structures from a single chemical building block is no longer a challenge. The preparation of DNA nanostructures is quick and accessible, see Sect. 11.1.3, and DNA is an abundant and inherently biocompatible material. DNA origami nanopores have previously been made and inserted into solid-state supports, Sect. 11.2.4. The groundwork for the creation of DNA-based synthetic membrane pores has thus already been carried out. The main remaining challenge was to achieve stable membrane insertion of the DNA construct. This has first been demonstrated by Langecker et al. who overcame the energy barrier for membrane insertion by attaching cholesterol tags to the DNA origami nanopore shown in Fig. 11.8a [61]. Recently, the design space of synthetic DNA membrane pores was expanded by creating significantly smaller and larger pores, Fig. 11.8b [189, 190]. Membrane anchoring of DNA nanopores has also been demonstrated with other hydrophobic modifications, including an ethane-phosphorothioate belt [191] and tetra-phenyl porphyrin tags [44], see Fig. 11.8c. Coarse-grained simulations provided a first quantitative investigation on the number of hydrophobic tags required to overcome the energy barrier for insertion [43, 44, 190].

In the following, we compare DNA membrane pores with the natural protein-based gold-standards. To do so, we test DNA membrane pores against the four characteristics of their natural counterparts presented in Sect. 11.3.1.

Membrane pores can have a variety of structures depending on their function Rapid prototyping of different nanoscale shapes is one of the most appreciated strengths of structural DNA nanotechnology. This is why it may seem surprising that the first DNA membrane pores all featured a 2 nm channel between six

Fig. 11.8 (continued) membrane-inserting DNA nanostructure to date, independently confirming its large conductance in the range of tens of nS. (Adapted from [190], https://pubs.acs.org/doi/abs/10.1021/acs.nanolett.6b02039. Further permissions related to the material excerpted should be directed to the ACS). (**e**) A membrane-spanning DNA duplex as the ultimately smallest DNA membrane pore. MD simulations reveal that ions flow through a toroidal pore at the DNA-lipid interface as illustrated in the sketch (left). Ionic current recordings reveal stable insertions (right). (Adapted from [189], https://pubs.acs.org/doi/abs/10.1021/acs.nanolett.6b02039. Further permissions related to the material excerpted should be directed to the ACS)

concentrically arranged DNA duplexes [44, 61, 160, 191]. The diameter of such a pore within the membrane has been confirmed with polyethylene glycol (PEG)-sizing experiments [160]. In the meantime, however, the structural variability of DNA membrane pores has been demonstrated with the creation of much smaller as well as much larger pores. A DNA channel made from four DNA duplexes had a nominal diameter of 0.8 nm, approaching the dimensions of natural ion channels [43]. Due to its simple design, this DNA channel self-assembled within a minute at RT. A membrane-inserting DNA origami funnel with a 6 nm construction was presented as the largest man-made pore to date [190]. It is wider than natural porins and comparable to the electrical diameter of the nuclear pore complex. Such large architectures are difficult to obtain with traditional chemical synthesis described in Sect. 11.3.2. Even though the intriguing complexity of natural membrane components cannot be matched, the diameters of synthetic pores made from DNA span the biologically relevant scales. Illustrations of the existing DNA membrane pore architectures are shown in Fig. 11.8b. Their functional characterisation follows in the next paragraphs.

Membrane pores span the lipid bilayer membrane TEM imaging, Fig. 11.8a [61], and confocal imaging of fluorescently labelled DNA nanopores [43, 190] revealed their attachment to lipid vesicles. Due to its fluorescent properties, porphyrin has been used for visualisation and membrane anchoring at the same time [44, 189]. The melting temperature of DNA nanostructures increases upon membrane insertion [43, 44] and the fluorescence emission of porphyrin shifts [44]. It should be noted, however, that these techniques can only confirm membrane attachment – functional studies are required to demonstrate that the DNA pores actually span the lipid bilayer. A study evaluating the relative proportions of membrane-spanning versus inactive DNA pores is still lacking.

Membrane pores mediate the rapid flux of ions without moving themselves Single-channel ionic current recordings are at the core of membrane pore characterisation – for both natural and synthetic pores. The ionic current passing through a single channel has for the first time been recorded by Neher and Sackman using their patch-clamp method [192]. A range of techniques have been developed to study membrane proteins ex vivo in synthetic membranes providing a simpler environment in which uncharacterised channels can be studied in isolation. All of them rely on the same measurement principle: A lipid membrane is supported across an aperture and an electric field is applied across the membrane inducing an ionic current. Magnitude, noise and fluctuations in the recorded ionic current allow for conclusions on the presence or absence of membrane pores, including their size, stability, and functional properties, such as gating, ligand-binding or ion selectivity [193]. Every lipid bilayer experiment starts with the formation of a lipid membrane. These bilayers are gigaohm resistors reducing the flow of ions across the aperture, its formation can be observed as a sudden decrease in the ionic current. The insertion of a membrane pore, on the other hand, causes the formation of transmembrane passage for ions and thus a stepwise increase in ionic current. Single-channel recordings of all DNA-based membrane pores revealed ohmic current-voltage characteristics at lower voltages and a relatively broad spread of conductances. The reported average conductance ranges from 0.25 nS for smaller pores [44] to 30 nS for the large DNA origami porin, see Fig. 11.8d [190]. Single-stranded DNA has been translocated through a 2 nm wide DNA pore [61]. The diameter of such a 2 nm DNA pore was confirmed with PEG-sizing experiments [160] and the ion flux was partially blocked with complementary single-stranded DNA ligands [194].

Membrane pores are often responsive to physical or chemical stimulation The functional characterisation of membrane pores also often relies on ionic current recordings. DNA membrane pores exhibited multiple conductance states reminiscent of gating observed for natural ion channels, independent of their architecture and the chemical nature of the membrane anchor. No gating has been reported for a recent DNA pore under the examined conditions [195]. Example traces for a range of different pores are shown in Fig. 11.9. Conductance steps can occur between different open states or between the conductance of the plain bilayer

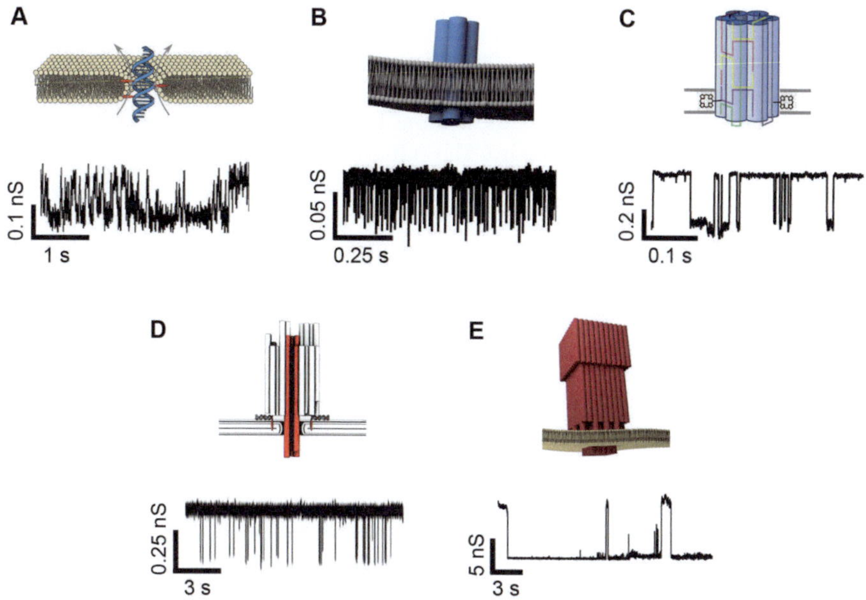

Fig. 11.9 Voltage-gating of DNA-based membrane pores. Sketch of the pore (top row) and gating trace (bottom row) of (**a**) the transmembrane spanning DNA duplex. (Adapted from [189], https://pubs.acs.org/doi/abs/10.1021/acs.nanolett.6b02039. Further permissions related to the material excerpted should be directed to the ACS); (**b**) the four-helix bundle. (Adapted with permission from [43]. Copyright 2015 American Chemical Society); (**c**) the scaffold-free six-helix pore. (Adapted with permission from [160], https://pubs.acs.org/doi/abs/10.1021/nn5039433. Further permissions related to the material excerpted should be directed to the ACS); (**d**) the DNA origami six-helix pore. (Adapted from [61], adapted with permission from AAAS), and (**e**) the large DNA origami porin. (Adapted from [190], https://pubs.acs.org/doi/abs/10.1021/acs.nanolett.6b02039. Further permissions related to the material excerpted should be directed to the ACS)

and the pore conductance, potentially caused by DNA pores flipping in and out of the membrane. Not surprisingly for such highly charged pores, the conductance states were shown to be voltage-dependent. At higher voltages, the pores were more likely to switch to a lower conductance state, or to leave the membrane entirely [61, 160]. Experiments [61] and MD simulations [196] suggest that dynamic motion of the mouth regions could be the origin of the gating.

MD simulations further revealed mechanosensitivity of DNA pores [197], specifically of the six-helix bundle [191]. This could explain why differences in the relative abundance of conductance states have been observed on different membrane systems [160]. Gating as a response to specifically designed DNA ligands has also been reported [194].

Despite the radically new approach towards the design of synthetic membrane pores, DNA-based pores already show the four main characteristics of their natural counterparts. Some of the challenges in the field, like the lack of common building blocks for diverse architectures, have successfully been addressed. The largest DNA pore even outperforms most biological pores in terms of its dimensions and conductance. The study of a membrane-spanning DNA duplex has revealed alternative conductance pathways through lipid membranes, which do not require the presence of a central physical channel – an observation that prompts questions about the role of such pathways in nature. Despite this significant progress, DNA membrane pores still trails the natural benchmarks when considering their structural and functional complexity. The full exploitation of the available toolbox for chemical functionalisation provides a route towards custom-designed stimuli-responsive DNA membrane pores. Using these technological advancements, the long-term goal of gaining the precision of nature should be achievable opening up exciting future avenues for DNA sensors, nanopores and ion channels.

Acknowledgements K.G. acknowledges funding from the Winton Programme for the Physics of Sustainability, Gates Cambridge and the Oppenheimer PhD studentship, U.F.K. from an ERC Consolidator Grant (Designerpores 647144) and Oxford Nanopore Technologies.

References

1. Watson JD, Crick FHC (1953) Molecular structure of nucleic acids. Nature 171:737–738
2. Holliday R (1964) A mechanism for gene conversion in fungi. Genet Res 5:282–304
3. Doniger J, Warner RC, Tessma I (1973) Role of circular dimer DNA in the primary recombination mechanism of bacteriophage S13. Nat New Biol 242:9–12
4. Seeman NC (1982) Nucleic acid junctions and lattices. J Theor Biol 99:237–247
5. Seeman NC, Kallenbach NR (1983) Design of immobile nucleic acid junctions. Biophys J 44:201–209
6. Yan H, Park SH, Finkelstein G, Reif JH, Labean TH (2003) DNA-templated self-assembly of conductive nanowires. Science 301:1882–1884
7. Zheng J, Birktoft JJ, Chen Y, Wang T, Sha R et al (2009) From molecular to macroscopic via the rational design of a self-assembled 3D DNA crystal. Nature 461:74–77

8. Winfree E, Liu F, Wenzler LA, Seeman NC (1998) Design and self-assembly of two-dimensional DNA crystals. Nature 394:539–544
9. Rothemund PWK, Papadakis N, Winfree E (2004) Algorithmic self-assembly of DNA Sierpinski triangles. PLoS Biol 2:2041–2053
10. Park SH, Pistol C, Ahn SJ, Reif JH, Lebeck AR et al (2006) Finite-size, fully addressable DNA tile lattices formed by hierarchical assembly procedures. Angew Chem Int Ed 45:735–739
11. Liu D, Wang M, Deng Z, Walulu R, Mao C (2004) Tensegrity: construction of rigid DNA triangles with flexible four-arm DNA junctions. J Am Chem Soc 126:2324–2325
12. Tesoro S, Göpfrich K, Kartanas T, Keyser UF, Ahnert SE (2016) Non-deterministic self-assembly with asymmetric interactions. Phys Rev E 94:1–8
13. Yin P, Hariadi RF, Sahu S, Choi HMT, Park SH et al (2008) Programming DNA tube circumferences. Science 321:824–826
14. Wei B, Dai M, Yin P (2012) Complex shapes self-assembled from single-stranded DNA tiles. Nature 485:623–627
15. Ke Y, Ong LL, Shih WM, Yin P (2012) Three-dimensional structures self-assembled from DNA bricks. Science 338:1177–1183
16. Yan H, LaBean TH, Feng L, Reif JH (2003) Directed nucleation assembly of DNA tile complexes for barcode-patterned lattices. Proc Natl Acad Sci 100:8103–8108
17. Shih WM, Quispe JD, Joyce GF (2004) A 1.7-kilobase single-stranded DNA that folds into a nanoscale octahedron. Nature 427:618–621
18. Rothemund PWK (2006) Folding DNA to create nanoscale shapes and patterns. Nature 440:297–302
19. Douglas SM, Dietz H, Liedl T, Högberg B, Graf F et al (2009) Self-assembly of DNA into nanoscale three-dimensional shapes. Nature 459:414–418
20. Dietz H, Douglas SM, Shih WM (2009) Folding DNA into twisted and curved nanoscale shapes. Science 325:725–730
21. Benson E, Mohammed A, Gardell J, Masich S, Czeizler E et al (2015) DNA rendering of polyhedral meshes at the nanoscale. Nature 523:441–444
22. Douglas SM, Marblestone AH, Teerapittayanon S, Vazquez A, Church GM et al (2009) Rapid prototyping of 3D DNA-origami shapes with caDNAno. Nucleic Acids Res 37:5001–5006
23. Said H, Schüller VJ, Eber FJ, Wege C, Liedl T et al (2013) M1.3 – a small scaffold for DNA origami. Nanoscale 5:284–290
24. Brown S, Majikes J, Martínez A, Girón TM, Fennell H et al (2015) An easy-to-prepare mini-scaffold for DNA origami. Nanoscale 7:16621–16624
25. Ke Y, Douglas SM, Liu M, Sharma J, Cheng A et al (2009) Multilayer DNA origami packed on a square lattice. J Am Chem Soc 131:15903–15908
26. Castro CE, Kilchherr F, Kim D-N, Shiao EL, Wauer T et al (2011) A primer to scaffolded DNA origami. Nat Methods 8:221–229
27. List J, Weber M, Simmel FC (2014) Hydrophobic actuation of a DNA origami bilayer structure. Angew Chem Int Ed 53:4236–4239
28. Funke JJ, Dietz H (2015) Placing molecules with Bohr radius resolution using DNA origami. Nat Nanotechnol 11:47–52
29. Liedl T, Högberg B, Tytell J, Ingber DE, Shih WM (2010) Self-assembly of three-dimensional prestressed tensegrity structures from DNA. Nat Nanotechnol 5:520–524
30. Andersen ES, Dong M, Nielsen MM, Jahn K, Lind-Thomsen A et al (2008) DNA origami design of dolphin-shaped structures with flexible tails. ACS Nano 2:1213–1218
31. Williams S, Lund K, Lin C, Wonka P, Lindsay S et al (2009) Tiamat: a three-dimensional editing tool for complex DNA structures. Springer, Heidelberg
32. Pan K, Kim D-N, Zhang F, Adendorff MR, Yan H et al (2014) Lattice-free prediction of three-dimensional structure of programmed DNA assemblies. Nat Commun 5:5578
33. Steffen C, Thomas K, Huniar U, Hellweg A, Rubner O et al (2010) TmoleX–a graphical user interface for TURBOMOLE. J Comput Chem 31:2967–2970

34. Martin TG, Dietz H (2012) Magnesium-free self-assembly of multi-layer DNA objects. Nat Commun 3:1103
35. Burns JR, Al-Juffali N, Janes SM, Howorka S (2014) Membrane-spanning DNA nanopores with cytotoxic effect. Angew Chem Int Ed 53:12466–12470
36. Göpfrich K, Zettl T, Meijering AEC, Hernández-Ainsa S, Kocabey S et al (2015) DNA-tile structures lead to ionic currents through lipid membranes. Nano Lett 15:3134–3138
37. Jungmann R, Liedl T, Sobey TL, Shih W, Simmel FC (2008) Isothermal assembly of DNA origami structures using denaturing agents. J Am Chem Soc 130:10062–10063
38. Sobczak J-PJ, Martin TG, Gerling T, Dietz H (2012) Rapid folding of DNA into nanoscale shapes at constant temperature. Science 338:1458–1461
39. Shaw A, Benson E, Högberg B (2015) Purification of functionalized DNA origami nanostructures. ACS Nano 9:4968–4975
40. Glasel J (1995) Validity of nucleic acid purities monitored by A260/A280 absorbance ratios. Biotechniques 18:62–63
41. Beer (1852) Bestimmung der Absorption des rothen Lichts in farbigen Flüssigkeiten. Ann Phys Chem 86:78–88
42. Green MR, Sambrook J (2012) Molecular cloning: a laboratory manual, 4th edn. Cold Spring Harbor Laboratory Press, New York
43. Göpfrich K, Zettl T, Meijering AEC, Hernández-Ainsa S, Kocabey S et al (2015) DNA-tile structures lead to ionic currents through lipid membranes. Nano Lett 15:3134–3138
44. Burns JR, Göpfrich K, Wood JW, Thacker VV, Stulz E et al (2013) Lipid-bilayer-spanning DNA nanopores with a bifunctional porphyrin anchor. Angew Chem Int Ed 52:12069–12072
45. Dunn KE, Dannenberg F, Ouldridge TE, Kwiatkowska M, Turberfield AJ et al (2015) Guiding the folding pathway of DNA origami. Nature 525:82–86
46. Holmes DL, Stellwagen NC (1990) The electric field dependence of DNA mobilities in agarose gels: a reinvestigation. Electrophoresis 11:5–15
47. Holmes DL, Stellwagen NC (1991) Estimation of polyacrylamide gel pore size from Ferguson plots of linear DNA fragments. II. Comparison of gels with different crosslinker concentrations, added agarose and added linear polyacrylamide. Electrophoresis 12:612–619
48. Pernodet N, Maaloum M, Tinland B (1997) Pore size of agarose gels by atomic force microscopy. Electrophoresis 18:55–58
49. Pluen A, Netti PA, Jain RK, Berk DA (1999) Diffusion of macromolecules in agarose gels: comparison of linear and globular configurations. Biophys J 77:542–552
50. Berne BJ, Pecora R (2000) Dynamic light scattering, 2nd edn. Dover Publications, Inc., New York
51. Göpfrich K, Li C-Y, Ricci M, Bhamidimarri SP, Yoo J et al (2016) Large-conductance transmembrane porin made from DNA origami. ACS Nano 10(9):8207–8214
52. Hansma HG, Revenko I, Kim K, Laney DE (1996) Atomic force microscopy of long and short double-stranded, single-stranded and triple-stranded nucleic acids. Nucleic Acids Res 24:713–720
53. Moreno-Herrero F, Colchero J, Baró AM (2003) DNA height in scanning force microscopy. Ultramicroscopy 96:167–174
54. Kuzuya A, Sakai Y, Yamazaki T, Xu Y, Komiyama, M (2011) Nanomechanical DNA origami 'single-molecule beacons' directly imaged by atomic force microscopy. Nat Commun 2:449
55. Arivazhagan R, Endo M, Hidaka K, Tao Tran PL, Teulade-Fichou M-P et al (2014) G-quadruplex-binding ligand-induced DNA synapsis inside a DNA origami frame. RSC Adv 4:6346–6355
56. Kocabey S, Kempter S, List J, Xing Y, Bae W et al (2015) Membrane-assisted growth of DNA origami nanostructure arrays. ACS Nano 9:3530–3539
57. Lyubchenko Y, Shlyakhtenko L, Ando T (2011) Imaging of nucleic acids with atomic force microscopy. Methods 54:274–283
58. Han D, Pal S, Liu Y, Yan H (2010) Folding and cutting DNA into reconfigurable topological nanostructures. Nat Nanotechnol 5:712–717
59. Williams DB, Carter CB (1996) Transmission electron microscopy. Plenum Press, New York

60. Bai X-C, Martin TG, Scheres SHW, Dietz H (2012) Cryo-EM structure of a 3D DNA-origami object. Proc Natl Acad Sci 109:20012–20017
61. Langecker M, Arnaut V, Martin TG, List J, Renner S et al (2012) Synthetic lipid membrane channels formed by designed DNA nanostructures. Science 338:932–936
62. Jungmann R, Avendaño MS, Woehrstein JB, Dai M, Shih WM et al (2014) Multiplexed 3D cellular super-resolution imaging with DNA-PAINT and exchange-PAINT. Nat Methods 11:313–318
63. Dai M (2017) DNA-PAINT super-resolution imaging for nucleic acid nanostructures. In: Ke Y, Wang P (eds) 3D DNA nanostructure: nethods and protocols. Humana Press, New York, pp 185–202
64. Fu J, Yang YR, Johnson-Buck A, Liu M, Liu Y et al (2014) Multi-enzyme complexes on DNA scaffolds capable of substrate channelling with an artificial swinging arm. Nat Nanotechnol 9:531–536
65. Voigt NV, Tørring T, Rotaru A, Jacobsen MF, Ravnsbæk JB et al (2010) Single-molecule chemical reactions on DNA origami. Nat Nanotechnol 5:200–203
66. Bell NAW, Keyser UF (2016) Digitally encoded DNA nanostructures for multiplexed, single-molecule protein sensing with nanopores. Nat Nanotechnol 11:645–651
67. Chen Y-J, Groves B, Muscat RA, Seelig G (2015) DNA nanotechnology from the test tube to the cell. Nat Nanotechnol 10:748–760
68. Stephanopoulos N, Freeman R, North HA, Sur S, Jeong SJ et al (2015) Bioactive DNA-peptide nanotubes enhance the differentiation of neural stem cells into neurons. Nano Lett 15:603–609
69. Douglas SM, Bachelet I, Church GM (2012) A logic-gated nanorobot for targeted transport of molecular payloads. Science 335:831–834
70. Kuzyk A, Schreiber R, Fan Z, Pardatscher G, Roller E-M et al (2012) DNA-based self-assembly of chiral plasmonic nanostructures with tailored optical response. Nature 483:311–314
71. Thacker VV, Herrmann LO, Sigle DO, Zhang T, Liedl T et al (2014) DNA origami based assembly of gold nanoparticle dimers for surface-enhanced Raman scattering. Nat Commun 5:3448
72. Kuzyk A, Yang Y, Duan X, Stoll S, Govorov AO et al (2016) A light-driven three-dimensional plasmonic nanosystem that translates molecular motion into reversible chiroptical function. Nat Commun 7:10591
73. Weller L, Thacker VV, Herrmann LO, Hemmig EA, Lombardi A et al (2016) Gap-dependent coupling of Ag-Au nanoparticle heterodimers using DNA origami-based self-assembly. ACS Photonics 3(9):1589–1595
74. Pal S, Deng Z, Ding B, Yan H, Liu Y (2010) DNA-origami-directed self-assembly of discrete silver-nanoparticle architectures. Angew Chem Int Ed 49:2700–2704
75. Schreiber R, Do J, Roller E-M, Zhang T, Schüller VJ et al (2014) Hierarchical assembly of metal nanoparticles, quantum dots and organic dyes using DNA origami scaffolds. Nat Nanotechnol 9:74–78
76. Tan SJ, Campolongo MJ, Luo D, Cheng W (2011) Building plasmonic nanostructures with DNA. Nat Nanotechnol 6:268–276
77. Tian Y, Zhang Y, Wang T, Xin HL, Li H et al (2016) Lattice engineering through nanoparticle–DNA frameworks. Nat Mater 15:654–661
78. Maune HT, Han S-P, Barish RD, Bockrath M, Lii WAG et al (2010) Self-assembly of carbon nanotubes into two-dimensional geometries using DNA origami templates. Nat Nanotechnol 5:61–66
79. Liu J, Geng Y, Pound E, Gyawali S, Ashton JR et al (2011) Metallization of branched DNA origami for nanoelectronic circuit fabrication. ACS Nano 5:2240–2247
80. Hernández-Ainsa S, Ricci M, Hilton L, Aviñó A, Eritja R et al (2016) Controlling the reversible assembly of liposomes through a multistimuli responsive anchored DNA. Nano Lett 16:4462–4466

81. Yao D, Li H, Guo Y, Zhou X, Xiao S et al (2016) A pH-responsive DNA nanomachine controlled catalytic assembly of gold nanoparticles. Chem Commun 52:7556–7559
82. Hemmig EA, Creatore C, Wünsch B, Hecker L, Mair P et al (2016) Programming light-harvesting efficiency using DNA origami. Nano Lett 16:2369–2374
83. Schmied JJ, Raab M, Forthmann C, Pibiri E, Wünsch B et al (2014) DNA origami-based standards for quantitative fluorescence microscopy. Nat Protoc 9:1367–1391
84. Yang Y, Wang J, Shigematsu H, Xu W, Shih WM et al (2016) Self-assembly of size-controlled liposomes on DNA nanotemplates. Nat Chem 8:476–483
85. Czogalla A, Kauert DJ, Franquelim HG, Uzunova V, Zhang Y et al (2015) Amphipathic DNA origami nanoparticles to scaffold and deform lipid membrane vesicles. Angew Chem Int Ed 54:6501–6505
86. Pedersen RO, Kong J, Achim C, LaBean TH (2015) Comparative incorporation of PNA into DNA nanostructures. Molecules 20:17645–17658
87. Pinheiro VB, Taylor AI, Cozens C, Abramov M, Renders M et al (2012) Synthetic genetic polymers capable of heredity and evolution. Science 336(6079):341–344
88. Pinheiro VB, Holliger P (2014) Towards XNA nanotechnology: new materials from synthetic genetic polymers. Trends Biotechnol 32:321–328
89. Yurke B, Turberfield AJ, Mills AP, Simmel FC, Neumann JL (2006) A DNA-fulled molecular machine made of DNA. Nature 128:10092–10102
90. Benenson Y, Paz-Elizur T, Adar R, Keinan E, Livneh Z et al (2001) Programmable and autonomous computing machine made of biomolecules. Nature 414:430–434
91. Zhang DY, Turberfield AJ, Yurke B, Winfree E (2007) Engineering entropy-driven reactions and networks catalyzed by DNA. Science 318:1121–1125
92. Bath J, Turberfield AJ (2007) DNA nanomachines. Nat Nanotechnol 2:275–284
93. Liu J, Cao Z, Lu Y (2009) Functional nucleic acid sensors. Chem Rev 109(5):1948
94. Chandrasekaran AR, Wady H, Subramanian HKK (2016) Nucleic acid nanostructures for chemical and biological sensing. Small 12:2689–2700
95. Torabi S-F, Lu Y (2011) Small-molecule diagnostics based on functional DNA nanotechnology: a dipstick test for mercury. Faraday Discuss 149:125–135
96. Chandrasekaran AR, Zavala J, Halvorsen K (2016) Programmable DNA nanoswitches for detection of nucleic acid sequences. ACS Sens 1:120–123
97. Koirala D, Shrestha P, Emura T, Hidaka K, Mandal S et al (2014) Single-molecule mechanochemical sensing using DNA origami nanostructures. Angew Chem Int Ed 53:8137–8141
98. Ke Y, Meyer T, Shih WM, Bellot G (2016) Regulation at a distance of biomolecular interactions using a DNA origami nanoactuator. Nat Commun 7:10935
99. You M, Peng L, Shao N, Zhang L, Qiu L et al (2014) DNA "nano-claw": logic-based autonomous cancer targeting and therapy. J Am Chem Soc 136:1256–1259
100. Tintore M, Gallego I, Manning B, Eritja R, Fabrega C (2013) DNA origami as a DNA repair nanosensor at the single-molecule level. Angew Chem Int Ed 52:7747–7750
101. Chen SX, Zhang DY, Seelig G (2013) Conditionally fluorescent molecular probes for detecting single base changes in double-stranded DNA. Nat Chem 5:782–789
102. Sannohe Y, Endo M, Katsuda Y, Hidaka K, Sugiyama H (2010) Visualization of dynamic conformational switching of the G-quadruplex in a DNA nanostructure. J Am Chem Soc 132:16311–16313
103. Miyake Y, Togashi H, Tashiro M, Yamaguchi H, Oda S et al (2006) MercuryII-mediated formation of thymine-HgII-thymine base pairs in DNA duplexes. J Am Chem Soc 128:2172–2173
104. Wang X, Yang C, Zhu S, Yan M, Ge S et al (2016) 3D origami electrochemical device for sensitive Pb2+ testing based on DNA functionalized iron-porphyrinic metal-organic framework. Biosens Bioelectron 87:108–115
105. Benenson Y, Gil B, Ben-Dor U, Adar R, Shapiro E (2004) An autonomous molecular computer for logical control of gene expression. Nature 429:423–429

106. Rudchenko M, Taylor S, Pallavi P, Dechkovskaia A, Khan S et al (2013) Autonomous molecular cascades for evaluation of cell surfaces. Nat Nanotechnol 8:580–586
107. Amir Y, Ben-Ishay E, Levner D, Ittah S, Abu-Horowitz A et al (2014) Universal computing by DNA origami robots in a living animal. Nat Nanotechnol 9:353–357
108. Modi S, Nizak C, Surana S, Halder S, Krishnan Y (2013) Two DNA nanomachines map pH changes along intersecting endocytic pathways inside the same cell. Nat Nanotechnol 8:459–467
109. Modi S, M G S, Goswami D, Gupta GD, Mayor S et al (2009) A DNA nanomachine that maps spatial and temporal pH changes inside living cells. Nat Nanotechnol 4:325–330
110. Keyser UF (2011) Controlling molecular transport through nanopores. J R Soc 8:1369–1378
111. Coulter WH (1953) Patent US2656508 A: means for counting particles suspended in a fluid
112. DeBlois RW, Bean CP (1970) Counting and sizing of submicron particles by the resistive pulse technique. Rev Sci Instrum 41:909–916
113. Kubitschek HE (1958) Electronic counting and sizing of bacteria. Nature 182:234–235
114. Bezrukov SM, Vodyanoy I, Parsegian VA (1994) Counting polymers moving through a single ion channel. Nature 370(6487):279–281
115. Song L, Hobaugh MR, Shustak C, Cheley S, Bayley H et al (1996) Structure of staphylococcal alpha-hemolysin, a heptameric transmembrane pore. Science 274:1859–1866
116. Kasianowicz JJ, Brandin E, Branton D, Deamer DW (1996) Characterization of individual polynucleotide molecules using a membrane channel. Proc Natl Acad Sci 93:13770–13773
117. Branton D, Deamer DW, Marziali A, Bayley H, Benner SA et al (2008) The potential and challenges of nanopore sequencing. Nat Biotechnol 26:1146–1153
118. Venkatesan BM, Bashir R (2011) Nanopore sensors for nucleic acid analysis. Nat Nanotechnol 6:615–624
119. Wanunu M (2012) Nanopores: a journey towards DNA sequencing. Phys Life Rev 9:125–158
120. Pennisi E (2012) Search for Pore-fection. Science 336:534–537
121. Meller A, Nivon L, Brandin E, Golovchenko J, Branton D (2000) Rapid nanopore discrimination between single polynucleotide molecules. Proc Natl Acad Sci 97:1079–1084
122. Bayley H, Cremer PS (2001) Stochastic sensors inspired by biology. Nature 413:226–230
123. Cherf GM, Lieberman KR, Rashid H, Lam CE, Karplus K et al (2012) Automated forward and reverse ratcheting of DNA in a nanopore at five angstrom precision. Nat Biotechnol 30:344–348
124. Fonslow BR, Stein BD, Webb KJ, Xu T, Choi J et al (2013) Reading DNA at single-nucleotide resolution with a mutant MspA nanopore and phi29 DNA polymerase. Nat Biotechnol 10:54–56
125. Clarke J, Wu H-C, Jayasinghe L, Patel A, Reid S et al (2009) Continuous base identification for single-molecule nanopore DNA sequencing. Nat Nanotechnol 4:265–270
126. Feng Y, Zhang Y, Ying C, Wang D, Du C (2015) Nanopore-based fourth-generation DNA sequencing technology. Genomics Proteomics Bioinformatics 13:4–16
127. Loose M, Malla S, Stout M (2016) Real time selective sequencing using nanopore technology. Nat Methods 13:751–754
128. Szalay T, Golovchenko JA (2015) De novo sequencing and variant calling with nanopores using PoreSeq. Nat Biotechnol 33:1–7
129. Loman NJ, Quick J, Simpson JT (2015) A complete bacterial genome assembled de novo using only nanopore sequencing data. Nat Methods 12:733–735
130. Daims H, Lebedeva E, Pjevac P, Han P, Herbold C et al (2015) Complete nitrification by Nitrospira bacteria. Nature 528:504–509
131. Quick J, Loman NJ, Duraffour S, Simpson JT, Severi E et al (2016) Real-time, portable genome sequencing for Ebola surveillance. Nature 530:228–232
132. Dunn A (2015) Sequencing DNA in the palm of your hand. NASA. Accessed 27 Jul 2016. https://www.nasa.gov/mission_pages/station/research/news/biomolecule_sequencer
133. Gu LQ, Braha O, Conlan S, Cheley S, Bayley H (1999) Stochastic sensing of organic analytes by a pore-forming protein containing a molecular adapter. Nature 398:686–690

134. Braha O, Gu LQ, Zhou L, Lu X, Cheley S et al (2000) Simultaneous stochastic sensing of divalent metal ions. Nat Biotechnol 18:1005–1007
135. Storm AJ, Chen JH, Ling XS, Zandbergen HW, Dekker C (2003) Fabrication of solid-state nanopores with single-nanometre precision. Nat Mater 2:537–540
136. Li J, Stein D, McMullan C, Branton D, Aziz MJ et al (2001) Ion-beam sculpting at nanometre length scales. Nature 412:166–169
137. Garaj S, Hubbard W, Reina A, Kong J, Branton D et al (2010) Graphene as a subnanometre trans-electrode membrane. Nature 467:190–193
138. Heerema SJ, Dekker C (2016) Graphene nanodevices for DNA sequencing. Nat Nanotechnol 11:127–136
139. Steinbock LJ, Otto O, Chimerel C, Gornall J, Keyser UF (2010) Detecting DNA folding with nanocapillaries. Nano Lett 10:2493–2497
140. Li W, Bell NAW, Herna S, Bujdoso R, Keyser UF (2013) Single protein molecule detection by glass nanopores. ACS Nano 7:4129–4134
141. Keyser UF, Koeleman BN, van Dorp S, Krapf D, Smeets RMM et al (2006) Direct force measurements on DNA in a solid-state nanopore. Nat Phys 2:473–477
142. Wanunu M, Meller A (2007) Chemically modified solid-state nanopores. Nano Lett 7:1580–1585
143. Kohli P, Harrell CC, Cao Z, Gasparac R, Tan W et al (2004) DNA-functionalized nanotube membranes with single-base mismatch selectivity. Science 305:984–986
144. Jovanovic-Talisman T, Tetenbaum-Novatt J, McKenney AS, Zilman A, Peters R et al (2009) Artificial nanopores that mimic the transport selectivity of the nuclear pore complex. Nature 457:1023–1027
145. Movileanu L (2009) Interrogating single proteins through nanopores: challenges and opportunities. Trends Biotechnol 27:333–341
146. Keyser UF (2016) Enhancing nanopore sensing with DNA nanotechnology. Nat Nanotechnol 11:106–108
147. Bell NAW, Keyser UF (2015) Specific protein detection using designed DNA carriers and nanopores. J Am Chem Soc 137:2035–2041
148. Kong J, Bell NAW, Keyser UF (2016) Quantifying nanomolar protein concentrations using designed DNA carriers and solid-state nanopores. Nano Lett 16:3557–3562
149. Bell NAW, Keyser UF (2016) Digitally encoded DNA nanostructures for multiplexed, single-molecule protein sensing with nanopores.Nat Nanotechnol 11:1–28
150. Bell NAW, Engst CR, Ablay M, Divitini G, Ducati C et al (2012) DNA origami nanopores. Nano Lett 12:512–517
151. Wei R, Martin TG, Rant U, Dietz H (2012) DNA origami gatekeepers for solid-state nanopores. Angew Chem Int Ed 124:4948–4951
152. Hall AR, Scott A, Rotem D, Mehta KK, Bayley H et al (2010) Hybrid pore formation by directed insertion of α-haemolysin into solid-state nanopores. Nat Nanotechnol 5:874–877
153. Hernández-Ainsa S, Keyser UF (2014) DNA origami nanopores: developments, challenges and perspectives. Nanoscale 6:14121–14132
154. Hernández-Ainsa S, Bell NAW, Thacker VV, Göpfrich K, Misiunas K et al (2013) DNA origami nanopores for controlling DNA translocation. ACS Nano 7:6024–6030
155. Bell NAW, Thacker VV, Hernández-Ainsa S, Fuentes-Perez ME, Moreno-Herrero F et al (2013) Multiplexed ionic current sensing with glass nanopores. Lab Chip 13:1859–1862
156. Steinbock LJ, Otto O, Skarstam DR, Jahn S, Chimerel C et al (2010) Probing DNA with micro- and nanocapillaries and optical tweezers. J Phys Condens Matter 22:454113
157. Hernández-Ainsa S, Misiunas K, Thacker VV, Hemmig EA, Keyser UF (2014) Voltage-dependent properties of DNA origami nanopores. Nano Lett 14:1270–1274
158. Haque F, Wang S, Stites C, Chen L, Wang C et al (2015) Single pore translocation of folded, double-stranded, and tetra-stranded DNA through channel of bacteriophage phi29 DNA packaging motor. Biomaterials 53:744–752
159. Leung C, Hodel AW, Brennan AJ, Lukoyanova N, Tran S (2017) Real-time visualization of perforin nanopore assembly. Nat Nanotechnol 12:467–473

160. Seifert A, Göpfrich K, Burns JR, Fertig N, Keyser UF et al (2014) Bilayer-spanning DNA nanopores with voltage-switching between open and closed state. ACS Nano 9:1117–1126
161. Bell NAW, Keyser UF (2014) Nanopores formed by DNA origami: a review. FEBS Lett 588:3564–3570
162. Li CY, Hemmig EA, Kong J, Yoo J, Keyser UF et al (2015) Ionic conductivity, structural deformation and programmable anisotropy of DNA origami in electric field. ACS Nano 9:1420–1433
163. Plesa C, Ananth AN, Linko V, Gulcher C, Katan A et al (2014) Ionic permeability and mechanical properties of DNA origami nanoplates. ACS Nano 8:35–43
164. Baker M (2010) Making membrane proteins for structures: a trillion tiny tweaks. Nat Methods 7:429–434
165. Knowles MR, Stutts MJ, Spock A, Fischer N, Gatzy JT et al (1983) Abnormal ion permeation through cystic fibrosis. Science 221:1067–1070
166. Feske S (2007) Calcium signalling in lymphocyte activation and disease. Nat Rev Immunol 7:690–702
167. Fernandez-Lopez S, Kim HS, Choi EC, Delgado M, Granja JR et al (2001) Antibacterial agents based on the cyclic D,L-alpha-peptide architecture. Nature 412:452–455
168. Czogalla A, Franquelim HG, Schwille P (2016) DNA nanostructures on membranes as tools for synthetic biology. Biophys J 110:1698–1707
169. Hille B (2001) Ion channels of excitable membranes. Palgrave Macmillan, Sunderland
170. Benz R (ed) (2004) Bacterial and eukaryotic porins WILEY-VCH Verlag GmbH & Co KGaA, Weinheim
171. Rout MP, Aitchison JD, Suprapto A, Hjertaas K, Zhao Y et al (2000) The yeast nuclear pore complex. J Cell Biol 148:635–652
172. Fyles TM (2007) Synthetic ion channels in bilayer membranes. Chem Soc Rev 36:335–347
173. Doyle DA, Morais Cabral J, Pfuetzner RA, Kuo A, Gulbis JM et al (1998) The structure of the potassium channel: molecular basis of K+ conduction and selectivity. Science 280:69–77
174. Hucho F, Weise C (2001) Ligand-gated ion channels. Angew Chem Int Ed 40:3100–3116
175. Jiang Y, Lee A, Chen J, Cadene M, Chait BT et al (2002) The open pore conformation of potassium channels. Nature 417:523–546
176. Carpenter EP, Beis K, Cameron AD, Iwata S (2008) Overcoming the challenges of membrane protein crystallography. Curr Opinion Struct Biol 18:581–586
177. White SH (1998–2019) Membrane proteins of known 3D structure, Stephen White laboratory, UC Irvine. Accessed 27 Jul 2016. https://blanco.biomol.uci.edu/mpstruc/
178. Sakai N, Matile S (2013) Synthetic ion channels. Langmuir 29:9031–9040
179. Pedersen CJ (1988) The discovery of crown ethers (Nobel address). Angew Chem Int Ed 27:1021–1027
180. Chimerel C, Field CM, Piñero-Fernandez S, Keyser UF, Summers DK (2012) Indole prevents Escherichia coli cell division by modulating membrane potential. Biochim Biophys Acta Biomembr 1818:1590–1594
181. Kennedy SJ, Roeske RW, Freeman AR, Watanabe AM, Besche HR (1977) Synthetic peptides form ion channels in artificial lipid bilayer membranes. Science 196:1341–1342
182. Wallace DP, Tomich JM, Eppler JW, Iwamoto T, Grantham JJ et al (2000) A synthetic channel-forming peptide induces Cl- secretion: modulation by Ca2+-dependent K+ channels. Biochim Biophys Acta Biomembr 1464:69–82
183. Tabushi I, Kuroda Y, Yokota K (1982) A,B,D,F-tetrasubstituted β-cyclodextrin as artificial channel compound. Tetrahedron Lett 23:4601–4604
184. Mamad-Hemouch H, Ramoul H, Abou Taha M, Bacri L, Huin C et al (2015) Biomimetic nanotubes based on cyclodextrins for ion-channel applications. Nano Lett 15:7748–7754
185. Fyles TM, Loock D, Zhou X (1998) A voltage-gated ion channel based on a bis-macrocyclic bolaamphiphile. J Am Chem Soc 7863:2997–3003
186. Carmichael VE, Dutton PJ, Fyles TM, James TD, Swan JA et al (1989) Biomimetic ion transport: a functional model of a unimolecular ion channel. J Am Chem Soc 111(2):767–769

187. Tanaka Y, Kobuke Y, Sokabe M (1995) A non-peptidic ion channel with K+ selectivity. Angew Chem Int Ed 34:693–694
188. Geng J, Kim K, Zhang J, Escalada A, Tunuguntla R et al (2014) Stochastic transport through carbon nanotubes in lipid bilayers and live cell membranes. Nature 514:612–615
189. Göpfrich K, Li C-Y, Mames I, Bhamidimarri SP, Ricci M et al (2016) Ion channels made from a single membrane-spanning DNA duplex. Nano Lett 16:4665–4669
190. Göpfrich K, Li C-Y, Ricci M, Bhamidimarri SP, Yoo J et al (2016) Large-conductance transmembrane porin made from DNA origami. ACS Nano 10:8207–8214
191. Burns JR, Stulz E, Howorka S (2013) Self-assembled DNA nanopores that span lipid bilayers. Nano Lett 13:2351–2356
192. Neher E, Sakmann B (1976) Single-channel currents recorded from membrane of denervated frog muscle fibres. Nature 260:799–802
193. Gutsmann T, Heimburg T, Keyser U, Mahendran KR, Winterhalter M (2015) Protein reconstitution into freestanding planar lipid membranes for electrophysiological characterization. Nat Protoc 10:188–198
194. Burns JR, Seifert A, Fertig N, Howorka S (2016) A biomimetic DNA-based channel for the ligand-controlled transport of charged molecular cargo across a biological membrane. Nat Nanotechnol 11:152–156
195. Krishnan S, Ziegler D, Arnaut V, Martin TG, Kapsner K et al (2016) Molecular transport through large-diameter DNA nanopores. Nat Commun 7:12787
196. Maingi V, Lelimousi M, Howorka S, Sansom MSP (2015) Gating-like motions and wall porosity in a DNA nanopore scaffold revealed by molecular simulations. ACS Nano 9:11209–11217
197. Yoo J, Aksimentiev A (2015) Molecular dynamics of membrane-spanning DNA channels: conductance mechanism, electro-osmotic transport, and mechanical gating. J Phys Chem Lett 6:4680–4687

Chapter 12
Bio Mimicking of Extracellular Matrix

Moumita Ghosh, Michal Halperin-Sternfeld, and Lihi Adler-Abramovich

Abstract Biomaterials play a critical role in regenerative strategies such as stem cell-based therapies and tissue engineering, aiming to replace, remodel, regenerate, or support damaged tissues and organs. The design of appropriate three-dimensional (3D) scaffolds is crucial for generating bio-inspired replacement tissues. These scaffolds are primarily composed of degradable or non-degradable biomaterials and can be employed as cells, growth factors, or drug carriers. Naturally derived and synthetic biomaterials have been widely used for these purposes, but the ideal biomaterial remains to be found. Researchers from diversified fields have attempted to design and fabricate novel biomaterials, aiming to find novel theranostic approaches for tissue engineering and regenerative medicine. Since no single biomaterial has been found to possess all the necessary characteristics for an ideal performance, over the years scientists have tried to develop composite biomaterials that complement and combine the beneficial properties of multiple materials into a superior matrix. Herein, we highlight the structural features and performance of various biomaterials and their application in regenerative medicine and for enhanced tissue engineering approaches.

Keywords Biomaterials · Extracellular matrix · Scaffolds · Peptides · Hydrogels · Supramolecular polymers

Abbreviations

3D	Three-dimensional
Al	Alumina
bFGF	Basic fibroblast growth factor

M. Ghosh · M. Halperin-Sternfeld · L. Adler-Abramovich (✉)
Department of Oral Biology, The Goldschleger School of Dental Medicine, Sackler Faculty of Medicine, Tel Aviv University, Tel Aviv, Israel
e-mail: LihiA@tauex.tau.ac.il

S. Perrett et al. (eds.), *Biological and Bio-inspired Nanomaterials*,
Advances in Experimental Medicine and Biology 1174,
https://doi.org/10.1007/978-981-13-9791-2_12

BMP-2	Bone morphogenic protein 2
CaP	Calcium phosphate
DOPA	3,4-dihydroxy-L-phenylalanine
ECM	Extracellular matrix
Fmoc	Fluorenylmethyloxycarbonyl
FmocFF	Fluorenylmethoxycarbonyl-diphenylalanine
GAGs	Glycosaminoglycans
GDNF	Glial cell derived neurotrophic factor
Mg	Magnesium
Mn	Manganese
MSCs	Mesenchymal stem cells
Na	Sodium
PCL	Poly(e-caprolactone)
PEG	Polyethylene glycol
PET	Polyethylene terephthalate
PGA	Polyglycolic acid
PGs	Proteoglycans
PLA	Polylactic acid
PLGA	Polylactic-co-glycolide
PMMA	Poly(methylmethacrylate)
PTFE	Poly(tetrafluoroethylene)
PU	Polyurethanes
RP	Rapid prototyping
Si	Silicon
TCP	Tricalcium phosphate
Ti	Titanium
UV	Ultraviolet
VEGF	Vascular endothelial growth factor
Zn	Zinc
Zr	Zirconium

12.1 Introduction

The term “biomaterial” was initially defined in the first Consensus Conference of the European Society for Biomaterials (ESB) in 1976 as “a nonviable material used in a medical device, intended to interact with biological systems”. However, it was later modified by the ESB and is currently defined as “a material intended to interface with biological systems to evaluate, treat, augment or replace any tissue, organ or function of the body”. The change in the definition indicates how the field of biomaterials has evolved over the years, from merely interacting with the body to influencing various biological processes, aimed at tissue regeneration. An alternative definition of a biomaterial is a substance that has been engineered to take a form which, alone or as part of a complex system, is used to direct, by controlling interactions with components of living systems, the course of any therapeutic or diagnostic procedure, in human or veterinary medicine [1].

The selection of biomaterials depends on their surface properties, mechanical properties and degradation behavior, which affect biological interactions including protein and peptide adhesion, cell adhesion, proliferation, and differentiation. The current approaches are widely applied to fabricate ideal biomaterials that will serve as scaffolds and that provide suitable matrices for cellular growth and attachment. Owing to these continual efforts, a wide range of biomaterials, natural or synthetic, organic or inorganic, has been proposed, aiming to produce functional replacement tissue for diseased organs, including artificial heart valves, hydroxyapatite-coated hip implants, three-dimensional (3D) scaffolds for dental applications, or to serve as a platform for drug delivery.

Throughout this chapter, we will present the various types of biomaterials and their applications, as well as the biological factors that underlie the design of biomaterials. We will also focus on the recent development of composite biomaterials for tissue engineering and will highlight current approaches for further development of novel biomaterials possessing superior biomedical and mechanical properties.

12.2 The Extracellular Matrix

The extracellular matrix (ECM) is the natural 3D microenvironment produced by the resident cells of every tissue and organ; it provides a physical scaffold in which cells are embedded. Additionally, it regulates cellular processes including growth, migration, differentiation, survival, homeostasis, and morphogenesis [2–4]. It is essential for maintaining the normal function of the tissue because it contains structural and functional molecules that permit communication between adjacent cells through transportation of water, electrolytes, nutrients, and metabolites. Another important role of the ECM in tissue homeostasis and development is to serve as a niche for stem cell differentiation whereby playing a most crucial role. The physical properties of the ECM, such as rigidity, porosity, insolubility, as well as its topography, which originate from the matrix components, determine the mechanical behavior of each tissue and the behavior of the residing cells [5, 6].

The metabolic adaptations of the cells in a specific matrix are governed by its composition and organization in response to alterations in the mechanical properties, pH, oxygen concentration, and other factors in the microenvironment [7]. This process of "dynamic reciprocity" is crucial for tissue development and homeostasis [2, 8].

The ECM is primarily composed of fiber-forming proteins such as collagens (mainly type I and type II) and elastic fibers, as well as non-fiber-forming proteins such as proteoglycans (PGs), glycosaminoglycans (GAGs), and soluble factors [9, 10] (Fig. 12.1). Collagens are involved, either directly or indirectly, in cell attachment and differentiation, and function as hemotactic agents. In addition, they are associated with other collagen fibers, as well as with ECM proteins and PGs and facilitate in developing a complex 3D matrix network [4]. PGs comprise one of the most important structural and functional macromolecules in tissues. They

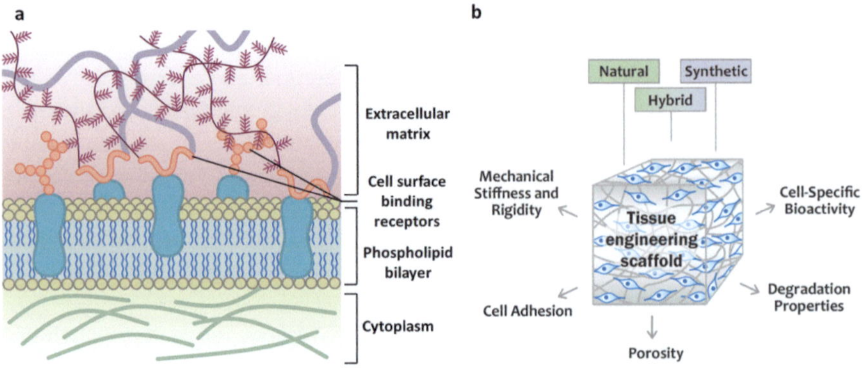

Fig. 12.1 ECM-mimicking biomaterials. (**a**) A model of the natural ECM, composed of proteins such as collagen (I, II, III, IV, and VII), laminin, and fibronectin, embedded in highly negatively charged polysaccharide-rich glycans, and its interactions with cell-surface receptors, including integrins, selectins, CD44, and syndecan, among others. (**b**) Composition-based classification of biomaterials into natural, synthetic, and hybrid, and a schematic representation of their properties

are composed of a core protein covalently attached to GAG chains of the same or a different type. They interact with various growth factors, cytokines and chemokines, cell surface receptors, and ECM molecules, either through core proteins or through GAG side chains. Thus, they play an important role in cell signaling, proliferation, migration, differentiation, apoptosis, and adhesion [11]. PGs are also important for organizing the ECM; thus, they contribute to the formation of the ECM biomimicking scaffold and to cell adherence and transportation within it. Moreover, the stem cell niche provided by the ECM is composed of both soluble factors and ECM macromolecules, which play a critical role in directing stem/progenitor cell differentiation fate [12, 13]. Providing a supportive medium for blood vessels, nerves, and lymphatics, as well as the diffusion of nutrients from the blood to the surrounding cells, the ECM is a natural scaffold. ECM-based materials are attractive candidates for scaffolds in regenerative medicine approaches to tissue reconstruction.

12.3 Classification of Biomaterials

Biomaterials can be classified into two main groups, namely, synthetic and natural. Synthetic biomaterials include metals, ceramics, non-biodegradable and biodegradable polymers, whereas naturally derived biomaterials include protein-based biomaterials (e.g., collagen, gelatin, and silk), polysaccharide-based biomaterials (e.g., cellulose, chitin/chitosan, and glucose), and decellularized tissue-derived biomaterials (e.g., decellularized heart valves, blood vessels, and liver) [14]. Herein, we will present the diversity of both types of biomaterials, providing a comprehensive understanding of their advantages and limitations, together with various architectural parameters of biomaterials important for tissue engineering (Fig. 12.1b).

12.4 Synthetic Biomaterials

12.4.1 Metallic Biomaterials

Metallic biomaterials have had the longest history among the various kinds of biomaterials [15, 16]. Owing to their excellent mechanical properties (modulus, strength, and ductility) and biocompatibility, they are widely used in various fields [17]. In cardiology, they are used in pacemakers, intravascular stents, and occlusion coils [18, 19], in orthopedics, metals are used in simple wires and screws, fracture fixation plates, and joint prosthesis (artificial joints) for hips, knees, shoulders, elbows, and more [20], and in dentistry, in the production of endosseous implants and dentures [21].

Titanium (Ti) and its alloys, stainless steel, and cobalt-chromium, are the three most widely used biomedical metals, mainly for supporting or fixation, especially in dentistry and orthopedics [22, 23]. They are popular primarily because of their biocompatibility and their ability to bear significant loads, withstand fatigue loading, and to undergo plastic deformation before failure. However, the long-term embedding of metal implants in the body's physiological environment may result in corrosion and decay of their surface [17]. In addition, since Ti metals and its alloys are bioinert, fibrous encapsulation may occur on the surface of such implants following implantation, preventing their direct attachment to the host bone [24]. This phenomenon may lead to implant loosening and the possible need for an additional operation. Using surface modification technologies, the physical and chemical properties of the medical metal surfaces can be modified, resulting in improved stability. In addition, surface modification allows functional materials to be incorporated into the metal surface; this endows the metals with improved biocompatibility, cell adhesion ability, drug delivery, bacterial growth inhibition, anti-tumor, anticoagulation, or other bio-functions [17].

Surface modification can be obtained by physical and chemical methods. Among the physical techniques, grinding, polishing, and sand blasting are commonly used to achieve specific surface topographies and roughness and to remove surface contamination. Other physical methods commonly used in fabricating films or coatings on biomedical metals include thermal spraying, plasma spraying, physical vapor deposition, and ion implantation. These methods alter the composition and properties of substrate surfaces, particularly for Ti and its alloys [25]. Alternatively, chemical methods can be applied using a chemical or electrochemical treatment, sol-gel, chemical vapor deposition, and biochemical modification [26]. Surface modification only changes the surface layer properties of the biomedical metals while preserving the mechanical properties of the metal devices, which are the primary advantages of biomedical metals. Indeed, instead of encapsulation by fibrous tissue following implantation of bio-inert biomedical metals, surface modification results in a chemical bond with the surrounding bony tissue [24].

Current developments in bone tissue engineering focus on surface modifications of biomedical metals to improve their bioactivity. However, surface modification methods can be greatly improved. For instance, it has been observed that hydroxyapatite coatings improve the osteointegration of metal implants via tight binding to the bone mineral phase, thus providing suitable conditions for favorable osteoblast adhesion and proliferation onto the implant surface. However, such coatings suffer from poor stability and tend to delaminate from the metal surface under mechanical stresses experienced by the implant, leading to its loosening. Other surface treatments applied recently are acid-alkali treatment and alkali-heat treatment, which make the surface more basic, thereby affecting cell proliferation [27].

Bio-functionalizing of biomedical metals via surface modification has been widely performed to make them more suitable for biomedical applications [23]. For example, developing new coatings for Ti and its alloys is one of the primary goals for the next generation of orthopedic and dental implants, aiming at improving their potential for osteointegration while retaining their mechanical properties.

Over the last decade, metallic alloys have been introduced as biodegradable metals in clinical cases requiring temporary support or fixation (such as plates and screws for bone fracture fixation and stents). Magnesium (Mg) alloys have been the most extensively studied biodegradable metallic materials for orthopedic applications due to their advantageous properties, such as low density, modulus of elasticity approaching that of cortical bone, and degradability via corrosion [23]. Fixation appliances composed of degradable biomaterials are often used to treat children, teenagers, and athletes, to avoid both future bone fractures caused by abnormal loading patterns, as well as a second surgery for fixation removal. The use of degradable implants also overcomes the limitations of permanent metallic biomaterials such as the stress shielding effect, which induces refracture caused by permanent implants, due to their higher Young's modulus compared with bone, as well as the release of toxic metallic ions or particles through corrosion, wear, and bacterial infection [28, 29].

Mg is often alloyed with alumina (Al), zinc (Zn), manganese (Mn), silicon (Si), and zirconium (Zr) in order improve its mechanical strength and corrosion resistance. In addition, different coatings, including bioceramics, polymers and composite layers, have been proposed to slow the corrosion process, enhance bone growth, and control biodegradation [30]. Mg implants have been widely applied to positively stimulate new bone formation, thereby assisting in bone fracture healing. As in permanent implants, surface modification affects the microstructure, mechanical properties, and corrosion behavior of the materials. Furthermore, in biodegradable metallic materials, it controls the degradation rate [31]. Therefore, the composition and the manufacturing route have to be carefully chosen to meet the requirements of each application.

To date, Mg-based fixators have been successfully applied in some low load-bearing fracture sites in patients without adverse clinical outcomes. This implicates the use of more Mg-based devices in other applications, such as developing novel hybrid systems containing Mg implants with traditional metal devices to fix fractures in high load-bearing sites [32].

12.4.2 Ceramic Biomaterials

Bioceramics are biocompatible ceramic compounds produced in porous or dense forms as well as powders, granulates, coatings, and injectable formulations that can be used alone or combined with additional organic or polymeric materials for soft and hard tissue replacement [33]. They can also be surface modified and designed accordingly to deliver biologically active substances aimed to repair, maintain and restore organs and tissues. They are hard, brittle materials with relatively poor tensile properties, but exhibit excellent compressive strength, high resistance to wear, and favorable low frictional properties in articulation.

Bioceramics can be classified into three major families, based on their tissue response: nearly inert (e.g., alumina and zirconia), bioactive (e.g., bioactive glass), and resorbable ceramics [e.g., β- and α-tricalcium phosphate (TCP)] [34]. Inert ceramics are widely used as femoral heads and acetabular cups for hip replacement as well as for developing dental implants. However, these materials are presently not being applied as scaffolds because their inertness triggers the formation of a thick "protective" fibrous capsule on the surface of the implant, thus preventing it from directly binding to the host bone tissue [34]. Since the ability to create a stable bond with the host tissue is of utmost importance in selecting bioceramics for scaffold production, bioactive and bioresorbable ceramics are used. Bioceramics also gradually degrade over time and are replaced by the natural host tissue, and can be completely removed once their task of serving as templates for new tissue generation has been completed [35].

Based on their microstructure, bioceramics can also be classified into crystalline ceramics, glass-ceramics, bioactive glasses, and composites. Crystalline ceramics, including calcium phosphate (CaP), are mainly used as scaffolds for bone tissue regeneration owing to their structure and chemical composition, showing similarity to the inorganic fraction of bones, and their osteoconductivity [36]. In addition, CaP exhibits bioactivity of osteointegration, an intimate physicochemical bond between the implant and bone [37]. Porous blocks of various sizes composed of CaP or Hydroxyapatite, a specific form of CaP with a ratio of 1:67, are currently used to repair small-sized bone defects [38, 39]. These blocks can be manipulated intraoperatively to match the size and contour of the existing defect. Hydroxyapatite is also used as a porous scaffold for orbital floor repair [40] and as a coating material of metallic joint prostheses and dental implants [41, 42]. The main drawback of all CaP scaffolds produced in a porous form is their low mechanical properties, such as brittleness and low fatigue strength, which largely limit their clinical application to low- or non-load-bearing parts of the skeleton.

Alumina is another crystalline ceramic that has been widely employed to fabricate components of hip and knee joint prostheses (femur head, acetabular cup, and tibial plate), largely due to alumina's high strength; thus, it is most suitable for load-bearing applications, excellent wear resistance, and bioinertness [43]. Porous alumina is primarily used for fabricating orbital implants (spherical porous scaffolds) for enucleated patients; this may allow for fibro-vascular ingrowth through the porous network and it remains in the patient's anophthalmic socket for an indefinite period, without undergoing degradation [14, 44].

Other forms of bioceramics are bioactive glasses, which are degradable biomaterials that have been widely used to restore osseous defects due to their unique ability to bond to host bone, stimulate new bone growth, and induce angiogenesis [45]. Their glass surface allows the formation of a biologically active layer of hydroxyapatite that provides the bonding interface with host tissues, whereas the dissolution products [e.g., Si, sodium (Na), Ca, P ions] stimulate the cells to produce new tissue [15]. Bioactive glasses can be fabricated by either the conventional melting-quenching routes or the sol-gel technique. Finally, the glasses can also be spun to fabricate glass fibers. In the last decade, the phosphate ones, in particular, have attracted increasing interest for soft-tissue engineering applications. They have been used as guides for muscle or nerve repair [46], as well as for fabricating glassy bone scaffolds [47], and for lung tissue engineering applications [48]. In the last decade, a new set of bioactive glasses, characterized by a highly ordered mesoporous texture, has been developed and studied as a smart platform for the controlled release of biomolecules, *in situ* therapy, and regenerative applications [49, 50].

Glass-ceramics represent the third form of bioceramics, exhibiting superior mechanical properties concerning their parent glasses, specifically their higher elastic modulus, hardness, failure strength, and wear resistance, and are often used to produce scaffolds.

Like most ceramic materials, a primary disadvantage of bioceramics is their low fracture toughness, which could limit their use in load-bearing applications. Furthermore, owing to their high stiffness, they may not be efficient for non-osseous applications, where adequate compliance with soft tissues is needed [51]. To overcome these drawbacks, composite scaffolds composed of bioceramics and polymers have been proposed. Typically, bioceramics are used as fillers or coatings for the polymer matrix to improve their mechanical properties. For instance, hydroxyapatite/polyethylene porous composites are currently used for repairing orbital floor fractures [52]. Another example is the use of bioceramics as a coating for metallic materials, such as stainless steel and titanium, for load-bearing prostheses. The bioceramics improve the biocompatibility and reduce the release of metal ions into the tissues [14]. Besides their "traditional" use for osseous defect repair, bioactive glasses and glass-ceramics in their non-porous form have been used for non-osseous applications, including wound healing and soft-tissue engineering, owing to their ability to bond to soft tissues and to elicit desirable biological responses, such as angiogenesis. Interestingly, recent studies have highlighted the suitability of bioactive glass particulates for (i) treating gastric ulcers, (ii) injectable radioactive glasses for killing cancer cells in liver tumors, (iii) glass-polymer composites for cardiac tissue engineering, (iv) regeneration of injuries at the tendon-to-bone interface, (v) fabricating a scaffold for wound dressing, and (vi) glass/polymer tubes for peripheral nerve regeneration [51, 53–55].

12.4.3 Synthetic Biodegradable Polymers

Synthetic biodegradable polymers have gained increasing attention as scaffolds in tissue engineering, drug delivery vehicles, biosensors, biomedical adhesives, and for replacing permanent devices in low-load bearing fracture sites [56–58]. Synthetic polymers are biocompatible, and their 3D structure and surface features can be tailored, resulting in highly controlled and consistent degradation properties and excellent reproducible mechanical and physical properties, such as tensile strength and elastic modulus [59]. In addition, they can be produced in large homogeneous quantities and have a long shelf time, as well as a lower price compared with biological scaffolds.

The current polymers used constructing 3D scaffolds for bone regeneration are saturated poly-a-hydroxy esters, including polylactic acid (PLA), polyglycolic acid (PGA), and their co-polymer polylactic-co-glycolide (PLGA) [60, 61]. Typical scaffold forms include meshes, fibers, sponges, and foams, which promote a uniform cell distribution, diffusion of nutrients, and growth of organized cell communities [59]. Different synthetic polymers exhibit varying physicochemical and mechanical properties and biodegradation rates. PGA is a hydrophilic polymer exhibiting a fast degradation rate, whereas PLA has a structure similar to PGA but has more hydrophobic features. The chemical properties of these polymers allow hydrolytic degradation by de-esterification. After degradation, the monomeric components of each polymer are eliminated by metabolic pathways. Owing to these properties, PLA and PGA have been widely applied in biomedical products and devices, such as degradable sutures [62]. The biochemical properties, such as a faster and more controllable biodegradation rate, the properties of PLGA, the co-polymer of these two polymers, are improved compared with the pure monomers [63]. Generally, for fabricating bone substitute constructs, the co-polymer PLGA is preferred compared to its constituent homopolymers because of its superior control of degradation by varying the ratio between its monomers [64]. However, although being biocompatible, the clinical application of pure PLGA for bone regeneration is limited by its poor osteoconductivity and suboptimal mechanical properties for load-bearing applications. Therefore, combining PLGA with other materials, such as ceramics and bioactive glass is preferred, to render it more biomimetic, thus enhancing bone regeneration [65]. PLA, PGA, and PLGA are also used as conduits for drug delivery [66] and as scaffolds for constructing artificial vessels due to their high porosity, which allows the diffusion of nutrients upon implantation and subsequent neovascularization, and their easy handling and fabrication into different shapes.

Other synthetic polymers have also been used in tissue engineering for a broad range of biomedical applications [67], including poly(trimethylene carbonate) in vascular tissue engineering [68], poly(methylmethacrylate) (PMMA) for producing intraocular lenses [69], poly(e-caprolactone) (PCL) for *in situ* tissue engineering of small diameter vascular prosthesis [70], polyethylene terephthalate (PET) (Dacron) for hepatic venous reconstruction [71], polyurethanes (PU) in heart valves, vascular grafts, catheters, and prostheses [72], polytetrafluoroethylene (PTFE) for small diameter vascular grafts [73], and membranes for guided tissue regeneration [74].

The mechanical properties of polymeric biomaterials largely depend on the molecular architecture of the macromolecules, i.e., the structure of the monomeric unit, the molecular weight of the polymer, and the morphology exhibited by the polymers [75].

12.5 Natural Biomaterials

Natural polymers can be considered as the first clinically used biodegradable biomaterials [76]. They can be classified primarily into proteins (e.g., silk, collagen, gelatin, fibrinogen, elastin, keratin, actin, and myosin), polysaccharides (e.g., cellulose, amylose, dextran, chitin, and glycosaminoglycans), or polynucleotides (e.g., DNA, RNA) [77]. Natural polymers are soft but flexible materials, which can adapt their shape to desired forms through a variety of molding and casting techniques. They contain specific molecular domains that can support cells at various stages of their development and promote rapid biological interactions of the scaffold with the surrounding host tissue [78], induce the deposition of cells and additional ECM, overall, accelerating angiogenesis [79]. Natural polymers thus exhibit several benefits over synthetic polymers, such as degradability, negligible toxicity, resemblance to native tissues/organs, and the ability to induce tissue remodeling. Owing to their similarity to the ECM as well as their biocompatibility and biodegradability, natural polymers have been developed as base materials for various applications in tissue engineering, including small intestinal submucosa, acellular dermis, bladder acellular matrix grafts, and amniotic membranes.

12.5.1 Collagen

Collagen constitutes approximately one-third of the protein of humans and two-thirds of the dry weight of skin; it plays a vital role in biological functions, such as tissue formation, cell attachment, and proliferation. The collagen protein is a helical polypeptide composed of repeating sequences of glycine, proline, hydroxyproline, and lysine. As an integral part of the ECM, collagen has great tensile strength for tissue growth; thus, it was extensively investigated for biomedical applications [80]. Collagen-based materials are considered to be favorable biomaterials for both cartilage and bone scaffolds, since collagen type II is the major component in articular cartilage, and collagen type I is the major component in bone. Based on the extent of their purification, collagen-based biomaterials can be classified into two categories: (i) decellularized collagen matrices that maintain the original tissue properties and the ECM structure; and (ii) more refined scaffolds developed by extraction, purification, and collagen polymerization [81]. Different types of collagen formulations have been developed for biomedical applications including gels, sponges, spheres, membranes, microfiber collagen scaffolds, and electrospun collagen nanofibrous scaffolds [82].

The clinical use of collagen began in 1881 as a biodegradable suture termed "catgut", a collagen-rich biomaterial prepared from the small intestine of sheep [83, 84]. Over the years, countless clinical innovations have extended the reach of collagen for the engineering and repair of soft and hard tissues [78], with different applications requiring different formulations. Collagen sponges serve as 3D cell culture scaffolds, wound dressings for severe burns, pressure sores, donor sites, and leg ulcers, artificial skin replacements, and as drug delivery carriers [82]. They also serve as effective scaffolds for applying exogenous growth factors [85–87] and antibiotics [88] to wounds. For example, collagen sponge scaffolds loaded with platelet-derived growth factor (PDGF) increase fibroblast influx into the wounds and enhance capillary formation, compared with control treatments [85]. These sponges can absorb large amounts of tissue exudate, adhere smoothly to a wet wound bed, and maintain a moist environment around the wound while protecting against mechanical trauma and bacterial infection. Collagen promotes cellular motility, and inflammatory cells actively invade the porous scaffold [88]. A highly vascularized granulation tissue develops, which stimulates the formation of new granulation tissue and epithelial layers. Commercial collagen sponges are insoluble protein forms derived from animals such as cows, horses, and pigs.

Collagen in the form of films and membranes can be used in wound healing [82], tissue engineering [89], and guided bone regeneration [90, 91]. Collagen-based wound dressings have long been used to treat burn wounds and ulcers. They are effective in accelerating wound healing by supporting an appropriate environment for fibroblast and keratinocyte proliferation. Collagen membranes can support the regeneration of periodontal tissues [92]. They have been widely used in periodontal and implant therapy as barriers for preventing the migration of epithelial cells and for encouraging wound repopulation by cells with regenerative capacity [90].

Recently, collagen has also been used as a carrier for drug delivery [93]. Collagen-based drug delivery systems consist of injectable microspheres based on gelatin (a degraded form of collagen), implantable collagen-synthetic polymer hydrogels, and interpenetrating networks of collagen and synthetic polymer collagen membranes [87]. In ophthalmology, composite collagen-alginate scaffolds are being investigated for sustained delivery of bioactive glial cell-derived neurotrophic factor (GDNF) to promote photoreceptor survival in inherited retinal degeneration [94]. Owing to their flowable, injectable, and biocompatible characteristics, collagen gels have also been utilized for reconstructing the liver, skin, blood vessels, and the small intestine [95].

The design and development of collagen-based scaffolds involve processing collagen solutions with other well-known biomolecules, such as elastin [96], GAGs [96], and chitosan [97]. Such biomaterials are produced by extracting and purifying collagen from natural tissues. Currently, techniques for generating new cross-links of collagen by covalent bonds to its amino and carboxyl groups are being developed to control the rate of degradation and absorption. These techniques include chemical cross-linking, physical cross-linking with ultraviolet light or dehydrothermal treatment, and cross-linking with enzymes.

12.5.2 Alginate

Alginate is a naturally occurring anionic polysaccharide obtained from calcium, magnesium, and sodium alginate salts found in the cell walls and intracellular spaces of different brown algae [98]. Alginate is a linear unbranched polymer composed of linked β-D-mannuronic acid (M units for mannuronic) and α L-guluronic acid (G units for guluronic) monomers along the polymer backbone. G units increase gel-forming and MG and M units increase the flexibility. A large number of M units, however, could induce immunogenicity [99]. Alginate, having biocompatibility, low toxicity, a relatively low cost, and mild gelation by the addition of divalent cations such as Ca^{2+}, has been extensively used for many biomedical applications [100]. It can be easily modified and fine-tuned to produce, for example, hydrogels, microspheres, microcapsules, sponges, foams, and fibers, which are widely used for wound healing, delivery of bioactive agents such as small chemical drugs and proteins, cell transplantation, and other applications.

Alginate wound dressings facilitate wound healing by maintaining a physiologically moist microenvironment and, minimal bacterial infection at the wound site [101]. They are usually produced by ionic cross-linking of an alginate solution with calcium ions to form a gel matrix, followed by the formation of freeze-dried porous sheets (i.e., foam), and fibrous non-woven dressings. Alginate dressings in the dry form typically induce wound fluid to re-gel, and the gels can resupply water to a dry wound, thus maintaining a physiologically moist microenvironment and minimizing bacterial infection at the wound site. These functions can also promote granulation tissue formation, rapid epithelialization, and healing.

Another medical application of alginate is the slow release of drug molecules, ranging from small chemical drugs to macromolecular proteins. These molecules can be released from alginate gels in a controlled manner, depending on the various types and methods of cross-linking, which employ a chemical or physical "barrier" to provide a controlled release of the drug. For example, alginate has been used for controlled and localized delivery of antineoplastic agents [102]. Alginate gels also enable simultaneous or sequential release of several drugs. Depending on the chemical structure of the drug and the mode of incorporation, a variety of drugs can be loaded into alginate-based gels allowing different release kinetics. For example, methotrexate does not interact with alginate and is rapidly released by diffusion, whereas doxorubicin can be covalently attached to alginate, enabling its release via chemical hydrolysis of the cross-linker. Mitoxantrone, ionically complexed to alginate, is only released after the gel has been dissociated [102]. Alginate is also widely used for delivery of protein drugs, since proteins can be incorporated into alginate-based formulations under mild conditions that minimize their denaturation, and the gels protect them from degradation until their release. Owing to the inherent porosity and hydrophilic nature of the gels, the release rate of proteins from alginate gels is rapid. Heparin-binding growth factors such as vascular endothelial growth factor (VEGF) or basic fibroblast growth factor (bFGF) exhibit reversible binding similar to alginate hydrogels, enabling a sustained and localized release [103, 104].

Like proteins, delivery of cell populations can also direct the regeneration or engineering of various tissues and organs in the body. Alginate gels are also utilized for cell transplantation in tissue engineering. In this approach, hydrogels serve as cell carriers to the desired site, localize them, and limit their migration away from the target site. They also provide space for new tissue formation, and control the structure and function of the engineered tissue by instructional signals that influence cellular behavior. Alginate scaffolds are actively explored for their ability to facilitate the regeneration of tissues and organs, including skeletal bone, skin, nerve, liver, and pancreas. Alginate microparticle and microfiber aggregated scaffolds have been produced via aggregation methods. Such a porous structure allows vascularization, oxygenation, cell migration, adhesion, and proliferation, which are necessary for bone tissue regeneration. Use of alginate-polymer nanocomposites for bone tissue regeneration has been extensively studied to enhance alginate biological performance. Specifically, an alginate-chitosan composite for bone tissue repair is one of the most studied materials. Such a hybrid scaffold for bone tissue engineering displayed high porosity, improved mechanical strength, and structural stability, and was shown to stimulate new bone formation and rapid vascularization [105]. In another study, a 3D porous alginate-chitosan scaffold for critical size calvarial defect repair in Sprague-Dawley rats was used. This scaffold supported undifferentiated mesenchymal stem cells (MSCs) in culture for 14 days. In addition, after 16 days, partial defect closure was observed, with the most significant defect closure ($71.56 \pm 19.74\%$) in the animal group treated with alginate-chitosan scaffolds in conjunction with bone morphogenic protein 2 (BMP-2) when compared to BMP-2 alone [106]. Recently, an injecatable composite alginate hydrogel was developed for bone regeneration, combining fluorenylmethoxycarbonyl-diphenylalanine (FmocFF), a self-assembling short aromatic peptide. The composite Alginate/FmocFF hydrogel exhibited a high storage modulus of approximately 10 kPa and facilitated adhesion, proliferation and osteogenic differentiation of MCT3T-E1 preosteoblasts [107].

A current challenge is matching the physical properties of alginate gels to the requirement of a particular application. The range of different available cross-linking strategies, using molecules with various chemical structures, molecular weights, and cross-linking functionality often leads to the formation of gels suitable for each application [108]. An alternative method is the utilization of self-assembling peptides, which enable to tailor the mechanical properties of the resulting gels [107, 109].

12.5.3 Cellulose

Cellulose is the most abundant polymer in the world, found in plant cell walls and in certain marine organisms, such as tunicates, and algae, such as Valonia. It is also produced by several bacteria, such as Acetobacter xylinum [110]. Cellulose is a linear polysaccharide composed of up to 15,000 D-glucose residues linked by β-(1 → 4)-glycosidic bonds. Based on the position of the hydrogen bonds between and within strands of units, different crystalline structures of cellulose were identified (cellulose I–IV).

Like other natural polymers, including alginates and chitosan, cellulose polymers exhibit good biocompatibility and wound healing characteristics. Cellulose is extensively used as a raw material for producing paper and cardboard products. In medical applications, it is often used in bioseparation, immobilized reaction, cell suspension culture, and as an adsorbent for sewage treatment. Different derivatives of cellulose, such as cellulose acetate, cellulose propionate, and cellulose acetate-butyrate are well known for immobilizing enzymes such as catalase, alcohol oxidase, and glucose oxidase, thereby improving their overall biocatalytic efficiency and storage stability. Microbial cellulose synthesized by acetobacterxylinum has been used as a wound healing agent. Cellulose is also suitable for use in micronerve surgery, as an artificial blood vessel suitable for microsurgery and as a candidate for capsule-based controlled drug delivery [111]. In dentistry, the cellulose membrane is used in guided bone regeneration as a space maintainer, enabling bone apposition underneath [112].

12.5.4 Chitin-Chitosan

Chitin is a white, hard, inelastic, nitrogenous polysaccharide consisting of (1 → 4)-β-N-acetyl-D-glucosamine units. It is found in the exoskeleton as well as in the internal structures of invertebrates, such as crustacean shell or insect cuticles as well as in some mushroom envelopes, the green algae cell wall, and yeast [113]. Chitosan, a partially deacetylated derivative of chitin, is composed of a mixture of N-acetyl-D-glucosamine and D-glucosamine, forming a linear polysaccharide, structurally similar to GAGs. Once dissolved, chitosan can be gelled by different methods including increasing the pH, extruding the solution into a non-solvent, by glutaraldehyde crosslinking, or ultraviolet (UV) light.

Chitosan is the second most abundant biosynthesized material, after cellulose. Chitosan can be easily processed to produce hydrogels, membranes, nanofibers, beads, micro/nanoparticles, scaffolds, foams, and sponges for various biomedical applications such as drug delivery, wound healing, and tissue engineering. The antifungal and bactericidal properties of chitosan, along with its permeability to oxygen, are often utilized for the treatment of wounds and burns. Chitosan forms a gel without any additive by neutralizing the chitosan amino group and thus, the repulsion between chitosan chains. Hydrogel formation occurs via hydrogen bonds, hydrophobic interactions, and chitosan crystallites.

Being biocompatible, non-toxic, non-allergenic, degradable, and mucoadhesive, chitosan can accelerate cell proliferation, making it a promising polymer for tissue engineering, especially for dental and bone implants, cartilage, and artificial skin [113].

Chitosan membranes have been extensively used as an artificial kidney membrane because of their suitable permeability and high tensile strength [114]. Chitosan is also used to fabricate contact lenses, as well as to control the release of different ophthalmic drugs from lens materials, due to its optical clarity, mechanical stability, sufficient optical correction, gas permeability—partially towards oxygen, as well as its wettability and immunological compatibility [115, 116].

12.6 Natural and Synthetic Composite Biomaterials

Composite biomaterials have become the focus of intense interdisciplinary research. Combining naturally derived biomaterials such as biopolymers, polysaccharides, or proteins with synthetic biomaterials such as metals or bioceramics provides more opportunities for designing "smart biomaterials". Strategies to produce "smart biomaterials" mainly aim to develop matrices that are instructive and inductive to cells, or that can stimulate target cell responses that are vital in the tissue regeneration process [117]. To fabricate composite structures, the bulk and surface properties are modified either chemically or physically, or by introducing therapeutic molecules such as drugs, proteins and genes within the structure to allow sustainable release, as well as time-dependent and sequential delivery of multiple biofactors. Another way to improve the multifunctionality of composite scaffolds is to induce responsiveness to internal or external stimuli, such as pH, temperature, ionic strength, and magnetism [117].

An extensive range of compositional variables is available by composite approach with a goal to mimic the physical properties of the native tissues. Tailoring of physical properties is based on polymer-polymer or polymer-inorganic compositions which targets both soft and hard tissue, respectively. In addition, nanotechnological advances in organizing/structuring biomaterial composites have focused on self-assembled biomaterials with or without a functional moiety in the nanostructure, as an approach to produce smart composites. Using this approach, hyaluronic acid was incorporated with the self-assembling FmocFF peptide and successfully formed a 3D hydrogel scaffold for cell culture and drug delivery without the use of cross-linking agents [118]. The combination of hyaluronic acid and FmocFF ideally mimics glycosaminoglycans and fibrous proteins (e.g., collagen, elastin, fibronectin and laminin), the two main macromolecules comprising the extracellular matrix, thus serving as an ideal microenvironment for cells. Another way of approach in the internal modulation to biomimicry is surface engineering with ECM molecules with full sequence of recombinant proteins or engineered peptides (short or oligopeptides) with key domains, or in a fused form that has multiple actions. The rationale is to provide biologically relevant environment to the surrounding cells, thereby aiding them in recognizing the biomaterial surface and enhancing target functions such as adhesion, proliferation, migration, and tissue differentiation.

12.7 Supramolecular Soft Biomaterials (Hydrogels)

Supramolecular hydrogels are part of the next generation of materials to enter the biomedical arena. These materials are composed of either polymeric or small molecular gelators and form 3D networks with suitable properties to entrap a large quantity of water within their entangled structures [119]. They are formed through intermolecular non-covalent interactions, including hydrogen bonds, π-π stacking and van der Waals interactions, thereby not requiring cross-linking agents. Hydrogels exhibit structural similarity to macromolecular-based components in the body and are considered biocompatible. Owing to their high biocompatibility and water content, they are highly appealing for biological applications. Amino acids and peptide-based hydrogelators are promising candidates to support 3D cells growth [120]. The structural and mechanical properties of the formed gels can be tailored by incorporating different amino acid types and sequences, or by mixing different peptides, rendering them to formulate bioactive hydrogels that can mimic the structure and function of native ECM [121]. Using a co-assembly approach, by mixing two short aromatic peptides, FmocFF and Fmoc-pentafluoro-phenylalanine, an ECM mimicking nanofibrous hydrogel with extraordinary mechanical properties was formed for potential use in bone tissue engineering [121]. Similarly, using the co-assembly approach, together with the incorporation of hydroxyapatite, a composite organic-inorganic hydrogel scaffold was fabricated allowing bone regeneration as well [122]. In the last two decades, different hydrogels have been comprehensively studied, focusing on designing, synthesizing, and using these materials for various biological and biomedical applications, particularly for tissue engineering and in regenerative medicine [123–132].

12.8 How to Design the Molecular Building Blocks for Hydrogels

12.8.1 Mimicking the Microarchitecture of the Native ECM with Engineered Scaffolds

Tissue engineering and regenerative approaches involve the use of 3D scaffolds, which serve as carriers for cells or growth factors to regenerate damaged tissues or organs. These scaffolds should mimic the ECM microenvironment, where cells interact and respond to mechanical cues received from the surrounding 3D environment. Specifically, native ECM presents cells with instructional signals that govern cellular behaviors, including proliferation, migration, and differentiation [133, 134]. This information can be encoded in a biochemical form via integrin binding sites or through the mechanical characteristics of the scaffolds, such as their stiffness and strain-hardening properties. Since these factors exert significant effects on cell behavior, there has been an intense interest to integrate such parameters into scaffold design strategies. For example, the presentation of specific cues regulating cell adhesion, such as tri-peptide Arginine-Glycin-Aspartic Acid (RGD) is widely employed to make synthetic scaffolds adhesive to cells [135]. In addition, mechanical cues are often integrated by altering the material's concentration or by crosslinking to control scaffold stiffness [136]. Hence, the scaffolds' material properties are of utmost importance in determining cellular response and fate (Fig. 12.2a).

12.8.2 Microarchitecture of Tissue-Engineered Scaffolds

Various scaffolds that are currently being used for 3D cell encapsulation present different structural environments to cells. These scaffolds can be classified into three fundamental categories: fibrous networks, crosslinked materials with nano-sized pores, and foam-like materials containing pores ranging from tens to hundreds of microns. Fibrous scaffolds consist of an entangled network of interacting fibers and include hydrogels such as reconstituted collagen and fibrin, as well as self-assembling peptides and many types of electrospun scaffolds. Because these fibers are much smaller than an individual cell, they can surround it to create a 3D environment similar to native stromal ECM. Crosslinked scaffolds with nanometer-sized pores, include gels composed of components such as matrigel or alginate, or synthetic polymers such as polyethylene glycol (PEG). These scaffolds surround cells with a small pore size if they are unable to enzymatically degrade them. A third scaffold group has pores that are larger than an individual cell. Although such scaffolds provide a 3D structure at a macro level, each cell experiences an environment that is similar to a 2D culture because of its large pore size.

Scaffold microarchitecture is not only guided by the particular material used, whereas the processing of the substance might also result in notable effects on the final structure. For example, large macro-pores around several hundred microns can be designed in a nanoporous scaffold through processing techniques such as particle leaching, freeze-drying, and gas foaming [137]. Alternatively, for spreading and migration of cells, proteolytic or photodegradable crosslinks can be incorporated into these nanoporous gels [126, 138]. Many polymers can also be fabricated with a fibrous architecture through electrospinning or other techniques. Thus, a single material can potentially be in any of the three structural categories based on how it is prepared [139]. Even the microarchitecture of inherently fibrous materials can change to a certain extent, based on the gelling conditions, since

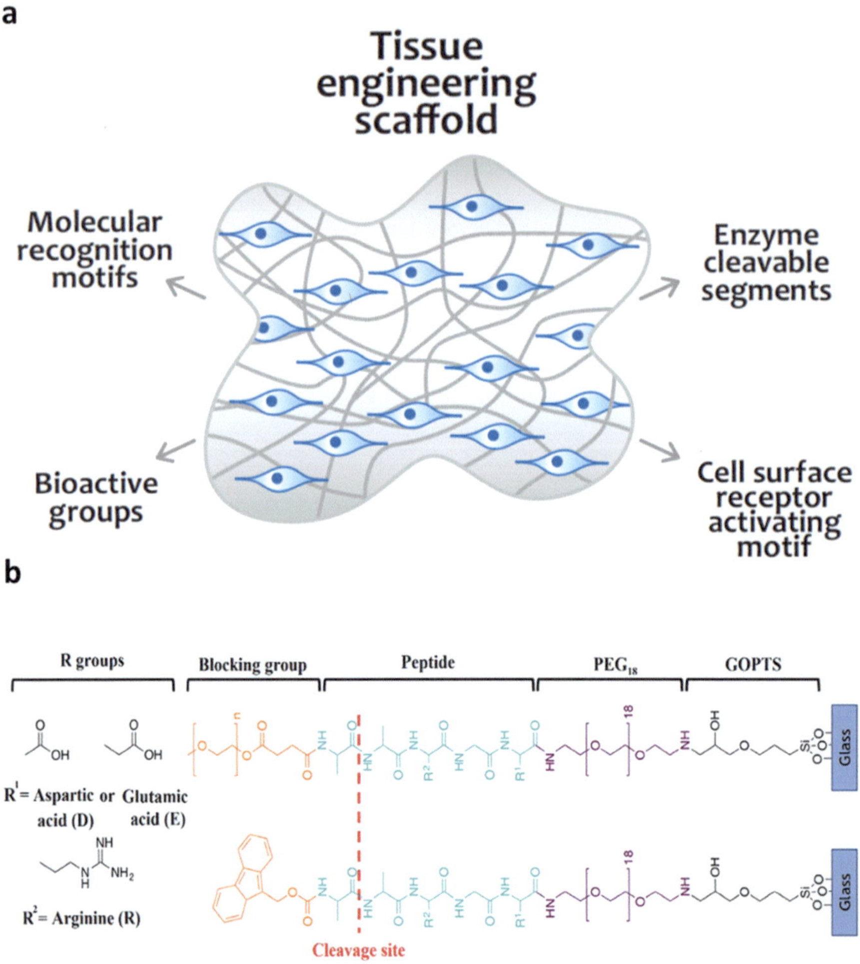

Fig. 12.2 Supramolecular biomaterial scaffolds. (**a**) A scheme describing various methods for activating receptors on cell surfaces and initiating intracellular signaling cascades using different dynamic, interchangeable, and reversible motifs for designing supramolecular biomaterial scaffolds. (**b**) A specific example of a supramolecular biomaterial scaffold with dynamic, enzyme cleavable surfaces for mesenchymal stem cell growth using Fmoc, PEG, and RGD peptide sequences as basic building blocks. Fmoc and PEG are used for blocking the RGD peptide, which is exposed upon the action of the elastase enzyme (the cleavage site is denoted by red dotted lines). (Reprinted and adapted with permission from Roberts JN, *ACS Nano*, 2016, 6667–6679)

temperature and pH can influence fiber diameter and length, or through crosslinking techniques that prevent fiber movement [140–142]. Therefore, one should be aware of such parameters and choose the appropriate processing technique when preparing scaffolds for different cellular applications.

While designing a hydrogel, the molecular scaffold should be decorated with biocompatible fragments of hydrophobic and hydrophilic units having the ability to control specific molecular interactions at the cell-material interface as well with water medium. In practice, the most desired and bioactive performances most suitable for the biomedical application properties include (i) transport properties (such as sustained release), (ii) tissue interactions (such as bioactivity), and (iii) chemical stability (such as degradability). Moreover, incorporating specific structural motifs such as molecular recognition motifs, bioactive groups, enzyme degradable segments, and motifs that can activate receptors on the cell surface and initiate intracellular signaling cascades helps in developing improved tissue engineering scaffolds having the desired properties (Fig. 12.2a).

12.9 Hydrogel Degradation

Scaffold degradation occurs as part of tissue remodeling and is controlled by physical or chemical processes and biological processes mediated by various agents, such as enzymes. The biodegradable scaffold gradually disintegrates over a predetermined period and is replaced by newly grown tissue from the adherent cells [143].

Strategies to control the degradation of hydrogels include the use of a peptide-polymer hydrogel design with cross-linking oligomers, which are substrates known for cell-secreted enzymes such as collagenases, gelatinases, and other matrix metalloproteases. These oligomeric sequences enable an *in vivo* resorption rate that coincides with the normal repair period. Several studies described the usage of PEG hydrogels containing degradable blocks to control degradation time based on various schemes [144–148].

12.10 Bioadhesion and Bioactivity

The term bioadhesion refers to the important property of cells and tissues adhering to hydrogels, enabling surgical repair and tissue regeneration. Bioadhesion is important for controlling hydrogel bioactivity, which is required to control specific biological events in the body such as endogenous cell recruitment, local morphogenesis, and controlled cell differentiation. The crucial factors, which govern these processes, are the presence of receptor-ligand complexes that mediate cell adhesion and mesenchymal migration, bound or soluble molecule interactions resulting in proteolytic biodegradation or transcriptional events that govern cell phenotype and other biophysical properties [129, 149, 150].

Hydrogel scaffolds can be incorporated with bioadhesive features by using linker molecules that induce covalent or non-covalent interactions between the implant and its surrounding cells. For example, Martino et al. [151] suggested that integrin-dependent cellular interactions with the ECM can be engineered to control stem cell fate. Integrin specificity was modulated by cell-adhesive oligopeptides derived from fibronectin central cell-binding domains (FNIII 9-10) [151]. Another method includes immobilizing fibronectin peptide fragments using a focused laser, thereby enabling one to guide and control axonal growth [152]. Mesenchymal stem cell adhesion to scaffolds can also be controlled by activating the scaffold surface using immobilized peptides that may be enzymatically activated to change their function [153]. The RGD peptide, which enables cell adherence when exposed by the elastase enzyme, can be masked by Fluorenylmethyloxycarbonyl (Fmoc) and PEG groups (Fig. 12.2b) [153]. Other typical examples of modifying hydrogels for bioadhesion include catechol-based modification t[3,4-dihydroxy-L-phenylalanine (DOPA)] of a PEG hydrogel implant to improve extra hepatic islet transplantation in cases of type I diabetes mellitus [154]. The hydrogel modification allows muscles to adhere to wet organic surfaces. Under oxidizing conditions, DOPA rapidly forms cross-links when bound, for example, to an end-functionalized star-PEG precursor. In addition, the catechol moiety forms strong covalent interactions with nucleophiles such as thiols and imidazoles found in the ECM.

12.11 3D Structures of Hydrogels

A wide range of hydrogels, based on synthetic or natural polymers, has been used regularly in medicine. For example, soft contact lenses are made from poly(hydroxyethyl methacrylic) acid, biological adhesives are made from reconstituted fibrin or albumin, biological fillers are made from hyaluronic acid, and alginate polysaccharide is used as a wound dressing. The poor geometry fidelity and limited mechanical properties of single network hydrogels have limited further biomedical applications such as artificial cartilage, muscle, and vascular. However, this drawback has led to the development of composite hydrogels, which exhibit improved mechanical properties. Calcium-alginate/poly(acrylamide) DN gel is an example of a composite hydrogel used as artificial cartilage [155]. It exhibits good geometric consistency with the tibia joint simulator as well as good resiliency and high shape retention, along with high stability under compression, stretching, and bending conditions [155] (Fig. 12.3a).

Currently, hydrogels have excellent potential for use in stem cell and cancer research, cell therapy, tissue engineering, immunomodulation, and *in vitro* diagnostics. Hydrogels have been extensively used in regenerative medicine and enable hierarchical organization of cells into tissue-like structures, where the architectural and molecular cues can be engineered with spatial and temporal presentations that mediate cellular behavior and fate. Hydrogel materials have been used for generating soft and hard tissue 3D implants. Hard tissue implants have been produced using rapid prototyping (RP) techniques. In addition, the RP method has been used for soft tissue 3D implant formation [156]. The plotting material was dispensed into a solution with a temperature below the gelation point of the

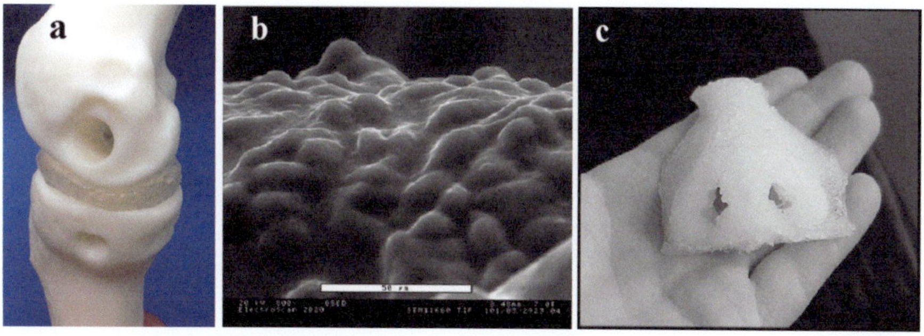

Fig. 12.3 3D structures of hydrogels. (**a**) Gel cartilage exhibits good geometric consistency with the tibia joint simulator. (Reprinted and adapted with permission from Wang J. et al., *J. Mater. Sci*, 2015, 5458–5465.) (**b**) ESEM micrograph of mouse fibroblasts on the surface of a strand of an agar scaffold and (**c**) Silicone model of the nose prepared by 3D plotting. (Both (**b**) and (**c**) were reprinted and adapted with permission from Landers R., *J. Mater. Sci*, 2002, 3107–3116)

hydrogel. It was also possible to seed and cultivate cells on the surface of these agar scaffolds, which were coated with hyaluronic acid to improve cell attachment [156] (Fig. 12.3b). The main feature of this RP technology is the 3D dispensing of liquid and pastes in liquid media [156] (Fig. 12.3c).

12.12 Conclusions

Since large variations exist in the material's composition, it is often difficult to determine the effectiveness of a particular biomaterial for tissue engineering and regeneration. It is evident that the mechanical conditions are of utmost importance to ensure engineering success. Hard materials provide compressive and torsional strength, but they are often poor at promoting bone tissue formation. Soft composites composed of natural polymers that more effectively promote cell expansion have a very low immunogenic potential, bioactive behavior, and ability to interact with the host tissue. However, they lack mechanical strength to withstand the forces typically observed in natural bone, which is the prime requisite for bone tissue engineering and tissue formation.

Most of the biomaterials available today do not ultimately meet all the demands that were discussed in this chapter, because cellular functions result from highly diverse interactions between cells, as well as between cells and biomaterials. The state of the art in biomaterial design has been continuously evolving over the past few decades, as the goals of biomedical engineering increase in complexity. Incorporating two polymers, natural and synthetic, or hybrid polymeric scaffolds combining natural and synthetic polymers, has attracted growing interest as part of the efforts to mimic the ECM of natural tissue. These new hybrid structures can present the full range of physicochemical properties and the processing techniques of synthetic polymers, as well as the biocompatibility and biological interactions of natural polymers, thereby combining the advantages of two different structures. In addition, biomaterials will not only serve as transplantation vehicles but also as tools that modulate cellular functions, thereby enhancing the endogenous regeneration response.

Acknowledgments We thank the support of the ISRAEL SCIENCE FOUNDATION (grant No. 1732/17) (L.A.A.). We thank Sharon Tsach for graphical assistance and the members of the Adler-Abramovich group for helpful discussions.

References

1. Williams DF (2009) On the nature of biomaterials. Biomaterials 30:5897–5909
2. Clause KC, Barker TH (2013) Extracellular matrix signaling in morphogenesis and repair. Curr Opin Biotechnol 24:830–833
3. Frantz C, Stewart KM, Weaver VM (2010) The extracellular matrix at a glance. J Cell Sci 123:4195–4200
4. Theocharis AD, Gialeli C, Hascall VC, Karamanos NK (2012) Extracellular matrix: a functional scaffold. In: Karamanos NK (ed) Extracellular matrix: Pathobiology and signaling. Walter de Gruyter, Berlin
5. Daley WP, Peters SB, Larsen M (2008) Extracellular matrix dynamics in development and regenerative medicine. J Cell Sci 121:255–264
6. Lu P, Takai K, Weaver VM, Werb Z (2011) Extracellular matrix degradation and remodeling in development and disease. Cold Spring Harb Perspect Biol 3(12):a005058
7. Bissell MJ, Hall HG, Parry G (1982) How does the extracellular matrix direct gene expression? J Theor Biol 99:31–68
8. Brown BN, Badylak SF (2014) Extracellular matrix as an inductive scaffold for functional tissue reconstruction. Transl Res 163:268–285
9. Hynes RO, Naba A (2012) Overview of the matrisome – an inventory of extracellular matrix constituents and functions. Cold Spring Harb Perspect Biol 4:a004903
10. Sharma A, Sharma NL, Lavy CB, Kiltie AE, Hamdy FC, Czernuszka J (2014) Three-dimensional scaffolds: An in vitro strategy for the biomimetic modelling of in vivo tumour biology. J Mater Sci 49:5809–5820
11. Theocharis AD, Skandalis SS, Gialeli C, Karamanos NK (2016) Extracellular matrix structure. Adv Drug Deliv Rev 97:4–27
12. Reilly GC, Engler AJ (2010) Intrinsic extracellular matrix properties regulate stem cell differentiation. J Biomech 43:55–62
13. Votteler M, Kluger PJ, Walles H, Schenke-Layland K (2010) Stem cell microenvironments – unveiling the secret of how stem cell fate is defined. Macromol Biosci 10:1302–1315
14. Baino F, Novajra G, Vitale-Brovarone C (2015) Bioceramics and scaffolds: a winning combination for tissue engineering. Front Bioeng Biotechnol 3:202
15. Cao W, Hench LL (1996) Bioactive materials. Ceram Int 22:493–507
16. Li HF, Zheng YF (2016) Recent advances in bulk metallic glasses for biomedical applications. Acta Biomater 36:1–20
17. Mahapatro A (2015) Bio-functional nano-coatings on metallic biomaterials. Mater Sci Eng C Mater Biol Appl 55:227–251
18. Lowe HC, Oesterle SN, Khachigian LM (2002) Coronary in-stent restenosis: current status and future strategies. J Am Coll Cardiol 39:183–193

19. Waksman R, Erbel R, Di Mario C, Bartunek J, De Bruyne B, Eberli FR, Erne P, Haude M, Horrigan M, Ilsley C, Bose D, Bonnier H, Koolen J, Luscher TF, Weissman NJ (2009) Early- and long-term intravascular ultrasound and angiographic findings after bioabsorbable magnesium stent implantation in human coronary arteries. JACC Cardiovasc Interv 2:312–320
20. Amin Yavari S, Van Der Stok J, Chai YC, Wauthle R, Tahmasebi Birgani Z, Habibovic P, Mulier M, Schrooten J, Weinans H, Zadpoor AA (2014) Bone regeneration performance of surface-treated porous titanium. Biomaterials 35:6172–6181
21. Padovani GC, Feitosa VP, Sauro S, Tay FR, Duran G, Paula AJ, Duran N (2015) Advances in dental materials through nanotechnology: facts, perspectives and toxicological aspects. Trends Biotechnol 33:621–636
22. Chen Q, Thouas GA (2015) Metallic implant biomaterials. Mater Sci Eng R 87:1–57
23. Xiao M, Chen YM, Biao MN, Zhang XD, Yang BC (2017) Bio-functionalization of biomedical metals. Mater Sci Eng C Mater Biol Appl 70:1057–1070
24. Goodman SB, Yao Z, Keeney M, Yang F (2013) The future of biologic coatings for orthopaedic implants. Biomaterials 34:3174–3183
25. Lüdecke C, Bossert J, Roth M, Jandt KD (2013) Physical vapor deposited titanium thin films for biomedical applications: reproducibility of nanoscale surface roughness and microbial adhesion properties. Appl Surf Sci 280:578–589
26. Kim HM, Miyaji F, Kokubo T, Nakamura T (1996) Preparation of bioactive Ti and its alloys via simple chemical surface treatment. J Biomed Mater Res 32:409–417
27. Kim SY, Kim YK, Park IS, Jin GC, Bae TS, Lee MH (2014) Effect of alkali and heat treatments for bioactivity of tio2 nanotubes. Appl Surf Sci 321:412–419
28. Manivasagam G, Suwas S (2014) Biodegradable mg and mg based alloys for biomedical implants. Mater Sci Technol 30:515–520
29. Sumner DR, Turner TM, Igloria R, Urban RM, Galante JO (1998) Functional adaptation and ingrowth of bone vary as a function of hip implant stiffness. J Biomech 31:909–917
30. Vasconcelos DM, Santos SG, Lamghari M, Barbosa MA (2016) The two faces of metal ions: from implants rejection to tissue repair/regeneration. Biomaterials 84:262–275
31. Hort N, Huang Y, Fechner D, Stormer M, Blawert C, Witte F, Vogt C, Drucker H, Willumeit R, Kainer KU, Feyerabend F (2010) Magnesium alloys as implant materials – principles of property design for mg-re alloys. Acta Biomater 6:1714–1725
32. Zhao D, Witte F, Lu F, Wang J, Li J, Qin L (2017) Current status on clinical applications of magnesium-based orthopaedic implants: a review from clinical translational perspective. Biomaterials 112:287–302
33. Best SM, Porter AE, Thian ES, Huang J (2008) Bioceramics: past, present and for the future. J Eur Ceram Soc 28:1319–1327
34. Hench LL (1999) Bioactive glasses and glass-ceramics, Materials science forum. Trans Tech Publications, Zurich, pp 37–64
35. Baino F, Vitale-Brovarone C (2011) Three-dimensional glass-derived scaffolds for bone tissue engineering: current trends and forecasts for the future. J Biomed Mater Res A 97:514–535
36. Legeros RZ (2002) Properties of osteoconductive biomaterials: calcium phosphates. Clin Orthop Relat Res 395:81–98
37. Anselme K (2000) Osteoblast adhesion on biomaterials. Biomaterials 21:667–681
38. Yuan H, Van Den Doel M, Li S, Van Blitterswijk CA, De Groot K, De Bruijn JD (2002) A comparison of the osteoinductive potential of two calcium phosphate ceramics implanted intramuscularly in goats. J Mater Sci Mater Med 13:1271–1275
39. Yuan H, Van Blitterswijk CA, De Groot K, De Bruijn JD (2006) Cross-species comparison of ectopic bone formation in biphasic calcium phosphate (BCP) and hydroxyapatite (HA) scaffolds. Tissue Eng 12:1607–1615
40. Levy RA, Chu TM, Halloran JW, Feinberg SE, Hollister S (1997) Ct-generated porous hydroxyapatite orbital floor prosthesis as a prototype bioimplant. AJNR Am J Neuroradiol 18:1522–1525

41. Rafieerad AR, Ashra MR, Mahmoodian R, Bushroa AR (2015) Surface characterization and corrosion behavior of calcium phosphate-base composite layer on titanium and its alloys via plasma electrolytic oxidation: a review paper. Mater Sci Eng C Mater Biol Appl 57:397–413
42. Xuereb M, Camilleri J, Attard NJ (2015) Systematic review of current dental implant coating materials and novel coating techniques. Int J Prosthodont 28:51–59
43. Rahaman MN, Yao A, Bal BS, Garino JP, Ries MD (2007) Ceramics for prosthetic hip and knee joint replacement. J Am Ceram Soc 90:1965–1988
44. Baino F, Perero S, Ferraris S, Miola M, Balagna C, Verne E, Vitale-Brovarone C, Coggiola A, Dolcino D, Ferraris M (2014) Biomaterials for orbital implants and ocular prostheses: overview and future prospects. Acta Biomater 10:1064–1087
45. Jones JR, Gentleman E, Polak J (2007) Bioactive glass scaffolds for bone regeneration. Elements 3:393–399
46. Vitale-Brovarone C, Novajra G, Lousteau J, Milanese D, Raimondo S, Fornaro M (2012) Phosphate glass fibres and their role in neuronal polarization and axonal growth direction. Acta Biomater 8:1125–1136
47. Gu Y, Huang W, Rahaman MN, Day DE (2013) Bone regeneration in rat calvarial defects implanted with fibrous scaffolds composed of a mixture of silicate and borate bioactive glasses. Acta Biomater 9:9126–9136
48. Tan A, Romanska H, Lenza R, Jones JR, Hench LL, Polak JM, Bishop A (2003) The effect of 58s bioactive sol-gel derived foams on the growth of murine lung epithelial cells. Key Eng Mater 240–242:719–724
49. Arcos D, Vallet-Regí M (2013) Bioceramics for drug delivery. Acta Mater 61:890–911
50. Baino F, Fiorilli S, Vitale-Brovarone C (2016) Bioactive glass-based materials with hierarchical porosity for medical applications: review of recent advances. Acta Biomater 42:18–32
51. Miguez-Pacheco V, Hench LL, Boccaccini AR (2015) Bioactive glasses beyond bone and teeth: emerging applications in contact with soft tissues. Acta Biomater 13:1–15
52. Tanner K (2010) Bioactive ceramic-reinforced composites for bone augmentation. J R Soc Interface 7:S541–S557
53. Day RM, Boccaccini AR, Shurey S, Roether JA, Forbes A, Hench LL, Gabe SM (2004) Assessment of polyglycolic acid mesh and bioactive glass for soft-tissue engineering scaffolds. Biomaterials 25:5857–5866
54. Rai R, Boccaccini AR, Knowles JC, Locke IC, Gordge MP, Mccormick A, Salih V, Mordon N, Keshavarz T, Roy I (2010) Fabrication of a novel poly (3-hydroxyoctanoate)/nanoscale bioactive glass composite film with potential as a multifunctional wound dressing. V international conference on times of polymers (top) and composites. AIP Publishing, pp 126–128
55. Tong SY, Wang Z, Lim PN, Wang W, Thian ES (2017) Uniformly-dispersed nanohydroxapatite-reinforced poly(ε-caprolactone) composite films for tendon tissue engineering application. Mater Sci Eng C 70(Part 2):1149–1155
56. Blum AP, Kammeyer JK, Rush AM, Callmann CE, Hahn ME, Gianneschi NC (2015) Stimuli-responsive nanomaterials for biomedical applications. J Am Chem Soc 137:2140–2154
57. Muskovich M, Bettinger CJ (2012) Biomaterials-based electronics: polymers and interfaces for biology and medicine. Adv Healthc Mater 1:248–266
58. Theato P, Sumerlin BS, O'Reilly RK, Epps TH 3rd (2013) Stimuli responsive materials. Chem Soc Rev 42:7055–7056
59. Dhandayuthapani B, Yoshida Y, Maekawa T, Kumar DS (2011) Polymeric scaffolds in tissue engineering application: a review. Int J Polym Sci 2011:1–19
60. Lin HR, Kuo CJ, Yang C-Y, Shaw SY, Wu YJ (2002) Preparation of macroporous biodegradable plga scaffolds for cell attachment with the use of mixed salts as porogen additives. J Biomed Mater Res 63:271–279
61. Mano JF, Sousa RA, Boesel LF, Neves NM, Reis RL (2004) Bioinert, biodegradable and injectable polymeric matrix composites for hard tissue replacement: state of the art and recent developments. Compos Sci Technol 64:789–817

62. Jagur-Grodzinski J (1999) Biomedical application of functional polymers. React Funct Polym 39:99–138
63. Gentile P, Chiono V, Carmagnola I, Hatton PV (2014) An overview of poly(lactic-co-glycolic) acid (plga)-based biomaterials for bone tissue engineering. Int J Mol Sci 15:3640–3659
64. Lanao RPF, Jonker AM, Wolke JG, Jansen JA, Van Hest JC, Leeuwenburgh SC (2013) Physicochemical properties and applications of poly (lactic-co-glycolic acid) for use in bone regeneration. Tissue Eng Part B Rev 19:380–390
65. Pan Z, Ding J (2012) Poly (lactide-co-glycolide) porous scaffolds for tissue engineering and regenerative medicine. Interface Focus 2:366–377
66. Bala I, Hariharan S, Kumar MN (2004) Plga nanoparticles in drug delivery: the state of the art. Crit Rev Ther Drug Carrier Syst 21:387–422
67. Griffith L (2000) Polymeric biomaterials. Acta Mater 48:263–277
68. Song Y, Kamphuis MMJ, Zhang Z, Sterk LMT, Vermes I, Poot AA, Feijen J, Grijpma DW (2010) Flexible and elastic porous poly(trimethylene carbonate) structures for use in vascular tissue engineering. Acta Biomater 6:1269–1277
69. Cassinelli C, Morra M, Pavesio A, Renier D (2000) Evaluation of interfacial properties of hyaluronan coated poly(methylmethacrylate) intraocular lenses. J Biomater Sci Polym Ed 11:961–977
70. Wang F, Mohammed A, Li C, Ge P, Wang L, King MW (2014) Degradable/non-degradable polymer composites for in-situ tissue engineering small diameter vascular prosthesis application. Biomed Mater Eng 24:2127–2133
71. Ara C, Akbulut S, Ince V, Aydin C, Gonultas F, Kayaalp C, Unal B, Yilmaz S (2015) Circumferential fence with the use of polyethylene terephthalate (Dacron) vascular graft for all-in-one hepatic venous reconstruction in right-lobe living-donor liver transplantation. Transplant Proc 47:1458–1461
72. Ding M, Li J, Tan H, Fu Q (2012) Self-assembly of biodegradable polyurethanes for controlled delivery applications. Soft Matter 8:5414–5428
73. Zhang J, Huang H, Ju R, Chen K, Li S, Wang W, Yan Y (2016) In vivo biocompatibility and hemocompatibility of a polytetrafluoroethylene small diameter vascular graft modified with sulfonated silk fibroin. Am J Surg 213(1):87–93
74. Gentile P, Chiono V, Tonda-Turo C, Ferreira AM, Ciardelli G (2011) Polymeric membranes for guided bone regeneration. Biotechnol J 6:1187–1197
75. Carson JS, Bostrom MP (2007) Synthetic bone scaffolds and fracture repair. Injury 38(Suppl 1):S33–S37
76. Nair LS, Laurencin CT (2007) Biodegradable polymers as biomaterials. Prog Polym Sci 32:762–798
77. Alexander H, Brunski JB, Cooper SL, Hench LL, Hergenrother R, Hoffman A, Kohn J, Langer R, Peppas N, Ratner B (1996) Classes of materials used in medicine. In: Biomaterials science: an introduction to materials in medicine. Academic, San Diego, pp 37–130
78. Nooeaid P, Salih V, Beier JP, Boccaccini AR (2012a) Osteochondral tissue engineering: scaffolds, stem cells and applications. J Cell Mol Med 16:2247–2270
79. Laschke MW, Harder Y, Amon M, Martin I, Farhadi J, Ring A, Torio-Padron N, Schramm R, Rücker M, Junker D (2006) Angiogenesis in tissue engineering: breathing life into constructed tissue substitutes. Tissue Eng 12:2093–2104
80. Vroman I, Tighzert L (2009) Biodegradable polymers. Materials 2:307
81. Gilbert TW, Sellaro TL, Badylak SF (2006) Decellularization of tissues and organs. Biomaterials 27:3675–3683
82. Chattopadhyay S, Raines RT (2014) Review collagen-based biomaterials for wound healing. Biopolymers 101:821–833
83. Lister J (1881) An address on the catgut ligature. Br Med J 1:183–185
84. Macewen W (1881) Clinical lectures on some points connected with the treatment of wounds. Br Med J 1:150–151
85. Lepisto J, Kujari H, Niinikoski J, Laato M (1994) Effects of heterodimeric isoform of platelet-derived growth factor PDGF-AB on wound healing in the rat. Eur Surg Res 26:267–272

86. Marks MG, Doillon C, Silver FH (1991) Effects of fibroblasts and basic fibroblast growth factor on facilitation of dermal wound healing by type I collagen matrices. J Biomed Mater Res 25:683–696
87. Rao KP (1995) Recent developments of collagen-based materials for medical applications and drug delivery systems. J Biomater Sci Polym Ed 7:623–645
88. Wachol-Drewek Z, Pfeiffer M, Scholl E (1996) Comparative investigation of drug delivery of collagen implants saturated in antibiotic solutions and a sponge containing gentamicin. Biomaterials 17:1733–1738
89. Li X, Xu J, Nicolescu C, Marinelli J, Tien J (2016) Generation, endothelialization, and microsurgical suture anastomosis of strong 1-mm-diameter collagen tubes. Tissue Eng Part A 23(7–8):335–344
90. Stoecklin-Wasmer C, Rutjes AW, Da Costa BR, Salvi GE, Juni P, Sculean A (2013) Absorbable collagen membranes for periodontal regeneration: a systematic review. J Dent Res 92:773–781
91. Wessing B, Urban I, Montero E, Zechner W, Hof M, Alandez Chamorro J, Alandez Martin N, Polizzi G, Meloni S, Sanz M (2016) A multicenter randomized controlled clinical trial using a new resorbable non-cross-linked collagen membrane for guided bone regeneration at dehisced single implant sites: interim results of a bone augmentation procedure. Clin Oral Implants Res 28(11):e218–e226
92. Pitaru S, Tal H, Soldinger M, Grosskopf A, Noff M (1988) Partial regeneration of periodontal tissues using collagen barriers. Initial observations in the canine. J Periodontol 59:380–386
93. An B, Lin YS, Brodsky B (2016) Collagen interactions: drug design and delivery. Adv Drug Deliv Rev 97:69–84
94. Wong FS, Wong CC, Chan BP, Lo AC (2016) Sustained delivery of bioactive GDNF from collagen and alginate-based cell-encapsulating gel promoted photoreceptor survival in an inherited retinal degeneration model. PLoS One 11:e0159342
95. Lee KY, Mooney DJ (2001) Hydrogels for tissue engineering. Chem Rev 101:1869–1879
96. Ellis DL, Yannas IV (1996) Recent advances in tissue synthesis in vivo by use of collagen-glycosaminoglycan copolymers. Biomaterials 17:291–299
97. Wu X, Black L, Santacana-Laffitte G, Patrick CW (2007) Preparation and assessment of glutaraldehyde-crosslinked collagen–chitosan hydrogels for adipose tissue engineering. J Biomed Mater Res A 81A:59–65
98. Tønnesen HH, Karlsen J (2002) Alginate in drug delivery systems. Drug Dev Ind Pharm 28:621–630
99. Kulseng B, Skjåk-Bræk G, Ryan L, Andersson A, King A, Faxvaag A, Espevik T (1999) Transplantation of alginate microcapsules: generation of antibodies against alginates and encapsulated porcine islet-like cell clusters1. Transplantation 67:978–984
100. Wee S, Gombotz WR (1998) Protein release from alginate matrices. Adv Drug Deliv Rev 31:267–285
101. Queen D, Orsted H, Sanada H, Sussman G (2004) A dressing history. Int Wound J 1:59–77
102. Bouhadir KH, Alsberg E, Mooney DJ (2001) Hydrogels for combination delivery of antineoplastic agents. Biomaterials 22:2625–2633
103. Lee KY, Peters MC, Mooney DJ (2003) Comparison of vascular endothelial growth factor and basic fibroblast growth factor on angiogenesis in SCID mice. J Control Release 87:49–56
104. Silva EA, Mooney DJ (2010) Effects of VEGF temporal and spatial presentation on angiogenesis. Biomaterials 31:1235–1241
105. Li Z, Ramay HR, Hauch KD, Xiao D, Zhang M (2005) Chitosan-alginate hybrid scaffolds for bone tissue engineering. Biomaterials 26:3919–3928
106. Florczyk SJ, Leung M, Li Z, Huang JI, Hopper RA, Zhang M (2013) Evaluation of three-dimensional porous chitosan-alginate scaffolds in rat calvarial defects for bone regeneration applications. J Biomed Mater Res A 101:2974–2983
107. Ghosh M, Halperin-Sternfeld M, Grinberg I, Adler-Abramovich L (2019) Injectable alginate-peptide composite hydrogel as a scaffold for bone tissue regeneration. Nanomaterials (Basel) 9:497

108. Lee KY, Mooney DJ (2012) Alginate: properties and biomedical applications. Prog Polym Sci 37:106–126
109. Gong X, Branford-White C, Tao L, Li S, Quan J, Nie H, Zhu L (2016) Preparation and characterization of a novel sodium alginate incorporated self-assembled Fmoc-FF composite hydrogel. Mater Sci Eng C Mater Biol Appl 58:478–486
110. Petersen N, Gatenholm P (2011) Bacterial cellulose-based materials and medical devices: current state and perspectives. Appl Microbiol Biotechnol 91:1277–1286
111. Eo MY, Fan H, Cho YJ, Kim SM, Lee SK (2016) Cellulose membrane as a biomaterial: from hydrolysis to depolymerization with electron beam. Biomater Res 20:16
112. Kim SM, Lee JH, Jo J, Lee SC, Lee SK (2005) Development of a bioactive cellulose membrane from sea squirt skin for bone regeneration-a preliminary research. J Korean Assoc Oral Maxillofac Surg 31:440–453
113. Croisier F, Jérôme C (2013) Chitosan-based biomaterials for tissue engineering. Eur Polym J 49:780–792
114. Lee SH, Lee JB, Bae MS, Balikov DA, Hwang A, Boire TC, Kwon IK, Sung H-J, Yang JW (2015) Current progress in nanotechnology applications for diagnosis and treatment of kidney diseases. Adv Healthc Mater 4:2037–2045
115. Behl G, Iqbal J, O'Reilly NJ, Mcloughlin P, Fitzhenry L (2016) Synthesis and characterization of poly(2-hydroxyethylmethacrylate) contact lenses containing chitosan nanoparticles as an ocular delivery system for dexamethasone sodium phosphate. Pharm Res 33:1638–1648
116. Silva D, Pinto LF, Bozukova D, Santos LF, Serro AP, Saramago B (2016) Chitosan/alginate based multilayers to control drug release from ophthalmic lens. Colloids Surf B Biointerfaces 147:81–89
117. Pérez RA, Won J-E, Knowles JC, Kim H-W (2013) Naturally and synthetic smart composite biomaterials for tissue regeneration. Adv Drug Deliv Rev 65:471–496
118. Aviv M, Halperin-Sternfeld M, Grigoriants I, Buzhansky L, Mironi-Harpaz I, Seliktar D, Einav S, Nevo Z, Adler-Abramovich L (2018) Improving the mechanical rigidity of hyaluronic acid by integration of supramolecular peptide matrix. ACS Appl Mater Interfaces 10:41883–41891
119. Dou XQ, Feng CL (2017) Amino acids and peptide-based supramolecular hydrogels for three-dimensional cell culture. Adv Mater 29:1604062
120. Hauser CA, Zhang S (2010) Designer self-assembling peptide nanofiber biological materials. Chem Soc Rev 39(8):2780–2790
121. Halperin-Sternfeld M, Ghosh M, Sevostianov R, Grogoriants I, Adler-Abramovich L (2017) Molecular co-assembly as a strategy for synergistic improvement of the mechanical properties of hydrogels. Chem Commun 53:9586–9589
122. Ghosh M, Halperin-Sternfeld M, Grigoriants I, Lee J, Nam KT, Adler-Abramovich L (2017) Arginine-presenting peptide hydrogels decorated with hydroxyapatite as biomimetic scaffolds for bone regeneration. Biomacromolecules 18:3541–3550
123. Benoit DS, Schwartz MP, Durney AR, Anseth KS (2008) Small functional groups for controlled differentiation of hydrogel-encapsulated human mesenchymal stem cells. Nat Mater 7:816–823
124. Cruise GM, Scharp DS, Hubbell JA (1998) Characterization of permeability and network structure of interfacially photopolymerized poly(ethylene glycol) diacrylate hydrogels. Biomaterials 19:1287–1294
125. Drury JL, Mooney DJ (2003) Hydrogels for tissue engineering: scaffold design variables and applications. Biomaterials 24:4337–4351
126. Khetan S, Guvendiren M, Legant WR, Cohen DM, Chen CS, Burdick JA (2013) Degradation-mediated cellular traction directs stem cell fate in covalently crosslinked three-dimensional hydrogels. Nat Mater 12:458–465
127. Liu Y, Chan-Park MB (2009) Hydrogel based on interpenetrating polymer networks of dextran and gelatin for vascular tissue engineering. Biomaterials 30:196–207
128. Lopérgolo LC, Lugão AB, Catalani LH (2003) Direct UV photocrosslinking of poly(n-vinyl-2-pyrrolidone) (PVP) to produce hydrogels. Polymer 44:6217–6222

129. Lutolf MP, Hubbell JA (2005) Synthetic biomaterials as instructive extracellular microenvironments for morphogenesis in tissue engineering. Nat Biotechnol 23:47–55
130. Nicodemus GD, Bryant SJ (2008) Cell encapsulation in biodegradable hydrogels for tissue engineering applications. Tissue Eng Part B Rev 14:149–165
131. Peppas NA, Hilt JZ, Khademhosseini A, Langer R (2006) Hydrogels in biology and medicine: from molecular principles to bionanotechnology. Adv Mater 18:1345–1360
132. Tibbitt MW, Anseth KS (2009) Hydrogels as extracellular matrix mimics for 3D cell culture. Biotechnol Bioeng 103:655–663
133. Guilak F, Cohen DM, Estes BT, Gimble JM, Liedtke W, Chen CS (2009) Control of stem cell fate by physical interactions with the extracellular matrix. Cell Stem Cell 5:17–26
134. Guo WH, Frey MT, Burnham NA, Wang YL (2006) Substrate rigidity regulates the formation and maintenance of tissues. Biophys J 90:2213–2220
135. Lowe SB, Tan VT, Soeriyadi AH, Davis TP, Gooding JJ (2014) Synthesis and high-throughput processing of polymeric hydrogels for 3d cell culture. Bioconjug Chem 25:1581–1601
136. Nemir S, West JL (2010) Synthetic materials in the study of cell response to substrate rigidity. Ann Biomed Eng 38:2–20
137. Levengood SKL, Zhang M (2014) Chitosan-based scaffolds for bone tissue engineering. J Mater Chem B 2:3161–3184
138. Kloxin AM, Kasko AM, Salinas CN, Anseth KS (2009) Photodegradable hydrogels for dynamic tuning of physical and chemical properties. Science 324:59–63
139. Madihally SV, Matthew HW (1999) Porous chitosan scaffolds for tissue engineering. Biomaterials 20:1133–1142
140. Carey SP, Kraning-Rush CM, Williams RM, Reinhart-King CA (2012) Biophysical control of invasive tumor cell behavior by extracellular matrix microarchitecture. Biomaterials 33:4157–4165
141. Hayen W, Goebeler M, Kumar S, Riessen R, Nehls V (1999) Hyaluronan stimulates tumor cell migration by modulating the fibrin fiber architecture. J Cell Sci 112(Pt 13):2241–2251
142. Lang NR, Skodzek K, Hurst S, Mainka A, Steinwachs J, Schneider J, Aifantis KE, Fabry B (2015) Biphasic response of cell invasion to matrix stiffness in three-dimensional biopolymer networks. Acta Biomater 13:61–67
143. Langer R (1994) Biodegradable polymer scaffolds for tissue engineering. Nat Biotechnol 12:689–693
144. Deforest CA, Polizzotti BD, Anseth KS (2009) Sequential click reactions for synthesizing and patterning three-dimensional cell microenvironments. Nat Mater 8:659–664
145. Kloxin AM, Kloxin CJ, Bowman CN, Anseth KS (2010) Mechanical properties of cellularly responsive hydrogels and their experimental determination. Adv Mater 22:3484–3494
146. Kraehenbuehl TP, Zammaretti P, Van Der Vlies AJ, Schoenmakers RG, Lutolf MP, Jaconi ME, Hubbell JA (2008) Three-dimensional extracellular matrix-directed cardioprogenitor differentiation: systematic modulation of a synthetic cell-responsive peg-hydrogel. Biomaterials 29:2757–2766
147. Lutolf MP, Lauer-Fields JL, Schmoekel HG, Metters AT, Weber FE, Fields GB, Hubbell JA (2003) Synthetic matrix metalloproteinase-sensitive hydrogels for the conduction of tissue regeneration: engineering cell-invasion characteristics. Proc Natl Acad Sci U S A 100:5413–5418
148. Mckinnon DD, Domaille DW, Brown TE, Kyburz KA, Kiyotake E, Cha JN, Anseth KS (2014) Measuring cellular forces using bis-aliphatic hydrazone crosslinked stress-relaxing hydrogels. Soft Matter 10:9230–9236
149. Discher DE, Mooney DJ, Zandstra PW (2009) Growth factors, matrices, and forces combine and control stem cells. Science 324:1673–1677
150. Hudalla GA, Murphy WL (2011) Biomaterials that regulate growth factor activity via bioinspired interactions. Adv Funct Mater 21:1754–1768
151. Martino MM, Mochizuki M, Rothenfluh DA, Rempel SA, Hubbell JA, Barker TH (2009) Controlling integrin specificity and stem cell differentiation in 2d and 3d environments through regulation of fibronectin domain stability. Biomaterials 30:1089–1097

152. Luo Y, Shoichet MS (2004) A photolabile hydrogel for guided three-dimensional cell growth and migration. Nat Mater 3:249–253
153. Roberts JN, Sahoo JK, Mcnamara LE, Burgess KV, Yang J, Alakpa EV, Anderson HJ, Hay J, Turner LA, Yarwood SJ, Zelzer M, Oreffo RO, Ulijn RV, Dalby MJ (2016) Dynamic surfaces for the study of mesenchymal stem cell growth through adhesion regulation. ACS Nano 10:6667–6679
154. Brubaker CE, Kissler H, Wang LJ, Kaufman DB, Messersmith PB (2010) Biological performance of mussel-inspired adhesive in extrahepatic islet transplantation. Biomaterials 31:420–427
155. Wang J, Wei J, Su S, Qiu J, Wang S (2015) Ion-linked double-network hydrogel with high toughness and stiffness. J Mater Sci 50:5458–5465
156. Landers R, Pfister A, Hübner U, John H, Schmelzeisen R, Mülhaupt R (2002) Fabrication of soft tissue engineering scaffolds by means of rapid prototyping techniques. J Mater Sci 37:3107–3116

Chapter 13
Bioinspired Engineering of Organ-on-Chip Devices

Li Wang, Zhongyu Li, Cong Xu, and Jianhua Qin

Abstract The human body can be viewed as an organism consisting of a variety of cellular and non-cellular materials interacting in a highly ordered manner. Its complex and hierarchical nature inspires the multi-level recapitulation of the human body in order to gain insights into the inner workings of life. While traditional cell culture models have led to new insights into the cellular microenvironment and biological control *in vivo*, deeper understanding of biological systems and human pathophysiology requires the development of novel model systems that allow for analysis of complex internal and external interactions within the cellular microenvironment in a more relevant organ context. Engineering organ-on-chip systems offers an unprecedented opportunity to unravel the complex and hierarchical nature of human organs. In this chapter, we first highlight the advances in microfluidic platforms that enable engineering of the cellular microenvironment and the transition from cells-on-chips to organs-on-chips. Then, we introduce the key features of the emerging organs-on-chips and their proof-of-concept applications in biomedical research. We also discuss the challenges and future outlooks of this state-of-the-art technology.

Keywords Bioinspired materials · Microfluidics · Organ-on-chip · Cellular microenvironment · Disease modeling · Drug testing

---

L. Wang · Z. Li · C. Xu
Division of Biotechnology, Dalian Institute of Chemical Physics, Chinese Academy of Sciences, Dalian, P. R. China

J. Qin (✉)
Division of Biotechnology, Dalian Institute of Chemical Physics, Chinese Academy of Sciences, Dalian, P. R. China

CAS Center for Excellence in Brain Science and Intelligence Technology, Chinese Academy of Sciences, Shanghai, China

Institute for Stem Cell and Regeneration, Chinese Academy of Sciences, Beijing, China

University of Chinese Academy of Sciences, Beijing, China
e-mail: jhqin@dicp.ac.cn

S. Perrett et al. (eds.), *Biological and Bio-inspired Nanomaterials*, Advances in Experimental Medicine and Biology 1174,
https://doi.org/10.1007/978-981-13-9791-2_13

Abbreviations

ADMET	Adsorption, distribution, metabolism, elimination and toxicity
BBB	Blood-brain-barrier
EBs	Embryonic body
ECM	Extracellular matrix
ECs	Endothelial cells
EMT	Epithelial-to mesenchymal
ESCs	Embryonic stem cells
FSS	Fluidic shear stress
iPSCs	Induced pluripotent stem cells
MEMS	Micro-electromechanical system
MSC	Mesenchymal stem cells
PDMS	Polydimethylsiloxane
PK/PD	Pharmacokinetics and pharmacodynamics
TEER	Trans-epithelial electrical resistance

13.1 Introduction

Appropriate model systems drive the development of biological and biomedical research. These model systems seek to recapitulate human physiology and pathology from the molecular level to the cellular, tissue and organ levels, thus providing insights into disease etiology, diagnostic therapeutics and disease prevention. *In vivo*, the body can be viewed as a variety of cellular and non-cellular materials interacting in a highly ordered manner. The complex and hierarchical nature of all living things inspires the multi-level recapitulation of the human body and development of biological model systems consisting of multiple cell types, with internal (cell-cell, cell-matrix) and/or external (e.g. cell-environment) interactions in a more relevant organ and multi-organ context.

Conventionally, animal models are often used to closely recapitulate human physiology in a variety of biomedical research areas, but they fail to faithfully mimic human responses due to the presence of confounding variables and differences between animal and human biology. It is also quite difficult to carry out real-time observation, utilization and throughput assays in animal models. While simplistic models, such as two dimensional (2D) monocultures of cells in a Petri-dish, have their merits to study the biological process of specific cell types, these formats often lack cell-cell and cell-matrix interactions that are necessary to maintain and define the specific phenotypes of cells. They also fail to mimic the cellular functions and intercellular communication that is present in tissues or organs. While three-dimensional (3D) cell aggregates and spheroid cultures can improve the cellular functions to some extent, they still lack many features that are critical for sustaining organ development and function, such as spatiotemporal biochemical cues, vascular perfusion, mechanical cues or multiple cell co-cultures. Obviously, most current model systems are far from being able to fully reconstitute functions spanning cellular, tissue and organ levels, but it is crucial to develop such biological systems that can address specific scientific questions in biomedical research.

Considerable advances in microfabrication and microfluidics technology have expanded our ability to culture cells in a tightly controlled complex cellular microenvironment in a spatiotemporal manner, thus mimicking the tissue microenvironment *in vivo*. The microfluidic culture platform can provide living cells with continuously perfused medium in microchannels at the microscale [1–5]. Integration of microfabrication with microfluidics technologies that enable precise control of dynamic fluid flow has made it possible to create cellular microenvironments that present cells (e.g. human cell lines, primary cells or stem cells) with appropriate organ-relevant chemical gradients and dynamic mechanical cues. These new capabilities have provided an impetus for the development of alternative cell-based *in vitro* models, which better mimic the complex structures and functional complexity of living organs, termed organs-on-chips [6–9]. These organs-on-chips combine microfluidics with bioengineering and cell biology, allowing study of the diverse biological processes and physiopathology of the human body in ways that are not possible using animal models and conventional 2D or 3D cell culture systems.

In this chapter, we first introduce the microfluidic culture systems that can offer dynamic cell cultures with enhanced capabilities. Then, we introduce the design considerations and key components required for engineering the cellular microenvironment using microfluidic chips. We further summarize the emerging transition from cells-on-a-chip to organs-on-a-chip, and the proof-of–concept applications of engineered organ-on-chip devices. We also discuss the challenges and future perspectives of this state-of–the-art technology.

13.2 Microfluidic Cell Culture System

Microfluidic technology allows precise manipulation of fluid in a microscale device created with technologies developed by the semiconductor industry and micro-electromechanical systems (MEMS). Advances in microfabrication and soft lithography have enabled microfluidic culture systems to reconstitute dynamic, controlled spatio-temporal physico-chemical microenvironments to mimic *in vivo* conditions, distinguishing them from the existing cell culture platforms.

Microfluidic culture devices are usually generated by soft lithography pioneered by Whitesides et al. using polydimethylsiloxane(PDMS) [10, 11]. The classic process of PDMS chip fabrication begins from a mould manufactured using photoresistor silicon [12, 13]. A mixture of silicone rubber and cross-linking agent is poured into the mould and is then easily peeled off the substrate after cross-linking. The PDMS block can be reversibly sealed to other substrates, such as glass and PDMS simply by direct contact. It is also easy to irreversibly bond the PDMS chip to a PDMS substrate, silicon or glass by plasma oxidizing the PDMS surface to form Si-O-Si bonds or by using PDMS as glue [14]. These features make it possible to fabricate multilayer microfluidic chips with flexible microstructure configurations and channels for compartmentalizing cell culture by simply stacking PDMS pieces to connect different layers [15, 16]. The high elasticity of PDMS enables the integration of pneumatic microvalves into chips to realize complicated cell manipulations on-chip [17–22]. The major superiority of PDMS is its permeability to oxygen and its biocompatibility, which are necessary for long-term cell culture in sealed microchambers or microchannels. The property of PDMS of being optically transparent is favorable for observing cellular behavior and detecting molecules expressed in cells using a brightfield or fluorescence microscope. Besides PDMS, other polymeric materials have been utilized for fabricating configurable microstructures (e.g. microwell, micropillars and microchannels) for microfluidic cell culture [23–28]. In addition, natural polymers, such as agarose, fibrin and collagen, can also be used for generating cell-laden microfluidic chips with the creation of an *in vivo*-like extra-cellular matrix (ECM). More recently, paper-based substrates, which have a fibrous structure analogous to native ECM, have been developed to fabricate chips with 3D microstructures to support the culture of cancer cells, stem cells and cardiomyocytes [29–34]. The diverse properties of these materials have broadened the applications of microfluidics in cell-based biomedical research.

Microfluidic systems were initially applied to cell culture by simply lining the cells in the microchannels or microchambers in a 2D manner with perfused medium in a controlled way. They are often used to study cell growth, proliferation, differentiation and in drug testing. Lately, microfluidic technology has gradually moved to create 3D cell culture models *in vitro* due to its capability to produce and manipulate micron-sized 3D spheroids in a high-throughput manner with great reproducibility [35–37]. Cell spheroid formation is mostly based on the self-assembly features of cells. Although some conventional methods can produce cellular spheroids, such as non-adhesive culture substrates and spinner systems, these methods fail to create uniform cellular spheroids rapidly with controlled size. Recently, microfluidic devices have been developed to produce uniformly sized cellular aggregates [38–40]. These microfluidic based 3D aggregates can be used for generation of multiple types of spheroids from cancer cells, liver cells, and adult stem cells, as well as embryonic bodies (EBs) from embryonic stem cells (ESCs) or induced pluripotent stem cells (iPSCs) [41–46]. The cellular spheroids formed by single or multiple cell types can promote cell-cell interactions via gap junctions between cells and thus reduce the distance between cells and the chemical signals secreted from adjacent cells via paracrine pathways. Thus, these 3D cellular spheroids are effective in creating a variety of functional microtissues *in vitro*. A dynamic culture model in a microdevice can also accelerate the proliferation and maturation of tissues.

Microfluidic cell-laden culture systems are an ideal platform for studying cell-ECM interactions as present in the tissue microenvironment. This approach can realize the spatial arrangements of different cellular spheroids via specific microchannel designs, facilitating the study of spheroid-spheroid, and spheroid-ECM interactions. Compared to existing 3D methods, microfluidic-based 3D cell-laden culture has many advantages, including controllable size, arbitrary shape, macro-tissue reproduction, and easy manipulation of different hydrogels [47–51]. Cell-laden gels as a building block have potential for construction of functional tissues or organs. Zhang et al. proposed a novel and straightforward strategy to produce shape-controlled collagen building blocks via a membrane-templated microdevice [52]. This strategy enables the collagen blocks to self-assemble into 3D tissue-like microstructures with spatial distributions of cell types. These studies open new opportunities to investigate the mechanisms of tissue or organ development and tissue engineering applications.

13.3 Microengineering the Cellular Microenvironment

In order to conduct reliable microfluidics-based cell cultures, it is critical to mimic the cellular microenvironment encountered in vivo (Fig. 13.1). To construct a biomimetic cellular microenvironment in vitro, complex and multi-purpose designs are required to integrate micro-fabricated substrates with microfluidics technologies and cell biology. The in vivo cellular microenvironment is composed of both biochemical and mechanical signals produced by cells and the ECM. These stimuli may

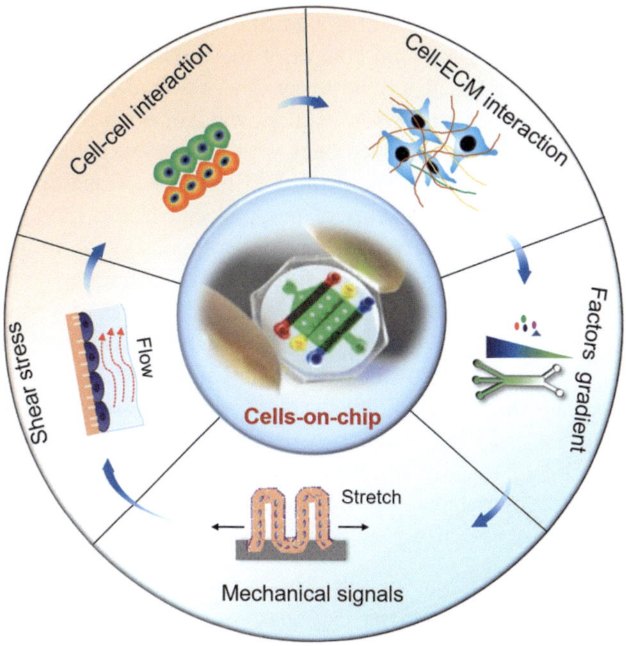

Fig. 13.1 Engineering the cellular microenvironment on chip. The microfluidic device provides cells with a controllable microenvironment, including biochemical and biophysical cues involved in maintaining cellular microenvironments

guide tissue organization and growth via orienting cell polarization and migration, balancing the growth and apoptosis, and regulating functional protein expression and cellular behavior to construct a functional and coordinated tissue. Cell-cell communications within cellular microenvironments share several common features, such as short communication distance between cells and other stimuli, continuous nutrient supply and waste removal, and synergistic actions of total cells to external stimuli. Microfluidic devices offer a powerful tool to reconstruct the cellular microenvironment via providing precise control of intercellular communication, as well as biochemical and biophysical cues that are necessary for the formation and development of tissues or organs.

13.3.1 Cell-Matrix Interaction

Extracellular matrix proteins secreted by different cell types provide important physical support for cells in the processes of tissue/organ formation [53, 54]. These matrix proteins direct cell fate and behavior via promoting cell-cell and cell-matrix interactions. The receptor proteins on cellular membranes can recognize the specific ligands on cells or extracellular matrix proteins and trigger intra-cellular signaling pathways. Microfluidic platforms exhibit the ability to integrate the ECM within microsystems to form gradients of nutrients, oxygen and soluble factors [55–57] via spatio-temporal control of the cellular microenvironment. For example, a microfluidic-based turning-assay device was designed to realistically mimic the microenvironment of neuronal growth cones in vivo by combining precise gradients of soluble guidance cues with surface-bound guidance signals. The surface-bound laminin gradient enabled to tune the polarity of the neuronal growth cone in response to gradients of neurotrophic factors [58]. Lanfer et al. fabricated aligned collagen matrices using a microfluidic set-up in order to study the effects of collagen on the growth and differentiation of mesenchymal stem cells (MSC) [59, 60]. Using microfabrication techniques, Chin et al. created an array of 10,000 microwells (coated by polyornithine and laminin) with 20–500 μm in diameter on a glass coverslip to study the combinatorial effects of growth factors and laminin protein on the proliferation and phenotypes of rat neural progenitor cells [61]. With these microdevices, the cell-matrix interactions could be studied *in vitro* in a biomimetic manner, which improves our understanding of the formation mechanism and development process of tissues/organs with anisotropic architecture.

13.3.2 Cell-Cell Interactions

Cell-cell interactions guide development and morphogenesis, as well as promote wound healing of tissues or organs because the body is composed of a variety of cell types working synergistically in organized structures. Commonly, cell-cell interactions under physiological conditions may happen either by direct contact relying on a tight cell-cell junction or by indirect contact via local diffusion of soluble factors or the system of endocrine regulation. It is feasible, with microfluidic technology, to manipulate and culture multiple cell types within a compartmentalized microdevice, which can facilitate the study of cell-cell interactions. Qin et al. developed a series of functional microfluidic chips to realise cellular co-cultures and investigate the interactions between different cell types. Two cell types, MSC and salivary gland cancer cells, were co-cultured in a compartmentalized PDMS microdevice fabricated with two separate chambers and PDMS pillar structures. This study demonstrated that MSCs could be recruited by cancer cells and this effect could be mediated by TGF-β, secreted by cancer cells [62]. It is well known that the first step of tumor cell metastasis is the transfer of circulating tumor cells across the vascular side in a tumor microenvironment. This work designed a 3D microfluidic cell co-culture model to investigate the interaction between cancer cells and endothelial cells, in which the device consists of a vessel-like cavity, endothelium and perivascular matrix containing chemokines [63]. This device enabled the modeling of tumor-cell metastasis in a dynamic manner and visualized observation of transendothelial invasion of cancer cells in real-time, something that is not possible by conventional methods.

The study of neurobiology requires the creation of cellular microenvironments containing multiple types of brain cells and biochemical cues in a precisely controlled manner. A microfluidic system with large open cell culture reservoirs was designed to generate neuronal microenvironments that enable to mimic axon transport and synapse formation via dynamic analysis. In this work, the motor neurons were co-cultured with C2C12-derived myotubes for substantial time periods on the devices in order to mimic the neuro-muscular junction [64]. This device provides a new platform to study the interaction between different cells, such as effector cells and target cells within the cellular microenvironment. Obviously, microfluidic culture systems offer advantages beyond existing methods by allowing to manipulate different cells in a flexible and organ relevant context.

13.3.3 Control of Biochemical Microenvironments

13.3.3.1 Gradients of Soluble Factors

In addition to cellular components, biochemical factors in the local tissue microenvironment may function as regulatory signals to guide various types of cellular behavior, such as cell growth and differentiation, migration and angiogenesis, by forming gradients of soluble molecules. It is quite difficult to generate physiologically relevant biochemical gradients on traditional 2D and 3D culture models.

The ability of microfluidic techniques for producing chemical gradients holds great promise for mimicking and investigating the role of spatially defined soluble microenvironments in guiding cellular behavior and adversity of biological responses [65–69]. Classical microfluidic gradients were generated by manipulating diffused mass transport across the interface between adjacently flowing liquid streams in microchannels under laminar flow at low Reynolds numbers [70, 71]. To validate the chemo-attraction of leukocytes under inflammatory stimuli, Han et al. created a 3D microfluidic cell culture device that generated spatially controllable and stable gradients of two chemo-attractants [72]. A microfluidic gradient chip containing hydrogel-incorporating chambers between surface-accessible microchannels has been utilized to investigate angiogenesis under growth factor gradients in 3D microenvironments [73]. Aside from these methods of gradient formation, relying on laminar flow, Torisawa et al. developed a novel microfluidic device to generate physiologically relevant biochemical gradients by patterning chemo-attractant-secreting (source) and chemo-attractant-scavenging (sink) cells in spatially defined locations inside microchannels [74]. This culture system with more physiological chemoattractant gradients can be broadly used for engineering tissue microenvironments for the study of complex intercellular communications.

13.3.3.2 Control of Oxygen Concentration

Oxygen gradients play a crucial role in maintaining homeostasis in specific tissues, promoting angiogenesis and inducing acute cellular response under inflammatory conditions. Cells in the human body can respond to a wide range of relative oxygen concentrations; for example, normal arterial oxygen content in human brain ranges from 5 to 10 ml/dl. However, the cells in conventional culture systems are usually maintained in atmospheric condition of approximately 20% O_2, which is higher than that in the body [75–77]. Microfluidic devices enable to control parameters of the cellular microenvironment and provide a unique opportunity to generate gaseous gradients with high spatio-temporal resolution [78–80]. A new type of microfluidic device capable of generating oxygen gradients for cell culture was developed based on spatially confined chemical reactions. This device requires a minimal amount of reagents and efficiently controls the oxygen gradients in cell cultures [81]. Derda et al. developed a paper-based cell culture system composed of stacked layers of paper that were impregnated with suspensions of cancer cells in an ECM hydrogel. This system enabled control of oxygen and nutrient gradients in a 3D microenvironment and allowed the analysis of molecular and genetic responses. The paper-supported gels provide a uniquely flexible platform to investigate cellular responses to 3D molecular gradients and to simulate tissue or organ level functions [32, 82].

13.3.4 Control of Biophysical Microenvironments

13.3.4.1 Fluid Flow-Induced Stress

Fluid flow exists ubiquitously in the human body, including in blood vessels, lymphatic vessels, and its role is mass transport and the distribution of soluble factors. The fluid flows in the body span a wide range of fluid velocities from 0.1 μm/s to 0.3 m/s [83, 84]. Different flow velocities may induce various responses of different cell types. It is advantageous for biomimetic microfluidic cell culture devices to enable the simulation and generation of fluidic shear stress (FSS) within microchannels in order to investigate the effects of FSS on cellular adhesion, growth, protein expression and morphology. It is beneficial to be able to reproduce physiologically relevant shear stresses and to study their roles in regulating the specific tissues or organs at levels relevant for organs [85–87]. One significant feature of microfluidic devices lies in their enabling of integrated bioassays. An integrated microfluidic perfusion chip was developed to simultaneously produce multiple-parameter FSS in order to study the effects of fluid flow stimuli on the fate of MSC, chondrocytes and Yes-associated protein (YAP) expression associated with the regulations of cell proliferation, survival, differentiation and organ size [88]. Wang et al. constructed a microfluidic-based vascular-like chip to mimic blood vessels *in vivo* with the aim to study the role of fluid flow in the arrangement of human iPSC-derived endothelial cells (ECs) [89]. In this study, the FSS promoted the arrangement of ECs comparing with static cultures. In addition, this work also mimicked vascular inflammatory reactions under flow conditions by analyzing the interactions of ECs and inflammatory monocytes and the response of ECs to the inflammatory factor TNF-α. The cells showed a physiologically relevant feature under fluidic conditions not possible with conventional cell culture systems. FSS can also modulate cellular behavior via a reorganization of the cytoskeleton (F-actin stress fibers) in cells [90]. Jang et al. developed a microsystem to investigate the effects of luminal FSS on the reorganization of the actin cytoskeleton of inner medullary collecting duct cells of the kidney as well as the translocation of water transport proteins anchoring on the cellular membrane [91, 92]. A biomimetic microfluidic chip was designed to mimic the flow of urine within the proximal tubule and to study the effects of complement C3a on the epithelial-to mesenchymal (EMT) of proximal tubular epithelial cells after being exposed to serum proteins [93].

Due to the microstructure design of the microfluidic devices, they can also be utilized for recapitulating the effects of fluid forces on modulating angiogenesis associated with tumor biology. To investigate the collective roles of fluid and soluble factors on endothelial sprouting, Song and Munn designed a microfluidic culture device that consists of two parallel microchannels lined with HUVEC and a central microchannel filled with collagen gel which separates the two parallel mcirochannels [94]. Multiple mechanical and chemical cues were generated in this microfluidic cell culture system to mimic the physiological microenvironment of ECs angiogenesis.

13.3.4.2 Tissue Mechanics

In addition to fluid shear stresses, cells also experience organ-specific mechanical cues, such as tensile and compressive forces under normal physiological and pathological conditions. A multilayer microfluidic device was designed to study the combinatorial effects of solid mechanical and surface tension stress induced by cyclic wall stretching and the propagation of the air-liquid interface in the alveoli of the lung [95]. This platform generated more physiologically relevant mechanical cues and provided more detailed information than previous models of ventilator-induced pulmonary injury which were based on cyclic stretching [96, 97] or air-liquid interface flow over the cells respectively [98, 99].

Previous studies of gut absorption and metabolism, using human intestinal epithelial cells (e.g. Caco-2), failed to reproduce most of the differentiated organ-specific features of living intestine by culturing these cells on plastic flasks or Transwell inserts as well as microfluidic chips [100–102]. The main reason is that epithelial cells in culture are not able to experience the natural mechanical stimuli including fluid flow and cyclic peristaltic motions in the normal intestine *in vivo*. Recently, Ingberet al. developed a multilayer microfluidic gut-on-a-chip device with a flexible, porous ECM-coated membrane between two PDMS chips containing microchannels. The intestinal epithelial cells were cultured on the membrane substrate in the microchannels and experienced physiological mechanical strains including trickling flow and cyclic mechanical distortion [103, 104]. These mechanical stimuli induced the cells to spontaneously form robust intestinal villi structures that hold the features of tight junctions between cells, coating with brush borders and mucus. The mechanical cues also promoted differentiation of the cells into four different cell lineages of the small intestine, including absorptive, mucus-secretory, enteroendocrine and Paneth, and produced a higher resistance epithelial barrier.

In addition, mechanical strains also influence the biological behaviors of stem cells. Gao et al. created a simple membrane-based microfluidic device to study the effects of cyclic tensile stress on the proliferation and differentiation of MSC [105]. This tensile stress was generated by deforming the elastic PDMS membrane sandwiched between the two layer PDMS chips via negative pressure. The MSCs cultured on the membrane were subjected to different magnitudes of tensile force. The results recapitulated the results that higher tensile stress could promote proliferation, osteogenesis and inhibit adipogenesis of MSCs, providing a new method using tensile stress to regulate the osteogenesis/adipogenesis balance in the development and damage repair of organs.

13.4 From Cells-on-Chip to Organs-on-Chips

In vitro cell culture systems can reflect some biological behavior and functions of cells. However, how cells as the basic building blocks of all living organisms can be made to assemble into functional tissues and organs *in vitro* has been, and remains, a major challenge of recent decades. Organ-on-a-chip technology rapidly advanced on the basis of integration of microfabrication and microfluidics technology that enable to recapitulate the dynamic and complex tissue microenvironment from the cellular to the tissue and organ level [106, 107]. Organs-on-chips have progressed to the point where it has become feasible to engineer the structural arrangements and functional complexity of living organs by culturing cells in microfluidic channels with multicellular microarchitecture, tissue-tissue interfaces, and biochemical/physical cues in an organ-relevant context (Fig. 13.2). It should benoted that this technology is not intended to engineer a whole living organ but rather to rebuild minimal functional units that represent some specific functions of tissues or organs [108]. In earlier work, the concept of organ-on-a-chip focused on a simple design with perfused microchannels or chambers containing one cell type (e.g., ECs, hepatocytes or intestinal epithelial cells) that represents the functions of one tissue type. Recently, more complex organ-on-chip devices were developed with multi-layer structures connected by porous membranes. Different cells were cultured on opposite sides of membranes to establish interfaces between different tissues (e.g. the blood-brain barrier (BBB) or the lung alveolar-capillary interface). These devices can combine simultaneously chemical signals (e.g. soluble factors) with physical cues (e.g. fluidic shear stress, mechanical pressure and cyclic mechanical strains). They can also be used for real-time analysis of organ-specific responses to drugs, toxins or other environmental perturbations and circulating immune cells. Nowadays, multiple types of organs-on-chips have been created for studying the various biological processes at physiological or pathological levels in ways that are not possible by traditional cell culture and animal models [109–112] (Table 13.1).

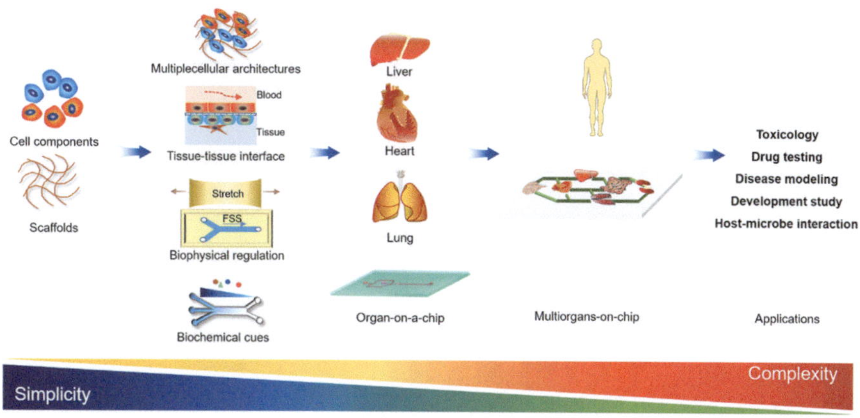

Fig. 13.2 Schematic illustrations of the biomimetic diagram for engineering organs-on-chips

Table 13.1 Summary of the presented organs-on-chip devices categorized by organ types, cell resources, chip features and applications

Organ	Cell source	Preparation process	Applications	References
Brain-BBB unit	Brain ECs (bEnd3) and astrocytic cells (C8D1A) or rat glial cells (C6) (M, R)	Porous polycarbonate sandwiched by PDMS chips with microchannels integrating TEER sensors	Physiological BBB models and toxicity of brain-targeting drugs	[159, 199]
	Brain ECs (R) and astrocytic cells (R)	Two independent vascular channel surrounding tissue compartment in center	Functional analysis	[200, 215]
Neurovascular unit	Neurons and astrocytes (R) HUVEC and cerebral microvascular ECs (H)	Two channels for 3D gels, Two channels for media 9 trapezoidal structures in gel channels	Functional analysis	[160]
Heart	iPSC-derived Cardiomyocytes(H);	Microcontact printing using PDMS stamps PDMS thin film	Disease model	[201]
	Cardiomyocytes (R)		Functional heart tissue; Drug testing	[150, 152, 202]
	Cardiomyocytes (M)	Tri-layer cell-laden hydrogel in a perfused PMDSchip	Functional heart tissue; drug testing	[145]
Lung Alveolar-capillary interface	Alveolar epithelial cells, pulmonary microvascular EC (H)	Two PDMS layers with microchannels sandwich a microporous PDMS membrane. Fluid flow, mechanical strain.	Pulmonary inflammation and infection using TNF-a and coli bacteria,toxico logy of nanoparticulat	[115, 116]
	Bronchial epithelial cells(Beas-2B)(H)	A porous silicon membrane is sandwiched between two PDMSmicrochannels	Pulmonary edema and drug toxicity drug testing Lung inflammation model Disease modeling (chronic obstructive pulmonary disease, COPD)	[203]
Lung small airway	Airway epithelial cells and lung microvascular ECs.	PDMS layers with channels separated by semiporous polyester membrane Fluid flow, mechanical strain.		[171]

Intestine	Intestinal epithelial cells (H)	Two PDMS layers with microchannels sandwich a microporous PDMS membrane. Fluid flow, mechanical strain.	Co-culture with microflora Intestinal villus differentiation absorption function host-microbe interaction	[102, 103, 104]
Kidney	Primary kidney inner medullary collecting duct(IMCD)cells (R)	Sandwiched assemble of PDMSmicrochannels, polyester porous membrane, PDMS well Fluid flow shear(1 dyn/cm2)	Physiological renal tubule model; drug screening	[91, 92]
	Primary renal proximal tubular epithelial cells(H) Renal proximal tubular cell line (HK-2)(H)	Fluid flow shear(0.2dyn/cm2) Flow rate (0.6 μl/min)	Drug transport and nephrotoxicity	[204] [93]
	Immortalized glomerular EC (GEnCs) and podocytes (MPC-5)(M)	Flow rate 0.001, 0.002 and 0.003 dyn/cm2	Kidney disease (proteinuric nephropathy) Kidney disease (hypertensive nephropathy)	[205]
Liver	Primary hepatocytes(R); hepatic stellate cell line(LX2)(H)	Multiple chambers reconfigurable coculture device with biosensors	Disease modeling	[206]
	Primary hepatocytes(R)	PDMS device containing multiplexed microchannels	Function analysis; Drug toxicity	[27]
	Primary hepatocytes(H&R)	PDMS chip containing central cell culture region and the outer flow channels with endothelial-like barrier consisted of a set of parallel channels	Function analysis	[126]
	Primary hepatocytes (R);primary adrenal medullary ECs (R); Bovine aortic ECs.	Two PDMS with channels sandwich a porous membrane. Flow perfusion.	Function analysis; Disease model(hepatitis)	[207]
Vasculature	Primary dermal microvascular EC (HDMECs)(H)	PDMS chip containing two compartments and three pump membrane.	Physiological vascular tissues	[208]
	Cordblood ECs (H); lung fibroblasts (H)	Long microchannel connects multiple microchambers	Structure and function analysis	[179, 209, 210]

(continued)

Table 13.1 (continued)

Organ	Cell source	Preparation process	Applications	References
Multiorgans Liver, lung, kidney, adipose	C3A; A549;HK-2;HPA	Multi-channel 3D microfluidic cell culture system (3D-μFCCS)under perfusing medium	Co-culture model	[169]
Neurospheres/liver	Undifferentiated NT2 (H); liver HepaRG cells (H)	PDMS chip containing two compartments and three pump membrane.	Toxicology	[165]
Heart, muscle, nerve, liver	HiPSC-derived cardiomyocytes(H); Skeletal Myofibers(H); Motoneurons(H);hiPSC- derived neurons (H); hepatocellular carcinoma	Two holders separate culture devices with pumpless. Continuous gravitational flow.	Organ-organ communication; Drug toxicity;	[211]
Intestine, liver, skin,kidney	HepG2/C3A(H); intestinal epithelial cells (H); HepaRG and primary hepatic stellate cells(HHSteC) (H); prepuce (H); proximal tubule cell line (RPTEC/TERT-1)(H)	Two PDMS layer with two fluid flow circuits with separate peristaltic on-chip micropump	ADME model	[168]
Intestine, liver	colorectal carcinoma cells(HCT-116 eGFP) (H);	Hang drop technique for fabrication of multi-cell microfluidic chip	Function analysis	[212]
Liver, tumor, Marrow	primary liver cells (R) HepG2/C3A (H); colon cancer cells (HCT-116)(H); Myeloblast cell line (Kasumi-1)(H)	Microscale cell culture (μCAA)	PK-PD model	[166]
Multiple heart chips	iPSC-derived cardiomyocytes (H)	Lego-like plug & play system(μOrgano)	Function analysis	[213]
Spleen	Red blood cells (H)	Device containing fast-flow and slow-flow channel and pillar matrix in slow-flow channel	Spleen filtering function	[214]

H human, *R* rat, *M* mouse

13.4.1 Bioengineering Organs on Chip

13.4.1.1 Lung on a Chip

The lung is the primary respiratory organ in humans. Much effort is devoted to save lives by improving lung health and preventing lung diseases in biomedical research. The alveolar-capillary interface is the fundamental functional unit of the living lung. Some methods developed to-date reproduce the geometry of the lung epithelial-endothelial interface via culturing epithelial and endothelial cells on opposite sides of a thin porous membrane. However, these models failed to mimic the mechanical microenvironment of a living breathing lung [113]. Tavana et al. developed a similar design to monitor the effect of pulmonary pressure on the cell culture system [114]. This model focused on the role that surfactants play in the damage of alveolar epithelial cells. The results demonstrated that the surfactant, Survanta, reduced the cell injuries induced by the liquid plug flow. This lung-on-a-chip system is helpful for the investigation of cellular and sub-cellular effects in airway reopening.

Huh et al. developed a microfluidic lung-on-a-chip device comprising two PDMS layers with microchannels separated by a nanoporous membrane with the purpose to analyze the effects of liquid plug flow on human epithelial cells [99]. Using a similar microdevice, this biomimetic microsystem was used to mimic the structural alveolar-capillary interface of the lung, and the physiological pulmonary respiratory movement via compression and expansion of the system [115]. In this system, human alveolar epithelial cells and pulmonary capillary endothelial cells were cultured on the opposite side of the flexible PDMS membrane that was used to reproduce physiological breathing. Two microchannels on both sides of the PDMS membrane acted as vacuum chambers to mimic breathing movements. The authors found that uptake of nanoparticles by the epithelial cells and endothelial cells increased and that the transport of nanoparticles into the underlying vascular channels was stimulated by the mechanicals constrains of air being pulled in and out of the device. The authors also introduced bacteria and inflammatory cytokines into the air and human bloodborne immune cells into the vascular channels in order to mimic the innate cellular response to pulmonary infection of bacterial origin. Another lung disease model was devised in order to mimic the development and progression of human pulmonary edema induced by the cancer drug interleukin-2 (IL-2) using the same lung-on-a-chip device [116]. Similar to previous chips, air was introduced into the upper alveolar channels and liquid was introduced into the lower vascular channels. The results demonstrated that IL-2 leaked into the alveolar channels increasingly under the cyclic mechanical strains and drug DSK2193874 could inhibit the leakage induced by IL-2. This human disease model-on-a-chip could bring research one step closer to predicting the efficacy of a new drug for pulmonary edema.

13.4.1.2 Gastrointestines on a Chip

The gut is one of the essential components for the maintainance of immune responses and for the natural development of the human body. It is necessary to construct *in vitro* cell-based gut models that aid in the study of the structural, mechanical, absorption functional, drug transportation and pathophysiological properties of the living gut [117]. Several models were developed using human intestinal tumor-derived Caco-2 cells to simulate the intestinal structure and function including Transwell filter inserts and miniaturized microfluidic devices [102, 118–120]. In other work, the researchers developed a new approach to rebuild the normal 3D microarchitecture of the intestinal lining *in vitro* via culturing intestinal epithelial cells on hydrogel substrates that were microfabricated to simulate the size, shape and density of human intestinal villi [121]. However, *these in vitro* models are not able to recapitulate the mechanical microenvironment of the gut, including peristaltic movements and intraluminal fluid flow of human living intestine that is crucial for normal organ physiology and pathology.

To overcome these limitations, Kim et al. demonstrated a functional human gut-on-a-chip microfluidic device composed of two microchannels separated by a porous flexible, clear PDMS membrane pre-coated with ECM and lined with Caco-2 cells [103] (Fig. 13.3). This model adopted fluid flow conditions at a low shear

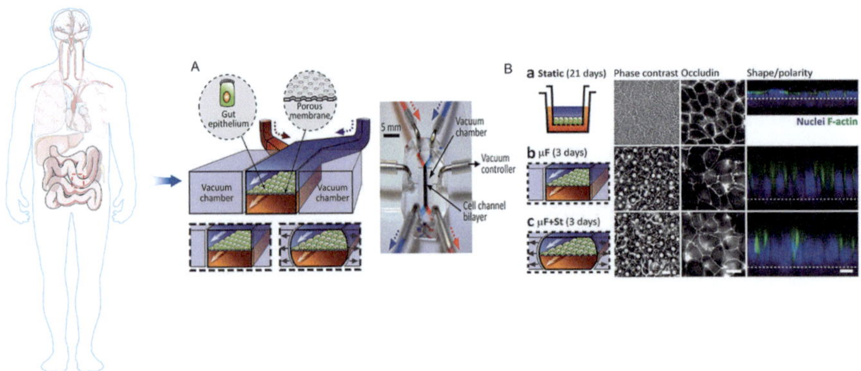

Fig. 13.3 Human gut-on-a-chip. (**A**) Scheme of a gut-on-a-chip device showing the flexible porous ECM-coated membrane lined by gut epithelial cells cross horizontally through the middle of the central microchannel, and full height vacuum chambers on both sides. (**B**) Morphology of the Caco-2 epithelial cells cultured in the static Transwell system for 21 days (*a*) versus in the gut-on-a-chip with microfluidic flow (30 μL h^{-1}; μF) without (*b*) or with (*c*) application of cyclic mechanical strain (10%; 0.15 Hz; μF + St) for 3 days. A scheme (left) shows the system layout; fluorescence views (center) show the distribution of the tight junction protein, occludin, in the epithelial monolayers; and the confocal fluorescence views (right) show a vertical cross section of the epithelium highlighting cell shape and polarity (nuclei in blue and F-actin in green). The regular array of small white circles in (*b*) and (*c*) are pores visible beneath the epithelial monolayer; the dashed white line indicates the top of the anchoring substrate (bar, 20 μm). Reproduced from [103] with permission

stress (0.002 $N{\cdot}m^{-2}$) and cyclic strain (10%; 0.15 Hz) simultaneously to mimic the intraluminal fluid flow and physiological peristaltic motions in living intestine. Caco-2 cells formed confluent polygonal epithelial monolayers with tight junction protein expression and exhibited differentiated intestinal barrier functions in the microfluidic device after only 3 days culture, which was less than that required on Transwell culture system (>21 days). The flow velocity also influenced the differentiation and polarization of epithelial cells. More interestingly, this study demonstrated for the first time that the planar columnar epithelium spontaneously deformed to form undulations and folds when Caco-2 cells were maintained in the device with the flow and cyclic strain for extended periods of time. These folds exhibited the morphology of normal intestinal villi formed by polarized columnar epithelium with basal nuclei and separated by crypts. In addition, the authors tried to culture the normal intestinal microbial flora, Lactobacillus rhamnosus GG, on the apical surface of differentiated intestinal epithelial monolayers under continuous flow and cyclic strain context. The results showed that normal intestinal microbes can be co-cultured for over 1 week on the apical surface of the cultured epithelium. Based on this gut-on-a-chip device, the same group further explored the structure and function of epithelium using the tumor-derived Caco-2 cells in detail [104]. The cells formed undulating structures containing basal proliferative crypts and the four types of differentiated epithelial cells (absorptive, mucus-secretory, enteroendocrine and Paneth) of normal intestinal villi. Furthermore, the intestinal models on chips behaved similarly to normal intestinal physiology, such as more efficient glucose reuptake, higher cytochrome P450-3A4 isoform activity and mucus production than previous static culture systems. Thus, this human gut-on-a-chip may play a critical role in studying the mechanical regulation of intestinal function and host-microbe symbiosis and evolution.

The stomach is another important organ in the digestive system with specific chemical and physical microenvironments. Gastric mucus serves in protecting the epithelial cells of the stomach wall from injures by the acidic digestive juices in the gastric lumen. Li et al. reported an *in vitro* microfluidic device that replicated a dynamic stomach acid barrier [122]. This study used native mucins obtained from pig gastric mucus and perfused continuously mucus liquid into the epithelial lumen. Several models of the mass transport were constructed to investigate the effects of H^+-mucin interaction on the diffusion of H^+ and acid penetration was monitored using the pH sensitive dye Oregon Green by live fluorescence microscopy. This work indicated that a continuously secreted mucin layer can hinder acid diffusion. This stomach-on-a-chip holds promise for the study of the barrier functions provided by the mucus layer and the interaction of the mucus layer and drugs.

13.4.1.3 Liver on a Chip

The liver is a major organ with multiple functions for protein synthesis, detoxification of various substrates, digestion and metabolic activities in the human body. Multiple liver-on-a-chip devices were fabricated to construct liver microsystem due to their ability to mimic the complex *in vivo* microenvironment [123–125]. To mimic the functional unit of the liver, Lee et al. developed a biologically inspired artificial liver sinusoid microdevice [126]. In this microfluidic chip, an endothelial-like barrier formed by microstructure served a mass transport function similar to the liver acinus. Primary rat and human hepatocytes were co-cultured in the configuration for 7 days without ECM coating. This model has also been applied to test the hepatotoxicity of diclofenac. With the advances of microfabrication technology, well-organized liver microtissues were created that simulated both the structural and physiological functions of the living liver in contrast to the conventional approaches using random co-culture methods. Micropatterned substrates can offer a suitable microenvironment to explore the cell-interactions in co-culture systems [127–130]. Although 2D patterning is convenient to control the spatial position of hepatocytes and other supporting cells, fresh primary hepatocytes cultured on 2D substrates rapidly, lose their capacity to proliferate, to form differentiated structures and their liver-specific functions compared to 3D culture microenvironments [131]. 3D culture systems can promote the hepatocytes' functions and maintain their differentiated properties for extended times *in vitro*, as well as mimic *in vivo* structural features including lobule and tubular architectures as closely as possible [132, 133]. Currently, several methods have been developed to construct 3D liver microtissues, including a spinner culture system, non-adhesive surfaces, pellet culture models, cell sheets and a hanging drop [134], but the microfluidic-based procedures may be more effective as they offer a convenient and straightforward platform in order to form uniformly sized and shaped 3D structures [135, 136].

Continuous perfusion is of great importance in 3D cell culture systems in order to maintain the long term function and viability of hepatocytes. Microfluidic technology enables the precise control mechanism for perfusing medium with nutrients and controlling its chemical composition not possible using traditional culture system. The fluid flow promoted and maintained the 3D tissue-like structure and cell-specific function of primary human hepatocytes and stem cell differentiation into hepatocytes [137–139]. Dash et al. reported that hepatocytes exhibited an enhanced capacity of metabolism of specific cytochrome P450 enzymes in a flow system compared with non-flow conditions [140]. Trietsch et al. developed a stratified 3D cell culture model incorporated in a microtiter plate format. The hepatocytes and fibroblasts were co-cultured in the way of mixed or side-by-side format to evaluate the toxicity of rifampicin. Furthermore, this device was used to study the invasion and aggregation of breast cancer cells [141]. To mimic the functional unit of the liver, a microfluidic culture device was created to rebuild the microscopic structure of the hepatic cords [142]. The asymmetric tip of the device with two separate compartments can house two cells side-by-side and the aligned hepatocytes can self-organize and form bile canaliculi along the cord-like microscale structure. Recently,

Esch et al. fabricated a low-cost cell culture device to culture 3D liver microtissues containing multiple liver cell types including primary hepatocytes, stellate cells, fibroblasts and Kupffer cells under fluid flow condition [143]. These 3D liver microtissues had good enzymatic activity and responded to bacterial lipoprotein. All these 3D liver microsystems represent an important step towards adoption of organ-on-a-chip technology for disease modeling and drug development.

13.4.1.4 Heart on a Chip

Microengineering cardiac tissue can potentially drive the development of efficacy and toxicity of drugs via reproducing crucial cardiac physiological features. Currently, several heart-on-chip systems have been explored to mimic physical, mechanical and biological functions of living heart which are not realized using conventional 2D culture systems [144–149]. Parker et al. designed a microfluidic device to culture rat cardiomyocytes using fibronectin-coated flexible elastomeric cantilevers cultured to form anisotropic muscular tissue [150]. The average systolic stress and diastolic stress were within the stress range published previously using isolated muscle strips [151]. These muscular thin films exhibited a chronotropic effect as the concentration of epinephrine increasing by dose-response curve experiments. This platform was further improved by combining a fluid flow control system and the platinum electrodes to collect accurate functional contractile response to a cardiac drug isoproterenol (β-adrenergic agonist) [152]. This high throughput heart-on-a-chip device is very useful for collecting large quantities of high quality functional data, testing cardiac drug and integration with other organs. Serena et al. reported a microfluidic device based on micropatterning techniques which can selectively treat areas of the cell array and perform multi-parametric assays [153].

Microfluidic technology can be integrated with stem cells to generate human derived cardiac tissues. The cardiac bodies derived from iPSCs were captured in niches along a perfusion microchannel in a microfluidic device. The cardiomyocyte clusters exhibited well-developed sarcomeres and cardiac protein markers (Nkx2.5 and Cx43) as well as *in vivo*-like cardiac function [154, 155]. In addition, a non-invasive recording method was introduced in this study to calculate the beating frequency of cardiac bodies using an ordinary microscope. These 3D cardiac bodies responded to different drugs rapidly in a similar manner to living heart tissue in a body. Other researchers explored different microfluidic approaches to form cell-laden hydrogels in order to evaluate the functionality of the cardiac tissue constructs. Aung et al. produced 3D perfused cardiomyoctes-laden hydrogels within a microfluidic chip and monitored the contractile stress of cardiac tissues in real-time and *in situ* measurements [156]. A cardiac microphysiological system was developed to recapitulate a minimal human cardiac microtissue in a central microchannel of the microfluidic device. This biomimetic system can drive the self-organization of human iPSC-derived cardiomyocytes into aligned 3D heart organoids which presented spontaneous beating at physiological level [157, 158]. The heart microtissues were also treated with multiple drugs at different concentrations. The results demonstrated that half maximal inhibitory concentration (IC50) and half maximal effective concentration (EC50) values were more consistent with the data on tissue scale references compared to studies at cellular level.

13.4.1.5 Blood-Brain-Barrier on Chip

The blood-brain-barrier (BBB) is a selective yet dynamic barrier between the central nervous tissues and the circulatory system, which is formed by microvascular ECs, pericytes and the perivascular end-feet of astrocytes. The dysfunction of this barrier is associated with brain tumors, Alzheimer's and neurodegenerative diseases. Reproducing the physiological characteristics and functional responses of the BBB in a reliable model will greatly promote the development of novel therapeutics for central nervous system diseases [199, 200]. To mimic the *in vivo* BBB, a microfluidic BBB (μBBB) was created by lining a fibronectin-coated polycarbonate membrane with brain microvascular endothelial cells and astrocytes on the opposite side of the membrane [159]. This BBB exhibited a well-developed biological function, including higher trans-epithelial electrical resistance (TEER) across the barrier, and being more impermeable to large molecules. A more complex microfluidic device of the neurovascular unit was created by co-culturing rat brain microvascular ECs and a mixture of three different cell types derived from rat brain (astrocytes, neurons and microglia) on opposite sides of a porous membrane [160]. The endothelial cells formed a good barrier and the neuronal cells fired inhibitory as well as excitatory potentials after having been cultured for 10 days. Moreover, a tissue-mimetic neuroinflammation model was also established on this chip by stimulating the ECs with inflammatory factor TNF-α for 6 h, leading to the significant activation of resident microglia and astrocytes on the neural side.

13.4.1.6 Multiple Organs on a Chip

The advances of single organ-on-a-chip technology highlight the importance of interaction between different cell types within the same functional unit of the organ. More complex microfluidic chips integrating tissue-tissue interfaces and compartmentalized microchambers connected with microchannels (e.g., blood-brain-barrier, lung alveolar-capillary barrier) make it possible to create multiple organs on a chip. These multi-organs on a chip are designed to maintain multiple cell types in different culture compartments in one device for long-term culture and integrate them into one system in order to mimic organ-organ interaction. In the past decade, several multiorgans-on-chip devices have been developed for potential use in disease modeling and drug toxicity screening applications. To mimic physiologically relevant organs realistically *in vivo*, some key aspects need to be considered to design the chip, such as the size of the organs, interactions between different organs and the organ volume-medium ratio (scaling of organ sizes and vascular flow *in vivo*). Wikswo et al. discussed scaling arguments about microengineering multi-organs on a chip to guide the design of a universal cell-culture medium without red blood cells [161]. It is mentioned that the prototype drug metabolized by the liver component could have therapeutic effects. Thus, the liver should be placed upstream of the drug target organs in designing the multi-organs on a chip.

The field of multiorgans-on-chips is explored and investigated for different purposes [162]. To study tissue cross-effects, liver and intestinal slices isolated from the same rat were cultured in one microdevice [163]. The liver and intestinal slices remained functional under flow conditions, indicating the potential of this chip for the study of inter-organ interactions. Wagner et al. cultured human primary hepatocytes and skin biopsies simultaneously in one microfluidic device for up to 28 days [164]. These studies demonstrated that the primary tissues from living body can maintain their functions *in vitro* under dynamic culture condition. Moreover, a 3D two-organs-on-chip device was created to support the survival of differentiated neurospheres and liver spheroids in a combined media circuit over 14 days [165]. In the study it was suggested that the two combined tissues were more sensitive to the 2, 5-hexanedione compared to respective single-tissue cultures, maybe due to the tissue-tissue interactions. As is well established, most orally adminstred drugs are absorbed and metabolised in the small intestine or the liver and excreted by the kidney. However, it is impossible to mimic these processes *in vitro* using conventional cell-based models.

A typical example of a three organs-on-chip device was designed with a 3D hydrogel culture in which three cell lines were cultured to assess metabolism-dependent cytotoxicity of anti-cancer drugs [166] (Fig. 13.4). In this microsystem, cancer cells, liver cells and myeloblasts were cultured in separate chambers representing cancerous tissue, liver and bone marrow. Tegafur, an oral prodrug of 5-fluorouracil, was introduced into this system to test the cytotoxicity to the three cell lines. Compared with conventional 96-well microtiter plates, the micro cell culture system enabled to reproduce the metabolism of Tegafur to 5-fluorouracil in the liver and the resulting death of cells caused by 5-fluorouracil, while cells in the 96-well microtiter plate did not display the same results. Esch et al. designed a microscale body-on-a-chip system containing gastrointestinal tract, liver and other tissues to mimic the oral uptake of nanoparticles. The nanoparticles were transported across the GI epithelium and interacted with liver cells resulting in the damage of the latter [167].

To improve the functions of multiorgans-on-chip, a four-organ-chip microphysiological system was proposed to maintain the functionality of four organs containing human intestine, liver, skin and kidney over 28 days [168]. Considering the size ratio of tissues on chip compared to human living organs, the intestine and skin tissues on chip are at a size 100,000-fold smaller than their counterpart organs. 3D liver microtissues, equivalent to ten liver lobules, were found to mimick liver function. Human proximal tubule cell lineage RPTEC/TERT-1 was used to form the proximal tubule barrier on a polymeric membrane to mimic the metabolite excretion. A peristaltic micropump integrated in this chip enables pulsed medium flow interconnecting the four organs through microchannels. This device considered

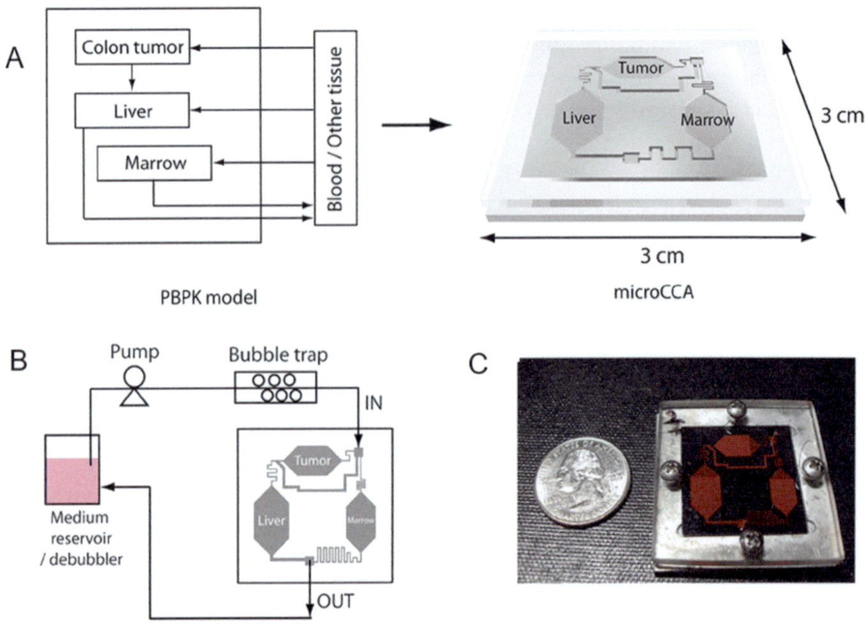

Fig. 13.4 Design of multi-ogans on a chip. (**a**) Liver-marrow-cancer on a chip. Hepatoma cells (HepG2/C3A), Myeblasts (Kasumi-1) and colon cancer cells (HCT-116) are embedded in a 3D hydrogel and cultured in separate chambers representing the liver, marrow and tumor respectively. Medium flows through the cell culture chambers via connecting channels with a pump mimicking blood flow. (**b**) Schematic diagram of operation setup of a single chip with medium recirculation. Medium is withdrawn from a reservoir, after circulating through the channels and chambers in a chip, medium goes back to the reservoir for recirculation. (**c**) A picture of an assembled Liver-marrow-cancer on a chip with red dye for visualization of channels and chambers. Reproduced from [166] with permission

the physiological fluid-to-tissue ratio which is important to accurately simulate the drug metabolism. This microsystem exhibited remarkably robust homeostasis and functionality of the four organ equivalents, presenting itself as a good platform for *in vitro* adsorption, distribution, metabolism, elimination (ADME) modeling. A new type of four organs-on-chip device was fabricated to culture four human cell types representing liver, lung, kidney and the adipose tissue, respectively [169]. A common culture medium is the key factor to be considered in the development of multiple cell type culture systems that should satisfy all types of cells. This is also the critical question for designing multi-organs on a chip in the future.

13.4.2 Integrated Analysis System

One of the most powerful opportunities for microfluidic organs-on-chip for biomedical applications is the capacity of integrated analysis system to monitor and control organs in real-time. Existing automated techniques enable us to monitor cellular proliferation, morphology and molecular changes in multiwell plates. However, it is difficult to monitor the complex multiple parameters on organ microsystems including gradients of chemical factors and of oxygen, fluid flow shear, mechanical cues, matrix composition and metabolites. Currently, several types of microsensors for real-time monitoring of organ conditions have provided measurements of basic cell culture parameters including oxygen, pH, temperature, and molecular composition (e.g., glucose and lactate). More specialized sensor measurements are widely used to monitor the trans-epithelial electrical resistance (TEER) which usually exists in biological barriers, such as the blood-brain barrier, intestinal epithelial-capillary barrier and the lung alveolar-capillary barrier. Recent efforts indicate that TEER measurements in organs-on-chip systems present specific challenges and are highly sensitive to slight changes in the cell barrier membrane [170]. Hence, the automated and precise measurements of TEER in a complex organ microsystem would increase reliability and offer key information on the organ condition and responses to different drugs.

13.5 Proof-of-Concept Applications of Organs-on-Chip

13.5.1 Disease Modeling

Disease modeling is one of the main applications of organs-on-chip technology, ranging from genetic diseases to infectious, cancerous and degenerative diseases, and is helpful for understanding the disease etiology, diagnostic strategies and effective therapies. Several proof-of-concept disease models have been reported to date.

13.5.1.1 Inflammatory-Related Diseases

Inflammation reactions are closely related to many diseases and can lead to serious pathological symptoms. Pneumonia is one of most common lung diseases, characterized by its complexity, acute onset and difficulty to control it. Establishing an *in vitro* model to study the disease mechanism and to be able to screen for efficient drugs is an urgent requirement in order to address this challenging disease. The breathing lung-on-a-chip, as mentioned above, recapitulated the epithelium-endothelium interface under fluid flow and cyclic mechanical strain conditions to mimic the functional unit of the lung. Inflammatory models were developed in this system by administrating immune activator TNF-α or bacteria which increased expression of surface ICAM-1 of endothelial cells and recruitment of human neutrophils [114]. Based on the lung-on-a-chip system, the same research group further mimicked a pulmonary edema model by the cancer drug interleukin-2 (IL-2) which can cause pulmonary vascular permeability and lung edema [115]. This lung disease model also recapitulated that the mechanical strain caused by the breathing motions promote increased vascular leakage and pulmonary edema induced by IL-2 but not the circulating immune cells. An organ-level lung small airway-on-a-chip model was created to study human chronic obstructive pulmonary disease (COPD) and drug response using patient and healthy lung microvascular endothelial cells and airway epithelial cells [171]. This model effectively rebuilt many properties of the structures and functions of human lung bronchioles and maintained them for weeks *in vitro* which is crucial for studying chronic disease *in vitro* including cell types and cilia structure. More importantly, interleukin-13 (IL-13) stimulated the epithelium leading to goblet cell hyperplasia, cytokine hypersecretion and ciliary functional decline of asthmatics. Using the robust in vitro lung-on-a-chip model, it is possible to screen for synergistic effects of lung epithelium and endothelium on cytokine secretion, discover new biomarkers of disease exacerbation and evaluate responses to anti-inflammatory drugs.

In addition, neuroinflammation was studied by a more complex microfluidic chip supporting the neurovascular unit containing endothelial barrier on one side of a porous membrane and three various brain cell types (astrocytes, neurons and microglia) on the other. Inflammatory factor, TNF-α, stimulated the vascular endothelium which activated adjacent microglia and astrocytes. This process is similar to what happens *in vivo* in situations such as neuroinfectious diseases [160].

13.5.1.2 Brain diseases on Chip

Neurodegenerative diseases such as Alzheimer's disease (AD) seriously threaten human health. In order to better understand disease etiology and to develop efficient treatment strategies, it is indispensable to seek a suitable *in vitro* model for exploring the mechanisms of brain diseases. Lee et al. presented a microfluidic device to create 3D neurospheroids which more closely simulate the *in vivo* brain microenvironment (Fig. 13.5) [172]. The uniform neurospheroids formed in concave microwell arrays are exposed to all directions and are able to interact. By introducing an interstitial level of flow using an osmotic micropump system, this model system demonstrated that the flow could promote neurospheroid growth, improved neural network formation compared with cultures under static conditions, as well as enabled long-term cultures without the need for peripheral devices. To test the potential of this microfluidic microsystem as an *in vitro* brain disease model, the authors added the amyloid-β peptide, which is generally viewed as the major contributor in AD, on 3D neurospheroids under interstitial flow conditions. The study indicated that the neurotoxic effects of amyloid-β led to a decline in neural cell viability and destroyed neural networks via inducing synaptic dysfunction. Taken together, this biomimetic brain-on-a-chip device recreates a 3D cytoarchitecture and interstitial flow of brain *in vivo*, as well as mimics the pathological changes of AD *in vitro*. This platform has great potential for investigating neurodegenerative disease pathology and treatment strategies as well as for drug testing applications.

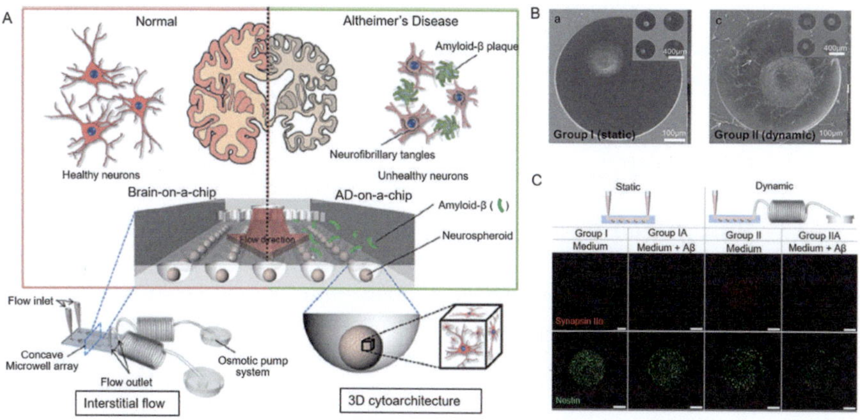

Fig. 13.5 Creating Alzheimer's disease modeling on chip. (**a**) Schematic diagram of a brain-on-a-chip device. A concave microwell array is utilized for homogeneous neurospheroid formation and a continuous flow mimicking interstitial flow is applied via an osmotic pump system. (**b**) Dynamic culture significantly promotes the growth and function of neurosopheroids. (**c**) Amyloid-β treatment via an osmotic micropump significantly reduced the viability of neurospheroids and caused a significantly more pronounced destruction of neural networks, compared to the amyloid-β treatment under static conditions. Reproduced from [172] with permission

13.5.1.3 Cancers on Chip

Cancers are a large family of diseases that seriously threaten human health and life. The development of cancers involves abnormal cell growth and invasion of other important organs. It is very helpful to establish *in vitro* models to study the biological behavior of cancer cells and anticancer drug effects. Currently, numerous groups have developed cancer-on-a-chip models by mimicking tumor microenvironments consisting of complex cell-cell and cell-matrix interactions, chemokine/cytokine gradients and biophysical cues [173–175]. Xu et al. fabricated a bilayer PDMS microfluidic chip to mimic the glioblastoma invasion into ECM under different concentrations of oxygen [176]. This work investigated the role of hypoxia and EMT in glioblastoma and hypoxia promoted the proliferation of the cancer cells and EMT-associated protein expression, and enhanced cell migration. The mechanism linking EMT and cancer cell behavior could be related to the Hypoxia-Inducible Factor 1α or 2α (HIF1α or HIF2α), indicating that developing inhibitors of HIFs may be a novel therapeutic drug. Li et al. demonstrated a high throughput microdevice containing tunable cell micro-niches, which performs flow-based analysis of large cell populations to evaluate various responses of lung adenocarcinoma cells to different ECM proteins and soluble factors [177]. This study indicated that tumor-cell growth is related to TGF-β and TGF-βR2 inhibitor drugs in a 3D matrix but not in a 2D culture. Angiogenesis of tumors is a key step of tumor metastasis towards distant target organs. Several works were performed to study the intravasation and extravasation of tumor cells using microfluidic devices [178, 179]. Other studies have utilized similar microfluidic technology to investigate the molecular mechanisms of cancer cell-immune cell crosstalk [180].

13.5.2 Drug Testing

13.5.2.1 Efficacy and Toxicity Testing

One challenge of drug development is the poor efficacy and unexpected toxicity in clinical trials caused by an absence of predicted therapeutic effects. The main reason for undesired outcomes is that existing approaches fail to accurately predict *in vivo* drug efficacy before clinical application. Human organ-on-a-chip systems that model human physiological and pathological functional units of living organs provide a promising tool to address the limitations of existing methods. They allow to reconstruct and pharmacologically modulate key aetiologies and clinical relevant responses at various levels of biological complexity and to test unanticipated off-target toxicity. As described above, a heart-on-a-chip composed of 20 rat cardiomyocyte thin films was utilized to evaluate the inotropic effects of the β-adrenergic agonist isoproterenol which are similar to those previously determined in rats [152]. A recent study reported a micro-engineered 3D model of EMT during cancer progression for testing drug efficacy [181]. This model demonstrated EMT-induced tumor dispersion and phenotypic changes by culturing lung tumor spheroids in a matrix gel close to microchannels inhabited with endothelial cells. In addition, 12 drugs, including approved drugs, as well as prospective drugs that are in the early discovery pipeline, were perfused into the vascular microchannels to evaluate the potential of this model as a drug screening platform. The ability of these drugs to inhibit EMT was tested by direct visualization of the cancer spheroids. The results showed that efficacious concentrations derived from the cancer-on-a-chip were higher than that from the 2D system, but are also closer to the range of effective drug concentrations determined in clinical trials. Several micro-engineered models of breast cancer and multiple organ models of uterine cancer, bone marrow and liver have also demonstrated a similar difference between effective drug concentrations determined by organs-on-chip platforms and those by conventional 2D culture systems [182, 183]. To investigate the communication between different brain regions, organotypic brain slices from rat hippocampus and entorhinal cortex were cultured in compartments interconnected by microchannels [184]. A glutamate receptor antagonist (kynurenic acid) was introduced to one microchannel in order to selectively inhibit the spontaneous electrical excitation of the treated brain tissue but not on the other brain slice. This design allows for the selective pharmacological administration of only one tissue and evaluation of its effects across the synaptic connection.

Recently, Qin et al. developed a novel *in vivo*-like 3D blood-brain barrier model that replicates the complex multicellular architecture, functions and mechanical properties of the BBB *in vivo* [185] (Fig. 13.6). The BBB model encompassed essential components, including primary brain microvascular endothelial cells,

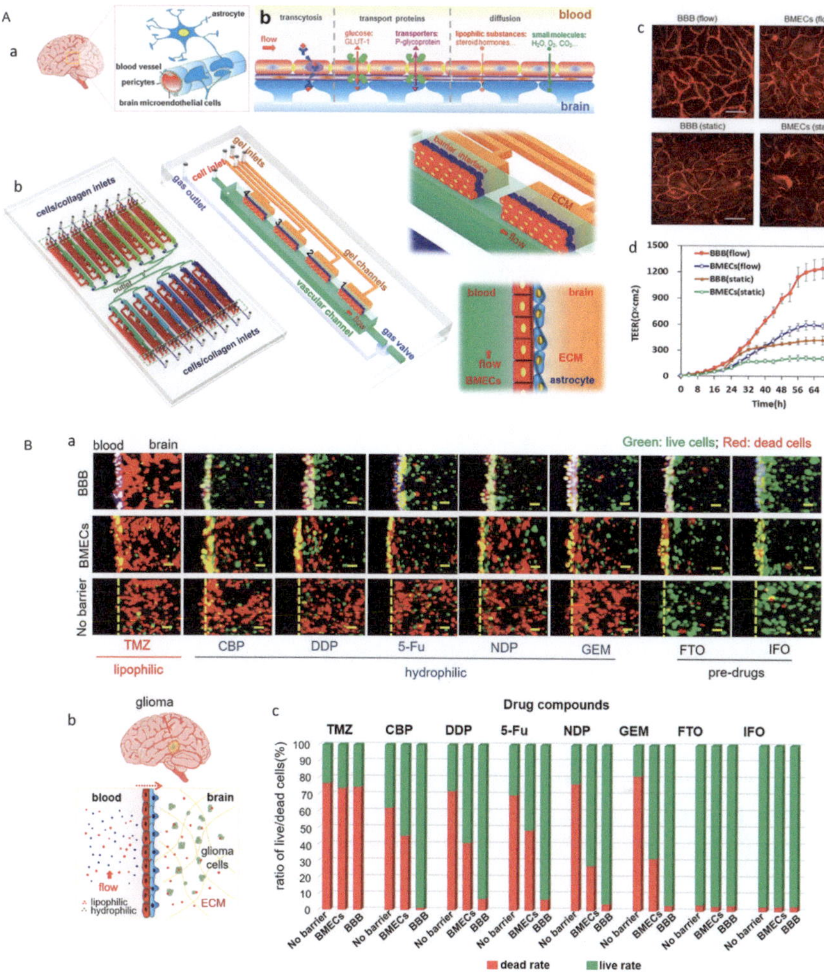

Fig. 13.6 Engineering 3D dynamic blood-brain-barrier (BBB) in brain tumor microenvironment on chip. (**A**-*a*) Cellular constituents and structural features of the BBB *in vivo*. (**A**-*b*) The details of the BBB microdevice are represented. It is composed of 16 independent units with fluidic channels, ECM gel channels, gas valve. The BMECs, astrocytes and 3D ECM gel formed the functional BBB under fluidic flow conditions. (**A**-*c*) Expression of adhesive protein, VE-cadherin, in integral BBB and BMECs under fluidic flow or static conditions. (**A**-*d*) TEER values in BBB and BMECs under fluidic flow and static conditions. (**B**-*a*) Cytotoxicity of eight chemotherapeutic agents to glioma cells (U87 cell line) in various barrier groups. (**B**-*b*) Schematic illustration of diffusion of lipophilic and hydrophilic drug compounds across the BBB *in vivo*. (**B**-*c*) Quantitative data of live/dead ratio of U87 cells induced by the different drugs introduced into the vascular channels of the BBB. Reproduced from [185] with permission

astrocytes and ECM which orchestrate to form a more stringent structure of a BBB (Fig. 13.6). In this microsystem, the TEER value in the BBB interface under dynamic culture condition is around 1300 $\Omega \times cm^2$, a value that far exceeds that reported in Transwell-based BBB models in a static context [186, 187]. Moreover, this BBB model was used to understand the interplay of the BBB and exogenous cancer cells in the course of brain metastasis in real-time. As is well-known, efficiently delivering a drug to the brain is still a challenge at present because of the selective permeability of the BBB to different exotic molecules. In this work, the *in vitro* BBB model exhibited a robust and realistic experimental result with regard to drug response of brain tumors including primary and secondary cancers and the high throughput design is very useful to screen multiple drugs simultaneously, which is not possible in other BBB models *in vitro* and in animal experiments. This physiological BBB microsystem provides a versatile and valuable platform for neuroscientific research, drug testing and pharmaceutical development.

Unanticipated adverse drug effects are another very common cause of the costly withdrawal of marketed drugs and clinical trial failures. Animal experiments often fail to reveal important toxic effects in humans and can lead to unnecessary rejection of drug candidates due to the genetic background and biological discrepancy and animal-specific pathways of toxicity. Organ-on-a-chip devices used in liver toxicity have been reported recently to analyze the metabolism of hepatotoxicity. For example, a microfluidic device was developed to culture human hepatocytes in order to monitor metabolic responses of hepatocytes to flutamide (a type of anticancer drug) and the hepatotoxic properties of flutamide and its active metabolite hydroxyflutamide. This study showed the metabolic signatures of toxic responses and described metabolic pathways induced by the drug and its metabolites in hepatotoxicity. The human liver-on-a-chip model leveraged in this work was influential in reducing biological noise inherent to *in vivo* metabolomics models, and indicating a potential source of hepatotoxicity-specific biomarkers. The secondary toxic effects of drug metabolites generated in the liver on other organs were also demonstrated by two organ liver-kidney organ-on-a-chip devices [188]. This study mimicked the systemic interplay between the two organs and recapitulated nephrotoxic responses to the ifosfamide metabolites produced by hepatocytes. More recently, drug-induced cardiotoxicity was also tested using an organ-on-a-chip platform by rebuilding and monitoring the contractile functions of heart muscle by means of quantitative analysis [189, 190].

13.5.2.2 Pharmacokinetic and Pharmacodynamic Studies

Pharmacokinetic and pharmacodynamic (PK/PD) models have been broadly used to analyze and predict the whole body response to drugs in a time-dependent manner by featuring the mechanistic basis of multi-organ interactions. Multiorgans-on-chips were developed to analyze the bioaccumulation, distribution, and toxicity of selected chemical compounds [191]. The toxicant naphthalene was converted by hepatocytes into its reactive metabolites which depleted the intracellular glutathione of lung cells as soon as circulating to the lung tissue chamber. Adipocytes differentiated from

3 T3-L1 in the fat tissue chamber moderated the glutathione depletion induced by naphthalene, but preferentially accumulated hydrophobic compounds. This study is the first model of adsorption, distribution, metabolism, elimination and toxicity (ADMET) that performed all of these functions on the same device. Over the past decade, many researchers have explored the utilization of organs-on-chips to investigate drug ADMET features, to support PK/PD modeling, and to evaluate drug efficacy [192–194]. Li et al. developed a new and multilayer organs-on-chip device to assess drug metabolism and its active metabolite drug efficacy and cytotoxicity in four organ-specific cells simultaneously representing the liver, breast cancer, lung tumor, and normal gastric cells [195]. In this study, the prodrug capecitabine (CAP) was first metabolized in a top liver tissue chamber with hepatocytes (HepG2) and its intermediate metabolites, 5′-deoxy-5-fluorocytidine (DFUR), was further metabolized into 5-fluorouracil (5-FU) by targeting cancer tissue and normal tissue cells (Fig. 13.7). This work recapitulated that the CAP exhibited strong cytotoxicity on breast and lung cancer cells, but not in normal gastric cells.

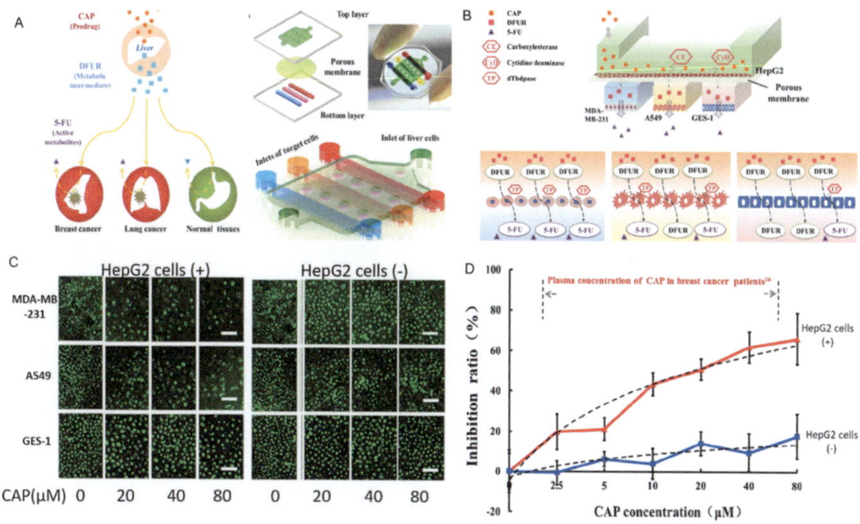

Fig. 13.7 Liver dependent drug metabolism on a multiorgan-on-chip. (**a**) Schematic diagram of organs-on-a-chip. The microfluidic chip consists of two layers separated by a porous membrane. The top layer was seeded with HepG2 cells representing the liver and the three channels on the bottom layer were cultured with breast cancer cells (MDA-MB231), lung cancer cells (A549) and normal tissue cells (GES-1). (**b**) Schematic illustration of drug metabolism on chip. CAP was introduced from the top layer and presented to HepG2 cells. CAP was metabolized by carboxylesterase and cytidine deaminase within HepG2 and transformed into DFUR. The DFUR were subsequently presented to the target cells on the bottom layer and transformed into cytoxic 5-FU by dThdpase expressed by target cells. Since the target cell lines express different levels of dThdpase, the cytotoxic effects vary. (**c**) Cytoxic effects of CAP on each cell type with or without HepG2. (**d**) Dose-dependent effects of CAP on the inhibition of breast cancer on chip. Reproduced from [195] with permission

All these studies demonstrate an opportunity for organ-on-a-chip devices in the field of drug development to simulate and predict the key physiological responses involved in drug metabolism, bioactivity, efficacy and toxicity, and increase likelihood of success in clinical trials.

13.5.3 Host-Microbe Interaction

The gut microbiome can deeply impact many aspects of human bio-behavior, such as mental activities, stress responses via brain-gut communication [196]. The balance of intestinal microbial diversity is crucial to maintain human health status and alteration of intestinal microbiota is related to many acute and chronic diseases including inflammatory bowel disease, diabetes, obesity and cardiovascular diseases [197].The gut microbial community is a dynamic ecosystem that can be influenced by many factors, such as food type, living condition and host genetics. Therefore, establishment of a stable host-microbe ecosystem *in vitro* is crucial for understanding the human intestinal diseases, regulating nutrient and drug absorption. However, existing 2D, static culture methods fail to rebuild functional intestinal structure and establish a stable symbiosis between host intestinal epithelium and a certain population of bacterial cells for extended time periods because bacterial overgrowth occurs rapidly within 1 day.

Alternatively, microfluidic devices can provide a desirable culture platform to co-culture the host cells and bacteria under dynamic conditions. The human gut-on-a-chip was developed using intestinal epithelial cells that grew into 3D villi stably on optically clear, microporous PDMS membranes sandwiched by two PDMS layers with parallel hollow microchannels. In this microsystem, multiple commensal microbes were directly co-cultured with epithelial cells for more than 1 week. This study recapitulated the individual contributors containing the peristalsis-associated mechanical deformations, gut microbiome and inflammatory cells to intestinal bacterial overgrowth and inflammation [198]. Lack of epithelial deformation generated by peristalsis-like motion led to bacterial overgrowth similar to that observed in inflammatory intestinal diseases. Most importantly, this *in vitro* intestinal model replicated results from past human and animal studies which demonstrated that antibiotic and probiotic treatment can suppress villus injury caused by pathogenic bacteria. The intestinal epithelial cells were also stimulated by immune cells and lipopolysaccharide endotoxin to produce four pro-inflammatory cytokines (IL-6, −8, −1β,and TNF-α) that can trigger villus injury and compromise intestinal barrier function. In future studies, gut-on-a-chip devices can be used to investigate the host-microbe interaction in a physiological or pathological context.

13.6 Conclusion and Outlooks

Novel bioengineered organs-on-chips platforms can provide tight environmental control of tissues and organs along with the physiological transport and signaling as present in the human body. These approaches offer unique advantages in studying the intercellular communications, tissue microenvironment, disease modeling and drug testing beyond the existing methods. These novel strategies complement traditional cell culture and animal model systems by providing the biomimetic niche of native tissues/organs, including multicellular architectures, tissue-tissue interfaces and spatio-temporal control of biochemical and biophysical cues. In addition, the compatibility of organs-on-chips with existing analysis methods and integrated biosensors has enabled real-time readout of crucial physiological and pathological information, thus extending their utility for biomedical applications.

While the emerging organs-on-chip technology has advanced our understanding of the cellular behaviors within the context of organ-relevant functions, much work remains to be done to fully mimic organ functions and to move toward a human-on-a-chip model with greater relevance for disease studies and drug discovery. Among these challenges, the cell source is one of the key issues for building organs-on-chips analogous with organs in the human body. Human primary cells are ideal resources to model human tissues/organs. To meet both ends, human iPSCs have been explored to rebuild organs-on-chips. Patient-specific iPSCs provide a promising cell source to engineer the disease-on-a-chip for personalized medicine. In order to study organ-organ interactions in multiorgans-on-chips, the identification of an appropriate common medium is also of great concern because the different cell types need various nutrient components when these cell lineages are assembled. To address this challenge, human plasma or serum might be a desirable substitute as supplements of culture medium for multiple organ-specific cell lineages.

In addition, the choice of material is another major concern in fabricating and designing the organs-on-chips. At present, PDMS is the most widely used material due to its ease of use, high transparency, air permeability and biocompatible features. However, PDMS can absorb small hydrophobic drugs, which might cause underestimation of drug toxicity and efficacy, thus limiting its practical utility in drug discovery. Chemical modifications of PDMS surfaces and other alternative materials are required to solve this type of problem. Furthermore, more powerful analysis and monitoring systems specifically designed or modified for microfluidic chips are required to provide comprehensive information of cell behavior, metabolism and drug responses from the organs-on-chips. The available methods, such as PCR analysis, *in vivo* imaging, two-photon confocal microscopy, and mass spectrometry can be further integrated with organs-on-chips for real-time monitoring, thus boosting the development of organ-on-a-chip technology for obtaining biological information. It will be necessary to develop new biomimetic microsystem, novel materials and appropriate cell sources that support the scale-up of organs-on-chips.

Although the development of engineered organs-on-chips is still in its infancy, its potential to model diseases and predict human response under physiological and pathological conditions is tremendous. The ongoing progress of this technology to simulate various cues in organ relevant microenvironments and to combine with stem cell biology and traditional biological assays will accelerate the pace toward creating a more realistic human-on-a-chip model system that is accessible, practical and robust for end users. Collectively, future success will require the collaborative efforts from academic investigators, bioengineers, industry and regulatory agencies to advance this technology to market.

Acknowledgments This research was supported by the Strategic Priority Research Program of the Chinese Academy of Sciences (XDA16020900, XDPB0305), National Nature Science Foundation of China (No. 91543121, 31671038, 81573394, 81803492), National Key R&D Program of China (No. 2017YFB0405400), Key Program of the Chinese Academy of Sciences (KFZD-SW-213), Innovation Program of Science and Research from the DICP, CAS (DICP TMSR201601).

References

1. Haeberle S, Zengerle R (2007) Microfluidic platforms for lab-on-a-chip applications. Lab Chip 7:1094–1110
2. Chin LK et al (2016) Imaging live cells at high spatiotemporal resolution for lab-on-a-chip applications. Lab Chip 16:2014–2024
3. Young EW, Beebe DJ (2010) Fundamentals of microfluidic cell culture in controlled microenvironments. ChemSoc Rev 39:1036–1048
4. Tehranirokh M et al (2013) Microfluidic devices for cell cultivation and proliferation. Biomicrofluidics 7:51502
5. Young EW, Simmons CA (2010) Macro- and microscale fluid flow systems for endothelial cell biology. Lab Chip 10:143–160
6. Bhatia SN, Ingber DE (2014) Microfluidic organs-on-chips. Nat Biotechnol 32:760–772
7. Sung JH et al (2013) Microfabricated mammalian organ systems and their integration into models of whole animals and humans. Lab Chip 13:1201–1212
8. Tseng P et al (2014) Research highlights: microtechnologies for engineering the cellular environment. Lab Chip 14:1226–1229
9. Williamson A (2013) The future of the patient-specific body-on-a-chip. Lab Chip 13:3471–3480
10. Xia Y, Whitesides GM (1998) Soft lithography. Annu Rev Mater Sci 1998:153–184
11. Mcdonald JC, Whitesides GM (2002) Poly(dimethylsiloxane) as a material for fabricating microfluidic devices. Acc Chem Res 35:491–499
12. Effenhauser CS et al (1997) Integrated capillary electrophoresis on flexible silicone microdevices: analysis of DNA restriction fragments and detection of single DNA molecules on microchips. Anal Chem 69:3451–3457
13. Duffy DC et al (1998) Rapid prototyping of microfluidic systems in poly(dimethylsiloxane). Anal Chem 70:4974–4984
14. Mcdonald JC et al (2000) Fabrication of microfluidic Systems in Poly(dimethylsiloxane). Electrophoresis 21:27–40
15. Wu HK et al (2003) Fabrication of complex three- dimensional microchannel systems in PDMS. J Am Chem Soc 125:554–559
16. Zhang Q et al (2012) A microfluidic-base device for study of transendothelial invasion of tumor aggregates in realtiem. Lab Chip 12(16):2837–2842

17. Unger MA et al (2000) Monolithic microfabricated valves and pumps by multilayer soft lithography. Science 288:113–116
18. Balagadde FK et al (2005) Long-term monitoring of Bacteria undergoing programmed population control in a microchemostat. Science 309:137–140
19. Huang B (2007) Counting low-copy number proteins in a single cell. Science 315:81–84
20. Wen H et al (2015) A droplet microchip with substance exchange capability for the developmental study of C. *elegans*. Lab Chip 15(8):1905–1911
21. Gao X et al (2009) Microvalves actuated sandwich immunoassay on an integrated microfluidic system. Electrophoresis 30(14):2481–2487
22. Shi W et al (2010) Droplet microfluidics for characterizing the neurotoxin-induced responses in individual Caenorhabditiselegans. Lab Chip 10(21):2855–2863
23. Ni M et al (2009) Cell culture on MEMS platforms: a review. Int J Mol Sci 10(12):5411–5441
24. Ma LA et al (2010) A porous 3D cell culture micro device for cell migration study. Biomed Microdevices 12(4):753–760
25. Shi Y et al (2015) Hypoxia combined with spheroid culture improves cartilage specific function in chondrocytes. Integr Biol (Camb) 7(3):289–297
26. Gottwald E et al (2007) A chip-based platform for the in vitro generation of tissues in three-dimensional organization. Lab Chip 7(6):777–785
27. Toh YC et al (2009) A microfluidic 3D hepatocyte chip for drug toxicity testing. Lab Chip 9(14):2026–2035
28. Choi J et al (2011) Wnt5a-mediating neurogenesis of human adipose tissue-derived stem cells in a 3D microfluidic cell culture system. Biomaterials 32(29):7013–7022
29. Cate DM et al (2015) Recent developments in paper-based microfluidic devices. Anal Chem 87(1):19–41
30. Wang L et al (2015) Human induced pluripotent stem cell-derived beating cardiac tissues on paper. Lab Chip 15(22):4283–4290
31. Mosadegh B et al (2014) Three-dimensional paper-based model for cardiac ischemia. Adv Healthc Mater 3(7):1036–1043
32. Derda R et al (2011) Multizone paper platform for 3D cell cultures. PLoS One 6(5):e18940
33. Park HJ et al (2014) Paper-based bioactive scaffolds for stem cell-mediated bone tissue engineering. Biomaterials 35(37):9811–9823
34. Mosadegh B et al (2015) A paper-based invasion assay: assessing chemotaxis of cancer cells in gradients of oxygen. Biomaterials 52:262–271
35. Walker GM et al (2004) Microenvironment design consideration for cellular scale studies. Lab Chip 4(2):91–97
36. Chung BG et al (2011) Microfluidic fabrication of microengineered hydrogels and their application in tissue engineering. Lab Chip 12(1):45–59
37. Ota H et al (2011) Microfluidic experimental platform for producing size-controlled three-dimensional spheroids. Sensors Actuators A Phys 169(2):266–273
38. Ma J et al (2016) Patterning hypoxic multicellular spheroids in a 3D matrix-a promising method for anti-tumor drug screening. Biotechnology 11(SI):127–134
39. Hardelauf H et al (2011) Microarrays for the scalable production of metabolically relevant tumour spheroids: a tool for modulating chemosensitivity traits. Lab Chip 11:419–428
40. Ruppen J et al (2015) Towards personalized medicine: chemosensitivity assays of patient lung cancer cell spheroids in a perfused microfluidic platform. Lab Chip 15:3076–3085
41. Kim C et al (2012) On-chip anticancer drug test of regular tumor spheroids formed in microwells by a distributive microchannel network. Lab Chip 12:4135–4142
42. Tekin H et al (2010) Stimuli-responsive microwells for formation and retrieval of cell aggregates. Lab Chip 10:2411–2418
43. Karimi M et al (2016) Microfluidic systems for stem cell-based neural tissue engineering. Lab Chip 16:2551–2571
44. No DY et al (2015) 3D liver models on a microplatform: well-defined culture, engineering of liver tissue and liver-on-a-chip. Lab Chip 15:3822–3837
45. Lee J et al (2016) A 3D alcoholic liver disease model on a chip. Integr Biol 8:302–308

46. Kim C et al (2011) 3-dimensional cell culture for on-chip differentiation of stem cells in embryoid body. Lab Chip 11:874–882
47. Khademhosseini A, Nichol JW (2009) Modular tissue engineering: engineering biological tissues from the bottom up. Soft Matter 5(7):1312–1319
48. Chung BG et al (2012) Microfluidic fabrication of microengineered hydrogels and their application in tissue engineering. Lab Chip 12:45–59
49. Yamada M et al (2015) Cell-sized condensed collagen microparticles for preparing microengineered composite spheroids of primary hepatocytes. Lab Chip 15:3941–3951
50. Yu Y et al (2014) Flexible fabrication of biomimetic bamboo-like hybrid microfibers. Adv Mater 26(16):2494–2499
51. Yue Y et al (2016) Simple spinning of heterogeneous hollow microfiber on Chip. Adv Mater 28(31):6649–6655
52. Zhang X et al (2015) Flexible fabrication of shape-controlled collagen building blocks for self-assembly of 3D microtissues. Small 11(30):3666–3675
53. Goldbrunner RH et al (1999) Cell-extracellular matrix interaction in glioma invasion. Acta Neurochir (Wien) 141(3):295–305
54. Ingber DE, Folkman J (1989) Mechanochemical switching between growth and differentiation during fibroblast growth factor-stimulated angiogenesis in vitro: role of extracellular matrix. J Cell Biol 109:317–330
55. Baker BM et al (2013) Microfluidics embedded within extracellular matrix to define vascular architectures and pattern diffusive gradients. Lab Chip 13(16):3246–3252
56. Choi NW et al (2007) Microfluidic scaffolds for tissue engineering. Nat Mater 6(11):908–915
57. Haessler U et al (2009) An agarose-based microfluidic platform with a gradient buffer for 3D chemotaxis studies for 3D chemotaxis studies. Biomed Microdevices 11(4):827–835
58. Joanne Wang C et al (2008) A microfluidics-based turning assay reveals complex growth cone responses to integrated gradients of substrate-bound ECM molecules and diffusible guidance cues. Lab Chip 8(2):227–237
59. Lanfer B et al (2008) Aligned fibrillar collagen matrices obtained by shear flow deposition. Biomaterials 29(28):3888–3895
60. Lanfer B et al (2009) The growth and differentiation of mesenchymal stem and progenitor cells cultured on aligned collagen matrices. Biomaterials 30(30):5950–5958
61. Chin VI et al (2004) Microfabricated platform for studying stem cell fates. Biotechnol Bioeng 88(3):399–415
62. Ma H et al (2012) Probing the role of mesenchymal stem cells in salivary gland cancer on biomimeticmicrodevices. Integr Biol (Camb) 4(5)):522–530
63. Zhang Q et al (2012) A microfluidic-based device for study of transendothelialinvasion of tumor aggregates in real-time. Lab Chip 12(16):2837–2842
64. Tong Z et al (2014) Engineering a functional neuro-muscular junction model in a chip. RSC Adv 4:54788–54797
65. Chung BG et al (2006) A microfluidic multi-injector for gradient generation. Lab Chip 6:764–768
66. Kim S et al (2010) Biological applications of microfluidic gradient devices. Integr Biol 2:584–603
67. Ye N et al (2007) Cell-based high content screening using an integrated microfluidic device. Lab Chip 7(12):1696–1704
68. Jeon NL et al (2000) Generation of solution and surface gradients using microfluidic systems. Langmuir 16:8311–8316
69. Li Y et al (2010) The effects of insulin-like growth factor-1 and basic fibroblast growth factor on the proliferation of chondrocytes embedded in the collagen gel using an integrated microfluidic device. Tissue Eng Part C Methods 16(6):1267–1275
70. Jeon NL et al (2002) Neutrophil chemotaxis in linear and complex gradients of interleukin-8 formed in a microfabricated device. Nat Biotechnol 20:826–830
71. Han S et al (2012) A versatile assay for monitoring in vivo-like transendothelial migration of neutrophils. Lab Chip 12(20):3861–3865

72. Shin Y et al (2011) In vitro 3D collective sprouting angiogenesis under orchestrated ANG-1 and VEGF gradients. Lab Chip 11:2175–2181
73. Jeong GS et al (2011) Sprouting angiogenesis under a chemical gradient regulated by interactions with an endothelial monolayer in a microfluidic platform. Anal Chem 83:8454–8459
74. Torisawa YS et al (2010) Microfluidic platform for chemotaxis in gradients formed by CXCL2 source-sink cells. Integr Biol 2:680–686
75. Dings J et al (1998) Clinical experience with 118 brain tissue oxygen partial pressure catheter probes. Neurosurgery 43:1082–1095
76. Evans SM et al (2004) Hypoxia is important in the biology and aggression of human glial brain tumors. Clin Cancer Res 10:8177–8184
77. Lo JF et al (2010) Oxygen gradient for open well cellular culture via microfluidic substrates. Lab Chip 10(18):2394–2401
78. Wang L et al (2013) Construction of oxygen and chemical concentration gradients in a single microfluidic device for studying tumor cell–drug interactions in a dynamic hypoxia microenvironment. Lab Chip 13(4):695–705
79. Oppegard SC, Eddington DT (2013) A microfabricated platform for establishing oxygen gradients in 3-D constructs. Biomed Microdevices 15(3):407–414
80. Yang W et al (2015) A novel microfluidic platform for studying mammalian cell chemotaxis in different oxygen environments under zero-flow conditions. Biomicrofluidics 9(4):044121
81. Chen YA et al (2011) Generation of oxygen gradients in microfluidic devices for cell culture using spatially confined chemicalreactions. Lab Chip 1(21):3626–3633
82. Derda R et al (2009) Paper-supported 3D cell culture for tissue-based bioassays. Proc Natl Acad Sci USA 106:18457–18462
83. Widmaier EP et al (2004) In: Fox SI (ed) Human physiology, 9th edn. McGraw-Hill, New York, pp 375–466
84. Griffith LG, Swartz MA (2006) Capturing complex 3D tissue physiology in vitro. Nat Rev Mol Cell Biol 7:211–224
85. Kim L et al (2006) Microfluidic arrays for logarithmically perfused embryonic stem cell culture. Lab Chip 6:394–406
86. Lu H et al (2004) Microfluidic shear devices for quantitative analysis of cell adhesion. Anal Chem 76:5257–5264
87. van der Meer AD et al (2009) Microfluidic technology in vascular research. J Biomed Biotechnol:823148
88. Zhong W et al (2013) Mesenchymal stem cell and chondrocyte fates in a multishear microdevice are regulated by yes-associated protein. Stem Cells Dev 22(14):2083–2093
89. Wang L et al (2016) Human induced pluripotent stem cells derived endothelial cells mimicking vascular inflammatory response under flow. Biomicrofluidics 10(1):014106
90. McCue S et al (2004) Shear-induced reorganization of endothelial cell cytoskeleton and adhesion complexes. Trends Cardiovasc Med 14(4):143–151
91. Jang KJ et al (2011) Fluid-shearstress-induced translocation of aquaporin-2 and reorganization of actin cytoskeleton in renal tubular epithelial cells. Integr Biol 3:134–141
92. Jang KJ et al (2010) A multi-layer microfluidic device for efficient culture and analysis of renal tubular cells. Lab Chip 10:36–42
93. Zhou M et al (2014) Induction of epithelial-to-mesenchymal transition in proximal tubular epithelial cells on microfluidic devices. Biomaterials 35(5):1390–1401
94. Song JW, Munn LL (2011) Fluid forces control endothelial sprouting. Proc Natl Acad Sci USA 108(37):15342–15347
95. Douville NJ et al (2011) Combination of fluid and solid mechanical stresses contribute to cell death and detachment in a microfluidic alveolar model. Lab Chip 11:609–619
96. Vlahakis NE et al (1999) Stretch induces cytokine release by alveolar epithelial cells in vitro. Am J Physiol Lung Cell Mol Physiol 277:L167–L173
97. Tschumperlin DJ et al (2000) Deformationinduced injury of alveolar epithelial cells: effect of frequency, duration, and amplitude. Am J Respir Crit Care Med 162:357–362

98. Bilek AM et al (2003) Mechanisms of surface-tensioninduced epithelial cell damage in a model of pulmonary airway reopening. J Appl Physiol 94:770–783
99. Huh D et al (2007) Acoustically detectable cellular-level lung injury induced by fluid mechanical stresses in microfluidic airway systems. Proc Natl Acad Sci USA 104:18886–18891
100. Hubatsch I et al (2007) Determination of drug permeability and prediction of drug absorption in Caco-2 monolayers. Nat Protoc 2(9):2111–2119
101. Kimura H et al (2008) An integrated microfluidic system for long-term perfusion culture and on-line monitoring of intestinal tissue models. Lab Chip 8(5):741–746
102. Imura Y et al (2009) A microfluidic system to evaluate intestinal absorption. Anal Sci 25(12):1403–1407
103. Kim HJ et al (2012) Human gut-on-a-chip inhabited by microbial flora that experiences intestinal peristalsis-like motions and flow. Lab Chip 12:2165–2174
104. Kim HJ, Ingber DE (2013) Gut-on-a-chip microenvironment induces human intestinal cells to undergo villus differentiation. Integr Biol (Camb) 5:1130–1140
105. Gao X et al (2011) A simple elastic membrane-based microfluidic chip for the proliferation and differentiation of mesenchymal stem cells under tensile stress. Electrophoresis 32(23):3431–3436
106. Oleaga C et al (2016) Multi-organ toxicity demonstration in a functional human in vitro system system composed of four organs. Sci Rep 6:20030
107. Whitesides GM (2006) The origins and the future of microfluidics. Nature 442(7101):368–373
108. El-Ali J et al (2006) Cells in chips. Nature 442(7101):403–411
109. Huh D et al (2012) Microengineered physiological biomimicry: organs-on-Chip. Lab Chip 12(12):2156–2164
110. van der Meer AD, van den Berg A (2012) Organs-on-chips: breaking the in vitro impasse. Integr Biol (Camb) 4(5):461–470
111. Ghaemmaghami AM et al (2012) Biomimetic tissues on a chip for drug discovery. Drug Discov Today 17(3–4):173–181
112. Huh D et al (2011) From three-dimensional cell culture to organs-on-chips. Trends Cell Biol 21(12):745–754
113. Steimer A et al (2005) Cell culture models of the respiratory tract relevant to pulmonary drug delivery. J Aerosol Med 18(2):137–182
114. Tavana H et al (2001) Epithelium damage and protection during reopening of occluded airways in a physiologic microfluidic pulmonary airway model. Biomed Microdevices 13(4):731–742
115. Huh D et al (2010) Reconstituting organ-level lung functions on a chip. Science 328(5986):1662–1668
116. Huh D et al (2012) A human disease model of drug toxicity-induced pulmonary edema in a lung-on-a-chip microdevice. Sci Transl Med 4(159):159ra147
117. Fagerholm U et al (1996) Comparison between permeability coefficients in rat and human jejunum. Pharm Res 13(9):1336–1342
118. Kim SH et al (2013) A microfluidic device with 3-d hydrogel villi scaffold to simulate intestinal absorption. J Nanosci Nanotechnol 13(11):7220–7228
119. Gan LSL, Thakker DR (1997) Applications of the Caco-2 model in the design and development of orally active drugs: elucidation of biochemical and physical barriers posed by the intestinal epithelium. Adv Drug Deliv Rev 23:77–98
120. Ramadan Q et al (2013) NutriChip: nutrition analysis meets microfluidics. Lab Chip 13(2):196–203
121. Sung JH et al (2011) Microscale 3-D hydrogel scaffold for biomimetic gastrointestinal (GI) tract model. Lab Chip 11(3):389–392
122. Li L et al (2012) A microfluidic in vitro system for the quantitative study of the stomach mucus barrier function. Lab Chip 12(20):4071–4079

123. Yoon ND et al (2015) 3D liver models on a microplatform: well-defined culture, engineering of liver tissue and liver-on-a-chip. Lab Chip 15(19):3822–3837
124. Bhise NS et al (2016) A liver-on-a-chip platform with bioprinted hepatic spheroids. Biofabrication 8(1):014101
125. Lee J et al (2014) Fabrication and characterization of microfluidic liver-on-a-chip using microsomal enzymes. Lab Chip 14(17):3290–3299
126. Lee PJ et al (2007) An artificial liver sinusoid with a microfluidic endothelial-like barrier for primary hepatocyte culture. Biotechnol Bioeng 97(5):1340–1346
127. Kang IK et al (2004) Co-culture of hepatocytes and fibroblasts by micropatterned immobilization of beta-galactose derivatives. Biomaterials 25(18):4225–4232
128. Zinchenko YS, Coger RN (2005) Engineering micropatterned surfaces for the coculture of hepatocytes and Kupffer cells. J Biomed Mater Res A 75(1):242–248
129. Ho CT et al (2013) Liver-cell patterning lab chip: mimicking the morphology of liver lobule tissue. Lab Chip 13(18):3578–3587
130. Ho CT et al (2006) Rapid heterogeneous liver-cell on-chip patterning via the enhanced field-induced dielectrophoresis trap. Lab Chip 6(6):724–734
131. Malinen MM et al (2014) Differentiation of liver progenitor cell line to functional organotypiccultures in 3D nanofibrillar cellulose and hyaluronan-gelatin hydrogels. Biomaterials 35(19):5110–5121
132. Lee KH et al (2011) Diffusion-mediated in situ alginate encapsulation of cell spheroids using microscale concave well and nanoporous membrane. Lab Chip 11(6):1168–1167
133. Gieseck RL 3rd et al (2014) Maturation of induced pluripotent stem cell derived hepatocytes by 3D-culture. PLoS One 9(1):e86372
134. Achilli TM et al (2012) Advances in the formation, use and understanding of multi-cellular spheroids. Expert Opin Biol Ther 12(10):1347–1360
135. Jun Y et al (2013) 3D co-culturing model of primary pancreatic islets and hepatocytes in hybrid spheroid to overcome pancreatic cell shortage. Biomaterials 34(15):3784–3794
136. Wong SF et al (2011) Concave microwell based size-controllable hepatosphere as a three-dimensional liver tissue model. Biomaterials 32(32):8087–8096
137. Goral VN et al (2010) Perfusion-based microfluidic device for three-dimensional dynamic primary human hepatocyte cell culture in the absence of biological or synthetic matrices or coagulants. Lab Chip 10(24):3380–3386
138. Miki T et al (2011) Hepatic differentiation of human embryonic stem cells is promoted by three-dimensional dynamic perfusion culture conditions. Tissue Eng Part C Methods 17(5):557–568
139. Lee SA et al (2013) Spheroid-based three-dimensional liver-on-a-chip to investigate hepatocyte-hepatic stellate cell interactions and flow effects. Lab Chip 13(18):3529–3537
140. Dash A et al (2013) Hemodynamic flow improves rat hepatocyte morphology, function, and metabolic activity in vitro. Am J Physiol Cell Physiol 304(11):C1053–C1063
141. Trietsch SJ et al (2013) Microfluidic titer plate for stratified 3D cell culture. Lab Chip 13(18):3548–3554
142. Nakao Y et al (2011) Bile canaliculi formation by aligning rat primary hepatocytes in a microfluidic device. Biomicrofluidics 5(2):22212
143. Esch MB et al (2015) Multi-cellular 3D human primary liver cell culture elevates metabolic activity under fluidic flow. Lab Chip 15(10):2269–2277
144. Tanaka Y et al (2007) A micro-spherical heart pump powered by cultured cardiomyocytes. Lab Chip 7:207–212
145. Aung A et al (2016) 3D cardiac μtissues within a microfluidic device with real-time contractile stress readout. Lab Chip 16(1):153–162
146. Morimoto Y et al (2016) Human induced pluripotent stem cell-derived fiber-shaped cardiac tissue on a chip. Lab Chip 16(12):2295–2301
147. Khademhosseini A et al (2007) Microfluidic patterning for fabrication of contractile cardiac organoids. Biomed Microdevices 9(2):149–157

148. Visone R et al (2016) Cardiac meets skeletal: what's new in microfluidic models for muscle tissue engineering. Molecules 21(9):piiE1128
149. Radisic M et al (2007) Biomimetic approach to cardiac tissue engineering. Philos Trans R Soc Lond Ser B Biol Sci 362:1357–1136
150. Grosberg A et al (2011) Ensembles of engineered cardiac tissues for physiological and engineered cardiac tissues for physiological and pharmacological study heart on a chip. Lab Chip 11(24):4165–4173
151. Effron MB et al (1987) Changes in myosin isoenzymes, ATPase activity, and contraction duration in rat cardiac muscle with aging can be modulated by thyroxine. Circ Res 60(2):238–245
152. Agarwal A et al (2013) Microfluidic heart on a chip for higher throughput pharmacological studies. Lab Chip 13:3599–3608
153. Serena E et al (2012) Micro-arrayed human embryonic stem cells-derived cardiomyocytes for in vitro functional assay. PLoS One 7(11):e48483
154. Kensah G et al (2013) Murine and human pluripotent stem cell-derived cardiac bodies form contractile myocardial tissue in vitro. Eur Heart J 34:1134–1146
155. Bergstrom G et al (2015) Stem cell derived in vivo-like human cardiac bodies in a microfluidic device for toxicity testing by beating frequency imaging. Lab Chip 15:3242–3249
156. Aung A et al (2016) 3D cardiac mutissues within a microfluidic device with real-time contractile stress readout. Lab Chip 16:153–162
157. Mathur A et al (2015) Human iPSC-based cardiac microphysiological system for drug screening applications. Sci Rep 5:8883
158. Mathur A et al (2016) In vitro cardiac tissue models: current status and future prospects. Adv Drug Deliv Rev 96:203–213
159. Booth R, Kim H (2012) Characterization of a microfluidic in vitro model of the blood-brain barrier (mBBB). Lab Chip 12:1784–1792
160. Achyuta AK et al (2013) A modular approach to create a neurovascular unit-on-a-chip. Lab Chip 13:542–553
161. Wikswo JP et al (2013) Scaling and systems biology for integrating multiple organs-on-a-chip. Lab Chip 13:3496–3511
162. Abaci HE, Shuler ML (2015) Human-on-a-chip design strategies and principles for physiologically based pharmacokinetics/pharmacodynamics modeling. Integr Biol (Camb) 7(4):383–391
163. vanMidwoud PM et al (2010) A microfluidic approach for in vitro assessment of interorgan interactions in drug metabolism using intestinal and liver slices. Lab Chip 10:2778–2786
164. Wagner I et al (2013) A dynamic multi-organ-chip for long-term cultivation and substance testing proven by 3D human liver and skin tissue co-culture. Lab Chip 13(18):3538–3547
165. Materne EM et al (2015) A multi-organ chip co-culture of neurospheres and liver equivalents-for long-term substance testing. J Biotechnol 205:36–46
166. Sung JH, Shuler ML (2009) A micro cell culture analog (microCCA) with 3-D hydrogel culture of multiple cell lines to assess metabolism-dependent cytotoxicity of anti-cancer drugs. Lab Chip 9(10):1385–1394
167. Esch MB et al (2014) Body-on-a chip simulation with gastrointestinal tract and liver tissues suggests that ingested nanoparticles have the potential to cause liver injury. Lab Chip 14(16):3081–3092
168. Maschmeyer I et al (2015) A four-organ-chip for interconnected long-term co-culture of human intestine, liver, skin and kidney equivalents. Lab Chip 15(12):2688–2699
169. Zhang C et al (2009) Towards a human-on-chip: culturing multiple cell types on a chip with compartmentalized microenvironments. Lab Chip 9:3185–3192
170. Odijk M et al (2015) Measuring direct current trans-epithelial electrical resistance in organ-on microsystem. Lab Chip 15(3):745–752
171. Benam KH et al (2016) Small airway-on-a-chip enables analysis of human lung inflammation and drug responses in vitro. Nat Methods 13(2):151–157

172. Park J et al (2015) Three-dimensional brain-on-a-chip with an interstitial level of flow and its application as an in vitro model of Alzheimer's disease. Lab Chip 15(1):141–150
173. Ma H et al (2013) Biomimetic tumor microenvironment on a microfluidic platform. Biomicrofluidics 7(1):11501
174. Kuo CT et al (2014) Modeling of cancer metastasis and drug resistance via biomimetic-nanocilial and microfluidics. Biomaterials 35(5):1562–1571
175. Choi Y et al (2015) A microengineered pathophysiological model of early-stage breast cancer. Lab Chip 15(16):3350–3357
176. Xu H et al (2015) Activation of hypoxia signaling induces phenotypic transformation of glioma cells: implications for bevacizumab antiangiogenic therapy. Oncotarget 6(14):11882–11893
177. Li CY et al (2013) Flow-based pipeline for systematic modulation and analysis of 3D tumor microenvironments. Lab Chip 13:1969–1978
178. Zervantonakis IK et al (2012) Three-dimensional microfluidic model for tumor cell intravasation and endothelial barrier function. Proc Natl Acad Sci USA 109:13515–13520
179. Moya ML et al (2013) In vitro perfused human capillary networks. Tissue Eng Part C Methods 19:730–737
180. Businaro L et al (2013) Cross talk between cancer and immune cells: exploring complex dynamics in a microfluidic environment. Lab Chip 13:229–239
181. Aref AR et al (2013) Screening therapeutic EMT blocking agents in a three-dimensional microenvironment. Integr Biol (Camb) 5:381–389
182. Vidi PA et al (2014) Disease-on-a-chip: mimicry of tumor growth in mammary ducts. Lab Chip 14:172–177
183. Tatosian DA, Shuler ML (2009) A novel system for evaluation of drug mixtures for potential efficacy in treating multidrug resistant cancers. Biotechnol Bioeng 103:187–198
184. Berdichevsky Y et al (2010) Building and manipulating neural pathways with microfluidics. Lab Chip 10:999–1004
185. Xu H et al (2016) A dynamic in vivo-like organotypic blood-brain barrier model to probe metastatic brain tumors. Sci Rep 6:36670
186. Lippmann ES et al (2012) Derivation of blood-brain barrier endothelial cells from human pluripotent stem cells. Nat Biotechnol 30:783–791
187. Deracinois B et al (2013) Glial-cell-mediated re-induction of the blood-brain barrier phenotype in brain capillary endothelial cells: a differential gel electrophoresis study. Proteomics 13:1185–1199
188. Choucha-Snouber L et al (2013) Investigation of ifosfamide nephrotoxicity induced in a liver–kidney co-culture biochip. Biotechnol Bioeng 110:597–608
189. McCain ML et al (2013) Recapitulating maladaptive, multiscale remodeling of failing myocardium on a chip. Proc Natl Acad Sci USA 110:9770–9775
190. Thavandiran N et al (2013) Design and formulation of functional pluripotent stem cell-derived cardiac microtissues. Proc Natl Acad Sci USA 110:E4698–E4707
191. ChouchaSnouber L et al (2013) Metabolomics-on-a-chip of hepatotoxicity induced by anticancer drug flutamide and its active metabolite hydroxyflutamide using HepG2/ C3a microfluidic biochips. Toxicol Sci 132:8–20
192. Shintu L et al (2012) Metabolomics-on-a-chip and predictive systems toxicology in microfluidic bioartificial organs. Anal Chem 84:1840–1848
193. Mahler GJ et al (2009) Characterization of a gastrointestinal tract microscale cell culture analog used to predict drug toxicity. Biotechnol Bioeng 104(1):193–205
194. Sung JH et al (2010) A microfluidic device for a pharmacokinetic-pharmacodynamic (PK-PD) model on a chip. Lab Chip 10(4):446–455
195. Li Z et al (2016) Assessment of metabolism-dependent drug efficacy and toxicity on a multilayer organs-on-a-chip. Integr Biol (Camb) 8(10):1022–1029
196. Cong X et al (2015) Early life experience and gut microbiome: the brain-gut-microbiota signaling system. Adv Neonatal Care 5(5):314–323

197. Dinan TG, Cryan JF (2012) Regulation of the stress response by the gut microbiota: implications for psychoneuroendocrinology. Psychoneuroendocrinology 37(9):1369–1378
198. Kim HJ et al (2016) Contributions of microbiome and mechanical deformation to intestinal bacterial overgrowth andinflammation in a human gut-on-a-chip. Proc Natl Acad Sci USA 113(1):E7–E15
199. Booth R, Kim H (2014) Permeability analysis of neuroactive drugs through a dynamic microfluidic in vitro blood-brain barrier model. Annals Biomed Engineering 42:2379–2391
200. Deosarkar SP et al (2015) A novel dynamic neonatal blood- brain barrier on a Chip. PLoS One 10(11):e0142725
201. Wang G et al (2014) Modeling the mitochondrial cardiomyopathy of Barth syndrome with induced pluripotent stem cell and heart-on-chip technologies. Nature Med 20:616–623
202. Alford PW et al (2010) Biohybrid thin films for measuring contractility in engineered cardiovascular muscle. Biomaterials 31(13):3613–3621
203. Punde TH et al (2015) A biologically inspired lung-on-a-chip device for the study of protein induced lung inflammation. IntegrBiol (Camb) 7(2):162–169
204. Jang KJ et al (2013) Human kidney proximal tubule-on-a-chip for drug transport and nephrotoxicity assessment. IntegrBiol (Camb). 5(9):1119–1129
205. Zhou M et al (2016) Development of a functional Glomerulus at the organ level on a Chip to mimic hypertensive nephropathy. Sci Rep 6:31771
206. Zhou Q et al (2015) Liver injury-on-a-chip: microfluidic co-cultures with integrated biosensors for monitoring liver cell signaling during injury. Lab Chip 15(23):4467–4478
207. Kang YB et al (2015) Liver sinusoid on a chip: long-term layered co-culture of primary rat hepatocytes and endothelial cells in microfluidic platforms. BiotechnolBioeng 112(12):2571–2582
208. Schimek K et al (2013) Integrating biological vasculature into a multi-organ-chip microsystem. Lab Chip 13(18):3588–3598
209. Hsu YH et al (2013) A microfluidic platform for generating large-scale nearly identical human microphysiological vascularized tissue arrays. Lab Chip 13(15):2990–2998
210. Wang X et al (2016) Engineering anastomosis between living capillary networks and endothelial cell-lined microfluidic channels. Lab Chip 16(2):282–290
211. Oleaga C et al (2016) Multi-organ toxicity demonstration in a functional human in vitro system composed of four organs. Sci Rep 6:20030
212. Frey O et al (2014) Reconfigurable microfluidic hanging drop network for multi-tissue interaction and analysis. Nat Commun 5:4250
213. Loskill P et al (2015) μOrgano: a Lego®-like Plug & Play System for Modular Multi-Organ-Chips. PLoS One 10(10):e0139587
214. Rigat-Brugarolas LG et al (2014) A functional microengineered model of the human splenon-on-a-chip. Lab Chip 14(10):1715–1724
215. Adriani G et al (2017) A 3D neurovascular microfluidic model consisting of neurons, astrocytes and cerebral endothelial cells as ablood-brain barrier. Lab Chip 17(3):448–459

MIX
Papier aus verantwortungsvollen Quellen
Paper from responsible sources
FSC® C105338

If you have any concerns about our products,
you can contact us on
ProductSafety@springernature.com

In case Publisher is established outside the EU,
the EU authorized representative is:
Springer Nature Customer Service Center GmbH
Europaplatz 3, 69115 Heidelberg, Germany

Printed by Libri Plureos GmbH
in Hamburg, Germany